Lecture Notes in Artificial Intelligence 10448

Subseries of Lecture Notes in Computer Science

More information about this series at http://www.springer.com/series/1244

Ngoc Thanh Nguyen · George A. Papadopoulos
Piotr Jędrzejowicz · Bogdan Trawiński
Gottfried Vossen (Eds.)

Computational Collective Intelligence

9th International Conference, ICCCI 2017
Nicosia, Cyprus, September 27–29, 2017
Proceedings, Part I

Springer

Editors
Ngoc Thanh Nguyen
Department of Information Systems, Faculty of Computer Science and Management
Wrocław University of Science and Technology
Wrocław
Poland

Bogdan Trawiński
Department of Information Systems, Faculty of Computer Science and Management
Wrocław University of Science and Technology
Wrocław
Poland

George A. Papadopoulos
Department of Computer Science
University of Cyprus
Nicosia
Cyprus

Gottfried Vossen
Department of Information Systems
University of Münster
Münster
Germany

Piotr Jędrzejowicz
Department of Information Systems
Gdynia Maritime University
Gdynia
Poland

ISSN 0302-9743 ISSN 1611-3349 (electronic)
Lecture Notes in Artificial Intelligence
ISBN 978-3-319-67073-7 ISBN 978-3-319-67074-4 (eBook)
DOI 10.1007/978-3-319-67074-4

Library of Congress Control Number: 2017952854

LNCS Sublibrary: SL7 – Artificial Intelligence

Printed on acid-free paper

This Springer imprint is published by Springer Nature
The registered company is Springer International Publishing AG
The registered company address is: Gewerbestrasse 11, 6330 Cham, Switzerland

Preface

This volume contains the proceedings of the 9th International Conference on Computational Collective Intelligence (ICCCI 2017), held in Nicosia, Cyprus, September 27–29, 2017. The conference was co-organized by the University of Cyprus, Cyprus and the Wrocław University of Science and Technology, Poland. The conference was run under the patronage of the IEEE SMC Technical Committee on Computational Collective Intelligence.

Following the successes of the 1st ICCCI (2009), held in Wrocław, Poland, the 2nd ICCCI (2010), in Kaohsiung, Taiwan, the 3rd ICCCI (2011), in Gdynia, Poland, the 4th ICCCI (2012), in Ho Chi Minh City, Vietnam, the 5th ICCCI (2013), in Craiova, Romania, the 6th ICCCI (2014), in Seoul, South Korea, the 7th ICCCI (2015), in Madrid, Spain, and the 8th ICCCI (2016), in Halkidiki, Greece, this conference continued to provide an internationally respected forum for scientific research in the computer-based methods of collective intelligence and their applications.

Computational Collective Intelligence (CCI) is most often understood as a sub-field of Artificial Intelligence (AI) dealing with soft computing methods that enable making group decisions or processing knowledge among autonomous units acting in distributed environments. Methodological, theoretical, and practical aspects of computational collective intelligence are considered as the form of intelligence that emerges from the collaboration and competition of many individuals (artificial and/or natural). The application of multiple computational intelligence technologies, such as fuzzy systems, evolutionary computation, neural systems, consensus theory, etc., can support human and other collective intelligence, and create new forms of CCI in natural and/or artificial systems. Three subfields of the application of computational intelligence technologies to support various forms of collective intelligence are of special interest but are not exclusive: semantic web (as an advanced tool for increasing collective intelligence), social network analysis (as the field targeted to the emergence of new forms of CCI), and multi-agent systems (as a computational and modeling paradigm especially tailored to capture the nature of CCI emergence in populations of autonomous individuals).

The ICCCI 2017 conference featured a number of keynote talks and oral presentations, closely aligned to the theme of the conference. The conference attracted a substantial number of researchers and practitioners from all over the world, who submitted their papers for the main track and seven special sessions.

The main track, covering the methodology and applications of computational collective intelligence, included: multi-agent systems, knowledge engineering and semantic web, social networks and recommender systems, text processing and information retrieval, data mining methods and applications, sensor networks and internet of things, decision support and control systems, and computer vision techniques. The special sessions, covering some specific topics of particular interest, included cooperative strategies for decision making and optimization, computational swarm intelligence, machine learning in medicine and biometrics, cyber physical systems in automotive area, internet of things - its relations and consequences, low resource language processing, and intelligent processing of multimedia in web systems.

We received in total over 240 submissions from 39 countries all over the world. Each paper was reviewed by 2–4 members of the International Program Committee of either the main track or one of the special sessions. We selected the 114 best papers for oral presentation and publication in two volumes of the Lecture Notes in Artificial Intelligence series.

We would like to express our thanks to the keynote speakers: Yannis Manolopoulos from the Aristotle University of Thessaloniki, Greece; Andreas Nürnberger from the Otto-von-Guericke University Magdeburg, Germany; Constantinos S. Pattichis from the University of Cyprus, Cyprus; and Sławomir Zadrożny from the Systems Research Institute of the Polish Academy of Sciences, Poland, for their world-class plenary speeches.

Many people contributed towards the success of the conference. First, we would like to recognize the work of the Program Committee co-chairs and special sessions organizers for taking good care of the organization of the reviewing process, an essential stage in ensuring the high quality of the accepted papers. The chairs of the workshops and special sessions deserve a special mention for the evaluation of the proposals and the organization and coordination of the work of seven special sessions. In addition, we would like to thank the PC members, of the main track and of the special sessions, for performing their reviewing work with diligence. We thank the Local Organizing Committee chairs, the publicity chair, the web chair, and the technical support chair for their fantastic work before and during the conference. Finally, we cordially thank all the authors, presenters, and delegates for their valuable contribution to this successful event. The conference would not have been possible without their support.

It is our pleasure to announce that the conferences of the ICCCI series continue a close cooperation with the Springer journal Transactions on Computational Collective Intelligence, and the IEEE SMC Technical Committee on Transactions on Computational Collective Intelligence.

Finally, we hope and intend that ICCCI 2017 will significantly contribute to the academic excellence of the field and lead to the even greater success of ICCCI events in the future.

September 2017

Ngoc Thanh Nguyen
George A. Papadopoulos
Piotr Jędrzejowicz
Bogdan Trawiński
Gottfried Vossen

Organization

Honorary Chairs

Pierre Lévy — University of Ottawa, Canada
Cezary Madryas — Wrocław University of Science and Technology, Poland
Costas Christophides — University of Cyprus, Cyprus

General Chairs

Ngoc Thanh Nguyen — Wrocław University of Science and Technology, Poland
George A. Papadopoulos — University of Cyprus, Cyprus

Program Chairs

Costin Badica — University of Craiova, Romania
Kazumi Nakamatsu — University of Hyogo, Japan
Piotr Jędrzejowicz — Gdynia Maritime University, Poland
Gottfried Vossen — University of Münster, Germany

Special Session Chairs

Bogdan Trawiński — Wrocław University of Science and Technology, Poland
Achilleas Achilleos — University of Cyprus, Cyprus

Doctoral Track Chair

George Pallis — University of Cyprus, Cyprus

Organizing Chair

Georgia Kapitsaki — University of Cyprus, Cyprus

Publicity Chair

Christos Mettouris — University of Cyprus, Cyprus

Keynote Speakers

Andreas Nürnberger	Otto von Guericke University Magdeburg, Germany
Yannis Manolopoulos	Aristotle University of Thessaloniki, Greece
Constantinos S. Pattichis	University of Cyprus, Cyprus
Sławomir Zadrożny	Systems Research Institute of the Polish Academy of Sciences, Poland

Special Sessions Organizers

1. *CSI 2017: Special Session on Computational Swarm Intelligence*

Urszula Boryczka	University of Silesia, Poland
Tomasz Gwizdałła	University of Lodz, Poland
Jarosław Wąs	AGH University of Science and Technology, Poland

2. *WebSys 2017: Special Session on Intelligent Processing of Multimedia in Web Systems*

Kazimierz Choroś	Wrocław University of Science and Technology, Poland
Maria Trocan	Institut Supérieur d'Électronique de Paris, France

3. *CPSiA 2017: Special Session on Cyber-Physical Systems in the Automotive Area*

Adam Ziębiński	Silesian University of Technology, Poland
Markus Bregulla	Technische Hochschule Ingolstadt, Germany
Rafal Cupek	Silesian University of Technology, Poland
Hueseyin Erdogan	Continental Ingolstadt, Germany
Daniel Grossman	Technische Hochschule Ingolstadt, Germany

4. *LRLP 2017: Special Session on Low-Resource Languages Processing*

Ualsher Tukeyev	al-Farabi Kazakh National University, Kazakhstan
Zhandos Zhumanov	al-Farabi Kazakh National University, Kazakhstan

5. *CSDMO 2017: Special Session on Cooperative Strategies for Decision-Making and Optimization*

Piotr Jędrzejowicz	Gdynia Maritime University, Poland
Dariusz Barbucha	Gdynia Maritime University, Poland

6. *IoT-RC 2017: Special Session on Internet of Things – Its Relations and Consequences*

Vladimir Sobeslav	University of Hradec Kralove, Czech Republic
Ondrej Krejcar	University of Hradec Kralove, Czech Republic

Peter Brida University of Žilina, Slovakia
Peter Mikulecky University of Hradec Kralove, Czech Republic

7. *MLMB 2017: Special Session on Machine Learning in Medicine and Biometrics*

Piotr Porwik University of Silesia, Poland
Alicja Wakulicz-Deja University of Silesia, Poland
Agnieszka Nowak-Brzezińska University of Silesia, Poland

International Program Committee

Muhammad Abulaish Jamia Millia Islamia (A Central University), India
Sharat Akhoury University of Cape Town, South Africa
Ana Almeida GECAD-ISEP-IPP, Portugal
Orcan Alpar University of Hradec Králové, Czech Republic
Bashar Al-Shboul University of Jordan, Jordan
Thierry Badard Laval University, Canada
Amelia Badica University of Craiova, Romania
Costin Badica University of Craiova, Romania
Hassan Badir Ecole Nationale des Sciences Appliquées de Tanger, Morocco
Dariusz Barbucha Gdynia Maritime University, Poland
Nick Bassiliades Aristotle University of Thessaloniki, Greece
Maria Bielikova Slovak University of Technology in Bratislava, Slovakia
Leon Bobrowski Bialystok University of Technology, Poland
Mariusz Boryczka University of Silesia, Poland
Urszula Boryczka University of Silesia, Poland
Abdelhamid Bouchachia Bournemouth University, UK
Peter Brida University of Žilina, Slovakia
Robert Burduk Wrocław University of Science and Technology, Poland
Krisztian Buza Budapest University of Technology and Economics, Hungary
Aleksander Byrski AGH University of Science and Technology, Poland
Jose Luis Calvo-Rolle University of A Coruna, Spain
David Camacho Universidad Autonoma de Madrid, Spain
Alberto Cano Virginia Commonwealth University, USA
Frantisek Capkovic Slovak Academy of Sciences, Slovakia
Richard Chbeir LIUPPA Laboratory, France
Shyi-Ming Chen National Taiwan University of Science and Technology, Taiwan
Amine Chohra Paris-East University (UPEC), France

Kazimierz Choroś	Wrocław University of Science and Technology, Poland
Mihaela Colhon	University of Craiova, Romania
Jose Alfredo Ferreira Costa	Universidade Federal do Rio Grande do Norte, Brazil
Boguslaw Cyganek	AGH University of Science and Technology, Poland
Ireneusz Czarnowski	Gdynia Maritime University, Poland
Paul Davidsson	Malmö University, Sweden
Tien V. Do	Budapest University of Technology and Economics, Hungary
Vadim Ermolayev	Zaporozhye National University, Ukraine
Nadia Essoussi	University of Carthage, Tunisia
Rim Faiz	University of Carthage, Tunisia
Faiez Gargouri	University of Sfax, Tunisia
Mauro Gaspari	University of Bologna, Italy
Janusz Getta	University of Wollongong, Australia
Daniela Gifu	University "Alexandru loan Cuza" of Iasi, Romania
Daniela Godoy	ISISTAN Research Institute, Argentina
Antonio Gonzalez-Pardo	Universidad Autonoma de Madrid, Spain
Manuel Grana	University of the Basque Country, Spain
Foteini Grivokostopoulou	University of Patras, Greece
Marcin Hernes	Wrocław University of Economics, Poland
Huu Hanh Hoang	Hue University, Vietnam
Tzung-Pei Hong	National University of Kaohsiung, Taiwan
Mong-Fong Horng	National Kaohsiung University of Applied Sciences, Taiwan
Frederic Hubert	Laval University, Canada
Maciej Huk	Wrocław University of Science and Technology, Poland
Dosam Hwang	Yeungnam University, South Korea
Lazaros Iliadis	Democritus University of Thrace, Greece
Agnieszka Indyka-Piasecka	Wrocław University of Science and Technology, Poland
Dan Istrate	Université de Technologie de Compiégne, France
Mirjana Ivanovic	University of Novi Sad, Serbia
Jaroslaw Jankowski	West Pomeranian University of Technology, Poland
Joanna Jędrzejowicz	University of Gdańsk, Poland
Piotr Jędrzejowicz	Gdynia Maritime University, Poland
Gordan Jezic	University of Zagreb, Croatia
Geun Sik Jo	Inha University, South Korea
Kang-Hyun Jo	University of Ulsan, South Korea
Jason Jung	Chung-Ang University, South Korea
Tomasz Kajdanowicz	Wrocław University of Science and Technology, Poland
Petros Kefalas	University of Sheffield International Faculty, CITY College, Greece

Rafał Kern	Wrocław University of Science and Technology, Poland
Marek Kisiel-Dorohinicki	AGH University of Science and Technology, Poland
Attila Kiss	Eotvos Lorand University, Hungary
Marek Kopel	Wrocław University of Science and Technology, Poland
Jerzy Korczak	Wrocław University of Economics, Poland
Jacek Koronacki	Polish Academy of Sciences, Poland
Leszek Kotulski	AGH University of Science and Technology, Poland
Ivan Koychev	University of Sofia "St. Kliment Ohridski", Bulgaria
Jan Kozak	University of Economics in Katowice, Poland
Adrianna Kozierkiewicz-Hetmańska	Wrocław University of Science and Technology, Poland
Bartosz Krawczyk	Virginia Commonwealth University, USA
Ondrej Krejcar	University of Hradec Králové, Czech Republic
Dalia Kriksciuniene	Vilnius University, Lithuania
Dariusz Król	Wrocław University of Science and Technology, Poland
Elzbieta Kukla	Wrocław University of Science and Technology, Poland
Julita Kulbacka	Wrocław Medical University, Poland
Marek Kulbacki	Polish-Japanese Academy of Information Technology, Poland
Piotr Kulczycki	Polish Academy of Sciences, Poland
Kazuhiro Kuwabara	Ritsumeikan University, Japan
Halina Kwaśnicka	Wrocław University of Science and Technology, Poland
Mark Last	Ben-Gurion University of the Negev, Israel
Nguyen Le Minh	Japan Advanced Institute of Science and Technology, Japan
Hoai An Le Thi	Université de Lorraine, France
Florin Leon	"Gheorghe Asachi" Technical University of Iasi, Romania
Edwin Lughofer	Johannes Kepler University Linz, Austria
Juraj Machaj	University of Žilina, Slovakia
Bernadetta Maleszka	Wrocław University of Science and Technology, Poland
Marcin Maleszka	Wrocław University of Science and Technology, Poland
Yannis Manolopoulos	Aristotle University of Thessaloniki, Greece
Urszula Markowska-Kaczmar	Wrocław University of Science and Technology, Poland
Adam Meissner	Poznań University of Technology, Poland
Ernestina Menasalvas	Universidad Politecnica de Madrid, Spain
Hector Menendez	Universidad Autonoma de Madrid, Spain

Jacek Mercik	WSB University in Wrocław, Poland
Radoslaw Michalski	Wrocław University of Science and Technology, Poland
Peter Mikulecky	University of Hradec Králové, Czech Republic
Alin Moldoveanu	University Politehnica of Bucharest, Romania
Javier Montero	Universidad Complutense de Madrid, Spain
Ahmed Moussa	Université Abdelmalek Essaadi, Morocco
Dariusz Mrozek	Silesian University of Technology, Poland
Kazumi Nakamatsu	University of Hyogo, Japan
Grzegorz J. Nalepa	AGH University of Science and Technology, Poland
Fulufhelo Nelwamondo	Council for Scientific and Industrial Research, South Africa
Filippo Neri	University of Napoli Federico II, Italy
Linh Anh Nguyen	University of Warsaw, Poland
Loan T.T. Nguyen	University of Warsaw, Poland
Adam Niewiadomski	Lodz University of Technology, Poland
Agnieszka Nowak-Brzezinska	University of Silesia, Poland
Alberto Nunez	Universidad Complutense de Madrid, Spain
Manuel Nunez	Universidad Complutense de Madrid, Spain
Tarkko Oksala	Aalto University, Finland
Mieczyslaw Owoc	Wrocław University of Economics, Poland
Marcin Paprzycki	Polish Academy of Sciences, Poland
Rafael Parpinelli	Santa Catarina State University, Brazil
Marek Penhaker	VSB -Technical University of Ostrava, Czech Republic
Isidoros Perikos	University of Patras, Greece
Marcin Pietranik	Wrocław University of Science and Technology, Poland
Elias Pimenidis	University of the West of England, UK
Piotr Porwik	University of Silesia, Poland
Radu-Emil Precup	Politehnica University of Timisoara, Romania
Ales Prochazka	University of Chemistry and Technology, Czech Republic
Paulo Quaresma	Universidade de Evora, Portugal
Mohammad Rashedur Rahman	North South University, Bangladesh
Ewa Ratajczak-Ropel	Gdynia Maritime University, Poland
Tomasz Rutkowski	University of Tokyo, Japan
Jose A. Saez	University of Salamanca, Spain
Virgilijus Sakalauskas	Vilnius University, Lithuania
Jose L. Salmeron	University Pablo de Olavide, Spain
Ali Selamat	Universiti Teknologi Malaysia, Malaysia
Natalya Shakhovska	Lviv Polytechnic National University, Ukraine
Andrzej Siemiński	Wrocław University of Science and Technology, Poland
Dragan Simic	University of Novi Sad, Serbia

Vladimir Sobeslav	University of Hradec Králové, Czech Republic
Stanimir Stoyanov	University of Plovdiv "Paisii Hilendarski", Bulgaria
Yasufumi Takama	Tokyo Metropolitan University, Japan
Zbigniew Telec	Wrocław University of Science and Technology, Poland
Diana Trandabat	University "Alexandru loan Cuza" of Iasi, Romania
Bogdan Trawinski	Wrocław University of Science and Technology, Poland
Jan Treur	Vrije Universiteit Amsterdam, Netherlands
Maria Trocan	Institut Supérieur d'Électronique de Paris, France
Krzysztof Trojanowski	Cardinal Stefan Wyszyński University in Warsaw, Poland
Ualsher Tukeyev	al-Farabi Kazakh National University, Kazakhstan
Olgierd Unold	Wrocław University of Science and Technology, Poland
Ventzeslav Valev	Bulgarian Academy of Sciences, Bulgaria
Bay Vo	Ho Chi Minh City University of Technology, Vietnam
Gottfried Vossen	ERCIS Muenster, Germany
Lipo Wang	Nanyang Technological University, Singapore
Izabela Wierzbowska	Gdynia Maritime University, Poland
Michal Wozniak	Wrocław University of Science and Technology, Poland
Sławomir Zadrożny	Polish Academy of Sciences, Poland
Drago Zagar	University of Osijek, Croatia
Danuta Zakrzewska	Lodz University of Technology, Poland
Constantin-Bala Zamfirescu	"Lucian Blaga" University of Sibiu, Romania
Katerina Zdravkova	Ss. Cyril and Methodius University in Skopje, Macedonia
Aleksander Zgrzywa	Wrocław University of Science and Technology, Poland
Adam Ziębiński	Silesian University of Technology, Poland

Program Committees of Special Sessions

CSI 2017: Special Session on Computational Swarm Intelligence

Urszula Boryczka	University of Silesia, Poland
Tomasz Gwizdałła	University of Łódź, Poland
Jarosław Wąs	AGH University of Science and Technology, Poland
Ajith Abraham	Scientific Network for Innovation and Research Excellence, USA
Andrew Adamatzky	University of the West of England, UK
Costin Badica	University of Craiova, Romania
Jan Baetens	Ghent University, Belgium
Mariusz Boryczka	University of Silesia, Poland
Wojciech Froelich	University of Silesia, Poland

Rolf Hoffmann	Technische Universität Darmstadt, Germany
Genaro Martínez	Computer Science Laboratory IPN, Mexico
Dariusz Pierzchała	Military University of Technology, Poland
Franciszek Seredyński	Cardinal Stefan Wyszyński University in Warsaw, Poland
Georgios Sirakoulis	Democritus University of Thrace, Greece
Rafał Skinderowicz	University of Silesia, Poland
Mirosław Szaban	Siedlce University of Science and Humanities, Poland
William Spataro	University of Calabria, Italy
Krzysztof Trojanowski	Cardinal Stefan Wyszyński University in Warsaw, Poland
Barbara Wolnik	University of Gdansk, Poland
Wojciech Wieczorek	University of Silesia, Poland

WebSys 2017: Special Session on Intelligent Processing of Multimedia in Web Systems

Kazimierz Choroś	Wrocław University of Science and Technology, Poland
Jarosław Jankowski	West Pomeranian University of Technology, Poland
Ondřej Krejcar	University of Hradec Kralove, Czech Republic
Tarkko Oksala	Helsinki University of Technology, Finland
Andrzej Siemiński	Wrocław University of Science and Technology, Poland
Jérémie Sublime	Institut Supérieur d'Électronique de Paris, France
Maria Trocan	Institut Supérieur d'Électronique de Paris, France

CPSiA 2017: Special Session on Cyber-Physical Systems in the Automotive Area

Markus Bregulla	Technische Hochschule Ingolstadt, Germany
Daniel Grossman	Technische Hochschule Ingolstadt, Germany
Dariusz Kania	Silesian University of Technology, Poland
Rafał Cupek	Silesian University of Technology Poland
Hueseyin Erdogan	Continental Ingolstadt, Germany
Damian Grzechca	Silesian University of Technology, Poland
Sebastian Budzan	Silesian University of Technology, Poland
Roman Wyżgolik	Silesian University of Technology, Poland
Krzysztof Tokarz	Silesian University of Technology, Poland
Marcin Fojcik	Western Norway University of Applied Sciences, Norway
Mirosław Łazoryszczak	West Pomeranian University of Technology, Poland
Grzegorz Ulacha	West Pomeranian University of Technology, Poland
Krzysztof Małecki	West Pomeranian University of Technology, Poland
Grzegorz Andrzejewski	University of Zielona Góra, Poland
Marek Drewniak	Aiut Sp. z o.o., Poland
Adam Ziębiński	Silesian University of Technology, Poland

LRLP 2017: Special Session on Low-Resource Languages Processing

Ualsher Tukeyev	al-Farabi Kazakh National University, Kazakhstan
Zhandos Zhumanov	al-Farabi Kazakh National University, Kazakhstan
Madina Mansurova	al-Farabi Kazakh National University, Kazakhstan
Altynbek Sharipbay	L.N. Gumilyov Eurasian National University, Kazakhstan
Rustam Musabayev	Institute of Information and Computational Technologies, Kazakhstan
Zhenisbek Assylbekov	Nazarbayev University, Kazakhstan
Jonathan Washington	Swarthmore College, USA
Djavdet Suleimanov	Institute of Applied Semiotics, Russia
Alimzhanov Yermek	al-Farabi Kazakh National University, Kazakhstan

CSDMO 2017: Special Session on Cooperative Strategies for Decision-Making and Optimization

Dariusz Barbucha	Gdynia Maritime University, Poland
Vincenzo Cutello	University of Catania, Italy
Ireneusz Czarnowski	Gdynia Maritime University, Poland
Joanna Jędrzejowicz	Gdansk University, Poland
Piotr Jędrzejowicz	Gdynia Maritime University, Poland
Edyta Kucharska	AGH University of Science and Technology, Poland
Antonio D. Masegosa	University of Deusto, Spain
Javier Montero	Complutense University, Spain
Ewa Ratajczak-Ropel	Gdynia Maritime University, Poland
Iza Wierzbowska	Gdynia Maritime University, Poland
Mahdi Zargayouna	IFSTTAR, France

IoT-RC 2017: Special Session on Internet of Things – Its Relations and Consequences

Ana Almeida	Porto Superior Institute of Engineering, Portugal
Jorge Bernardino	Polytechnical Institute of Coimbra, Spain
Peter Brida	University of Žilina, Slovakia
Ivan Dolnak	University of Žilina, Slovakia
Josef Horalek	University of Hradec Kralove, Czech Republic
Ondrej Krejcar	University of Hradec Kralove, Czech Republic
Goreti Marreiros	Porto Superior Institute of Engineering, Portugal
Peter Mikulecký	University of Hradec Kralove, Czech Republic
Juraj Machaj	University of Žilina, Slovakia
Marek Penhaker	VSB Technical University of Ostrava, Czech Republic
José Salmeron	Universidad Pablo de Olavide of Seville, Spain
Ali Selamat	Universiti Teknologi Malaysia, Malaysia
Vladimir Sobeslav	University of Hradec Kralove, Czech Republic
Stylianakis Vassilis	University of Patras, Greece
Petr Tucnik	University of Hradec Kralove, Czech Republic

MLMB 2017: Special Session on Machine Learning in Medicine and Biometrics

Nabendu Chaki	University of Calcutta, India
Robert Czabański	University of Silesia, Poland
Rafał Deja	Academy of Business in Dąbrowa Górnicza, Poland
Michał Dramiński	Polish Academy of Sciences, Poland
Adam Gacek	Institute of Medical Technology and Equipment, Poland
Marina Gavrilova	University of Calgary, Canada
Manuel Grana	Computer Intelligence Group, Spain
Michał Kozielski	Silesian University of Technology, Poland
Marek Kurzyński	Wrocław University of Technology, Poland
Dariusz Mrozek	Silesian University of Technology, Poland
Bożena Małysiak-Mrozek	Silesian University of Technology, Poland
Agnieszka Nowak-Brzezińska	University of Silesia, Poland
Nobuyuki Nishiuchi	Tokyo Metropolitan University, Japan
Piotr Porwik	University of Silesia, Poland
Małgorzata Przybyła-Kasperek	University of Silesia, Poland
Roman Simiński	University of Silesia, Poland
Dragan Simic	University of Novi Sad, Serbia
Ewaryst Tkacz	Silesian University of Technology, Poland
Alicja Wakulicz-Deja	University of Silesia, Katowice, Poland

Additional Reviewers

Ben Brahim, Afef
Filonenko, Alexander
Holmgren, Johan
Le, Hoai Minh
Liutvinavicius, Marius
Meditskos, Georgios
Mihailescu, Radu-Casian
Mls, Karel
Montero, Javier
Phan, Duy Nhat
Piasny, Lukasz
Schomm, Fabian
Thilakarathne, Dilhan
Vascak, Jan
Zając, Wojciech

Contents – Part I

Knowledge Engineering and Semantic Web

Mapping the Territory for a Knowledge-Based System 3
Ulrich Schmitt

A Bidirectional-Based Spreading Activation Method for Human Diseases Relatedness Detection Using Disease Ontology 14
Said Fathalla and Yaman Kannot

Semantic Networks Modeling with Operand-Operator Structures in Association-Oriented Metamodel 24
Marek Krótkiewicz, Marcin Jodłowiec, and Krystian Wojtkiewicz

Knowledge Integration in a Manufacturing Planning Module of a Cognitive Integrated Management Information System 34
Marcin Hernes and Andrzej Bytniewski

The Knowledge Increase Estimation Framework for Ontology Integration on the Relation Level 44
Adrianna Kozierkiewicz-Hetmańska and Marcin Pietranik

Particle Swarm of Agents for Heterogenous Knowledge Integration. 54
Marcin Maleszka

Design Proposal of the Corporate Knowledge Management System. 63
Ivan Soukal and Aneta Bartuskova

Dipolar Data Integration Through Univariate, Binary Classifiers 73
Leon Bobrowski

Intelligent Collective: The Role of Diversity and Collective Cardinality 83
Van Du Nguyen, Mercedes G. Merayo, and Ngoc Thanh Nguyen

RuQAR: Querying OWL 2 RL Ontologies with Rule Engines and Relational Databases 93
Jarosław Bąk and Michał Blinkiewicz

The Efficiency Analysis of the Multi-level Consensus Determination Method 103
Adrianna Kozierkiewicz-Hetmańska and Mateusz Sitarczyk

Collective Intelligence Supporting Trading Decisions on FOREX Market. . . . 113
Jerzy Korczak, Marcin Hernes, and Maciej Bac

Social Networks and Recommender Systems

Testing the Acceptability of Social Support Agents in Online Communities . . . 125
Lenin Medeiros and Tibor Bosse

Enhancing New User Cold-Start Based on Decision Trees Active Learning by Using Past Warm-Users Predictions . . . 137
Manuel Pozo, Raja Chiky, Farid Meziane, and Elisabeth Métais

An Efficient Parallel Method for Performing Concurrent Operations on Social Networks . . . 148
Phuong-Hanh Du, Hai-Dang Pham, and Ngoc-Hoa Nguyen

Simulating Collective Evacuations with Social Elements . . . 160
Daniel Formolo and C. Natalie van der Wal

Social Networks Based Framework for Recommending Touristic Locations . . . 172
Mehdi Ellouze, Slim Turki, Younes Djaghloul, and Muriel Foulonneau

Social Network-Based Event Recommendation . . . 182
Dinh Tuyen Hoang, Van Cuong Tran, and Dosam Hwang

Deep Neural Networks for Matching Online Social Networking Profiles . . . 192
Vicentiu-Marian Ciorbaru and Traian Rebedea

Effect of Network Topology on Neighbourhood-Aided Collective Learning . . . 202
Lise-Marie Veillon, Gauvain Bourgne, and Henry Soldano

A Generic Approach to Evaluate the Success of Online Communities . . . 212
Raoudha Chebil, Wided Lejouad Chaari, and Stefano A. Cerri

Considerations in Analyzing Ecological Dependent Populations in a Changing Environment . . . 223
Kristiyan Balabanov, Robinson Guerra Fietz, and Doina Logofătu

Automatic Deduction of Learners' Profiling Rules Based on Behavioral Analysis . . . 233
Fedia Hlioui, Nadia Aloui, and Faiez Gargouri

Predicting the Evolution of Scientific Output . . . 244
Antonia Gogoglou and Yannis Manolopoulos

Data Mining Methods and Applications

Enhanced Hybrid Component-Based Face Recognition . . . 257
Andile M. Gumede, Serestina Viriri, and Mandlenkosi V. Gwetu

Enhancing Cholera Outbreaks Prediction Performance in Hanoi, Vietnam Using Solar Terms and Resampling Data 266
Nguyen Hai Chau

Solving Dynamic Traveling Salesman Problem with Ant Colony Communities 277
Andrzej Siemiński

Improved Stock Price Prediction by Integrating Data Mining Algorithms and Technical Indicators: A Case Study on Dhaka Stock Exchange 288
Syeda Shabnam Hasan, Rashida Rahman, Noel Mannan, Haymontee Khan, Jebun Nahar Moni, and Rashedur M. Rahman

A Data Mining Approach to Improve Remittance by Job Placement in Overseas 298
Ahsan Habib Himel, Tonmoy Sikder, Sheikh Faisal Basher, Ruhul Mashbu, Nusrat Jahan Tamanna, Mahmudul Abedin, and Rashedur M. Rahman

Determining Murder Prone Areas Using Modified Watershed Model 307
Joytu Khisha, Naushaba Zerin, Deboshree Choudhury, and Rashedur M. Rahman

Comparison of Ensemble Learning Models with Expert Algorithms Designed for a Property Valuation System. 317
Bogdan Trawiński, Tadeusz Lasota, Olgierd Kempa, Zbigniew Telec, and Marcin Kutrzyński

Multi-agent Systems

Multiagent Coalition Structure Optimization by Quantum Annealing 331
Florin Leon, Andrei-Ştefan Lupu, and Costin Bădică

External Environment Scanning Using Cognitive Agents 342
Marcin Hernes, Anna Chojnacka-Komorowska, and Kamal Matouk

OpenCL for Large-Scale Agent-Based Simulations 351
Jan Procházka and Kamila Štekerová

A Novel Space Filling Curves Based Approach to PSO Algorithms for Autonomous Agents 361
Doina Logofătu, Gil Sobol, Daniel Stamate, and Kristiyan Balabanov

Multiplant Production Design in Agent-Based Artificial Economic System . . . 371
Petr Tucnik, Zuzana Nemcova, and Tomas Nachazel

Role of Non-Axiomatic Logic in a Distributed Reasoning Environment 381
Mirjana Ivanović, Jovana Ivković, and Costin Bădică

Agent Having Quantum Properties: The Superposition States and the Entanglement 389
Alain-Jérôme Fougères

Sensor Networks and Internet of Things

A Profile-Based Fast Port Scan Detection Method 401
Katalin Hajdú-Szücs, Sándor Laki, and Attila Kiss

Sensor Network Coverage Problem: A Hypergraph Model Approach 411
Krzysztof Trojanowski, Artur Mikitiuk, and Mateusz Kowalczyk

Heuristic Optimization of a Sensor Network Lifetime Under Coverage Constraint 422
Krzysztof Trojanowski, Artur Mikitiuk, Frédéric Guinand, and Michał Wypych

Methods of Training of Neural Networks for Short Term Load Forecasting in Smart Grids 433
Robert Lis, Artem Vanin, and Anastasiia Kotelnikova

Scheduling Sensors Activity in Wireless Sensor Networks 442
Antonina Tretyakova, Franciszek Seredynski, and Frederic Guinand

Application of Smart Multidimensional Navigation in Web-Based Systems 452
Ivan Soukal and Aneta Bartuskova

WINE: Web Integrated Navigation Extension; Conceptual Design, Model and Interface 462
Ivan Soukal and Aneta Bartuskova

Real-Life Validation of Methods for Detecting Locations, Transition Periods and Travel Modes Using Phone-Based GPS and Activity Tracker Data 473
Adnan Manzoor, Julia S. Mollee, Aart T. van Halteren, and Michel C.A. Klein

Adaptive Runtime Middleware: Everything as a Service 484
Achilleas P. Achilleos, Kyriaki Georgiou, Christos Markides, Andreas Konstantinidis, and George A. Papadopoulos

Decision Support & Control Systems

Adaptive Neuro Integral Sliding Mode Control on Synchronization of Two Robot Manipulators 497
Parvaneh Esmaili and Habibollah Haron

Ant-Inspired, Invisible-Hand-Controlled Robotic System to Support Rescue Works After Earthquake 507
Tadeusz Szuba

Estimation of Delays for Individual Trams to Monitor Issues in Public Transport Infrastructure 518
Marcin Luckner and Jan Karwowski

Novel Effective Algorithm for Synchronization Problem in Directed Graph 528
Richard Cimler, Dalibor Cimr, Jitka Kuhnova, and Hana Tomaskova

Bimodal Biometric Method Fusing Hand Shape and Palmprint Modalities at Rank Level 538
Nesrine Charfi, Hanene Trichili, and Basel Solaiman

Adaptation to Market Development Through Price Setting Strategies in Agent-Based Artificial Economic Model 548
Petr Tucnik, Petr Blecha, and Jaroslav Kovarnik

Efficacy and Planning in Ophthalmic Surgery – A Vision of Logical Programming 558
Nuno Maia, Manuel Mariano, Goreti Marreiros, Henrique Vicente, and José Neves

A Methodological Approach Towards Crisis Simulations: Qualifying CI-Enabled Information Systems 569
Chrysostomi Maria Diakou, Angelika I. Kokkinaki, and Styliani Kleanthous

Multicriteria Transportation Problems with Fuzzy Parameters 579
Barbara Gładysz

Author Index 589

Contents – Part II

Cooperative Strategies for Decision Making and Optimization

Gene Expression Programming Ensemble for Classifying Big Datasets 3
Joanna Jędrzejowicz and Piotr Jędrzejowicz

Shapley Value in a Priori Measuring of Intellectual Capital Flows. 13
Jacek Mercik

MDBR: Mobile Driving Behavior Recognition Using Smartphone Sensors. . . 22
Dang-Nhac Lu, Thi-Thu-Trang Ngo, Hong-Quang Le, Thi-Thu-Hien Tran, and Manh-Hai Nguyen

Adaptive Motivation System Under Modular Reinforcement Learning for Agent Decision-Making Modeling of Biological Regulation 32
Amine Chohra and Kurosh Madani

Computational Swarm Intelligence

Simulated Annealing for Finding TSP Lower Bound 45
Łukasz Strąk, Wojciech Wieczorek, and Arkadiusz Nowakowski

A Cellular Automaton Based System for Traffic Analyses on the Roundabout 56
Krzysztof Małecki, Jarosław Wątróbski, and Waldemar Wolski

The Swarm-Like Update Scheme for Opinion Formation 66
Tomasz M. Gwizdałła

A Comparative Study of Different Variants of a Memetic Algorithm for ATSP 76
Krzysztof Szwarc and Urszula Boryczka

Improving ACO Convergence with Parallel Tempering 87
Rafał Skinderowicz

Modeling Skiers' Dynamics and Behaviors. 97
Dariusz Pałka and Jarosław Wąs

Genetic Algorithm as Optimization Tool for Differential Cryptanalysis of DES6 107
Kamil Dworak and Urszula Boryczka

Machine Learning in Medicine and Biometrics

Edge Real-Time Medical Data Segmentation for IoT Devices with Computational and Memory Constrains 119
Marcin Bernas, Bartłomiej Płaczek, and Alicja Sapek

A Privacy Preserving and Safety-Aware Semi-supervised Model for Dissecting Cancer Samples 129
P.S. Deepthi and Sabu M. Thampi

Decision Fusion Methods in a Dispersed Decision System - A Comparison on Medical Data 139
Małgorzata Przybyła-Kasperek, Agnieszka Nowak-Brzezińska, and Roman Simiński

Knowledge Exploration in Medical Rule-Based Knowledge Bases. 150
Agnieszka Nowak-Brzezińska, Tomasz Rybotycki, Roman Simiński, and Małgorzata Przybyła-Kasperek

Computer User Verification Based on Typing Habits and Finger-Knuckle Analysis 161
Hossein Safaverdi, Tomasz Emanuel Wesolowski, Rafal Doroz, Krzysztof Wrobel, and Piotr Porwik

ANN and GMDH Algorithms in QSAR Analyses of Reactivation Potency for Acetylcholinesterase Inhibited by VX Warfare Agent 171
Rafael Dolezal, Jiri Krenek, Veronika Racakova, Natalie Karaskova, Nadezhda V. Maltsevskaya, Michaela Melikova, Karel Kolar, Jan Trejbal, and Kamil Kuca

Multiregional Segmentation Modeling in Medical Ultrasonography: Extraction, Modeling and Quantification of Skin Layers and Hypertrophic Scars 182
Iveta Bryjova, Jan Kubicek, Kristyna Molnarova, Lukas Peter, Marek Penhaker, and Kamil Kuca

Cyber Physical Systems in Automotive Area

Model of a Production Stand Used for Digital Factory Purposes 195
Markus Bregulla, Sebastian Schrittenloher, Jakub Piekarz, and Marek Drewniak

Improving the Engineering Process in the Automotive Field Through AutomationML 205
Markus Bregulla and Flavian Meltzer

Enhanced Reliability of ADAS Sensors Based on the Observation of the Power Supply Current and Neural Network Application 215
Damian Grzechca, Adam Ziębiński, and Paweł Rybka

ADAS Device Operated on CAN Bus Using PiCAN Module for Raspberry Pi 227
Marek Drewniak, Krzysztof Tokarz, and Michał Rędziński

Obstacle Avoidance by a Mobile Platform Using an Ultrasound Sensor 238
Adam Ziebinski, Rafal Cupek, and Marek Nalepa

Monitoring and Controlling Speed for an Autonomous Mobile Platform Based on the Hall Sensor 249
Adam Ziebinski, Markus Bregulla, Marcin Fojcik, and Sebastian Kłak

Using MEMS Sensors to Enhance Positioning When the GPS Signal Disappears 260
Damian Grzechca, Krzysztof Tokarz, Krzysztof Paszek, and Dawid Poloczek

Application of OPC UA Protocol for the Internet of Vehicles 272
Rafał Cupek, Adam Ziębiński, Marek Drewniak, and Marcin Fojcik

Feasibility Study of the Application of OPC UA Protocol for the Vehicle-to-Vehicle Communication 282
Rafał Cupek, Adam Ziębiński, Marek Drewniak, and Marcin Fojcik

Data Mining Techniques for Energy Efficiency Analysis of Discrete Production Lines 292
Rafal Cupek, Jakub Duda, Dariusz Zonenberg, Łukasz Chłopaś, Grzegorz Dziędziel, and Marek Drewniak

Internet of Things - Its Relations and Consequences

Different Approaches to Indoor Localization Based on Bluetooth Low Energy Beacons and Wi-Fi 305
Radek Bruha and Pavel Kriz

Towards Device Interoperability in an Heterogeneous Internet of Things Environment 315
Pavel Pscheidl, Richard Cimler, and Hana Tomášková

Hardware Layer of Ambient Intelligence Environment Implementation 325
Ales Komarek, Jakub Pavlik, Lubos Mercl, and Vladimir Sobeslav

Lightweight Protocol for M2M Communication 335
Jan Stepan, Richard Cimler, Jan Matyska, David Sec, and Ondrej Krejcar

Wildlife Presence Detection Using the Affordable Hardware Solution and an IR Movement Detector 345
Jan Stepan, Matej Danicek, Richard Cimler, Jan Matyska, and Ondrej Krejcar

Text Processing and Information Retrieval

Word Embeddings Versus LDA for Topic Assignment in Documents 357
Joanna Jędrzejowicz and Magdalena Zakrzewska

Development of a Sustainable Design Lexicon. Towards Understanding the Relationship Between Sentiments, Attitudes and Behaviours 367
Vargas Meza Xanat and Yamanaka Toshimasa

Analysing Cultural Events on Twitter 376
Brigitte Juanals and Jean-Luc Minel

A Temporal-Causal Model for Spread of Messages in Disasters 386
Eric Fernandes de Mello Araújo, Annelore Franke, and Rukshar Wagid Hosain

One Approach to the Description of Linguistic Uncertainties 398
Nikita Ogorodnikov

Complex Search Queries in the Corpus Management System 407
Damir Mukhamedshin, Olga Nevzorova, and Aidar Khusainov

Entropy-Based Model for Estimating Veracity of Topics from Tweets 417
Jyotsna Paryani, Ashwin Kumar T.K., and K.M. George

On Some Approach to Integrating User Profiles in Document Retrieval System Using Bayesian Networks 428
Bernadetta Maleszka

Analysis of Denoising Autoencoder Properties Through Misspelling Correction Task 438
Karol Draszawka and Julian Szymański

New Ontological Approach for Opinion Polarity Extraction from Twitter 448
Ammar Mars, Sihem Hamem, and Mohamed Salah Gouider

Study for Automatic Classification of Arabic Spoken Documents 459
Mohamed Labidi, Mohsen Maraoui, and Mounir Zrigui

"*Come Together!*": Interactions of Language Networks and Multilingual Communities on Twitter 469
Nabeel Albishry, Tom Crick, and Theo Tryfonas

Bangla News Summarization 479
Anirudha Paul, Mir Tahsin Imtiaz, Asiful Haque Latif, Muyeed Ahmed, Foysal Amin Adnan, Raiyan Khan, Ivan Kadery, and Rashedur M. Rahman

Low Resource Language Processing

Combined Technology of Lexical Selection in Rule-Based Machine Translation 491
Ualsher Tukeyev, Dina Amirova, Aidana Karibayeva, Aida Sundetova, and Balzhan Abduali

New Kazakh Parallel Text Corpora with On-line Access 501
Zhandos Zhumanov, Aigerim Madiyeva, and Diana Rakhimova

Design and Development of Media-Corpus of the Kazakh Language 509
Madina Mansurova, Gulmira Madiyeva, Sanzhar Aubakirov, Zhantemir Yermekov, and Yermek Alimzhanov

Morphological Analysis System of the Tatar Language 519
Rinat Gilmullin and Ramil Gataullin

Context-Based Rules for Grammatical Disambiguation in the Tatar Language 529
Ramil Gataullin, Bulat Khakimov, Dzhavdet Suleymanov, and Rinat Gilmullin

Computer Vision Techniques

Evaluation of Gama Analysis Results Significance Within Verification of Radiation IMRT Plans in Radiotherapy 541
Jan Kubicek, Iveta Bryjova, Kamila Faltynova, Marek Penhaker, Martin Augustynek, and Petra Maresova

Shape Classification Using Combined Features 549
Laksono Kurnianggoro, Wahyono, Alexander Filonenko, and Kang-Hyun Jo

Smoke Detection on Video Sequences Using Convolutional and Recurrent Neural Networks 558
Alexander Filonenko, Laksono Kurnianggoro, and Kang-Hyun Jo

Intelligent Processing of Multimedia in Web Systems

Improved Partitioned Shadow Volumes Method of Real-Time Rendering Using Balanced Trees 569
Kazimierz Choroś and Tomasz Suder

Online Comparison System with Certain and Uncertain Criteria Based on Multi-criteria Decision Analysis Method 579
Paweł Ziemba, Jarosław Jankowski, and Jarosław Wątróbski

Assessing and Improving Sensors Data Quality in Streaming Context 590
Rayane El Sibai, Yousra Chabchoub, Raja Chiky, Jacques Demerjian, and Kablan Barbar

Neural Network Based Eye Tracking 600
Pavel Morozkin, Marc Swynghedauw, and Maria Trocan

Author Index 611

Knowledge Engineering and Semantic Web

Mapping the Territory for a Knowledge-Based System

Ulrich Schmitt(✉)

University of Stellenbosch Business School, PO Box 610, Bellville 7535, South Africa
schmitt@knowcations.org

Abstract. Although many powerful applications are able to locate vast amounts of digital information, effective tools for selecting, structuring, personalizing, and making sense of the digital resources available to us are lacking. As a result, the opportunities to connect and empower knowledge workers are severely limited.

In recognizing these constraints, predictions of the 'Next Knowledge Management (KM) Generation' focus on nurturing personal and social settings and on utilizing existing and creating new knowledge. Levy even envisages a decentralizing KM revolution that gives more power and autonomy to individuals and self-organized groups. But, such promising scenarios have not materialized yet. It might be time to follow Pollard's suggestion of going back to the original premise and promise of KM and start again - but this time from the bottom up by developing processes, programs, and tools to improve knowledge workers' effectiveness and sense-making. As part of an ongoing design science research (DSR) project, this paper contributes to prior publications by synthesizing renowned computer-based methods of collective intelligence to provide a visual meta-perspective of a novel personal knowledge management (PKM) concept and prototype application. In focusing on time, space, and causality, the bottom-up approach taken, pictures the relevant personal and organizational knowledge spaces as a substitute for the intangible KM territory and provides a guiding map for knowledge workers and KM education.

Keywords: Knowledge management · Personal knowledge management · Design science research · Knowledge worker · Information space · Knowcations

1 Role of Metaphors and Visuals for Design Science and PKM

Knowledge is an abstract concept with no clearly delineated structure and no 'real world' referent. To give it structure and make it comprehensible, metaphors allow mapping 'real world' things we are familiar with onto the concept of knowledge. Regardless of whether metaphors are used as a thinking device or to support dialogue and education, their potential "for communicating and stimulating creativity may be further enhanced when combined with visuals" [1, 2]. Thus, this paper's purpose is to visualize the Personal and Organizational Knowledge Spaces so Knowledge Workers can be guided by a map that substitutes for the intangible KM territory.

N.T. Nguyen et al. (Eds.): ICCCI 2017, Part I, LNAI 10448, pp. 3–13, 2017.
DOI: 10.1007/978-3-319-67074-4_1

Personal and organizational KM could have been so much more focused on supporting Knowledge Workers, if Bush's vision of the 'Memex[1]' had materialized already [3–6]. This lapse - exaggerated by the relentless rising abundance of information which is threatening the very attention our finite cognitive capabilities are able to master - constitutes probably today's biggest threat to individual and collective intelligence, performance, and development. As a seven decades old inspiring idea never realized, the 'Memex' represents, however, the as-close-as-it-gets ancestor of the novel Personal Knowledge Management (PKM) concept and prototype system-in-progress[2].

The PKM approach is firmly rooted in Design Science Research (DSR)[3], and the metaphors and visualizations presented aim for 'Theory Effectiveness', a DSR notion which characterizes a theory that is incrementally and iteratively designed to be purposeful in terms of its utility and content but also in its communication to an audience [23]. Accordingly, prior posters, papers, and articles have reported to and received feedback from a wide range of disciplinary conferences and journals (see www.researchgate.net/profile/Ulrich_Schmitt2 or mini-abstracts [18]). With access to them assured, this paper focuses on the mapping of space (three-dimensional ecosystems), time (workflows), and causality (learning cycles) to synthesize underlying methodologies and to illustrate the PKM Systems (PKMS) functionalities, objects, and repositories.

2 Employing Metaphors and Maps in Knowledge Management

"The function of theory-building is not to build general, accurate and or simple theory per se but to produce purposeful theory." As such, effective theory-building is a design process that seeks and continuously evaluates the theory's utility and embraces visuals as a means of communication by highlighting those theoretic features that are appropriate for a purpose and its audience [24].

[1] In 1945, Vannevar Bush (then President Truman's Scientific Research Director) imagined the 'Memex', a hypothetical sort of mechanized private file/desk/library-device. It is supposed to act as an enlarged intimate supplement to one's memory, and enables an individual to store, recall, study, and share the "inherited knowledge of the ages". It would have facilitated the addition of personal records, communications, annotations, and contributions, but, above all, the recording of non-fading trails of one's individual "interests through the maze of materials available" - all easily accessible and sharable with the 'Memexes' of acquaintances [3].

[2] Prior publications have elaborated on the scope of anticipated PKM outcomes and the appealing opportunities they provide for stakeholders engaged in the context of curation [7, 8], education [9–11], research [12, 13], development [14, 15], experience management [16], business and entrepreneurship [17–19]. Further papers assessed the PKMS potential against established criteria as a disruptive innovation [20] and as a general-purpose-technology [13] and pointed out missing capabilities and the PKM affordances in need of being conferred [21].

[3] DSR guidelines and methodologies are meant to supplement the reactive behavioral (natural) science paradigm with the proactive design science paradigm in order to support researchers in creating innovative IT artefacts that extend human and social capabilities and meet desired outcomes [22].

Eppler reinforces these notions for the field of KM in his classification of knowledge maps by defining their essential role as graphic overviews and references of knowledge-related content that add further knowledge and serve knowledge management-related purposes, including facilitating [25]. Visuals also assist us to overcome the textual constraint of providing merely linear accounts of a nonlinear world [26]. This paper endeavors to map PKMS's contexts of space, time, and causality which are themselves "only metaphors of knowledge, with which we explain things to ourselves" [27].

The PKMS Map is based on Boisot's three-dimensional Information-Space or I-Space Model (Fig. 1) which originally depicts the dynamic flow of knowledge assets following a Social Learning Cycle (SLC) through six phases [28]. To accommodate the PKMS contexts and motivations[4], the I-Space Model has been modified by dividing its codification axis – in line with Popper's Three Worlds notion and the Digital Ecosystems introduced [22] - into distinct sections (from left to right): uncodified (emotions or intuition based tacit knowledge), codified (rationality based explicit information), PKMS captured (intellectual, social, and emotional capital), and PKMS shared knowledge.

The two left sections represent the physical objects (hosts and vectors) embodied in the external environment with their relationships and effects. Hosts are members of society and institutions (embodying the tacit uncodified knowledge of the collective human mind set) and must possess the potential capacity to elaborate on knowledge and to perform those cognitive tasks that we refer to as "understanding". A knowledge worker interacts with them through field research via observations and/or interviews. Vectors comprise the technological ecosystem of physical bodies and artefacts as objects of their encapsulated knowledge as well as the extelligence ecosystem with its encoded content stored in physical knowledge containers (e.g. books, posters, or digital files). A knowledge worker interacts through desk research via re-engineering, analysis, or reading.

Five of the many renowned KM notions instrumental to the PKMS thinking have also been appropriately positioned within these sections to demonstrate their synergies with the novel PKMS approach. While Kolb's Learning Model [30] concentrates on the individual, Boisot's Social Learning Cycle occupies the full space of the abstraction-codification-diffusion framework. Wierzbicki's and Nakamori's Nanatsudaki Model [31] (which integrates the SECI Spiral [32]) further differentiates between intuition and

[4] In parallel to the prototype development, the PKMS design process and its methodological design elements have been validated against the systems thinking techniques of the transdiscipline of Informing Science (IS) [29] and the accepted general DSR research guidelines alluded to [22]. Rather than to justify the research paradigm of the PKMS project in an ad hoc and fragmented manner with each new paper, the dedicated articles present the IS and DSR perspectives comprehensively as evidence of their relevance, utility, rigor, and publishability. Their conclusions emphasize PKM's status as a 'wicked' problem (ill-defined; incomplete, contradictory, changing requirements; complex interdependencies) where the information needed to understand the challenges depends upon one's idea for solving them. Accordingly, a chain of meta-arguments addressed the central ideas of the PKMS concept (incorporating notions of complexity and Popper's three worlds) leading to the development of a PKM framework made up of six Digital Ecosystems referring to the particular spaces of technology, extelligence. knowledge workers, institutions, society, and the ideosphere.

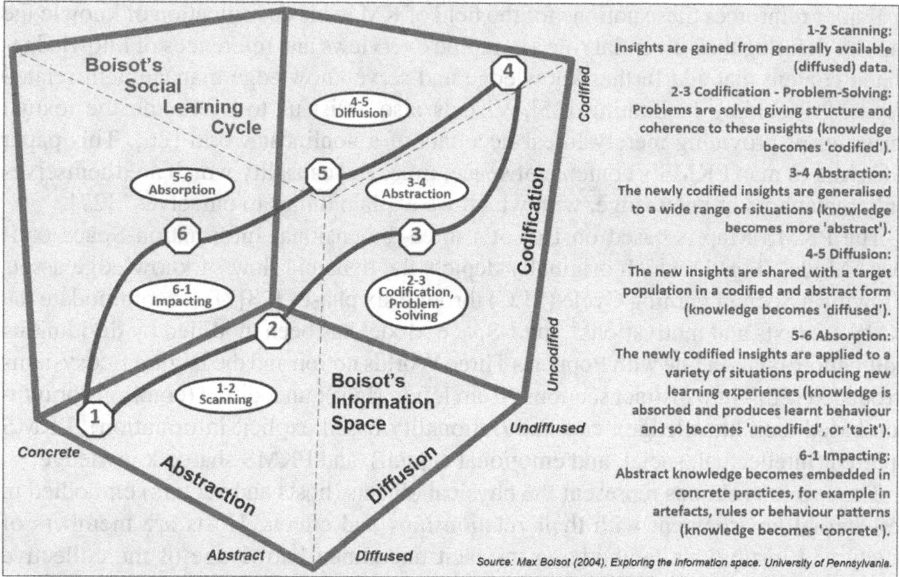

Fig. 1. Boisot's Information-Space (I-Space) and Social Learning Curve (SLC) [28]

emotions (tacit knowledge) and between individual and group perspectives, as discussed in a prior paper [33]. For further illustration, the knowledge assets suggested by Boisot and Nonaka [28, 32] have been placed accordingly.

The middle section represents an individual's human mind as an atomic constituent of potentially many collective mindsets (e.g. teams, guilds, professions, institutions, cultures, societies) and his/her knowledge worker ecosystem. It depicts Pirolli's and Card's Notional Model of the Sensemaking Loop for Intelligence Analysis, an empirical descriptive study in the context of expertise and work [34, 35]. Sensemaking "suggests an active processing of information to achieve understanding" which "involves not only finding information but also requires learning about new domains, solving ill-structured problems, acquiring situation awareness, and participating in social exchanges of knowledge" [36]. However, the future of work and knowledge societies is said to be based on the notion that the knowledge and skills of a knowledge worker are portable and mobile. Accordingly, the PKM affordance presented [21] would finally enable individuals - moving from one project or responsibility to another - to take their personal version of a KM system (able to be continually maintained and updated) with them wherever they choose to go, engage, share, or collaborate.

The two right sections represent the PKMS system. The left section focuses on the technological means to support individuals' mobility and autonomy via decentralized PKMS devices termed 'Knowcations®'. Its extended flows, functionalities, and loops incorporate Pirolli's and Card's model with some of their terminology amended to better fit the PKM concept.

The foraging loop is shown in red, the sensemaking loop in blue, and green arrows depict the workflow loop of additional PKMS support functions and services; all three loops substitute for Boisot's original Social Learning Cycle.

The right section pictures the supporting centralized repository (termed WHOMER™ for 'World Heritage of Memes Repository' [8]) as a voluntarily shared, global digital cloud-based library for the grass-roots, decentralized, networked, autonomous, personal devices and capacities, envisaged to constitute "the elementary process that makes possible the emergence of the distributed processes of collective intelligence, which in turn feed it" via creative conversations [38].

The three distinct areas in the right section represent the PKMS community with their decentralized PKMS workflows and devices with their access to the cloud-based PKMS repository. The underlying PKMS concept substantially deviates from the current document-centric KM systems and, instead, is based on the capturing, storing, and repurposing of basic information structures (ideas or memes[5] [39]) and their relationships (to create knowledge assets and other archetypal reconstructions thereof) rather than storing and referencing them the conventional way in their containers only (e.g. book, paper, report) It, thus, follows Bush's notion of 'Associative Indexing' [3] and supports Usher's concept of 'Cumulative Synthesis', a process-oriented performative account of innovation [41]. Further reasons are the better traceability of knowledge and the reducing of redundancies from the ever-increasing information abundance alluded to, a rationale evaluated in the DSR-related paper [22] and publication case [42].

3 A Knowledge Map Visualizing Knowledge in Space and Time

The following subsections provide an updated and considerably expanded account [33, 43] of the PKMS workflow loops and repositories depicted in Fig. 2.

3.1 Foraging Loop: Knowledge Identification, Acquisition, and Preservation

Memes and ideas are uncovered via field research (1H) by interacting with Hosts in form of conversations, interviews, and/or observations. The PKMS user extracts the relevant information accessible to him/her and stores it (1H&RH) in a preliminary Shoe Box or Case File '*Authorship*'. The outcome resembles a disorganized pile in need of further examination. Furthermore, references (e.g. contact details, date, place) are stored (RH: 2P) in the knowledge base '*Profiles*' and linked to the respective memes in '*Authorship*' to support their searchability, findability, and referencing later.

Memes are also uncovered via desk research (1V) but are usually packaged or absorbed in complex ways inside larger vectors/sources (e.g. books, files). The user selects and collects the relevant content accessible to him/her and adds them (1V&RV)

[5] Memes were originally described by Dawkins [40] as units of cultural transmission or imitation (e.g. ideas, tunes, catch-phrases, skills, technologies). They are (cognitive) information-structures that evolve over time through a Darwinian process of variation, selection and transmission with their longevity being determined by their environment.

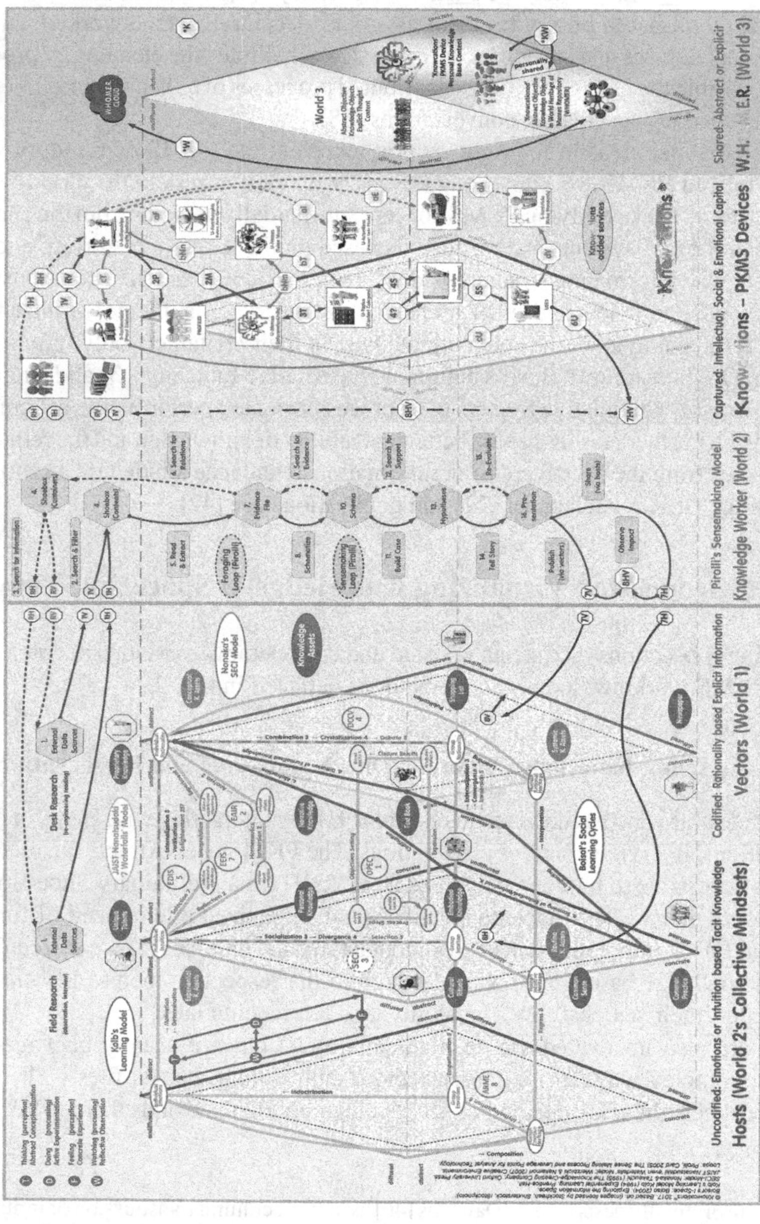

Fig. 2. PKM Information Space representing the External and PKM System Environment [37] (complementary high-resolution download from www.researchgate.net/publication/317551944)

to the Case File '*Authorship*'. Again, references (e.g. origins, authors, publishers, contact details, titles, formats, licenses) are stored (RV:2P) in '*Profiles*' as above.

Based on his/her interests, knowledge and editorial literacy, the user consciously chooses suitable content gathered from the Case File '*Authorship*', captures their memes' intended original messages and stores them (2M) in a dedicated Knowledge Base '*Memes*'. To ease re-usability any meme captured and codified ideally should be in an atomic state which might require the variation and replication of an original meme in a creative manner. To enhance accessibility, the user qualifies a meme by linking it to a multi-dimensional classification system, made up of pre- or user-defined abstract Meme Types (e.g. area, concept, process, tool) or topics (e.g. decision methods, logistics, ecology) to be stored (3T) in a Schema Base '*Topics*'. To represent the user's social capital and other's social relationships, hosts (individuals, teams, communities, organizations) can be linked to each other, and further qualified according to their research/project-related, industrial, service-oriented, and geographic settings and relevant documents to be stored in '*Profiles*'

3.2 Sensemaking Loop: Knowledge Goals, Development, Diffusion, Use

During the authoring process, the accumulated meme pool is scanned to activate appropriate 'meme candidates' for composing a planned Script (e.g. article, lecture, or presentation) to be stored (4S) in a Scripts and Hypotheses Base '*Scripts*'. Any gaps will be filled with a provisional 'known-unknown' meme as a reminder for field and desk research or creative work to be carried out (4?). Any finalized script can be converted (5U) into a presentation, pdf or paper version and stored in a Report or Presentation Base '*Uses*' for publication and wider distribution (6U) in order to become part of the public world extelligence (Hosts & Vectors) or to be shared in the WHOMER library.

The novel insights are shared via newly codified vectors (7V) by publication and diffusion or via oral presentation or discussion from mind-to-mind (7H). They might get absorbed by people and can become personal or organizational extelligence and lead to new learning experiences and behaviors. Subsequently, the abstract knowledge absorbed might make an impact by becoming embedded in concrete practices, either in codified formats such as documents or products (8V) or uncodified formats such as unwritten rules or patterns of behavior (8H).

The sharing and absorbing activities (7HV, 8HV) have been addressed by Nonaka and Takeuchi [32] in their groundbreaking SECI-Model. A recent paper demonstrates and visualizes its considerable synergies with the novel PKMS approach. While the SECI Model promotes individual and collective real-world learning processes in the Socializing-Externalizing-Combining-Internalizing-cycle, the PKMS's meme-based workflows are following the reverse ICES-order allowing for the PKMS support of [organizational] OKMS cycles [20, 21].

3.3 PKMS Support Functions: Knowledge Measurement and Management

The iterative process described [steps 1*–6*] is driven by the user's intentions, self-understanding, and self-reflection; an 'Extended PKM Ignorance Matrix' and 'PKM Value Chain' have been presented and visualized to further guide this endeavor [29].

While individual performance histories (own and others) are recorded and tracked via the '*Profiles*' base, three further bases are reflecting on the user's emotional capital by supporting the projects in progress: The '*Forethoughts*' Base stores longer-term plans and objectives and related thoughts and responses, the '*Intentions*' Base deals with shorter term tasks and diaries, and the '*Evaluations*' Base records, for example, feed-backs, reflections, and references to personal assets.

Connected in a feedback cycle (aF-aI-aE), they allow for the scheduling and monitoring of progress via dedicated fields for comments and to-do-lists regarding any entity in the repositories. This way, outstanding 'work to be carried out' (meme gaps to be filled) is evaluated against progress, and decisions can be made or revised (e.g. if already captured memes have to be utilized, further suitable external memes need to be found, new memes have to be self-authored or re-purposed from personal repositories). The entries are 'private' (not to be published) memes linked to particular memes to contain, for example, annotations, further ideas, feed-backs from colleagues or peers.

While the Pirelli/Card Model focusses on intelligence analysis [34], the PKMS Model also encompasses creating New Memes '*Nemes*' based on own ideas (bNn) or on modifying captured memes (bNm) which can like captured memes linked to (bT) entries in the '*Topics*' Base. Supporting evidence of any relationships can be stored (cT) separately in the '*Testimonials*' Base and attached (cU) to the '*Uses*' Base to back up any statements or claims made in the user's output to be published.

Any finalized script published can be retained in the knowledge base in a format with all reference links kept intact and instant access to the underlying information-rich contributing memes, their sources and alternate uses. This type of digital document can be transferred to the WHOMER Library and - if the content is appropriate for the purpose - is also ideal for storage (dY) in a Benchmarking/Standards Base '*Yardsticks*' able to feed forward (dA) to related subsequent projects and activities by providing templates, samples, best-practice methods, proven heuristics, regulations, tutorials, evaluation criteria, or trial assessments and in support of experience management [16].

4 Conclusions and the Road Ahead

"While today we have many powerful applications for locating vast amounts of digital information, we lack effective tools for selecting, structuring, personalizing, and making sense of the digital resources available to us" [44], a shortcoming of recent developments in ICT [45]. The novel PKM concept and prototype system-in-progress offers a solution. It merges distinctive voluntarily shared knowledge objects/assets of diverse disciplines into a single unified digital knowledge repository allowing for concretizing Popper's abstract *World 3* [22]. Any shared meme becomes available for learning and personalized curation as well as reusable in new contexts. On the one hand, these features provide the means to tackle the widening opportunity divides by affording individual knowledge workers with continuous life-long support from trainee, student, novice, or mentee towards professional, expert, mentor, or leader. On the other hand, they add transparency and momentum to the digital asset production and value creation and, with it, to the evolution of knowledge at the personal, institutional, and societal level. In a co-evolutionary PKMS-OKMS context, they would strengthen the absorptive capacity, ambidexterity, and resulting dynamic capability of organizations considerably, not at the expense of disinterested employees but as a means to motivate them by serving their self-interests [18].

As suggested in the introduction, the visualizations presented are, hence, of particular importance not only for forthcoming educational activities but also for successfully communicating the PKM concept (and the shortfalls of traditional KM) to a diverse portfolio of audiences. To further widen its appeal, the PKM Concept's innovative features and educational philosophies are currently aligned to an established Learning Management System [11].

Further publications and posters are also under review or planned addressing a PKMS Sustainability Vision, demonstrations and tutorials/workshops, and how the PKMS concept compares to an alternative PKM approach based on semantic technologies [46], and also how it can make use of and further add to semantic web technologies. After completing the test phase of the prototype, its transformation into a viable PKMS device application and a cloud-based WHOMER server based on a rapid development platform and a noSQL-database is estimated to take 12 months.

References

1. Andriessen, D.: On the metaphorical nature of intellectual capital: a textual analysis. J. Intellect. Capital **7**(1), 93–110 (2006)
2. Andriessen, D., Kliphuis, E., McKenzie, J., Van Winkelen, C.: Pictures of knowledge management, developing a method for analysing knowledge metaphors in visuals. Electron. J. Knowl. Manag. **7**(4), 405–415 (2009)
3. Bush, V.: As we may think. Atl. Mon. **176**(1), 101–108 (1945)
4. Davies, S.: Still building the Memex. Commun. ACM **53**(2), 80–88 (2011)
5. Osis, K., Gaindspenkis, J.: Modular personal knowledge management system and mobile technology cross-platform solution towards learning environment support. In: Annual International Conference on Virtual and Augmented Reality in Education (VARE), pp. 114–124 (2011)
6. Schmitt, U.: Overcoming the seven barriers to innovating personal knowledge management systems. In: Proceedings of the International Forum on Knowledge Asset Dynamics (IFKAD), Matera, Italy, 11–13 June 2014, pp. 3662–3681 (2014)
7. Schmitt, U.: Supporting digital scholarship and individual curation based on a meme-and-cloud-based personal knowledge management concept. Acad. J. Sci. (AJS) **4**(1), 220–237 (2015)
8. Schmitt, U.: Towards a 'world heritage of memes repository' for tracing ideas, tailoring knowledge assets and tackling opportunity divides: supporting a novel personal knowledge management concept. Int. J. Technol. Knowl. Soc. Annu. Rev. **10**, 25–44 (2015)
9. Schmitt, U., Butchart, B.A.H.: Making personal knowledge management part and parcel of higher education programme and services portfolios. J. World Univ. Forum **6**(4), 87–103 (2014)

10. Schmitt, U.: Redefining knowledge management education with the support of personal knowledge management devices. In: Uskov, V.L., Howlett, R.J., Jain, L.C. (eds.) Smart Education and e-Learning 2016. SIST, vol. 59, pp. 515–525. Springer, Cham (2016). doi: 10.1007/978-3-319-39690-3_46
11. Schmitt, U., Saade, R.G.: Taking on opportunity divides via smart educational and personal knowledge management technologies. In: 12th International Conference on e-Learning (ICEL), Orlando, USA, 1–2 June 2017 (2017)
12. Schmitt, U.: Knowledge management as artefact and expediter of interdisciplinary discourses. In: Proceedings of 9th International Multi-Conference on Society, Cybernetics and Informatics (IMSCI), Orlando, USA, 12–15 July 2015, pp. 92–98 (2015)
13. Schmitt, U.: Knowledge management systems as an interdisciplinary communication and personalized general-purpose technology. J. Syst. Cybern. Inform. **13**(6), 28–37 (2015)
14. Schmitt, U.: Making sense of e-skills at the dawn of a new personal knowledge management paradigm. In: Proceedings of the 2014 e-Skills for Knowledge Production and Innovation Conference, Cape Town, South Africa, 17–21 November 2014, pp. 417–447 (2014)
15. Schmitt, U.: Personal knowledge management for development (PKM4D) framework and its application for people empowerment. In: Elsevier Procedia Computer Science (International Conference on KM (ICKM), Vienna, 10–11 October 2016, vol. 99, pp. 64–78 (2016)
16. Schmitt, U.: The logic of use and functioning of personal KM-supported experience management. In: 7th German Workshop on Experience Management (GWEM) within the 9th Conference on Professional Knowledge Management (ProWM), Karlsruhe, 5–7 April 2017 (2017)
17. Schmitt, U.: Quo vadis, knowledge management: a regeneration or a revolution in the making? J. Inf. Knowl. Manage. (JIKM), **14**(4) (2015)
18. Schmitt, U.: Tools for exploration and exploitation capability: towards a co-evolution of organizational and personal knowledge management systems. Int. J. Knowl. Cult. Change Manag. Annu. Rev. **15**, 23–47 (2016)
19. Schmitt, U.: Strengthening SMEs impact and sustainability with the support of personal knowledge management systems and concepts. In: 15th International Entrepreneurship Forum (IEF), Venice, Italy, 14–16 December 2016, pp. 73–91 (2016)
20. Schmitt, U.: Utilizing the disruptive promises of personal knowledge management devices for strengthening organizational capabilities of innovativeness and leadership. Presented Paper at the 5th Ashridge International Research Conference (AIRC5), Berkhamsted, UK, 3–5 July 2016 (2016)
21. Schmitt, U.: Devising enabling spaces and affordances for personal knowledge management system design. Inform. Sci. Int. J. Emerg. Transdiscipl. (InformingSciJ) **20**, 63–82 (2017)
22. Schmitt, U.: Design science research for personal knowledge management system development—revisited. Inform. Sci. Int. J. Emerg. Transdiscipl. (InformingSciJ) **19**, 345–379 (2016)
23. O'Raghallaigh, P., Sammon, D., Murphy, C.: The design of effective theory. Syst. Signs Actions **5**(1), 117–132 (2011)
24. O'Raghallaigh, P., Sammon, D., Murphy, C.: Map-making informing a framework for effective theory-building. All Sprouts Content, Paper 371 (2010)
25. Eppler, M.J.: A process-based classification of knowledge maps and application examples. Knowl. Process Manag. **15**(1), 59–71 (2008)
26. Mintzberg, H.: Developing theory about the development of theory. In: Smith, K.G., Hitt, M.A. (eds.) Great Minds in Management: The Process of Theory Development, pp. 355–372. Oxford University Press, New York (2005)
27. Nietzsche, F.: The philosopher: reflections on the struggle between art and knowledge (1872). https://archive.org/details/StruggleBetweenArtAndKnowledge

28. Boisot, M.: Exploring the information space: a strategic perspective on information systems. Working Paper Series WP04-003, University of Pennsylvania (2004)
29. Schmitt, U.: Putting personal knowledge management under the macroscope of informing science. Int. J. Emerg. Transdiscipl. (InformingSciJ) **18**, 145–175 (2015)
30. Kolb, D.A.: Experiential Learning: Experience as the Source of Learning and Development. Prentice Hall, Englewood Cliffs (1984)
31. Wierzbicki, A.P., Nakamori, Y.: Creative Environments: Issues of Creativity Support for the Knowledge Civilization Age. Springer, Berlin (2007)
32. Nonaka, I., Takeuchi, H.: The Knowledge-Creating Company. Oxford University Press, Oxford (1995)
33. Schmitt, U.: Knowcations—positioning of a meme and cloud-based personal second generation knowledge management system. In: Skulimowski, A.M.J., Kacprzyk, J. (eds.) Knowledge, Information and Creativity Support Systems: Recent Trends, Advances and Solutions. AISC, vol. 364, pp. 243–257. Springer, Cham (2016). doi: 10.1007/978-3-319-19090-7_19
34. Pirolli, P., Card, S.: The sensemaking process and leverage points for analyst technology. In: Proceedings of International Conference on Intelligence Analysis (2005)
35. Schmitt, U.: Innovating personal knowledge creation and exploitation. In: 2nd Global Innovation and Knowledge Academy Conference (GIKA), Valencia, Spain, 9–11 July 2013 (2013)
36. Pirolli, P., Russell, D.M.: Introduction to this special issue on sensemaking. Hum. Comput. Interact. **26**(1–2), 1–8 (2011)
37. Schmitt, U.: PKM information space representing the external and personal knowledge management system environment. Poster supporting key note address at the 9th Conference on Professional Knowledge Management (Workshop "Flexible Knowledge Practices and the Digital Workplace"—FKPDW), Karlsruhe, Germany, 5–7 April 2017 (2017)
38. Levy, P.: The Semantic Sphere 1. Wiley, New York (2011)
39. Schmitt, U.: The Significance of memes for the successful formation of autonomous personal knowledge management systems. In: Kunifuji, S., Papadopoulos, G.A., Skulimowski, A.M.J., Kacprzyk, J. (eds.) Knowledge, Information and Creativity Support Systems. AISC, vol. 416, pp. 409–419. Springer, Cham (2016). doi:10.1007/978-3-319-27478-2_29
40. Dawkins, R.: The Selfish Gene. Paw Prints, Clermont (1976)
41. Usher, A.P.: A History of Mechanical Inventions, Revised edn. Courier Corporation, North Chelmsford (2013)
42. Schmitt, U.: How this paper has been created by leveraging a personal knowledge management system. In: 8th International Conference on Higher Education (ICHE), Tel Aviv, Israel, 16–18 March 2014, pp. 22–40 (2014)
43. Schmitt, U.: The Significance of 'Ba' for the successful formation of autonomous personal knowledge management systems. In: Kunifuji, S., Papadopoulos, G.A., Skulimowski, A.M.J., Kacprzyk, J. (eds.) Knowledge, Information and Creativity Support Systems. AISC, vol. 416, pp. 391–407. Springer, Cham (2016). doi:10.1007/978-3-319-27478-2_28
44. Kahle, D.: Designing open educational technology. In: Vijay Kumar, M.S., Iiyoshi, T. (eds.) Opening Up Education, pp. 27–46. MIT Press, Cambridge (2009)
45. Schmitt, U.: Shortcomings of the web of documents and data for managing personal knowledge and collaboration. In: 1st International Conference on Next Generation Computing Applications (NextComp), Mauritius, 19–21 July 2017. To be published in IEEE Xplore (2017)
46. Völkel, M.: Personal Knowledge Models with Semantic Technologies. BoD–Books on Demand, Norderstedt (2011)

A Bidirectional-Based Spreading Activation Method for Human Diseases Relatedness Detection Using Disease Ontology

Said Fathalla[1,2(✉)] and Yaman Kannot[3]

[1] Enterprise Information Systems (EIS), University of Bonn, Bonn, Germany
fathalla@cs.uni-bonn.de

[2] Faculty of Science, Alexandria University, Alexandria, Egypt

[3] Software Engineer, Alexandria, Egypt
yaman.kce@gmail.com

Abstract. There is a numerous demand for a standard representation of the ubiquitous available information on the web. Developing an efficient algorithm for traversing large ontologies is a key challenge for many semantic web applications. This paper proposes spreading activation over ontology method based on bidirectional search technique in order to detect the relatedness between two human diseases. The aim of our work is to detect disease relatedness by considering semantic domain knowledge and description logic rules to identify diseases relatedness. The proposed method is divided into two phases: Semantic Matching and Disease Relatedness Detection. In Semantic matching phase, diseases in submitted query are semantically identified in the ontology graph. In Disease relatedness detection phase, disease relatedness is detected by running a bidirectional-based spreading activation algorithm and return the related path (set of diseases) if so. In addition, the classification of these diseases is provided as well.

Keywords: Bidirectional search · Disease ontology · Semantic web · Spreading activation

1 Introduction

The use of ontologies in the field of health informatics has become a mainstream activity within bioinformatics due to the vast growing of healthcare system. In bioinformatics, ontology is used for representing and organizing medical vocabularies. Spreading Activation (SA) could be run on semantic networks and could also be used for information retrieval process [2]. Spreading activation is appropriate to run on incomplete and large graphs. It runs on a graph structure that comprises a set of nodes connected by edges in which concepts are nodes with an activation value and the relations between them are represented by edges. An activation value is assigned to each node in the graph and then the algorithm spreads to the nodes with the higher activation value. The algorithm runs in a set of iterations and terminates when a stopping condition is reached. The output is a list of

N.T. Nguyen et al. (Eds.): ICCCI 2017, Part I, LNAI 10448, pp. 14–23, 2017.
DOI: 10.1007/978-3-319-67074-4_2

activated nodes within each iteration. For each iteration or cycle, there are three substantial actions: (1) The list of nodes is expanded by adding adjacent nodes (all nodes which have links to the nodes in the list), (2) The activation value at each node in the list is recomputed based on the activation value of the node itself and the weight of links which exist between other nodes, and (3) The list is filtered by excluding the nodes with activation values less than a given threshold. Below a group of definitions related to the proposed methodology:

- *Semantic Relation*: A semantic relation between any two concepts in the ontology is one of the relations among the set of semantic relations $\sum =$ {Hypernym, Hyponymy, Synonymy},
- *Hyponym, Hypernym*: a hyponym is a word whose semantics is a specific meaning of another word which called its Hyperonym or its Hypernym. For instance, vaccinia and smallpox are all Hyponyms of viral infectious disease (their Hypernym),
- *Co-hyponyms*: Given a concept C has two hyponyms A and B, then A and B are identified as co-hyponyms.

This research investigates the question of whether two human diseases are related to each other by any means and what is the relation between them if exists. For instance, is there a relatedness between *Vasculogenic Impotence* and *Transvestism*? If so, what is the relatedness? A striking feature of finding a relatedness between diseases is that physicians can treat patients not only based on symptoms they suffer but they can treat the real cause of this disease which may be related to another disease that causes these symptoms. Therefore, physicians can treat the real cause disease, not the symptoms. For instance, Gallstones disease (Cholelithiasis) may be caused by *Hemolytic Anemia* disease so crushing gallstones is not a solution or treatment because the stones will develop again [24]. Therefore, the objectives of detecting the relatedness between diseases are[1]:

- *Causality*: A disease may occur due to the existence of another disease. For example, *Hereditary Spherocytosis* dieases is an autosomal preponderant anomaly of erythrocytes that causes gallstones. Pigmented gallstones occur in approximately half of untreated patients,
- *Complications of diseases*: one disease may increase the complications of another. For example, Diabetes and HCV hepatitis,
- *Treatments prescription*: Treatments may differ when there is a relation between two diseases.

The remainder of the article is structured as follows: We present an overview on related work in Sect. 2. The disease ontology which is used as a semantic knowledge base for the proposed method is described in Sect. 3. We present the proposed method in Sect. 4. The workflow of the proposed method is presented in Sect. 5. The proposed method is illustrated by using a running example in Sect. 6. The conclusion and the directions for future work are outlined in Sect. 7.

[1] Thank you Dr. Diaa Elsayed, Gastroenterology and Hepatology specialist, for the Counseling.

2 Related Work

There are a numerous amount of literature on biomedical knowledge management and medical decision making due to the explosion of biomedical knowledge over the last recent years. Therefore, biomedical knowledge available on the web is growing considerably as most of the biomedical research papers are published online. Semantic matching is used to expose information which is semantically related to structured data based matching concepts not keyword-based [12]. Abundant assorted frameworks and algorithms of semantic matching have been proposed so far such as [13,21,25]. Some examples of individual approaches addressing the matching problem can be found in [6,7]. In their cutting edge paper of 2010, Ngo, Cao, and Le [20] proposed an ontology-based vector space model for semantic annotation and semantic search by combines different ontologies. It takes advantage of ontological features of both named entities and WordNet vocabularies and develops a spreading activation algorithm for query expansion. As anticipated, their experiments evince that their model is better than the solely keyword-based model and also better than the ones using only WordNet or named entities. In addition, it merges various ontologies to improve the semantic search process. De Maio et al. [5] proposed a project named ODINO which uses a fuzzy knowledge approach for disease diagnosis that supports medical decision-making. ODINO has three main features which are a faceted search of diseases through taxonomy constraints, disease catalog browsing, and preliminary medical diagnosis. In the field of bioinformatics, there are a lot of research has been made to represent medical information as a semantic knowledgebase for further processing by semantic applications. Due to the existence of many medical ontologies, it is important to reuse and integrate ontologies to establish suitable mappings between their concepts. Therefore, a lot of research in ontology matching and integration [8,23] have been done in the recent years. Shvaiko and Euzenat [23] surveyed the state of the art of ontology matching and addressed some worthy challenges for ontology matching techniques. In addition, they analyzed the results of recent ontology matching evaluations. Some of the famous medical-related ontologies are:

- *Human Disease Ontology (DO)*[2] is a standardized biomedical ontology which contains a considerable number of disease terminologies,
- *Vaccine Ontology (VO)*[3] is a biomedical ontology which contains more than 2000 terms and relationships for vaccines and vaccinations,
- *Infectious Disease Ontology (IDO)*[4] compromises a set of ontologies which represent infectious diseases,
- *Ontology for Biomedical Investigations (OBI)*[5] is an integrated ontology for the concepts which belong to life-science and clinical practice,

[2] http://www.disease-ontology.org.
[3] http://www.violinet.org/vaccineontology.
[4] http://infectiousdiseaseontology.org/page/Main_Page.
[5] http://obi-ontology.org/page/Main_Page.

3 The Disease Ontology

The main purpose of developing ontology is to use it as a semantic knowledge base for identifying concepts in a specific domain and to share a data semantics among software agents so that it becomes machine understandable [10]. Köhler et al. [17] and Croft et al. [4] discuss in their research that the human disease data is a cornerstone of biomedical research. Therefore, there's an enormous need for a consistent representation of human disease for robust data analysis [18]. Creating a biomedical knowledgebase in the form of ontologies creates a rigid knowledgebase for semantic annotation of biomedical data through defined concepts and relations connecting them [14]. The disease ontology (DO) has been selected to be used as the semantic knowledgebase for the representation of human diseases. The objective of using DO is to provide the biomedical community with a convenient, reusable and robust knowledgebase of human disease concepts [16]. Major enhancements to the DO database since 2012 has been made including: the content of DO has had several revisions, including the addition of 32% of all terms. The Disease Ontology database has been updated to the latest ontology as of March 2, 2017. The DO project has had a considerable influence on the development of biomedical resources, as evidenced by 307 Google Scholar citations (as of April 14, 2017) to DO's paper [22] published in 2012. We have used the disease hierarchy in DO to infer disease relatedness. Furthermore, synonyms of diseases are also used in semantic matching phase. For instance, *Carotenemia* disease has exact synonym *Hypercarotinemia*.

4 Methodology

The methodology of the proposed work is divided into two phases: Semantic matching and Disease relatedness detection.

In Semantic matching phase, diseases in submitted query are semantically identified in the ontology graph. The output of this phase is whether these diseases are found in the ontology or not. If the disease has been identified then, the Uniform Resource Identifier (URI) of both diseases is retrieved. In the Disease relatedness detection phase, the URI of each disease is passed to the relatedness detector to find whether they are related or not. If they are related, the algorithm returns the set of diseases that connect them in the path from the first to the second and the classification of both diseases as well.

4.1 Semantic Matching

One of the most common approaches to perform semantic matching for determining the semantic similarity between concepts in an ontology is the node-based approach [3] which we used in this work. Semantic matching technique is used to identify candidate diseases in the disease ontology. Concept disambiguation is performed by querying WordNet [11] and DO. Each disease name in the query is first disambiguated into concepts using vector space models [3], representing concepts as vectors of features in a k-dimensional space where k is the number of pertinent keywords for each disease. In other words, each disease in the query is represented by a vector of pertinent keywords found through WordNet and DO. Pertinent keywords are *hyponyms*, *direct-hypernyms*, co-hyponyms, and *synonyms* of the disease. One-level Hyponyms and direct-hypernyms are retrieved from DO and synonyms are retrieved from both DO and WordNet. Direct-hypernyms are one-level up of a disease node in DO hierarchy and Hyponyms are its siblings. After sense disambiguation, the proposed matching algorithm returns the URIs of matched diseases. It is assumed that, when searching for a concept, it is also important to match synonyms of that concept. For example, the synonyms of the disease *Hyperuricemia* are *Hyperuricaemia* and *Uricacidemia*. Therefore, diseases describing these concepts are retrieved as well. Figure 1 shows the semantic matching process cycle.

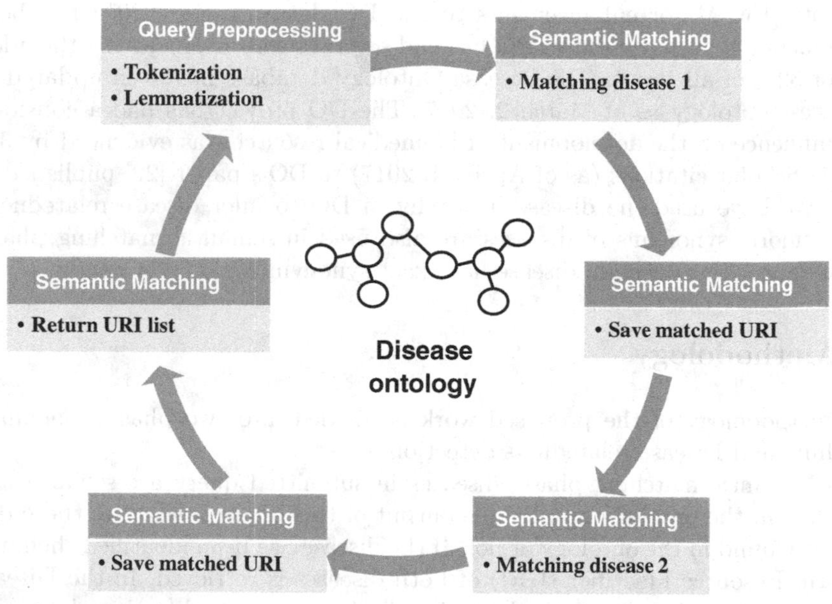

Fig. 1. The semantic matching process cycle.

4.2 Disease Relatedness Detection

In disease relatedness detection phase, diseases relatedness is detected by using bidirectional spreading activation on ontology graph which consists of a set of finite cycles/iterations. Checking for termination conditions is performed in each cycle. We are considering only hierarchical relations for nodes activation. The worthiness of bidirectional search is the speed and it requires less memory [19]. Figure 2 depicts the relatedness detector components. The algorithm starts with two initial nodes which are the two diseases and the follows the following steps:

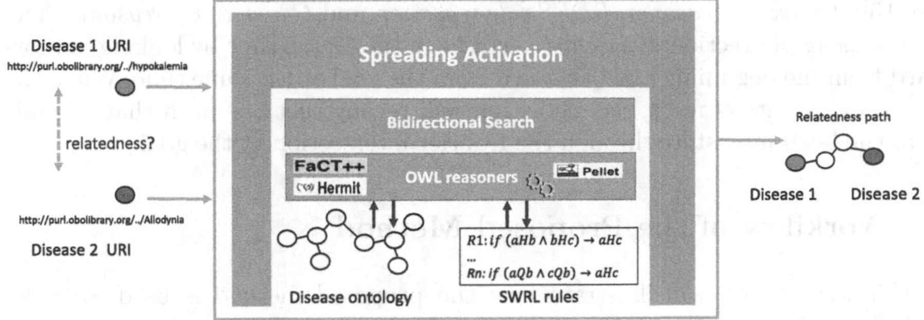

Fig. 2. Relatedness detector components.

1. Assume all nodes in the graph have activation values of zero and starting nodes have activation value of 1.0,
2. For each Link l_{ij} connecting the source node n_i with target node n_j, compute $a_j = \sum_{i=1}^{n} a_i w_{i,j}$ where n is the number of nodes connected to n_j and $w_{i,j}$ is the relatedness weight,
3. If a candidate node takes an activation value exceeds 1.0, then set its new activation value is set to 1.0. Likewise, set the activation value of the candidate node to 0.0 if it takes a value below 0.0. Nodes in the disease ontology receive the highest value of $w_{i,j}$ if they are one of the pertinent keywords of the current activated node ($w_{i,j} = 0.8$ for hyponyms and $w_{i,j} = 0.1$ for direct-hypernyms) because they build the taxonomy using "is-a" relation. A low value of $w_{i,j} = 0.0$ is assigned to any other relation which is not effective in this case,
4. A significance threshold ($F = 0.8$) determines whether to include the activated node to the output list,
5. Activated node will not be considered in the next cycles,
6. Nodes with activation value exceeds the threshold F are marked as activated on the next cycle,
7. The procedure terminates when a node is reached from more than one path (sexual disorder in the running example shown in Fig. 3).

An ambiguity may arise if two homonyms are used in the search query but there is no relatedness between them. For example, *Stewart* in *Stewart-Treves syndrome*, (a chronic lymphedema disease), is different from the one in *Stewart-Bluefarb syndrome.* The latter is a type of *acro angiodermatitis* which was described independently by Stewart as well as by Bluefarb and Adams on the legs of patients with Arterio-venous malformations [1]. Semantically matched diseases could be found using Jena reasoners. One feature of Jena is the support of different reasoners, which infer additional knowledge. Jena inference subsystem contains various inference engines or reasoners. These reasoners are used to check ontology consistency and allow additional facts to be inferred from instance data and class descriptions. The predefined reasoners included in the Jena distribution are [15]: *Transitive reasoner*, *RDFS rule reasoner*, and *Generic rule reasoner*. The idea of using bidirectional search is to cut back the search time by looking out forward from the beginning and backward from the goal at the same time. When the two search frontiers meet, the algorithm will reconstruct one path that extends from the beginning state through the frontier intersection to the goal.

5 Workflow of the Proposed Method

In this section, we will illustrate how the proposed method is used over the underlying disease ontology. User submitted query is automatically processed in the following steps:

1. User submits the two diseases as a string using the system interface.
2. Perform pre-processing on input, a detailed discussion of pre-processing falls outside the scope of this paper:
   (a) *Intelligent Tokenization and Stop list elimination*: This step includes tokenizing the query stream by breaking it down into understandable segments. Then, eliminate all stop words. The intelligence here is that the method does not blindly remove all stop words like traditional techniques but using a concept tokenization technique in which the keyword is taken with the preposition after it as a concept. If that concept has a match in the ontology, then this proposition will not be removed otherwise it will be removed,
   (b) *Stemming*: Stemming removes word suffixes: both inflectional suffixes (-s, -es, -ed) and derivational suffixes (-able, -ability) are stemmed [9].
3. Formulate semantic query using SPARQL query language and executes it using Jena-embedded query engine (ARQ) against the ontology which performs the semantic matching process described in the methodology section,
4. Diseases relatedness is detected by using bidirectional-based spreading activation on ontology graph if the diseases found in the matching process. The output of this process is a set of diseases that builds the path between the two diseases. In addition, the classification of these diseases could be detected by performing one more cycle of the spreading process which could be the parent disease (if available) in the hierarchy.

As soon as these steps have been carried out, a set of diseases that may connect the two initially submitted diseases are displayed to the user and the classification of them is displayed as well.

6 Running Example

As a running example in this paper, suppose a patient already has *Vasculogenic impotence* disease and the physician discovered that he/she got recently *Transvestism* so the question is does these two diseases is related to each other by any means? The answer to this question will support the medical decision of the physician for treatment prescription and whether there are any complications caused by one because of the other. In this case, the physician should submit a query includes two diseases: *Vasculogenic impotence* and *Transvestism*. As shown in Fig. 3, the relatedness between *Vasculogenic impotence* and *Transvestism* diseases are detected by the intersection node *sexual disorder*. The algorithm runs in a set of cycles. When the algorithm detects an intersection node, it stops. One more cycle can be performed to get one more up-level disease connecting both which could be considered as a classification of both diseases.

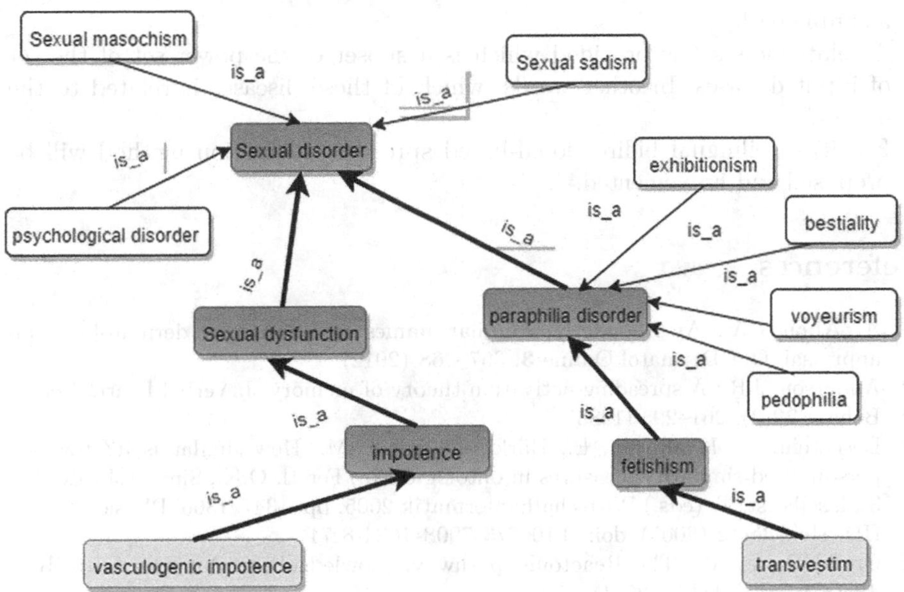

Fig. 3. Partial view of the disease ontology shows the relatedness path (marked in bold) between *Vasculogenic Impotence* and *Transvestism* diseases.

7 Conclusion and Future Work

In this paper, we proposed a Bidirectional-Based Spreading Activation Method in order to detect the relatedness between two human diseases. This relatedness can be detected by running spreading activation algorithm using bidirectional search methodology on a large disease ontology. One key feature of the proposed method is to identify whether two human diseases are related to each other. Moreover, identify related diseases that may connect them to a common path. As a result, detecting the relatedness between diseases helps in finding out the real cause of a disease (in case it is caused by another disease), decrease the prognosis of a disease by treating the other disease which increases the complications of that disease and helps in treatments prescriptions. Therefore, physicians can treat the main cause, not just the symptoms. Consequently, we believe that the proposed method will assist physicians and will support medical decision-making by considering semantic domain knowledge to infer diseases relatedness. This idea could also be applied to VO ontology to find out relations between vaccines and also in Gene ontology to find out relations between genes. Our main line of future research involves:

- Extending our approach to integrate different biomedical ontologies using and ontology matching and integration services,
- Detecting the relatedness between more than two diseases using multiple goal search algorithms and provide the relatedness graph between them not only a simple path,
- A relatedness set is provided which is a subset of the power set of the set of input diseases. In other words, which of these diseases is related to the others?
- Finally, a bilingual bidirectional-based spreading activation method will be proposed and implemented.

References

1. Al Aboud, A., Al Aboud, K.: Similar names and terms in dermatology; an appraisal. Our Dermatol Online **3**, 367–368 (2012)
2. Anderson, J.R.: A spreading activation theory of memory. J. Verbal Learn. Verbal Behav. **22**(3), 261–295 (1983)
3. Bernstein, A., Kaufmann, E., Bürki, C., Klein, M.: How similar is it? towards personalized similarity measures in ontologies. In: Ferstl, O.K., Sinz, E.J., Eckert, S., Isselhorst, T. (eds.) Wirtschaftsinformatik 2005, pp. 1347–1366. Physica-Verlag HD, Heidelberg (2005). doi:10.1007/3-7908-1624-8_71
4. Croft, D., et al.: The Reactome pathway knowledgebase. Nucleic Acids Res. **42**(D1), D472–D477 (2014)
5. De Maio, C., et al.: Fuzzy knowledge approach to automatic disease diagnosis. In: 2011 IEEE International Conference on Fuzzy Systems (FUZZ), pp. 2088–2095. IEEE (2011)
6. Dhamankar, R., et al.: iMAP: discovering complex semantic matches between database schemas. In: Proceedings of the 2004 ACM SIGMOD International Conference on Management of Data, pp. 383–394. ACM (2004)
7. Do, H.-H., Rahm, E.: COMA: a system for flexible combination of schema matching approaches. In: Proceedings of the 28th International Conference on Very Large Data Bases, pp. 610–621. VLDB Endowment (2002)
8. Euzenat, J., Meilicke, C., Stuckenschmidt, H., Shvaiko, P., Trojahn, C.: Ontology alignment evaluation initiative: six years of experience. In: Spaccapietra, S., et al. (eds.) Journal on Data Semantics XV. LNCS, vol. 6720, pp. 158–192. Springer, Heidelberg (2011). doi:10.1007/978-3-642-22630-4_6
9. Fan, Y., Huang, X., An, A.: York university at TREC 2006: enterprise email discussion search. In: TREC 2006 (2006)
10. Fathalla, S.M., Hassan, Y.F., El-Sayed, M.: A hybrid method for user query reformation and classification. In: 2012 22nd International Conference on Computer Theory and Applications (ICCTA), pp. 132–138. IEEE (2012)
11. Fellbaum, C.: WordNet. Wiley Online Library, New York (1998)

12. Giunchiglia, F., Yatskevich, M., Shvaiko, P.: Semantic matching: algorithms and implementation. In: Spaccapietra, S., et al. (eds.) Journal on Data Semantics IX. LNCS, vol. 4601, pp. 1–38. Springer, Heidelberg (2007). doi:10.1007/978-3-540-74987-5_1
13. Guo, J., et al.: Semantic matching by non-linear word transportation for information retrieval. In: Proceedings of the 25th ACM International on Conference on Information and Knowledge Management, pp. 701–710. ACM (2016)
14. Hoehndorf, R., Dumontier, M., Gkoutos, G.V.: Evaluation of research in biomedical ontologies. Brief. Bioinform. **14**(6), 696–712 (2013)
15. Jena, A.: Reasoners and rule engines: jena inference support. The Apache Software Foundation (2013)
16. Kibbe, W.A., et al.: Disease ontology 2015 update: an expanded and updated database of human diseases for linking biomedical knowledge through disease data. Nucleic Acids Res. **43**(D1), D1071–D1078 (2015)
17. Köhler, S., et al.: The human phenotype ontology project: linking molecular biology and disease through phenotype data. Nucleic Acids Res. **42**(D1), D966–D974 (2014)
18. LePendu, P., Musen, M.A., Shah, N.H.: Enabling enrichment analysis with the human disease ontology. J. Biomed. Inform. **44**, S31–S38 (2011)
19. Li, H., Xu, J., et al.: Semantic matching in search. Found. Trends R Inf. Retriev. **7**(5), 343–469 (2014)
20. Ngo, V.M., Cao, T.H., Le, T.M.: Combining named entities with wordnet and using query-oriented spreading activation for semantic text search. In: 2010 IEEE RIVF International Conference on Computing and Communication Technologies, Research, Innovation, and Vision for the Future (RIVF), pp. 1–6. IEEE (2010)
21. Qin, Y., Yao, L., Sheng, Q.Z.: Approximate semantic matching over linked data streams. In: Hartmann, S., Ma, H. (eds.) DEXA 2016. LNCS, vol. 9828, pp. 37–51. Springer, Cham (2016). doi:10.1007/978-3-319-44406-2_5
22. Schriml, L.M., et al.: Disease ontology: a backbone for disease semantic integration. Nucleic Acids Res. **40**(D1), D940–D946 (2012)
23. Shvaiko, P., Euzenat, J.: Ontology matching: state of the art and future challenges. IEEE Trans. Knowl. Data Eng. **25**(1), 158–176 (2013)
24. Trotman, B.W., et al.: Studies on the pathogenesis of pigment gallstones in hemolytic anemia: description and characteristics of a mouse model. J. Clin. Invest. **65**(6), 1301 (1980)
25. Wu, Z., et al.: An efficient Wikipedia semantic matching approach to text document classification. Inf. Sci. **393**, 15–28 (2017)

Semantic Networks Modeling with Operand-Operator Structures in Association-Oriented Metamodel

Marek Krótkiewicz[1], Marcin Jodłowiec[2], and Krystian Wojtkiewicz[1(✉)]

[1] Department of Information Systems,
Wrocław University of Science and Technology, Wrocław, Poland
krystian.wojtkiewicz@gmail.com

[2] Institute of Computer Science, Opole University of Technology, Opole, Poland

Abstract. Semantic networks are nowadays one of the most frequently used knowledge representation method. In this paper authors present a novel approach towards the definition and semantics of semantic networks with the use of predefined primitives as key structure elements. The presented solution is a part of Semantic Knowledge Base project, in which it is used to store complex information such as facts and rules. This approach aims at increased expressiveness of knowledge representation with the higher level of clarity of message. It introduces not only a unique duality of nodes types, namely operators and operands, but also provides mechanisms such as multiplicity, quantifiers or modifiers possible to apply to each and every of network elements.

Keywords: Semantic networks · Semantic Knowledge Base · Partitioned semantic nets · Association-oriented database modeling

1 Introduction

The article presents issues from the knowledge representation area. Knowledge is understood as information with the ability to interpret it. This means that the data and their semantics must be one indivisible and unequivocal whole, i.e. the basis for operating in the area of knowledge. While the data is considered in the area of databases where data structures (metastructure or intensional part) and the data (extensional part) exist. The data does not itself have any semantics. Their semantics depends on the form of a model representing a part of reality called the problem domain. Database modeling involves the creation of data structures, that in the best possible way enable storage and processing of data. Database models should show the utmost compatibility with the logic of the system that implements the problem domain. Compatibility between database layer and the logic of the system can be considered in several aspects. The primary purpose of the database is to provide a permanent storage for data, which can then be read by the logic of the system for processing. Apart from technical

N.T. Nguyen et al. (Eds.): ICCCI 2017, Part I, LNAI 10448, pp. 24–33, 2017.
DOI: 10.1007/978-3-319-67074-4_3

issues, the focus should be on the compatibility of data storage structures with the data structures used by the system logic.

A significant problem is the acquisition of knowledge presented in the natural language. For obvious reasons, this language may not be compatible with any system of inference. The issue of knowledge representation is also problematic, since natural language grammar depends heavily on language being chosen. Regardless of the natural language choice, each of them is built in a very distant way from knowledge representation methods. Much more problematic is the fact that the natural languages operate in the area of terms rather than concepts. Relationship *term* $\longleftrightarrow$ *concept* has *many-to-many* multiplicity, and choosing the right concept for a specific term is in general ambiguous. The text itself does not contain semantics, it is merely a sequence of characters that can be grouped in terms. By sentence parsing you can try to determine the grammatical relationships between terms, which define their roles in the analyzed sentence. In order to reproduce the original semantics of phrases stored in the natural languages complex systems are being created using dictionaries, thesauri, corpora or analyzing closer and further contexts. The transformation of the information stored in natural language into a form acceptable be knowledge representation system is a very complex task, however, necessary for making any knowledge processing, which is a key task of knowledge-based systems. Knowledge bases, unlike databases, should by the assumption allow exploration of knowledge in a free manner. It means that database applications have their functional scope determined at the design stage. In principle, they require human factor in order to interpret the data. Database systems themselves do not serve to produce new information form existing, especially not in an autonomous way.

For above reasons it is important to build complex knowledge representation systems taking into account the specific nature of knowledge, which will be stored in it. In particular it applies to e.g. conceptual structures similar to the object-oriented approach or semantic networks, which are based on the concept of graphs, etc. In order to fulfill the above postulate the data structure tailored to the nature of knowledge should be modeled. In the proposed solution the structure based on specialized, interrelated modules is used. In this study, the main focus is put on the Extended Semantic Network Module of SKB (ESNM^{SKB}) [1]. It is a completely separate element with respect to the language layer, which is dedicated to a separate module. Both modules share the information contained in the ontological core and in particular the central point of the system namely the concept (CONCEPT). In the area of database layer, SKB system was modeled in Association-Oriented Database Metamodel (AODB) [2]. With this embodiment the possibility is provided to distinguish the data and relationships between them, the independent inheritance among data containers and relationships, direct modeling of n-ary relationships, multiplicities on the side of the relationship, the mutual dependence of life time of the relationships and elements bonded by it, and many more efficient mechanisms for modeling complex structures and their constraints.

In this article, the authors focused on the presentation of their approach to semantic networks modeling. The primary contribution is the introduction of *operator-operand* structure. It allows to extend the classical approach to the semantic network as a directed graph in which each link is a binary term, into more general structure based on n-ary links with named roles. Section 2 presents an overview Semantic Knowledge Base, i.e. system in which the method was implemented. The next section provides an overview of state of the art in semantic network modeling, and the assumptions that turned into features of adopted solution. The Sect. 4 is dedicated to description of contribution, a structure representing implementation of the idea, and the evaluation presented as examples. The last section is a summary.

2 SKB Overview

SKB consist of separate modules [1,3] modeled in AODB [2]:

- Structural Module (SM^{SKB})
  - Ontological Core Module (OCM^{SKB})
  - Relationships Module (RM^{SKB})
  - Cyclic Value Ranges Module ($CVRM^{SKB}$)
- Dimension & Space Module (DSM^{SKB})
- Extended Semantic Network Module ($ESNM^{SKB}$)
- Behavioral Module (BM^{SKB})
- Linguistic Module (LM^{SKB})

The OCM^{SKB} task is to store information about concepts and basic relationships between them. The most important conceptions in this module include relations: *class-feature*, *class-instance*, *property-feature-value*, *set-instance*, *concept-connection*, *concept-relationship*. Other types of relations are performed by RM^{SKB}.

The RM^{SKB} module is able to store information about relationships between the concepts, both in intensional and extensional point of view. The basic conceptual structure of this module is a hypergraph. It represents a compound in which sets of concepts can perform specific roles. Roles are represented by hyperedges. Moreover, the roles may have hierarchical structure, which allows building of aspect-oriented links. That type of relationships represent the situation in which it is possible to selectively determine the role played by the given concept depending on the aspect in which it is considered.

The $CVRM^{SKB}$ module is designed to store information about the values which may be cyclic or could be defined in the form of value ranges. This is especially helpful in case of information that express time, because of the frequent use of cyclic time points and time intervals, which may also be cyclic.

The DSM^{SKB} module has the task of representing knowledge in the sense of spaces in which they can apply to individual elements of SKB. This applies to many levels of detail, from single concepts, through relations, to complex expressions stored in semantic networks. The main idea of this module is based on building spaces that consist of dimensions.

The $ESNM^{SKB}$ module allows storage of complex facts based on the *operator-operand* structure with a number of extensions, such as cardinality constraints, modifiers or quantifiers.

The BM^{SKB} module is used for behavioral knowledge representation which relates to description of elements changes in the dimension of time. The concept of this module is based on the process approach, in which the components being timers define timing of the system. Each event related to a timer causes a specific action that might affect the status of the modeled process. The process model is a fixed structure, in which the parameters are dynamic.

The LM^{SKB} module addresses the linguistic aspects of knowledge representation system. It is not a essential element of the system itself. However, due to the need for communication with the outside world, system should operate on terms rather than on concepts that are existing internally in the system. Concepts in SKB have unique identifiers and have no direct reference to the human conceptual sphere. The linguistic module is a communication interface allowing to express concepts and relationships between them in the form of natural language terms.

3 Semantic Networks

Semantic Network is a network that represents semantic relations (edges) between concepts (nodes). Edges (like relations) can be both directed and undirected (bidirectional). Edges may be binary, as in the classical semantic networks, or n-ary like e.g. in $ESNM^{SKB}$. Edges can have one predefined character for example in the case of semantic networks describing taxonomy, meronymy, etc. Edges may also belong to a predefined set containing their types. In the most general case, it is possible to define any edge type, but then they should have a type that allows for the definition of their role in the relationship, namely the relationship semantics. Nodes can be concepts form OCM^{SKB} or any other elements that are the associated elements. It is also possible to specify node type, which allows a more precise definition of semantics. Such elements of $ESNM^{SKB}$ as cardinality constraints, modifiers, quantifiers, represent a significant semantic capacity expansion of semantic network.

3.1 Related Work

Modeling of the structures with the semantic network nature dates back to antiquity. According to J. Sowa [4] the oldest semantic network is a graph called the Tree of Porphyry drawn in the third century AD by the Greek philosopher Porphyry showing taxonomy of concept types. Graph, which provided the philosopher comment to Aristotle Categories presented concepts as nodes while the edges were the IS-A relationship, both in the context of a generalization-specialization compound as well as class-instance. In the computer science interests semantic networks have grown their popularity in the 60s of the twentieth century. with the concept of semantic memory presented by Quillian's Teachable Language Comprehender (TLC) [5,6]. Quillian's model was based on assigning a words (terms) to nodes of semantic network which represent concepts. The concepts had their inherent properties (has-a), as well as connectivity to other concepts through links in order to store IS-A compounds. Quillian concept has become the standard for the semantic network in the area of knowledge engineering and many researchers have followed the same mainstream. They are modifying it, expanding and designing new graphic languages and text formats for the exchange of information about stored knowledge. Hendrix's partitioned semantic network theory [7,8] seems worth mentioning here. It introduced the distinction between a single concept and the whole system of concepts which constitutes a kind of a separate subnet. It derives from the fact that the network elements can be fused together in a space on an equal level and consequently be treated as independent nodes. Another conception worth mentioning is a variation of a semantic network i.e. inheritance networks proposed by Brachman [9] and further developed by other researchers [10–12]. These networks are directed acyclic graphs and model generalization-specialization relationship with its semantic consequences in the form of inheritance of features. Another idea presented and accepted as one of leading formal definition of semantic networks are Conceptual Graphs by J. Sowa [13]. They are used to represent formulas in first order logic in the form of directed labeled graphs.

There are various applications of semantic networks, i.e. visualisation of economical and financial knowledge [14]. The idea of the semantic networks as a way of knowledge representation is not merely a conceptual tool. There are a number of abandoned as well as still developed projects built on the base of this theory. One of the most well-known systems based on semantic networks is SNePS [15] by S. Shapiro. SNePS is a hybrid system using knowledge representation based on defined logic frames and semantic networks. The use of semantic networks in SNePS system involves the representation of sentences in the form of directed, labeled graph. Helbig [16] developed a MultiNet paradigm based on an expanded concept of the semantic network. This system is based on the terms, allowing to assign them the concept in the form of nodes connected by edges, performing relationships predefined in the system. Its main objectives are the possibility of being a interlingua for natural language processing systems, and thus, a layer of semantic annotation of natural languages. Very well documented and interesting project with rich set of semantic relations is Universal Networking Language [17]. At the present time an extremely popular is a mainstream of the Semantic Web [18]. Despite the similarity of names it is a distinct trend in relation to the development of the concept of the semantic networks. Semantic Web primarily uses two languages: RDF and OWL. RDF allows you to represent knowledge in the form of SOP triples, which define a common semantic network.

3.2 Key Features of $ESNM^{SKB}$

The following assumptions have been made and fulfilled while designing of semantic network model of the $ESNM^{SKB}$: 1. formal universal semantic network semantics, 2. unambiguous method of relationship interpretation, e.g. ISA arc can be: (a) A is kind of B, (b) A is subset of B, (c) A is part of B, (d) A is instance of B, due to ambiguity of term "*is*" in many natural languages. Some natural languages are able to cope with this issue, but the majority allows multiple meanings. 3. freedom in definition of relations. In majority of systems the relations are narrowed to closed set of few or several relation types, 4. ability of quantification, 5. logical operators (including negation), 6. credibility, 7. time representation, 8. uncertain and incomplete information, 9. definition of cardinality constraints.

4 Operand-Operator Constructions in ESNM^{SKB}

4.1 Syntax and Semantics

The proposed grammar and semantics of the knowledge representation system has a number of very important features. They are particularly important for the precision of knowledge expression as well as for possibility to build knowledge processing systems.

The basic structure of the *operator-operand*:

$$r(p_1, p_2, ..., p_n) \tag{1}$$

where r – operator, p_k – operand, $k \in 1 \ldots n$ provides a natural approach to n-ary compound, as a primary element of facts representation. Since each operator may also be used as an operand it allows to create complex networks expressing facts with complex structure. The operator may be set as *Activity* or *State*.

$$r = Activity|State \tag{2}$$

Activity requires intentional action to be taken. This means that the ultimate cause of the activity must be an action taken by the actor. The original reason for taking activity may be independent of the actor.

State is characterized by, uninterrupted for the duration of the particular state, having a property set defined for a given concept. Wherein the values describing the property may be defined as a range.

The operands can perform the following pre-defined roles: *Actor*, *Co-operator*, *Object*, *Number*, *Owner*, *Attribute*, *Relationship*, *Adverbial of Manner*, *Source*, *Target*, *Tool*. Both operators and operands perform strictly defined roles in the Semantic Network. Assigning specific concepts (CONCEPT) to certain roles provides semantic networks the ability to express concrete facts.

$$\begin{aligned} role = Actor|Co\text{-}operator|Object|Number|Owner|Attribute| \\ Relationship|Adverbial\ of\ Manner|Source|Target|Tool \end{aligned} \tag{3}$$

This conception has been extended by multiplicity assigned to link between operand and operator, which provides the ability to determine the size restrictions (cardinality constraint). Both operators and operands may have modifiers, which clearly define the concept performing a specific role.

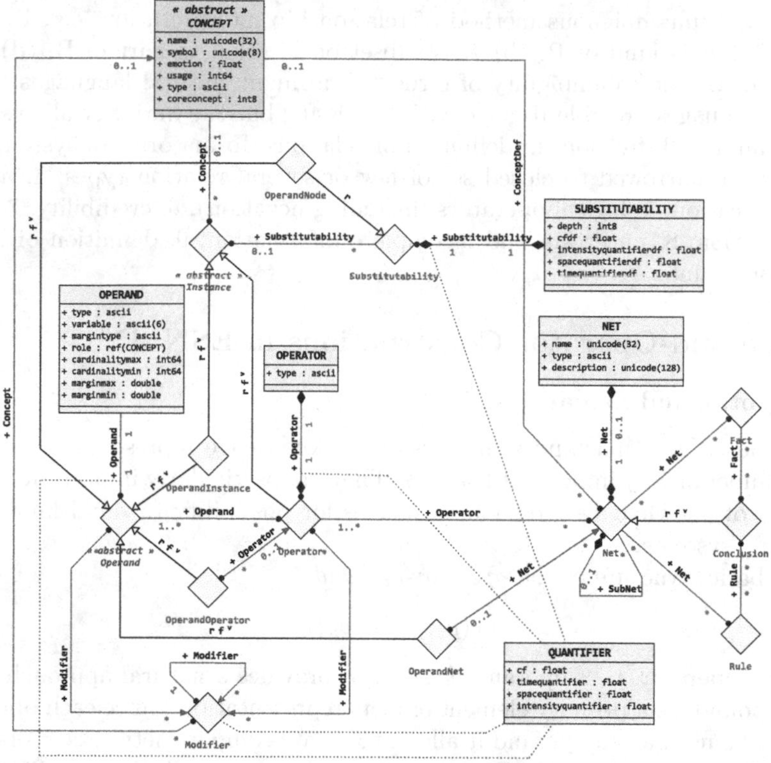

Fig. 1. Intensional part of ESNM^{SKB} module in AODB

4.2 Structure

Diagram in the Fig. 1 shows a structure of ESNM^{SKB} model presented in Association-Oriented Modeling Language AML [2]. The syntax and semantics of AML was described in publications [1,19]. Rectangles represent collections of data, while the diamonds represent the associations which containers of compounds. Continuous lines having a closed circle indicate the role of the associations, while a solid line with an open arrow at the end is a symbol of a generalization. Operand was modeled as a combination of Operand association with a collection OPERAND through the role Operand. It should be noted, however, that the association Operand is abstract, that is to say cannot have instances and can only be a source of inheritance for real, specific sets of associations: OperandNode, OperandInstance, OperandOperator, OperandNet. Each of these associations model specific type of operand, respectively, when the operand is: a specific network node, an instance of the concepts defined in the OCM^{SKB}, the operator and separate semantic network. The operator has been modeled analogously through the association Operator connected by the role Operator with the collection OPERATOR. In addition, the relationship *operand-operator* has its emanation as a role Operand in association Operator with a multiplicity of 1 ... * to the side of the bound element (operand) and * on the other side.

4.3 Examples

Lets consider an exemplary sentence "*John owns the car*", where operand "*car*" can be further redefined by the modifier "*big*", which will modify the sentence to "*John owns a big car*". Complementing this sentence with appropriate cardinality constraint leads to the sentence "*John owns one big car*". Modifiers can also be recursively modified, e.g. in the sentence "*John owns one very big car.*" Another important element influencing the semantic capacity are Quantifiers. In ESNM^{SKB} quantifiers cf - certainty factor, tq - time quantifier, sq - space quantifier and q - intensity quantifier are defined. Figure 2 presents a conceptual representation of the semantic network of ESNM^{SKB} for the sentence: "*John owns one very big car*" along with its complementary quantifiers.

The strength of semantic networks lays in the fact that it can consist of more than one node. The information contained in a set of nodes can be read in many different ways depending on the starting point of graph linearization algorithm. Figure 3 is a representation of a set of facts. It is a complex structure with a quite detailed semantics. Simplified representation of this network is as follows: 1. Ford is a car. 2. John owns one very big car. 3. John works in New York. 4. John goes very fast to New York by very big car. 5. New York is a city.

However, it should be noted that it lacks lots of details. Considering again the part of the network presented on Fig. 3 we can distinguish following facts, which approximate natural language representation is as follows:

Certainly ($cf = 1$) John owns *intensively ($iq = 0.9$)* one *probably ($cf = 0.5$)* car.

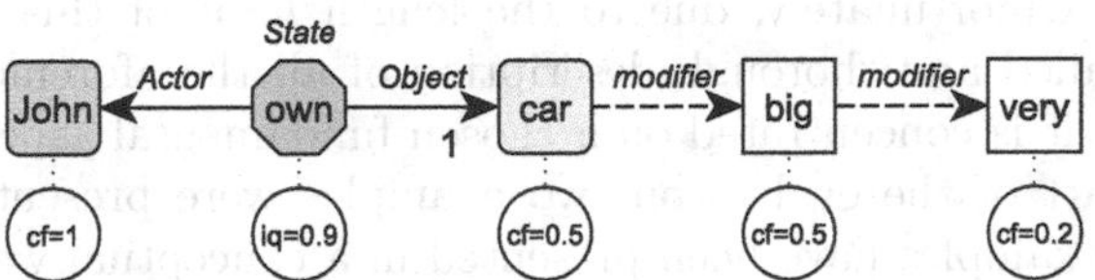

Fig. 2. The diagram depicting the Semantic Network representation of sentence "*John owns one very big car*"

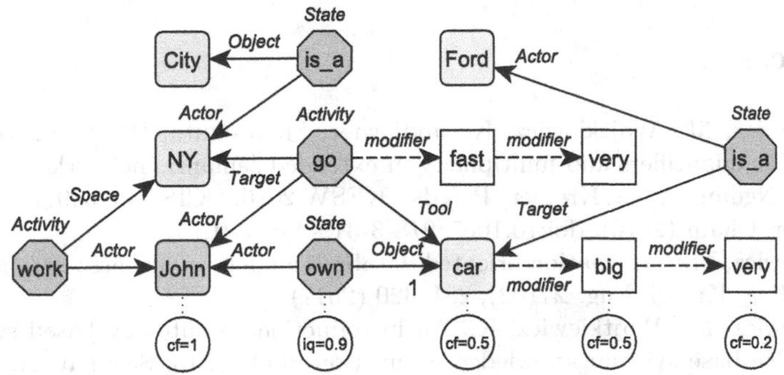

Fig. 3. The diagram depicting the Semantic Network representation of complex fact

At the same time we know that:

A car is *probably (cf = 0.5)* big and *probably (cf = 0.5 * 0.2) very big.*

The above example shows that the semantic networks presented by authors has high expressiveness, meaning that lots of facts and information can be presented using little elements.

The proposed solutions significantly extend the classical approach to semantic networks. It should be emphasized that the network stored in ESNM^{SKB} does not require conversion for permanent storage in the database layer. Is does do not require a text representation, which would be parsed in order to determine its grammatical structure. Facts are stored directly in the association-oriented database, the structure of which is shown in the Fig. 1.

5 Summary

The presented approach to semantic networks modeling is the basis for the solution adopted in of the Semantic Knowledge Base modules, namely Extended Semantic Networks Module. Its main goal is to directly support n-ary links with named roles, which constitute an enhancement in comparison to the previously explored solutions. An important aspect of the presented *operator-operand* structure is the fact that it was implemented in Association-Oriented Database as a basic construct of ESNM^{SKB} constituting SKB module in which facts and rules are stored and processed. This module supports a number of other features that expand the semantic capacity of the semantic networks, and which are indicated in the Sect. 3.2. Unfortunately, due to the length limit of this paper, it does not contain detailed and thorough description of all the aforementioned properties, therefore, it is concentrated on a chosen fundamental aspect of the basic structure. As part of the evaluation two examples were presented. To ensure readability, the examples have been presented in a conceptual way, rather than as an object diagram in AODB. As part of the future works the authors intend to focus on completing the SKB data processing methods as well as inference algorithms.

References

1. Krótkiewicz, M., Wojtkiewicz, K., Jodłowiec, M., Pokuta, W.: Semantic knowledge base: quantifiers and multiplicity in extended semantic networks module. In: Ngonga Ngomo, A.-C., Křemen, P. (eds.) KESW 2016. CCIS, vol. 649, pp. 173–187. Springer, Cham (2016). doi:10.1007/978-3-319-45880-9_14
2. Krótkiewicz, M.: Association-oriented database model – n-ary associations. Int. J. Softw. Eng. Knowl. Eng. **27**(02), 281–320 (2017)
3. Krótkiewicz, M., Wojtkiewicz, K.: An introduction to ontology based structured knowledge base system: knowledge acquisition module. In: Selamat, A., Nguyen, N.T., Haron, H. (eds.) ACIIDS 2013. LNCS, vol. 7802, pp. 497–506. Springer, Heidelberg (2013). doi:10.1007/978-3-642-36546-1_51

4. Sowa, J.F.: Principles of Semantic Networks: Explorations in the Representation of Knowledge. Morgan Kaufmann, San Mateo (2014)
5. Collins, A.M., Quillian, M.R.: Retrieval time from semantic memory. J. Verbal Learn. Verbal Behav. **8**(2), 240–247 (1969)
6. Quillian, M.R.: The teachable language comprehender: a simulation program and theory of language. Commun. ACM **12**(8), 459–476 (1969)
7. Hendrix, G.G.: Encoding knowledge in partitioned networks. In: Findler, N.V. (ed.) Associative Networks: Representation and Use of Knowledge by Computers, pp. 51–92. Academic Press, New York (1979)
8. Bundy, A., Wallen, L.: Partitioned semantic net. In: Bundy, A., Wallen, L. (eds.) Catalogue of Artificial Intelligence Tools, p. 89. Springer, Heidelberg (1984). doi:10.1007/978-3-642-96868-6_172
9. Brachman, R.J., Levesque, H.J., Reiter, R.: Knowledge Representation. MIT Press, Cambridge (1992)
10. Horty, J.F., Thomason, R.H.: Boolean extensions of inheritance networks. In: Proceedings of the Eighth National Conference on Artificial Intelligence, AAAI 1990, vol. 1, pp. 633–639. AAAI Press (1990)
11. Selman, B., Levesque, H.J.: The tractability of path-based inheritance. In: IJCAI, pp. 1140–1145. Citeseer (1989)
12. Thirunarayan, K., Kifer, M., Warren, D.S.: On the declarative semantics of inheritance networks. In: IJCAI (1989)
13. Sowa, J.F.: Conceptual graphs. In: van Harmelen, F., Lifschitz, V., Porter, B. (eds.) Handbook of Knowledge Representation, vol. 3(07), pp. 213–237. Elsevier, Amsterdam (2008)
14. Dudycz, H.: Usability of business information semantic network search visualization. In: Proceedings of the Mulitimedia, Interaction, Design and Innnovation, MIDI 2015, Warsaw, Poland, pp. 13:1–13:9, 29–30 June 2015
15. Shapiro, S.C.: An introduction to SNePS 3. In: Ganter, B., Mineau, G.W. (eds.) ICCS-ConceptStruct 2000. LNCS, vol. 1867, pp. 510–524. Springer, Heidelberg (2000). doi:10.1007/10722280_35
16. Helbig, H., Gnörlich, C.: Multilayered extended semantic networks as a language for meaning representation in NLP systems. In: Gelbukh, A. (ed.) CICLing 2002. LNCS, vol. 2276, pp. 69–85. Springer, Heidelberg (2002). doi:10.1007/3-540-45715-1_6
17. Cardeñosa, J., Gelbukh, A., Tovar, E.: Universal networking language: advances in theory and applications. Res. Comput. Sci. **12**, 1–443 (2005)
18. Shadbolt, N., Berners-Lee, T., Hall, W.: The semantic web revisited. IEEE Intell. Syst. **21**(3), 96–101 (2006)
19. Jodłowiec, M., Krótkiewicz, M.: Semantics discovering in relational databases by pattern-based mapping to association-oriented metamodel—a biomedical case study. In: Piętka, E., Badura, P., Kawa, J., Wieclawek, W. (eds.) Information Technologies in Medicine. AISC, vol. 471, pp. 475–487. Springer, Cham (2016). doi:10.1007/978-3-319-39796-2_39

Knowledge Integration in a Manufacturing Planning Module of a Cognitive Integrated Management Information System

Marcin Hernes(✉) and Andrzej Bytniewski

Wrocław University of Economics, ul. Komandorska 118/120, 53-345 Wrocław, Poland
{marcin.hernes,andrzej.bytniewski}@ue.wroc.pl

Abstract. One of the important functions of the operation of integrated management information systems, including multi-agent systems, is to properly planning the production. Due to the different production planning strategies (methods) and the company's limited production capacity, the agents running in the system may generate different versions of the production plans. In other word, agents' knowledge may differ. The final version may be selected by the system user, however, it should be noted that this is a time-consuming process, and there is a risk of the user choosing the worst version. The better solution is to automatically integrate the agents' knowledge and to determine one version of the plan presented to the user. The aim of this paper is to develop a consensus algorithm that will allow integrating manufacturing plans generated by different agents, and present one solution (that is very close to these plans, but not necessarily one of them) to user.

Keywords: Integrated management information systems · Manufacturing planning · Multi-agent systems · Cognitive agents

1 Introduction

Integrated management information systems, including multi-agent systems, support enterprises in business processes realization, especially in planning, control, monitoring, manufacturing and linking the business to the environment. One of the important aspects of the operation of such systems, due to the proper satisfaction of customers' needs to obtain flexibility and competitive advantage, is to properly planning the manufacturing. The system module responsible for this task allows (based on customer's orders and demand forecasts) to prepare a plan for the production of goods. Decision making in an enterprise, including manufacturing plans, is most often carried out under uncertainty and risk conditions, since the effects of a decision can only be predicted with a certain probability. Due to the different production planning strategies (methods) and the company's limited production capacity, the agents running in the system may generate different versions of the manufacturing plans. In other word, agents' knowledge may differ. The final version may be selected by the system user, however, it should be noted that this is a time-consuming process, and there is a risk of the user choosing the worst

N.T. Nguyen et al. (Eds.): ICCCI 2017, Part I, LNAI 10448, pp. 34–43, 2017.
DOI: 10.1007/978-3-319-67074-4_4

version. The better solution is to automatically integration of agents' knowledge and determines one version of the plan presented to the user.

The aim of this paper is to develop a consensus algorithm that will allow integrating manufacturing plans, generated by different agents, and presented one solution (that is very close to it, but not necessarily one of them) to user. This algorithm will be implemented in manufacturing planning module in a Cognitive Integrated Management Information System (CIMIS). This will, in consequence, increase the efficiency of production planning by reducing the time it takes to set a target production plan, change the plan in real time, and reduce the risk associated with this process.

2 Related Works

Planning of manufacturing is supported by different methods and tools. For example, paper [1] presents using Bayesan network for manufacturing planning. This solution allows minimizing the maintenance and inspection costs by reducing downtime whilst optimizing inspection intervals. Paper [2] presents a nonlinear mixed integer programming model to minimize the average total cost per unit product subject to constraints such as satisfying customer demand in various geographic regions, relationships between supply flows and demand flows within the physical configuration, and the production limitation of different size plants. In paper [3], a framework based on an integer linear programming (based on real time product data collected by Internet of things) is proposed for the purpose of planning and making precise production. The work [4] describes mixed integer linear model for integrating production and distribution (determined as NP-hard problem). The authors developed a decomposition method based on successive sub problem solving method. The problem is decomposed to a set of sub problems. After transformation, the sub problems are solved with an approach based on traveling repairman with profit. A knowledge-based intelligent decision system for production planning and decision is presented by [5]. This system detects and avoids collisions for multi-axis multi-tool machines by using knowledge from the vision sensor, manufacturing process, and machining data.

Knowledge integration in multi-agent systems is performed by using different methods. For example negotiations [6], or deduction-calculation methods [7] can be used. Negotiations enable effective integration of knowledge by reaching a compromise, however they require exchanging a large number of messages between agents, which results in decreased efficiency of the multi-agent system. The deduction-calculation methods (e.g. ones based on the theory of games, classical mechanics, or methods of choice) enable one to obtain a great computational capacity of a system, however they do not guarantee a proper result of knowledge integration [8]. In order to eliminate the presented problems, consensus methods may be applied which enable integration of knowledge in real time and guarantee reaching a good compromise at a lower level of risk, which may consequently lead to selecting decisions producing profits satisfactory for a decision maker [9].

3 Manufacturing Planning Module in CIMIS

The CIMIS had been widely described in [10]. In this paper (due to page limit), the manufacturing planning module, which is a component of manufacturing management sub-system, will be characterized.

The production process in the enterprise is aimed primarily at fulfilling the requirements of customers. Production is aimed at the execution of specific orders, moving away from being "stocked" to avoid unnecessary storage costs for finished goods. Consequently, this shortens the life cycle of the products and the production is more or less monolithic or even individual [4]. The correct implementation of the production planning process, according to [11], consists of five basic stages:

- set goals,
- analyze the situation,
- develop an action plan (proper planning),
- develop a timetable for action,
- make a decision on implementation.

Due to the turbulence of the environment, it is necessary, for decision makers, to make quick and effective decisions, which requires rapid analysis of large amounts of information performed by IT systems.

At present, it is pursuing the full integration of IT systems supporting all areas of business operations. Therefore, in CIMIS the manufacturing planning module is included in the manufacturing management subsystem architecture. This module performs functions such as:

- preparing a production plan,
- controlling the implementation of this plan,
- taking into account available resources and production capacity,
- simulation of plans.

The manufacturing planning module supports production planning on three levels [12]:

- strategic - achieving the company's main goals and mission,
- tactical - defining the ways of implementing strategic plans, achieving the goals set in the tactical plan,
- operational - defining tactical plans, meeting the objectives set out in the operational plan.

The considered module performs three groups of tasks [1]:

1. Basic tasks, including poor inventory production control, short flow times and high delivery deadlines.
2. Ancillary tasks, which include supplying the company with materials, tools and interference response.
3. Information tasks. The module should provide information relating to, inter alia, to the number and parameters of orders waiting before the machining system and in production, the flow time (completion of orders) or the capacity utilization rate.

Performing the above tasks is related to the implementation in the module, among others, such functions as [13–15]:

- planning with limited resources,
- "forward", "backward" and bi-directional planning,
- effective planning of the time of completion of operations based on current data from production departments,
- detection of delayed tasks or tasks that have to start before the date of release of production resources,
- modification the batch of sizes.

Several agents are running in the module. Each agent uses different production planning methods such as [9, 12, 15, 16]:

- linear and dynamic programming,
- planning and scheduling production by determining the degree of urgency of orders and priority rules (using heuristic decision rules),
- event-oriented scheduling - just in time,
- scheduling production using Gantt graphs,
- planning using network methods,
- planning and scheduling production through operational research,
- artificial intelligence methods (expert systems, genetic algorithms, neural networks).

The variety of production planning methods often leads to the generation of different versions of plans by particular agents. The system user (the decision maker) needs one version of the plan that will be implemented in a specific time frame. It is extremely difficult to choose such a version because of the need to analyze all versions generated by the system. The time spent on this analysis is very limited, as the system generates further versions of the plans (the turbulent environment forces changes in plans to be close to the real ones). User cannot rely on intuition or experience, as there is the risk of choosing a non-optimal plan. There is thus inconsistency of knowledge in manufacturing plans that will be characterized in the next part of the paper.

4 Inconsistency of Knowledge in Manufacturing Plans

Inconsistency of knowledge in manufacturing plans in CIMIS arises from the use, by agents, of different methods for production planning at the same time (described in the previous section of this paper). The manufacturing plans generated by the agents are represented in the form of a specific structure. Therefore, the inconsistency occurs when agents assign different values to the same objects and attributes. If we assume that the technological line allows us to start order at the same time, then the following production plans may alternatively be generated as an example of an inconsistency:

- to manufacture 2t product p1 starting production at 10.00 am, planned finish - 11.00 am, planned cost: 100,
- to manufacture 3t product p1 starting production at 10.30 pm, planned finish time - 13:30, planned cost: 200,

- to manufacture 7t product p1 starting production at 11.00 am planned finish time - 12.00, planned cost: 400.

Thus, the inconsistency affects both attributes like "quantity" and "time" as well as "cost" attribute (planned technical cost of manufacture). In this example, in each plan, the time allocated for a unit of product (for example, tons) is different (this may be due, for example, to taking into account different planning method or another configuration of the production line), so it is difficult to determine which of these plans is optimal.

Integration of agents' knowledge allows presenting to user a saturated version of production plan that is a compromise between alternate versions generated by the different agents. This will give the user a satisfactory low risk solution.

Next part of the paper presents consensus algorithm proposed to integration of agents' knowledge.

5 Consensus of Manufacturing Plans

Determining a consensus with regard to manufacturing plans consists of several stages. The structure of the production plans generated by the agents must be thoroughly developed first, and then calculate the distance between the individual plans. Determining a consensus

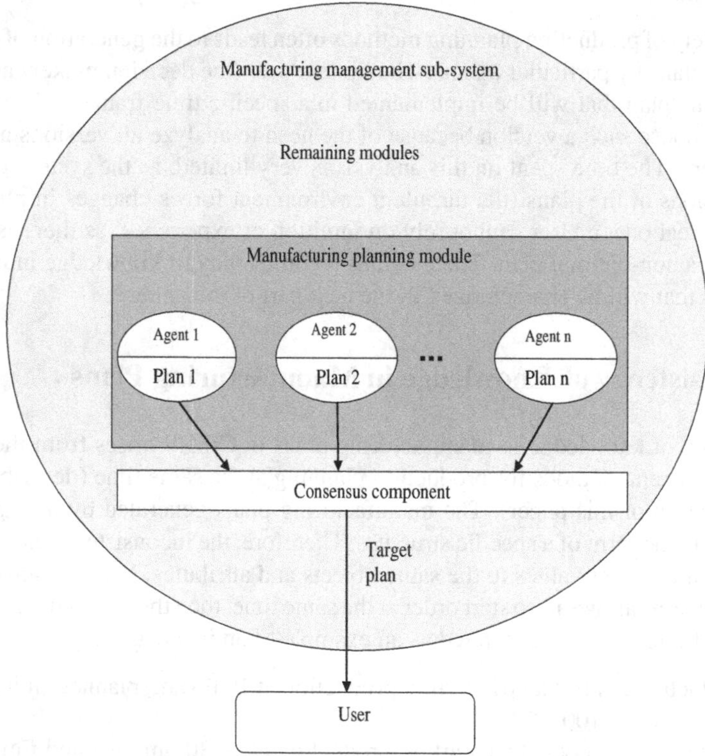

Fig. 1. Conception of using a consensus method in manufacturing planning module.

is to design a plan so that the sum of the distances between that plan (consensus) and the individual alternative plans generated by the agents, is minimal. Figure 1 presents a conception of using consensus method in manufacturing planning module

This conception assumes that the agents running in manufacturing planning module generate different versions of manufacturing plans represented by knowledge structures (defined in the next part of the paper). The consensus determining algorithms are executed automatically when the agents generate different alternative versions of the manufacturing plans. The results obtained using the consensus algorithm, are presented to the user, who can make decisions on the production management.

The structure of agents' knowledge representation is defined as follows:

Definition 1. Let set of products $O = \{p_1, p_2, \cdots, p_n\}$.

The agents' knowledge related to manufacturing plan is represented as follows:

$$x = \{\langle p_1, dts_1, dtf_1, i_1, c_1\rangle, \langle p_2, dts_2, dtf_2, i_2, c_2\rangle, \ldots, \langle p_n, dts_n, dtf_n, i_n, c_n\rangle\}$$

Where:

$dts_1, dts_2, \ldots, dts_n$ - date and time of planned starting manufacturing of products $p_1, p_2, \cdots, p_n$,

$dtf_1, dtf_2, \ldots, dtf_r$ - date and time of planed finishing manufacturing of product $p_1, p_2, \cdots, p_n$

$i_1, i_2, \ldots, i_n$ - the amount of product $p_1, p_2, \ldots, p_n$ (the size of the order),

$c_1, c_2, \ldots, c_n$ - the cost of manufacturing of product $p_1, p_2, \ldots, p_n$.

We can define the distance function between structures. It must be noted that calculation of a distance between structures may be based on calculation and summation of distances between individual elements of those structures. For the purpose of defining time distance between two dates, we use following function:

Definition 2. Distance ϑ between two dates dt_1 and dt_2 is called the function:

$$\vartheta(dt_1, dt_2) = |dt_1 - dt_2| \tag{1}$$

In considering the distance between number of the product and the costs of transport, it can be use the function presented in many papers (e.g. Nguyen 2008) specifying the distance between real numbers:

Definition 3. The distance between numbers x, y belonging to the sequence composed with Q real numbers is called following function:

$$\chi(x, y) = \frac{1}{Q}|x - y|. \tag{2}$$

The distance between two structures of manufacturing plans is defined as follow:

Definition 4. The distance Ψ between two structures:

$$\begin{aligned} W^{(1)} &= \{\langle p_1^{(1)}, dts_1^{(1)}, dtf_1^{(1)}, i_1^{(1)}, c_1^{(1)}\rangle, \ldots, \langle p_n^{(1)}, dts_n^{(1)}, dtf_n^{(1)}, i_n^{(1)}, c_n^{(1)}\rangle\} \\ W^{(2)} &= \{\langle p_1^{(2)}, dts_1^{(2)}, dtf_1^{(2)}, i_1^{(2)}, c_1^{(2)}\rangle, \ldots, \langle p_n^{(2)}, dtf_n^{(2)}, dte_n^{(2)}, i_n^{(2)}, c_n^{(2)}\rangle\} \end{aligned} \tag{3}$$

is called following function:

$$\Psi(W^{(1)}, W^{(2)}) = \sum_{j=1}^{n} (\vartheta(dts_j^{(1)}, dts_j^{(2)}) + \vartheta(dtf_j^{(1)}, dtf_j^{(2)}) + \chi(i_j^{(1)}, i_j^{(2)}) + \chi(c_j^{(1)}, c_j^{(2)}))$$

The postulated method of distance calculation may be employed in development the consensus algorithm in order to knowledge integration. The consensus is determining on the basis of set of structures of agents' knowledge related to manufacturing plans, called the profile, defined as follows:

Definition 5. The profile W = {W$^{(1)}$, W$^{(2)}$, …, W$^{(K)}$} is called set K structures, such that:

$$\begin{aligned} W^{(1)} &= \{\langle p_1^{(1)}, dts_1^{(1)}, dtf_1^{(1)}, i_1^{(1)}, c_1^{(1)}\rangle, \ldots, \langle p_n^{(1)}, dts_n^{(1)}, dtf_n^{(1)}, i_n^{(1)}, c_n^{(1)}\rangle\}, \\ W^{(2)} &= \{\langle p_1^{(2)}, dts_1^{(2)}, dtf_1^{(2)}, i_1^{(2)}, c_1^{(2)}\rangle, \ldots, \langle p_n^{(2)}, dts_n^{(2)}, dtf_n^{(2)}, i_n^{(2)}, c_n^{(2)}\rangle\} \\ &\vdots \\ W^{(K)} &= \{\langle p_1^{(K)}, dts_1^{(K)}, dtf_1^{(K)}, i_1^{(K)}, c_1^{(K)}\rangle, \ldots, \langle p_n^{(K)}, dts_n^{(K)}, dtf_n^{(K)}, i_n^{(K)}, c_n^{(K)}\rangle\} \end{aligned}$$

Algorithm for consensus determining is as follows:

Algorithm 1

Data: Profile $W = \{W(1), W(2), \ldots\ W(K)\}$ consists of K structures of agents` knowledge related to manufacturing plans.

Result: Consensus

$CON = \{\langle CON(p_1), CON(dts_1), CON(dtf_1), CON(i_1), CON(c_1)\rangle, \ldots, \langle CON(p_n), CON(dts_n), CON(dtf_n), CON(i_n), CON(c_n)\rangle\}$ **for profile W.**

START

1: *Let* $\forall CON(x) \in CON.CON(x) = \emptyset$.

2: j:=1.

3: g:=dts_j.

4: Determine pr(g).

5: $l_i^1 = (K+1)/2$, $l_i^2 = (K+2)/2$.

6: Determine g^y where $l_j^1 \leq g^y \leq l_j^2$.

$CON(g) = 0$;

7: If $g = dts_j$ then $g := dtf_j$, go to: 4.

If g= dtf_j then g:=i_j, go to: 4.

If g= i_j then g:=c_j, go to: 4.

If g=c_j then go to: 8.

8: g:=p_j.

9: $g \in CON(g) \Leftrightarrow \min(\chi(CON(c_j), c_j))$.

10: If g=p_j then j:=j+1.

11: If $j \leq n$ then go to: 3.

If $j > n$ then END.

END.

Computational complexity is $O(n^2K)$.

6 Discussion and Experiment Results

Developed algorithm allows for determining consensus of manufacturing plans. Considering the example presented in Sect. 4, a consensus is to produce 3t of product p1 (minimum distance sum is 1 + 0 + 4 = 5) at 10.30 (minimum distance sum is 30 + 0 + 30 = 60 min) to 12.00 (minimum distance sum is 60 + 90 + 0 = 150 min), the cost is 200 (the minimum distance sum is 100 + 0 + 200 = 300).

Developed algorithm was implemented in manufacturing planning module in CIMIS. Next, the research experiment has been done, witch following assumption:

- consensus have been determined on the basis of manufacturing plans generated by 7 agents (they uses planning methods described in Sect. 3), on the basis of simulated data,
- the time of production and cost of production was assumed as performance measures and as a measure of risk the average coefficient of variation was used,
- the calculation has been repeated 100 times.

Analyzing the results of verification one can notice that two agents have generated better results (the lower average costs and the shorter production times) than results generated using the consensus algorithm. However five agents have generated worse results. While analyzing the risk connected with choice of manufacturing plan, it has been observed that the use of the consensus algorithm enables executing the process with the lowest level of risk. It can be said then that plans generated by the integration component are characterized by a low level of fluctuation of production costs and times. The phenomenon may positively affect the stability of a company's financial liquidity, and it help maintain continuity of production (low fluctuation of production times lowers the risk of downtime).

7 Conclusions

Correct production management is a very important element of the functioning of a production enterprise. Formerly, it was hand-made by man. Today, due to the turbulent nature of the modern economy, production management without the use of information systems is impossible. These systems make it possible to manage production with optimum utilization of available production resources, which obviously has an impact on the ability of companies to adapt to market demand. However, they function effectively only if they allow a dynamic response to market needs. The variety of production planning methods often leads to a situation in which the system generates different versions of alternative solutions, among which not all are optimal. It is very difficult for you to choose the best plan, as it requires thorough analysis of all versions. This takes time and significantly lowers the dynamics and thus the effectiveness of production management. Nor is there any certainty that even after the analysis, the user will choose a good plan. Often, due to time constraints related to the continuity of the production process, analysis is not possible. Using consensus methods to resolve a conflict of plans and to designate one version later presented to the user can lead to a shorter time to designate the final version as well as to reduce the risk of choosing an incorrect version. This paper presents preliminary results of researches. Future works may involve integration of planning a production capacity plans or integrating an orders realization process by using a consensus method.

References

1. Jones, B., Jenkinson, I., Yang, Z., Wang, J.: The use of Bayesian network modelling for maintenance planning in a manufacturing industry. Reliab. Eng. Syst. Saf. **95**(3), 267–277 (2010)

2. Hsu, C.I., Li, H.C.: An integrated plant capacity and production planning model for high-tech manufacturing firms with economies of scale. Int. J. Prod. Econ. **118**(2), 486–500 (2009)
3. Fang, C., Liu, X., Pei, J., Fan, W., Pardalos, P.M.: Optimal production planning in a hybrid manufacturing and recovering system based on the internet of things with closed loop supply chains. Int. J. Oper. Res. **16**(3), 543–577 (2016)
4. Azadian, F., Murat, A., Chinnam, R.B.: Integrated production and logistics planning: contract manufacturing and choice of air/surface transportation. Eur. J. Oper. Res. **247**(1), 113–123 (2015)
5. Ahmad, R., Tichadou, S., Hascoet, J.Y.: A knowledge-based intelligent decision system for production planning. Int. J. Adv. Manuf. Technol. **89**(5), 1717–1729 (2017)
6. Dyk, P., Lenar, M.: Applying negotiation methods to resolve conflicts in multi-agent environments. In: Zgrzywa, A. (ed.) Multimedia and Network Information systems, MISSI 2006. Oficyna Wydawnicza PWr, Wrocław (2006)
7. Barthlemy, J.P.: Dictatorial consensus function on n-trees. Math. Soc. Sci. **25**, 59–64 (1992)
8. Maleszka, M., Mianowska, B., Nguyen, N.T.: A method for collaborative recommendation using knowledge integration tools and hierarchical structure of user profiles. Knowl.-Based Syst. **47**, 1–13 (2013)
9. Hernes, M., Sobieska-Karpińska, J.: Application of the consensus method in a multi-agent financial decision support system. Inf. Syst. e-Bus. Manag. **14**(1), 167–185 (2016)
10. Hernes, M.: A Cognitive Integrated Management Support System for Enterprises. In: Hwang, D., Jung, J.J., Nguyen, N.-T. (eds.) ICCCI 2014. LNCS, vol. 8733, pp. 252–261. Springer, Cham (2014). doi:10.1007/978-3-319-11289-3_26
11. Biermann, D., Gausemeier, J., Heim, H.P., Hess, S., Petersen, M., Ries, A., Wagner, T.: Planning and optimisation of manufacturing process chains for functionally graded components—part 2: case study on self-reinforced thermoplastic composites. Prod. Eng. Res. Dev. **9**(3), 405–416 (2015)
12. Krumeich, J., Werth, D., Loos, P.: Prescriptive control of business processes. Bus. Inf. Syst. Eng. **58**(4), 261–280 (2016)
13. Camstar Systems Inc: Advanced Manufacturing Execution System for Solar Manufacturing (2007). http://www.camstar.com/products-services/enterprise-platform/camstar-manufacturing/. Accessed 04 Feb 2016
14. Bytniewski, A. (ed.): Architecture of Integrated Management System. Wydawnictwo UE we Wrocławiu, Wrocław (2015)
15. Hong, T.P., Peng, Y.C., Lin, W.Y., Wang, S.L.: Empirical comparison of level-wise hierarchical multi-population genetic algorithm. J. Inf. Telecommun. **1**(1), 66–78 (2017)
16. Zhang, J.: Multi-Agent-Based Production Planning and Control. Wiley, New York (2017)

The Knowledge Increase Estimation Framework for Ontology Integration on the Relation Level

Adrianna Kozierkiewicz-Hetmańska(✉) and Marcin Pietranik

Faculty of Computer Science and Management,
Wroclaw University of Science and Technology,
Wybrzeze Wyspianskiego 27, 50-370 Wroclaw, Poland
{adrianna.kozierkiewicz,marcin.pietranik}@pwr.edu.pl

Abstract. The task of integration of sets of data or knowledge (regardless the choice of its representation) can be very daunting procedure, requiring a lot of computational resources and time. Authors claim that it is beneficial to develop a formal framework which could be used to estimate the profitability of the integration, ideally even before the integration even occurs. Therefore, a set of algorithms for such estimation of the increase of knowledge concerning relation level of ontology integration is proposed.

1 Introduction

One of the most common tasks related to knowledge management concerns its integration, which can be understood as a process of unification of a set of different and independent knowledge sources into one, consistent representation of the combined knowledge of the collective. This involves not only providing a summary of available information, but also resolving any conflicts which may entail inconsistencies and therefore, result with unreliable knowledge base. On the other hand, a new knowledge may appear as on outcome of a synergy - the unified, integrated collective knowledge may contain more information than a sum of its parts. To represent such knowledge a plethora of different methods and frameworks can be found in the literature. In our research we have focused on using ontologies as a knowledge representation. Their structure (according to [10]) can be expressed using a notion of "*ontology stack*" consisting of levels of concepts, relations and instances which express increasing level of abstraction of knowledge expressed within a particular ontology. The task of their integration can be formally defined as follows: *for given n ontologies* $O_1, O_2, ..., O_n$ *one should determine an ontology* O^* *which is the best representation of given input ontologies.*

This paper is devoted to the knowledge increase estimation framework serving as one of the quality assessment measures of ontology integration. This notion is related to answering the question about how much knowledge has been gained thanks to the performed integration and can be useful in a variety of applications like the one presented in [7]. Until now we have developed methods of estimating

N.T. Nguyen et al. (Eds.): ICCCI 2017, Part I, LNAI 10448, pp. 44–53, 2017.
DOI: 10.1007/978-3-319-67074-4_5

this knowledge increase during the integration of ontologies on the level of concepts and instances [5,6]. Due to the limitation of this paper only measures for concepts' relation level will be considered. To illustrate its usefulness we propose simple algorithms for ontology concepts' relations' integration, that are not a part of our framework.

Due to the limited space, the paper focuses only on the integration of concepts relations. The paper is organised as follows. Section 2 contains an overview of related works. Section 3 serves as an introduction to ontologies and basic notions used throughout the rest of the paper. In Sect. 4 the developed algorithms are described. This is followed by Sect. 5 that contains a variety of different use cases in which proposed measures may be useful. The last section is a summary and a brief description of authors' upcoming research plans.

2 Related Works

For assessing the integration process many authors [1,4] use popular measures like: completeness, precision, accuracy, consistency, relevance and reliability. However, the described functions have one, serious defect- all of them require the integration to be performed and only after it if completed they can be used to evaluate the obtained results. Other authors like [3] have considered ontology quality from the philosophical point of view where the data quality is defined and called as "fitness for use".

A more interesting solution has been presented in [13]. The overall model of ontology quality analysis has been proposed. Authors have defined the two types of metrics: schema and instance. Both are dedicated to the relation level of ontology. The relationship richness has reflected the diversity of relations and placement of relations in the ontology. Relationship richness classified to instance metric, reflecting how much of the properties in each class in the schema is actually being used at the instances level.

On the other hand, authors of [2] fit into the modern approach of treating everything as a service, by introducing a notion of Ontology as a Service (OaaS). To illustrate OaaS, they propose a sub-ontology extraction and merging, where a set of sub-ontologies are extracted from various input ontologies. Then extracted sub-ontologies are integrated to form a final ontology to be used by the user. However, authors do not propose any method of estimating a profitability of such process.

In [12] authors have presented a set of similarity measures between ontologies. Authors have distinguished two layers view of ontologies: lexical and conceptual. The relation overlap based on the geometric mean value of how similar their domain and range concepts are have been determined. This measure has reflected the accuracy that two relations match. Despite that authors have experimentally demonstrated the utility of their methods, the proposed measures are not able to estimate the potential knowledge increase during the integration process.

Lozano-Tello and Gómez-Pérez [8] have proposed the complex framework called Ontometric. Authors have defined a taxonomy of 160 characteristics, that provides an outline able to choose and to compare existing ontologies. However, they have not been clearly presented and we suppose that Ontometric is not able to assess the growth of knowledge after adding a new ontology to the existing set.

Authors of [11] propose to approach ontology integration (also referred to as merging) as a task of ontology aggregation understood as a social choice. In other words, as a problem of aggregating the input of the procedure into an adequate collective decision. However, no estimation of the knowledge gained thanks to such collective approach has been given.

3 Basic Notions

In our framework a pair (A, V) denotes a real world, where A is a set of attributes that can be used to describe objects taken from some universe of discourse and V is a set of these attributes valuations. Formally $V = \bigcup_{a \in A} V_a$ where a domain of an attribute a is denoted as V_a. Ontology is a tuple:

$$O = (C, H, R^C, I, R^I) \tag{1}$$

where C is a set of concepts, H is concepts' hierarchy, R^C is a set of relations between concepts $R^C = \{r_1^C, r_2^C, ..., r_n^C\}$, $n \in N$, $r_i \subset C \times C$ for $i \in [1, n]$, I is a finite set of instances' identifiers and $R^I = \{r_1^I, r_2^I, ..., r_n^I\}$ denotes a set of relations between concepts' instances such that a relation r_j^C denotes a set describing possible connections between instances of some concepts from the set C and r_j^I are those connections actually materialised. In other words - relations from R^C define what objects can be connected with each other, while R^I defines what is connected. For example, in some ontology a set R^C may contain relations *is_sister* and *is_brother*, while R^I my contain definitions that John is a brother of Jane, and Jennifer is a sister of David.

Concepts taken from the set C are defined as $c = (id^c, A^c, V^c, I^c)$, where id^c is an identifier of a concept c, A^c is a set of its attributes, V^c is a set attributes domains (formally: $V^c = \bigcup_{a \in A^c} V_a$) and I^c is a set of particular concepts' instances. For short, we write $a \in c$ to denote that the attribute a belongs the the concept's c set of attributes A^c. An ontology is called *(A, V)-based* if the condition $\forall_{c \in C} ((A^c \subseteq A) \wedge (V^c \subseteq V))$

Concepts' instances are formally defined as a pair $i = (id^i, v_c^i)$, where id^i is its identifier and v_c^i is a function with a signature: $v_c^i : A^c \to V^c$. Referring to the consensus theory [9], the function v_c^i may by interpreted as a tuple of type A^c.

A set of instances from the base ontology definition (from the Eq. 1) is denoted below:

$$I = \bigcup_{c \in C} \{id^i | (id^i, v_c^i) \in I^c\} \tag{2}$$

we write $i \in c$ to denote a fact that the concept c contains an instance with an identifier i.

We define an auxiliary function Ins^{-1} that generates a set of concepts to which an instance with some identifier belongs. It has the signature $Ins^{-1} : I \to 2^C$ and is defined below:

$$Ins^{-1}(i) = \{c | c \in C \wedge i \in c\} \tag{3}$$

To simplify set operations we also define a set *Ins(c)* which contains only identifiers of instances assigned to concept c. Formally it can be defined as $Ins(c) = \{id^i | (id^i, v_c^i) \in I^c\}$.

L_s^R is a sublanguage of the sentence calculus and is used within a function that assigns semantics of relations from the set R^C. This function has a signature $S_R : R^C \to L_s^R$. As a consequence, we can define formal criteria for relationships between relations:

- *equivalency* between relations r and r' (denoted as $r \equiv r'$) occurs only if a sentence $S_R(r) \iff S_R(r')$ is a tautology
- a relation r' is more general than the relation r (denoted as $r' \leftarrow r$) if a sentence $S_R(r) \implies S_R(r')$ is a tautology
- *contradiction* between relations r and r' (denoted as $r \sim r'$)occurs only if a sentence $\neg(S_R(r) \wedge S_R(r'))$ is a tautology

The hierarchy of concepts (denoted in Eq. 1 as H) may be treated as a distinguished relation between concepts. Thus, $H \subset C \times C$. A pair of concepts $c_1 = (id^{c_1}, A^{c_1}, V^{c_1}, I^{c_1})$ and $c_2 = (id^{c_2}, A^{c_2}, V^{c_2}, I^{c_2})$ may be included within it (which will be denoted using a symbol $\leftarrow$), stating that c_2 is more general than c_1 ($c_2 \leftarrow c_1$), only if all of the following postulates are met:

1. $|A^{c_1}| \geq A^{c_2}$
2. $\forall a' \in A^{c_2} \exists a \in A^{c_1} : (a \equiv a') \vee (a' \leftarrow a)$
3. $Ins(c_1) \subseteq Ins(c_2)$

As previously stated, relations from the set R^C define which objects can be connected, while R^I defines what is actually connected. In our framework, to denote this fact, we will use the same index of relations taken from both sets. Therefore, a relation $r_j^I \in R_I$ contains only pairs of concepts' instances that are connected by a relation denoted as $r_j^C \in R^C$. A set of formal criteria that both sets must comply to is given below:

1. $r_j^I \subseteq \bigcup_{(c_1,c_2) \in r_j^C} (Ins(c_1) \times Ind(c_2))$
2. $(i_1, i_2) \in r_j^I \implies \exists (c_1, c_2) \in r_j^C : (c_1 \in Ins^{-1}(i_1)) \wedge (c_2 \in Ins^{-1}(i_2))$ which states that two instances may be in a relation with each other only if there is a relation connecting concepts they belong to
3. $(i_1, i_2) \in r_j^I \implies \neg \exists r_k^I \in R^I : ((i_1, i_2) \in r_k^I) \wedge (r_j^C \sim r_k^C)$ which describes that fact that two instances cannot be connected by two relations that have been defined as contradicting with each other using relations' semantics S_R
4. $(i_1, i_2) \in r_j^I \wedge \exists r_k^I \in R^I : r_k^C \leftarrow r_j^C \implies (i_1, i_2) \in r_k^I$ which states that if two instances are connected by some relation and there exists a more general relation, then these two instance are also connected by this relation

4 The Quantity of Knowledge on the Relations' Level of Ontologies

4.1 Overview of Integration Algorithms

The estimation of knowledge increase require information about how the two or more ontologies are integrated on relational level. In our work we assume the approach presented in Algorithm 1. It is conducted for two ontologies and any other new ontology can be iteratively added to the previous result. It is based on simple sum of parts of integrated ontologies and the only additional step is removing a redundant equivalent relations while preserving the knowledge about which concepts have been connected by discarded relations. At first, we also considered to remove relations that are a generalisation of other relations, but according to considerations from Sect. 3 discarding such relation may cause the loss of knowledge about connected concepts which may not meet requirements to participate in more specific relation. For example, coexisting more general relation "*is_family*" along with a relation "*is_mother*" should not entail its removal, due to the fact that it also expresses connections other than motherhood.

Algorithm 1. Concept relations integration

Require: Set of input ontologies: $O_1 = (C_1, H_1, R_1^C, I_1, R_1^I), O_2 = (C_2, H_2, R_2^C, I_2, R_2^I), ..., O_m = (C_m, H_m, R_m^C, I_m, R_m^I)$;

Set $R^* = \bigcup_{i=1}^{m} R_i^C$;
for all $(r, r') \in R^* \times R^*$ **do**
  if $r \equiv r'$ **then**
    $r = r \cup r'$;
    $R^* = R^* \setminus \{r'\}$;
  end if
end for

The integration of hierarchies in Algorithm 2 is different. It is not the integration of input hierarchies, but a process of generating a new taxonomy concepts that are a result of the ontology integration on concept level. The algorithm utilises criteria described in Sect. 3 and (as it will be further described in next section) it may create new relations that were not present in any of the input ontologies.

4.2 Algorithms for Knowledge Increase Estimation

The Algorithm 3 contains a procedure of calculating knowledge increase gained thanks to the integration of relations between concepts. It consists of three main steps, first of which being calculating the increase of knowledge coming from broadening the scope of equivalent relations. This situation refers to the fact that two equivalent relations may contain different pairs of concepts, and the

Algorithm 2. Hierarchy integration

Require: The integrated ontology $O^* = (C^*, H^*, R^{C^*}, I^*, R^{I^*})$ created from a set of input ontologies: $O_1 = (C_1, H_1, R_1^C, I_1, R_1^I), O_2 = (C_2, H_2, R_2^C, I_2, R_2^I), ..., O_m = (C_m, H_m, R_m^C, I_m, R_m^I)$;

Set $H^* = \phi$
for all $(c, c') \in C^* \times C^*$ **do**
  if $(c \leftarrow c')$ **then**
    $H^* = H^* \cup \{(c, c')\}$;
  end if
end for

eventual value of the increase of knowledge should reflect such supplementation. Second part of the algorithm concerns the integration of two relations, one being more general than the other. A naive approach would discard such relation, but this could entail a potential loss of knowledge because not all of the concepts in broader relation could participate in more specific interaction (e.g. not all parenting is a maternity). The last part reflects the situation in which two relations have nothing in common with each other, therefore the increase of knowledge can be maximal.

Algorithm 3. Knowledge increase during relations' integration

Require: A set of input ontologies: $O_1 = (C_1, H_1, R_1^C, I_1, R_1^I), O_2 = (C_2, H_2, R_2^C, I_2, R_2^I), ..., O_m = (C_m, H_m, R_m^C, I_m, R_m^I)$;

Set $R^* = \bigcup_{i=1}^{m} R_i^C$;
Set $R_U = R^* \times R^*$;
Set $\omega = |R^*|$;
Set $\Delta_R = 0$;
for all $(r, r') \in R_U$ **do**
  if $r \neq r'$ **then**
    if $r \equiv r'$ **then**
      $\Delta_R = \Delta_R + (1 - \frac{|r \cap r'|}{|r \cup r'|})$;
      $R_U = R_U \setminus \{(r', r)\}$
      $\omega = \omega - 1$
    else if $r \leftarrow r'$ **then**
      $\Delta_R = \Delta_R + \frac{|r \cap r'|}{|r|}$;
    else
      $\Delta_R = \Delta_R + 1$;
    end if
  end if
end for
return $\frac{\Delta_R}{\omega}$

Due to the fact that hierarchies are a specific kind of relations we claim that the increase of knowledge during the integration of ontologies should be calculated separately using Algorithm 4. Equation 1 states that hierarchies are subsets of the Cartesian product of sets of concepts, so in the first step the algorithm checks if any of the integrated taxonomies are entirely included in the other one. If this is the case then the knowledge increase coming from origin ontologies is equal to 0. Otherwise the algorithm calculates ordinary Jaccard distance. This serves as an indication of how much knowledge has been gained thanks to strict integration of two ontologies and is denotes as δ_H^-. Due to the fact that the integration of hierarchies may result in new connections between concepts (utilising criteria from Sect. 3) the algorithm should handle such situation, because it may highly influence the final result. This is done by calculating the value δ_H^+ in the penultimate step of the algorithm. The final result is a simple sum of δ_H^- and δ_H^+. Obviously, the final value may be higher than 1 which represents the fact that the completely new knowledge (that has not existed in the partial ontologies) has been created as a result of the integration. This situation is discussed further in the next section of the article.

Algorithm 4. Knowledge increase during hierarchy integration

Require: The integrated ontology $O^* = (C^*, H^*, R^{C^*}, I^*, R^{I^*})$ created from two input ontologies: $O_1 = (C_1, H_1, R_1^C, I_1, R_1^I), O_2 = (C_2, H_2, R_2^C, I_2, R_2^I)$;

if $H_1 \subseteq H_2 \vee H_2 \subseteq H_1$ **then**
  $\delta_H^- = 0$;
else
  $\delta_H^- = 1 - \frac{|H_1 \cap H_2|}{|H_1 \cup H_2|}$;
end if
$\delta_H^+ = \frac{|H^* \setminus (H_1 \cup H_2)|}{|H_1 \cup H_2|}$
$\delta_H = \delta_H^+ + \delta_H^-$
return δ_H

5 Uses Case Scenarios for Hierarchy and Relation Integration

5.1 Hierarchy Integration

Let us illustrate by simple examples how the degree to which the knowledge increases is calculated in case of hierarchy integration (see Fig. 1). In the first case (Fig. 1A) the integrated ontologies are quite different. The fact that for these two ontologies any of the hierarchies is included in the other one entails that $\delta_H^- = 1$. After the integration, any of the new hierarchies is added and any of the old hierarchies is not replaced or removed. Therefore, $\delta_H^+ = 0$ because of the cardinality of a set $H* = H_1 \cup H_2$. Eventually, we obtain $\delta_H = 1$ and we can say that during the integration process on the relation level we doubled the knowledge we had.

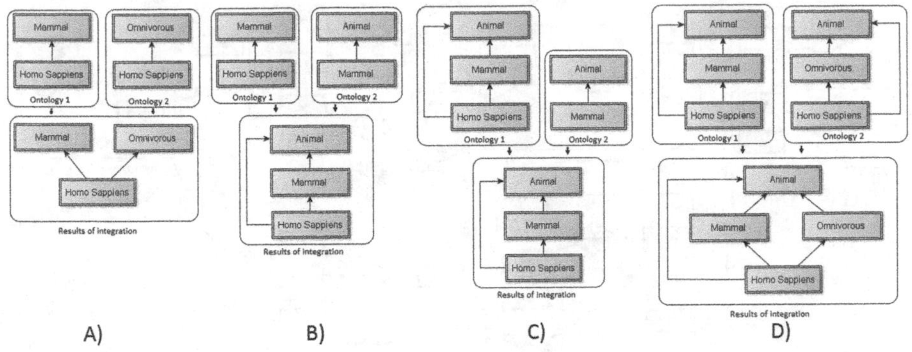

Fig. 1. Examples of ontologies integration at hierarchy level

In the second case (Fig. 1B), input ontologies seem to be very similar to the previous one. As in the previous example, any hierarchy is included in the other one, therefore $\delta_H^- = 1$. However, $H * \backslash H_1 \cup H_2 = 1$, then $\delta_H^+ = \frac{1}{2}$ and $\delta_H = \frac{3}{2}$. In this situation we "create" a new knowledge during the integration. If we consider inputs separately, we only know that *a Homo Sapiens* is *a Mammal* and *a Mammal* is *an Animal*. After the integration, we additionally know that each *Homo Sapiens* is also *an Animal*, therefore, we have found out something new. This knowledge has not been included in any of input ontologies. In this point of view, the presented integration process is very beneficial.

The next case (Fig. 1C) presents a situation where the whole set of hierarchy of Ontology 2 is included within a hierarchy of Ontology 1, formally $H_2 \subseteq H_1$. Therefore, $\delta_H^- = \delta_H^+ = 0$ and the integration process neither increases nor decreases the knowledge about that instance ($\Delta_H = 0$). Eventually, we can say that the integration of these two ontologies in not beneficial from the knowledge increase point of view.

The last example (Fig. 1D) is the most complex. The hierarchies of input ontologies are not included in each other, however they are some common parts i.e. each *Homo Sapiens* is *an Animal*. In this case $\delta_H^- = 1 - \frac{1}{5} = \frac{4}{5}$ and $\delta_H^+ = 0$ because of the cardinality of $H* = H_1 \cup H_2$ (no new knowledge has been created). Eventually, we get $\Delta_H = \frac{4}{5}$.

5.2 Integration of Concepts' Relations

Figure 2 presents some use case scenarios of the integration of concepts' relations. In the first one (Fig. 2A) two input ontologies (with two different relations) are integrated. It is easy to calculate that the potential knowledge increase in this case is maximal and equal to 1.

The second case (Fig. 2B) represents the situation where relation from ontology 2 is more general than the relation in ontology 1: "*is parent*" $\leftarrow$ "*is mother*". The common part of inputs ontologies is pair: (woman, boy) and (woman, girl). The more general relation can not be replaced by more detailed because it cause

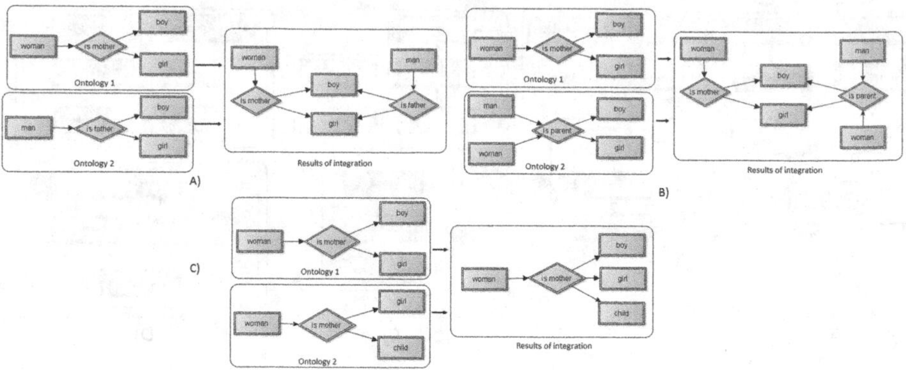

Fig. 2. Examples of ontologies integration at relation level

the lost information that man is parent of boy and man is parent of girl. Therefore $\Delta_R = \frac{3}{2}$. Due to the fact that $\omega = 2$, the final knowledge increase is equal to $\frac{3}{4}$.

In the last example (Fig. 2C) relations in inputs ontologies are equivalency. The "new knowledge" is contained only in the second ontology and it is pair *(woman, child)*. Therefore, $|r \cap r'| = 1$ and $|r \cup r'| = 3$, so $\Delta_R = \frac{2}{3}$ and $\omega = 1$. Finally, we can say that the ontology integration on the relation level increases our knowledge by $\frac{2}{3}$.

6 Future Works and Summary

This paper is a straight continuation of our previous research [5,6], which addressed the problem of ontology integration on the concept and instance levels. This article is devoted to the problem of ontology integration on the concepts' relation level. Due to the limited space we have omitted the integration of relations that exist not only on a "schema" level of concepts, but actually describe interactions of instances of concepts.

The main contribution of the paper is a set of algorithms of ontology integration on relation level. These algorithms distinguish the integration into the procedure of combining relations between concepts and the integration of concept's hierarchies (which in our opinion entail too many consequences and restrictions to be integrated using the same algorithm as relations). The article also contains a set of algorithms that can be used to evaluate the valency of the performed integration. Every procedure is carefully analysed and the outcomes are described with illustrative examples.

In the future, we plan to extend the framework of knowledge increase integration of aforementioned relations between instances. We also plan to investigate the usefulness of our ideas in other knowledge or data representation methods, such as federated data warehouses.

References

1. Bobrowski, M., Marré, M., Yankelevich, D.: Measuring data quality. Universidad de Buenos Aires. Report. 1999:99–002 (1999)
2. Flahive, A., Taniar, D., Rahayu, W.: Ontology as a Service (OaaS): a case for sub-ontology merging on the cloud. J. Supercomput. **65**, 185–216 (2013). doi:10.1007/s11227-011-0711-4
3. Frank, A.U.: Data quality ontology: an ontology for imperfect knowledge. In: Winter, S., Duckham, M., Kulik, L., Kuipers, B. (eds.) COSIT 2007. LNCS, vol. 4736, pp. 406–420. Springer, Heidelberg (2007). doi:10.1007/978-3-540-74788-8_25
4. Geisler, S., Weber, S., Quix, C.: An ontology-based data quality framework for data stream applications. In: 16th International Conference on Information Quality, pp. 145–159 (2011)
5. Kozierkiewicz-Hetmańska, A., Pietranik, M.: The knowledge increase estimation framework for ontology integration on the concept level. J. Intell. Fuzzy Syst. **32**(2), 1161–1172 (2017). doi:10.3233/JIFS-169116
6. Kozierkiewicz-Hetmańska, A., Pietranik, M., Hnatkowska, B.: The knowledge increase estimation framework for ontology integration on the instance level. In: Nguyen, N.T., Tojo, S., Nguyen, L.M., Trawiński, B. (eds.) ACIIDS 2017. LNCS, vol. 10191, pp. 3–12. Springer, Cham (2017). doi:10.1007/978-3-319-54472-4_1
7. Le, D.H., Dang, V.T.: Ontology-based disease similarity network for disease gene prediction Vietnam (2016). doi:10.1007/40595-016-0063-3
8. Lozano-Tello, A., Gómez-Pérez, A.: Ontometric: a method to choose the appropriate ontology. J. Database Manage. **2**(15), 1–18 (2004)
9. Nguyen, N.T.: Advanced Methods for Inconsistent Knowledge Management. Springer, London (2008). doi:10.1007/978-1-84628-889-0
10. Pietranik, M., Nguyen, N.T.: A multi-atrribute based framework for ontology aligning. Neurocomputing **146**, 276–290 (2014). doi:10.1016/j.neucom.2014.03.067
11. Porello, D., Endriss, U.: Ontology merging as social choice: judgment aggregation under the open world assumption. J. Logic Comput. **24**(6), 1229–1249 (2014)
12. Maedche, A., Staab, S.: Measuring similarity between ontologies. In: Gómez-Pérez, A., Benjamins, V.R. (eds.) EKAW 2002. LNCS, vol. 2473, pp. 251–263. Springer, Heidelberg (2002). doi:10.1007/3-540-45810-7_24
13. Tartir, S., Arpinar, I.B., Moore, M., Sheth, A.P., Aleman-Meza, B.: OntoQA: metric-based ontology quality analysis (2005). http://lsdis.cs.uga.edu/library/download/OntoQA.pdf

Particle Swarm of Agents for Heterogenous Knowledge Integration

Marcin Maleszka(✉)

Faculty of Computer Science and Management,
Wroclaw University of Science and Technology,
Wyb. Wyspianskiego 27, 50-370 Wroclaw, Poland
marcin.maleszka@pwr.edu.pl

Abstract. There is an ever increasing number of sources that may be used for knowledge processing. Often this requires dealing with heterogeneous knowledge and current methods become inadequate in these tasks. Thus it becomes important to develop better general methods and tools, or methods tailored to specific problems. In this paper we consider the problem of knowledge integration in a group of social agents. We use approaches based on particle swarm optimization – without the optimization component – to model the diffusion of information in a group of social agents. We present a short description of the theoretical model – a modification of PSO heuristics. We also conduct an experiment comparing this approach to previously researched models of knowledge integration in a group of social agents.

Keywords: Knowledge integration · Multiagent system · Collective knowledge · Knowledge diffusion · Multiagent system

1 Introduction

Integration of heterogeneous knowledge from different sources is becoming more and more common with increasing knowledge processing in modern information society. It may occur in decision making, information retrieval and various other applications. Heterogeneous knowledge may be understood as a specific case of inconsistent knowledge, so some of the methods developed for the latter type of problems should be possible to be adapted to the former. In our overall research we deal with different situations related to integration of knowledge over time, which may be treated as parallel to knowledge diffusion problems. After considering different types of communication, knowledge integration in members of a collective and structures of the collective, in this paper we consider an expanded model that is more focused on heterogeneous knowledge.

We consider a collective of agents, each with some knowledge that may be of various structure, in a task of determining the overall knowledge of this collective. We split this task into three parts: (a) division of the collective into more homogenous groups (b) knowledge diffusion in groups with small inter-group

N.T. Nguyen et al. (Eds.): ICCCI 2017, Part I, LNAI 10448, pp. 54–62, 2017.
DOI: 10.1007/978-3-319-67074-4_6

interference (c) integration of knowledge from representatives of each group. The second of these parts uses particle swarm approach and some of our previous research. The third part is a single task of centralized knowledge integration, where knowledge is heterogeneous but preprocessed to facilitate the integration process.

This paper is organized as follows: in Sect. 2 we provide short descriptions of other works relevant to ours; in Sect. 3 we present the description of the proposed model of heterogeneous knowledge integration; Sect. 4 presents our approach to verify this model in some basic situations; in Sect. 5 we provide some concluding remarks and detail possible further applications of our research.

2 Related Works

While this paper is focused on heterogeneous knowledge integration, it is a part of a larger research covering multi-agent systems, decentralized systems, asynchronous communication protocols, collective knowledge and more. This section covers various research that is relevant to ours, including other publications relevant to this paper.

The basis of this overall research is the consensus theory and the observation that consensus may change over time. This is somewhat similar to multi-agent approach to continuous time consensus (e.q. autonomous robots, network systems [16]), but is more focused on determining the final knowledge of some collective – thus the problem of finite time consensus in that area (whether the agents will reach consensus at all). In these areas of research the continuous or finite time consensus are used for attitude alignment, flocking, formation control, negotiations, measuring of stability of multi-agent systems and more [2]. This may also be considered in centralized or decentralized agent systems [6] and to determine whether a new agent adds sufficient knowledge to a collective to be a justified expenditure [13] – that is the problem of determining whether a collective may have some optimal number of experts for decision making systems. This overall area of research is in turn based on a mathematical model developed in [1,9] to solve a single practical problem and then further extended to methods of determining median knowledge. Since those publications, this approach has been proven to be useful for conflict solving, knowledge integration and knowledge inconsistency resolution. The postulates and knowledge functions are, among others: reliability, consistency, Condorcet consistency, general consistency, proportion, unanimity, quasi-unanimity, simplification, 1-optimality and 2-optimality [12]. In this paper we use the last two, which are used to determine medians of different knowledge states – a state with minimal distance to all other elements, and a state with minimal square distance to all other elements. We also use the basic approach to knowledge integration by simply using the sum of all knowledge. All these are the basis for developing various integration algorithms for different knowledge structures or practical applications.

Our research also incorporates ideas from previous publications in the area of centralized and decentralized communication between agents. The basic idea for centralized systems is some supervisor agent that gathers data from all other agents and makes decision, e.q. in [4] this is a decision making system for a whole traffic system in some area, where each single agent monitors a single area (crossroads) or single functionality (traffic lights). In decentralized systems the agents communicate their knowledge to other agents, but make decisions on their own, e.q. in [14] this is done for surveillance. There are also hybrid approaches, where agents communicate between each other, but the overall decision is still made by some supervisor, e.q. in [11] this is done to monitor power grid.

We also consider research done in social networks area, especially as related to knowledge diffusion and integration aspects. This types of methods are proposed to improve teaching process, knowledge dissemination in companies, etc. [8]. The authors of [3] show that determining strong ties in a social network would help with improvements to group results. In this paper we base a large part of our model on work done in [10]. The authors use Boids algorithm [15] as a basis for simulating a community of homogenous agents and explore the potential of their model for online advertising (recommendations).

3 Particle Swarm Model of an Agent System

In our previous research we used the following model of decentralized asynchronous communication between agents [7]:

- Each agent has a list of agents (*friends*) that he will communicate to more often.
- In irregular time intervals, each agent has a chance to start communication with a random agent of the whole population or a random agent of its *friends*.
- The communication between agents is unidirectional - once an agent sends a message, he does not expect a reply or confirmation of receipt.
- Upon receiving the knowledge, an agent may use different strategies to make use of it, e.g. substituting own knowledge, gathering a list of knowledge states and integrating them to create his new state of knowledge, etc.

This decentralized model worked quite well in many aspects, but it was unable to deal with heterogeneous sources of knowledge. To cover this gap, we expand our model by ideas derived from the work done in [10]. Towards this purpose we divide the whole collective of agents into smaller groups (each more homogeneous than the whole collective), where each group follows some leader and each member (including the leader) can change their knowledge state. Once each group decides on its position, we integrate only the knowledge of each leader. To complete this whole procedure, we require the following functionalities: (a) division of collective into subgroups, (b) group leader selection, (c) knowledge changes in groups, (d) interaction between different groups, (e) top-level integration. Each functionality will be described below in subsequent sections.

3.1 Division of Collective

As previously stated, the intention is for each subgroup to be more homogenous than the whole collective. We consider the problem in terms of dimensionality – i.e. different knowledge structures are orthogonal in some knowledge state (e.q. logical knowledge and ontological knowledge). In that case we may group collective members based on distance. It is relatively simple to determine the distance in terms of one dimension, but calculating distance between orthogonal knowledge is non-trivial. We simplify this by using Manhattan distance, that is for two orthogonal knowledge states A and B in knowledge dimensions a and b, the distance between them is:

$$d(A, B) = d_a(A, \oslash) + d_b(\oslash, B), \quad (1)$$

where d_a is the distance measure for dimension a and d_b is distance measure for dimension b.

With distance measure defined, we use one of known clustering method (several are tested in Sect. 4 for best results). As may be noted, this approach will lead to subgroups having lower number of dimensions for its members knowledge than the whole collective. Thus additional increase of homogeneity will occur.

3.2 Leader Selection

The leader of each group is selected initially as the best representative of that group, but as the knowledge of each member of the group may change over time at some point a different member may become a better representative. Thus we are continuously monitoring the position of virtual best representative (which should remain mostly unchanged in a group) and its distance to other group members. The leader of the group is always the member that is closest (most similar) to this virtual representative knowledge.

Over time the leader of each group influences all group members, making them slowly change their knowledge to be more similar to his own.

3.3 Inter-group Interaction

As previously stated, each group is selected based on similarity of its knowledge. Different groups have little similarity, but the overall collective was created towards common purpose. Thus we introduce the possibility that leaders of each group can influence each other, thus in turn influencing their followers. This occurs during the parallel diffusion of knowledge in each group.

Following the social model presented in [10], we assume that there is small repulsion between different group leaders. This works to make groups more distinct and with properly tuned parameters, helps with increasing homogeneity of the group.

3.4 Diffusion of Knowledge in Groups

Inside the groups we use particle swarm approach to changes in member knowledge. In an optimization problem this would mean trying to determine the best solution by exploring the solution space. In an integration problem, this means that a single integrated solution should be found and that each member of the group should change its own knowledge state to that integrated solution. Usually this problem is solved by gathering all knowledge from all the members, calculating its mean (integrated value) and sending it back to all members as their new solution. As previously stated, this is not always a satisfactory solution, especially in situations where the knowledge is heterogeneous or inconsistent. Our previous research in [7] has shown that when group members communicate in a decentralized and asynchronous manner, the collective knowledge is often better. In particular, when members gather a few opinions before integration, the results for the whole collective were shown to be best.

Following that idea, we adopt it to the particle swarm approach: each member of the group communicates with several randomly selected other members of the group, sending them his own state of knowledge. The receivers wait until they gather at least a predetermined number of inputs, then integrate the result. This is the first component of the new knowledge state. The others components are the attraction toward team leader and inertia (previous own knowledge state). The new knowledge state of the agent is the weighted average of these values.

3.5 Leader Knowledge Integration

After several iterations of knowledge diffusion, the leaders of each group are newly determined knowledge states of existing members of the collective. They are similar to mean values of integrated knowledge of each group, but are not required to be identical. The integrated knowledge of the collective is determined based only on those leaders. On this level we use simple centralized approach, that is each leader sends its knowledge state to some central supervisor, which then uses these states to determine the collective knowledge of the whole collective. This may be done e.q. by calculating a mean or summarizing the knowledge, depending on the exact type of problem. Overall each group should be very different, but complementary to others – thus the preference is summarization. As such each group would give partial knowledge and only the sum of it would be the correct result. We further test this in the simulation environment in Sect. 4.

4 Experimental Evaluation

In our previous research we focused on two approaches to verification of our models: a multi-agent simulation environment and a distributed system integrating different weather forecasts into a single prediction. While in this paper we also use these two approaches, it must be noted that weather forecasts have a more homogenous structure of knowledge and do not fully reflect the model proposed in this paper. On the other hand, by using the same system we are able to compare it to the previously suggested models.

The simulation environment is based on JADE agent framework [5]. The centralized nature of this framework works for gathering data on the functioning of agents and allows simulation of distributed system (a collective with distributed asynchronous communication), but any practical application of our work would need some other agent environment. We use identical *Social Agents* to represent different users or different sources of heterogeneous data. In the current simulation we use only multi-attribute and logic knowledge, but the attribute groups may be different for different agents. In simulation the agents generate random structure and value of knowledge upon initialization (e.q. a single integer in some range, a set of logical values). After division of collective based on several methods tested below, the communication between agents occurs in irregular intervals (each time moment, there is some probability that the agent will communicate). The communication occurs according to the description in previous Section. After some number of iterations, we gather the knowledge of the group leaders (several agents from the collective) and integrate it using one of several approaches.

In Table 1 we present combinations of different methods were tested on the datasets, with final collective knowledge compared to expert determined knowledge (i.e. the initial knowledge is generated randomly and we compare the result of integration by the expert with the result of integration using our approach).

The observed results show that the difference between methods used are minimal, but the best ones were achieved with using k-means clustering, O1 integration method in agents (determining the virtual representative) and summarizing the knowledge of group leaders. The differences are very small, so for

Table 1. Results of collective knowledge integration using the present approach in comparison with expert determined knowledge of the collective, for different integration and clustering methods. For each set of main parameters, the weights used in inter-group knowledge diffusion phase were independently tuned.

Agents	Clustering	Final integration	Result comp. to expert know.
100, using O1	K-means	O1	0.82
100, using O1	Hierarchical	O1	0.83
100, using O2	K-means	O1	0.81
100, using O2	Hierarchical	O1	0.80
100, using O2	K-means	O2	0.80
100, using O2	Hierarchical	O2	0.79
100, using O2	K-means	Sum	0.82
100, using O2	Hierarchical	Sum	0.83
100, using O1	K-means	Sum	0.86
100, using O1	Hierarchical	Sum	0.85

other types of knowledge structures other methods may be better. This means that the approach must be tuned for each specific problem.

The second experiment conducted was applying the proposed approach to our long-standing test bed of a weather prediction system [7]. This application is only marginally fitting to this approach, as the knowledge integrated is quite homogenous – most sources provide all information about weather forecasts (temperature, rain, cloud covers, etc.) and only some sources miss part of these information. After tuning the particle swarm approach, we determined that using O1 integration in agents, hierarchical clustering and O2 integration in leaders was the best for this problem. The comparison of results with other methods we tested for weather prediction is presented in Table 2.

Table 2. All observed runs of the weather prediction system in all variants: basic dominant value (B. Dominant), centralized consensus (Cen. Cons.), decentralized consensus (Dec. Cons.), decentralized consensus with *friend* relation (D.-S. Cons.), two new variants of dominant value approach: full decentralized voting in source layer (Dom. Dec.) and voting interchangeable with consensus in source layer (Dom. Mix.), particle swarm approach (PSO).

System-run	MAE	Comp. w/Best Src.	Comp. w/Worst Src.	Comp. w/Avg. Src.
B. Dominant (IV-V '15)	1,857	89%	17% better	3% better
Cen. Cons. (IV-V '15)	2, 018	82%	7% better	95%
Cen. Cons. (X '15)	2, 132	75%	2% better	90%
Dec. Cons. (X '15)	1, 984	83%	9% better	97%
Cen. Cons. (IV '16)	1, 991	85%	6% better	93%
Dec. Cons. (IV '16)	1, 994	85%	6% better	93%
D.-S. Cons.(IV-V '16)	1, 989	85%	6% better	93%
B. Dominant (X '16)	1,993	89%	15% better	99%
Cen. Cons. (X '16)	1,956	87%	12% better	97%
Dec. Cons. (X '16)	1,931	88%	14% better	98%
D.-S. Cons. (X '16)	1,933	88%	14% better	98%
Dom. Dec. (X '16)	1,898	90%	16% better	*equal*
Dom. Mix. (X '16)	1,892	90%	16% better	1% better
B. Dominant (IV '17)	2,271	90%	13% better	98%
Cen. Cons. (IV '17)	2,256	91%	14% better	99%
Dec. Cons. (IV '17)	2,211	92%	16% better	1% better
D.-S. Cons. (IV '17)	2,232	92%	15% better	*equal*
Dom. Dec. (IV '17)	2,260	90%	14% better	98%
Dom. Mix. (IV '17)	2,191	93%	17% better	2% better
PSO (IV '17)	2,208	93%	16% better	1% better

The comparison of results for the weather prediction data in April 2017 shows that the particle swarm approach is comparable with the better of previously tested methods. As stated, it is not a problem best suited for this method, but even in this case the results are promising.

5 Conclusions

In this paper we have described a model of heterogeneous knowledge integration based on a particle swarm of identical social agents. The model works by first dividing the collective into several groups that are more homogenous. Following this, agents in each group influence each other with overall knowledge of the group changing only minimally. There is also a slight interaction with other groups. Each group has a leader, that is the agent closest to the virtual centroid of this groups knowledge. The final knowledge of the whole collective is determined by integrating the knowledge of these leaders after several cycles of communication within the groups. The evaluation of this approach was performed both in a simulation environment and in a prototype system of a practical application. Both show that the approach is feasible and promising – in the practical application its results were comparable to the best of other tested methods. We intend to further study this approach with regards to more complex knowledge structures in other practical applications. It may be also interesting to study this approach as a model of knowledge diffusion of groups of heterogeneous agents.

Acknowledgment. This research was co-financed by Polish Ministry of Science and Higher Education grant.

References

1. Barthelemy, J.P., Janowitz, M.F.: A formal theory of consensus. Siam J. Discrete Math. **4**, 305–322 (1991)
2. Bhat, S.P., Bernstein, D.S.: Finite-time stability of continuous autonomous systems. Siam J. Control Optim. **38**(3), 751–766 (2000)
3. De Montjoye, Y.-A., Stopczynski, A., Shmueli, E., Pentland, A., Lehmann, S.: The strength of the strongest ties in collaborative problem solving. Scientific reports 4, Nature Publishing Group (2014)
4. Iscaro, G., Nakamiti, G.: A supervisor agent for urban traffic monitoring. In: IEEE International Multi-disciplinary Conference on Cognitive Methods in Situation Awareness and Decision Support (CogSIMA), pp. 167–170. IEEE (2013)
5. JADE: Java Agent Development Framework. http://jade.tilab.com/
6. Li, S., Dua, H., Lin, X.: Finite-time consensus algorithm for multi-agent systems with double-integrator dynamics. Automatica **47**, 1706–1712 (2011)
7. Maleszka, M.: Observing collective knowledge state during integration. Expert Syst. Appl. **42**(1), 332–340 (2015)
8. Maleszka, M., Nguyen, N.T., Urbanek, A., Wawrzak-Chodaczek, M.: Building educational and marketing models of diffusion in knowledge and opinion transmission. In: Hwang, D., Jung, J.J., Nguyen, N.-T. (eds.) ICCCI 2014. LNCS (LNAI), vol. 8733, pp. 164–174. Springer, Cham (2014). doi:10.1007/978-3-319-11289-3_17
9. McMorris, F.R., Powers, R.C.: The median procedure in a formal theory of consensus. Siam J. Discrete Math. **14**, 507–516 (1995)
10. Morzy, M., Kruk, T.: Particle swarm as a model for community formation in social networks. In: Proceedings of Network Intelligence Conference (ENIC) 2016, pp. 40–47. IEEE (2016)

11. Nagata, T., Sasaki, H.: A multi-agent approach to power system restoration. IEEE Trans. Power Syst. **17**(2), 457–462 (2002)
12. Nguyen, N.T.: Advanced Methods for Inconsistent Knowledge Management. Springer, London (2007)
13. Nguyen, V.D., Nguyen, N.T.: An influence analysis of the inconsistency degree on the quality of collective knowledge for objective case. In: Nguyen, N.T., Trawiński, B., Fujita, H., Hong, T.-P. (eds.) ACIIDS 2016. LNCS (LNAI), vol. 9621, pp. 23–32. Springer, Heidelberg (2016). doi:10.1007/978-3-662-49381-6_3
14. Peterson, C.K., Newman, A.J., Spall, J.C.: Simulation-based examination of the limits of performance for decentralized multi-agent surveillance and tracking of undersea targets. In: International Society for Optics and Photonics, SPIE Defense+ Security, p. 90910F (2014)
15. Reynolds, C.W.: Flocks, herds and schools: a distributed behavioral model. ACM SIGGRAPH Comput. Graph. **21**(4), 25–34 (1987)
16. Ren, W., Beard, R.W., Atkins, E.M.: A survey of consensus problems in multi-agent coordination. In: American Control Conference, 2005, Proceedings of the 2005, pp. 1859–1864. IEEE (2005)

Design Proposal of the Corporate Knowledge Management System

Ivan Soukal[1] and Aneta Bartuskova[2(✉)]

[1] Department of Economics, Faculty of Informatics and Management, University of Hradec Kralove, Rokitanskeho 62, 500 03 Hradec Kralove, Czech Republic
ivan.soukal@uhk.cz

[2] Faculty of Informatics and Management, Center for Basic and Applied Research, University of Hradec Kralove, Rokitanskeho 62, 500 03 Hradec Kralove, Czech Republic
aneta.bartuskova@uhk.cz

Abstract. This paper presents a proposal of the knowledge management system for managing and refining corporate knowledge. This information system is web-based to allow online immediate access. Corporate knowledge is organized to enable efficient extraction and use in business processes. It can be stored in a form of the course (organized by time sequence) or repository (i.e. collection of resources), depending on the characteristics of the particular knowledge. The system´s proposal will be demonstrated on the principle of several facilitators and many system users (employees). The proposed system can be just the same used for educational purposes, especially in lifelong learning.

Keywords: Corporate knowledge · Collective intelligence · Knowledge management system · Information system · Knowledge representation

1 Introduction

Corporate education is emerging as one of the most influential, dynamic and effective means for retaining existing employees or attracting new ones as well as facilitate innovation [1]. In this paper we propose a knowledge management system for managing and refining corporate knowledge. This knowledge is organized and intended to be further visualized to enable efficient knowledge extraction and use in business processes. The managed knowledge can be in a form of the process or course (organized by time sequence) or repository (i.e. collection of resources). In this paper we will use term "process-based" (or "course-based" when more relevant) and "repository-based" systems. Schemes, which are presented throughout this paper, are specified for processes rather than collections. This is because in corporate environment the knowledge is usually associated with particular business processes. These processes are best represented in a form of course-based knowledge. In case of collections, organization by time sequence would be simply replaced by other mean of organization. The system is designed to be web-based to allow online immediate access. Regarding users, we expect several facilitators and many users of the system (employees) in our proposal, however it can be adapted to custom needs. By the role of facilitator is meant selected supervising

N.T. Nguyen et al. (Eds.): ICCCI 2017, Part I, LNAI 10448, pp. 63–72, 2017.
DOI: 10.1007/978-3-319-67074-4_7

employee or associate, who is in charge of selected business process. The role of regular user is a representation of regular employee, who is participating in various business processes.

In the case of this particular model, presented in this paper, the first hierarchical level of resources is composed by knowledge representation of business processes. Each process then creates its own repository, which is the second level of the hierarchy. These repositories can be further divided by categories into groups. Throughout the chapter, there are comments regarding possible extensions or modifications of this basic model, which can be carried out in the implementation.

2 Conceptual Foundation

Our system´s proposal is based on the shortcomings of commonly used systems for knowledge management systems and educational systems, which are very similar in the need to organize knowledge and make it accessible for its users. They both support the processes of traditional knowledge management such as creation, transfer and storage/ retrieval [2]. According to some authors, e-learning and knowledge management function as complements and components critical to learning [3], which has a great role in corporate knowledge management.

Peng et al. presented a knowledge management system which would support web-based learning in higher education [4]. In this system, students showed their will to use the system to select useful course resources for themselves because it is easy to integrate the course resources to their own resources [4]. Similarly, we expect that employees would appreciate the possibility to store the official corporate materials along with their own knowledge. By doing so in our proposed system, at the same time they will contribute to building and improving the corporate knowledge.

Each of the discussed issue will be labelled for later reference, following this format: *{xxx}*, placed after the relevant issue. As the main limitations of process-based systems such as learning management systems were identified [5]:

- (all system users) an insufficient personalization support for:
  - organization of processes / courses *{per1}*
  - organization of resources inside processes/courses *{per2}*
  - annotation mechanisms (tags, notes, comments,..) *{per3}*
- (employees / students) the lack of visualization for distinguishing:
  - the process/course basic structure *{mng1}*
  - between important and optional resources *{mng2}*
- (facilitators / instructors) the system´s interface does not encourage:
  - regular revising of existing content *{mng3}*
  - disposing of outdated content *{mng4}*

Information overload is very common; we now live in a world of abundance where editing and curating become more crucial than ever [7]. However user interface of these systems usually does not encourage this desirable behaviour. Actions that are made easy by system are more likely to occur, while those that have barriers are less likely to [8].

The above mentioned issues were gathered by thorough literature study and personal experience with several knowledge and learning management systems. Furthermore we have conducted an analysis of 15 computer science courses in system used in University of Hradec Kralove, Czech Republic. The most common way to organize content in analysed course-based knowledge was:

- by category - primary content division, also in sidebar navigation, sections usually partially reflect content types as categories *{org1}*
- by continuum (time) - secondary content division, usually used in lectures section, which are divided into sessions *{org2}*

The analysis of selected knowledge repositories also identified several issues. We have chosen well-known learning object repository Ariadne Foundation [9] and Multimedia Educational Resource for Learning and Online Teaching [10] for the thorough analysis and encountered these issues:

- an inefficient and confusing hierarchy and labelling of categories *{res1}*
- search results are outdated or do not exist anymore - resources are not further managed (revised, updated, deleted) *{res2}*
- metadata are missing, inaccurate, wrong or redundant - resources are not sufficiently described *{res3}*

The following table (Table 1) presents the proposed areas of improvement for identified issues of educational and knowledge-management systems. This will be used as a guideline for the new system proposal. The personalization requirement is further the prerequisite for the extension - collaborative knowledge building.

Table 1. The proposed areas of improvement

	Support for issues
{ORG} organization of navigation items and resources	{mng1} {mng2} {org1} {org2} {res1}
{MNG} management of resources	{mng3} {mng4} {res2} {res3}
{PER} means of personalization (in organization and management)	{per1} {per2} {per3}

3 The Core Management System

This chapter is devoted to the proposal of the core management system. In relation to the proposed areas of improvement, this chapter is primarily focused on the means of personalization *{PER}*. The management of resources is naturally also covered, however mainly the fundamentals such as CRUD. The solution for identified issues regarding management *{MNG}* will be subject of later chapters, especially Sect. 4 "Collaborative knowledge building". The core model assumes many facilitators (one facilitator per one business process) and many users - employees (who can attend any number of these courses).

3.1 Core Functions

The core functions are represented by key activities, which users can perform in the system. As the primary division of corporate content were selected businesses process, as was already mentioned in the introduction to this chapter. Each process in a form of respective course then creates its own repository, which can be further organized into groups of resources - or simply categories. Individual resources or pieces of knowledge present separate entries in the system.

Regarding personalization, the first issue to consider is the extent of possible personalization. In current systems, user can often only "read" the content. While learning business processes, employees are forced to create their own repository of materials, either as a local copy of official materials or they are accumulating their own resources or the combination of both. The lack of personalization support, which was discussed in Sect. 6.1, can be divided into these two main categories:

- organization of resources (regrouping, adding, deleting) *{per1}, {per2}*
- modification of resources (annotation, highlighting, notes) *{per3}*

The first category of personalization (organization of content) is reflected on the key activities list. These are: CRUD course, CRUD category and CRUD entry from the facilitator´s role and CRUD custom category, CRUD custom entry and customize entry from the common user´s role. The second category (modification of content) is contained in "update" function from the "CRUD" acronym. "CRUD" is an abbreviation for four basic functions of persistent storage: create, read, update, delete.

Organizing is usually a subjective process, when done by ambiguous schemes, and language used for labelling is also often ambiguous [6]. Therefore users should be allowed not only to add and modify files, but also change labels or the position of files and thus adapt the environment to their needs. With this possibility, they could e.g. relocate selected resources to the separate group (category) for easier access.

3.2 Main Segments of the System

The proposed system has two main segments. The first one we would call "core repository", which in forms the central knowledge base. This repository provides access to all business processes with all resources to both facilitators and employees. Employee´s access can be restricted according to his enrolment or other requirements, this depends on the actual implementation of the system. Employee´s access to the core repository is "read-only", as is the facilitator´s access to the course/process of other facilitator. Again, according to actual implementation there could be several facilitators with rights to manage single business process, however we will stick to the basic model. The facilitator can edit his course directly in the core repository, edit categories and individual entries. The extent of editing is characterized by CRUD. These changes would affect the view of core repository for every user of the system.

The second segment consists of individual user´s accounts, which is in fact the implementation of employee´s personalization. It is not precluded that also a facilitator can have his personalized account. However it is desirable that all employees could benefit from facilitator´s management of resources, therefore it should take place in the core repository. In the scope of personalized user´s account and selected business process, the employee can add his own content to the course, update the existing resource or delete the extra resources. Of course, these changes would take effect only in the individual user´s view, not in the core repository of resources. As far as an employee does not personalize the business process, he will only "read" the core repository. When the employee starts manipulating content, the changes are logged, new content is stored, and his view of the course becomes the personalized course. However users should retain the option to display the original course as it is stored in the core repository.

The possibilities of personalized organization of corporate content are depicted in (Fig. 1). The diagram consists of the following use cases:

- user "Facilitator A" created "Business process A", within this course he created three categories (groups) "Rules", "Partners" and "Evaluation" and within these categories he added several entries in each;
- user "Facilitator B" created "Business process B" and analogically any number of categories and entries;
- user "User 1" has access to both courses in the core repository;

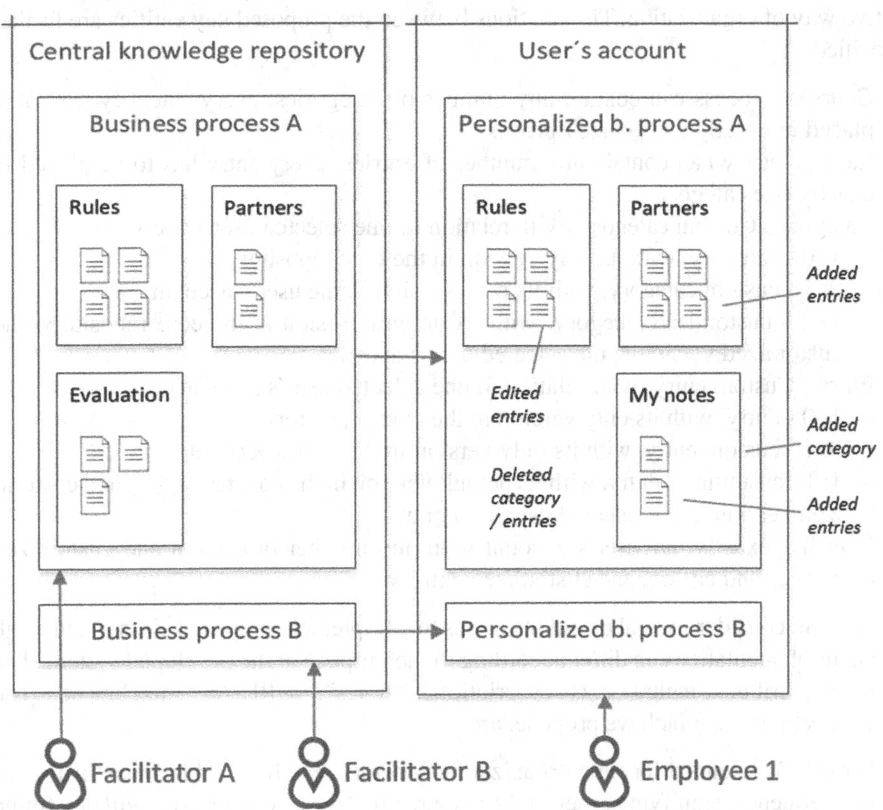

Fig. 1. Personalized organization of knowledge

- when "User 1" changes anything within the scope of any course, the personalized version of the respective course is created;
- user "User 1" created his personalized course by deleting "Evaluation" category, creating custom "My notes" category, creating custom entries in "Partners" and "My notes" category and customizing entries in "Rules"

The resources are for simplicity presented as files in folders, labels were selected arbitrarily for demonstration. For illustration only three users are showed in the schema. However the proposed system supports unlimited number of both roles.

3.3 Information Architecture

Several key classes are considered in the system, from the object-oriented programming point of view. These are: "course", "category", "entry" and "topic" in the core repository. There is only one instance of the core repository, however there are many user accounts. These accounts are represented by the following classes: "user", "custom category" and "custom entry". It is important to note that the "custom entry" in the diagram can signify both the new "custom entry" and the customization of existing entry. The concept of "custom category" is analogical. The management of entries by topic presents an alternative way of organization. The relations between the proposed key entities are further specified:

- Course / process can contain any number of categories, every category has to be placed in exactly one course / process
- Each category can contain any number of entries, every entry has to be placed in exactly one category
- Category : Custom category (with relation to one selected user´s account)
  - 1: 0 - category, with its only version in the core repository
  - 0: 1 - custom category, with its only version in the user´s account
  - 1: 1 - customized category, with its default version in the core repository and customized version in the selected user´s account
- Entry : Custom entry (with relation to one selected user´s account)
  - 1: 0 - entry, with its only version in the core repository
  - 0: 1 - custom entry, with its only version in the user´s account
  - 1: 1 - customized entry, with its default version in the core repository and customized version in the selected user´s account
- User has exactly one user´s account with any number of custom and customized categories and custom and customized entries

We can consider several useful attributes to complement proposed classes, although actual implementation can differ according to the purpose of the developed system. The common attributes include e.g. "description", "notes", "URL" or "attachment". The more special ones, which we propose, are:

- "priority", intended for time organization scheme - can be filled in by an instructor as a sequence, signifying order of the resource in the process. Priority will determine horizontal arrangement of the navigation list of resources. An employee can switch views between organization by "priority" (common version from the core repository) or "user priority" (personalized).

- "user rating" - continuum organization scheme, by which individual users can impose effect on the core repository, as average rating of the respective resource. Again, user should be able to switch views between organization by "average rating" (core repository) and "user rating" (personalized).
- "user progress" - an additional personalization technique

The system can implement various custom attributes, which can users add to the resources in order to personalize the experience. User progress, expressed as a number or percentage, could - similarly as user rating - function as a continuum organization scheme. The resources in personalized view of the business process could be then organized according to the user´s progress with the resource.

The ideas for possible variables as organization schemes are summarized in the above (Table 2). An addition to the classical organization schemes are tags with variables "topics" and "user topics". Tags were identified as the categorical organization schemes with cross-listing.

Table 2. Proposed variables as organization schemes

Org. scheme	Variable used by the org. scheme	Personalized variable[a]
Alphabet	Title (of entry)	User title (of custom entry)
Time	Priority (of entry)	User priority (of custom entry)
	Created/modified time stamp	–
Category	Category	Custom category
Continuum	Average rating	User rating
	–	User progress
Tags	Topics	User topics

[a]Custom entry/category is the same as the customized entry/category

4 Collaborative Knowledge Building

The proposals in this chapter will primarily deal with limitations in the area of management of resources *{MNG}*. By collaborative knowledge building is meant the cooperation of facilitators and employees alike in order to keep resources correct and up-to-date. This is very desirable especially in fields of business, which are under continuous progressive development such as medicine or computer science, where stored and used knowledge needs to be regularly updated. The proposed model´s purpose is to promote regular revising of existing content *{mng3}* and disposing of outdated content *{mng4}*. The lacks of such leads to usability issue such as that resources are not managed or sufficiently described *{res2}*, *{res3}*.

4.1 Extended Functions

In order to utilize collaborative knowledge building, the list of key activities would be extended by these use cases: “approve and add custom entry”, “reject custom entry” and “show new custom entries”. Collaborative knowledge building in the selected course in the proposed system is formed by cooperation of one facilitator and many employees participating in that business process. The actual implementation of the system can of course benefit from cooperation of more facilitators or possibly across more processes. Contributions from users are formed by their personalization; custom and customized entries but also deleted entries. These are accessible to the facilitator as a list of modifications from all employees of the business process managed by him. The facilitator can then accept new entries or modifications to the core repository, if he considers it an improvement. If not, the modifications will remain in the particular employee´s account and the core repository will remain unchanged for the rest of the employees. This way the moderated improvement of the core repository such as central knowledge base would be managed.

4.2 Process of Knowledge Building

The arrangement proposed in the previous section facilitates a follow-up of personalization - knowledge building. Users can access business processes from the core repository and they can personalize them. The “group repository” should reflect all performed personalization, including the new content, modified content, information about deleted content and changed labels and position of the content. Individual single-purposed personalization is this way transformed into a reverse process, spontaneous collective knowledge building. Facilitators receive feedback from users and as we assume it will inspire them to revise regularly existing content and to dispose of outdated content, which was one of the main discussed issues.

This idea is depicted in (Fig. 2). Employees can access knowledge about business processes from the core repository. They can personalize knowledge representation of business processes, by which they refine existing content and add new content. This process leads to spontaneous knowledge building, which can be used by facilitators (original creators) as a form of feedback. Ideally facilitators should be inspired to refine the business processes based on users’ personalization. Collaborative building of resources presents a viable solution of knowledge management. In order to transform personalized knowledge management into a collaborative activity, facilitator is intended to have access to personalized and newly created entries of users, which he can evaluate, approve and add to the central repository.

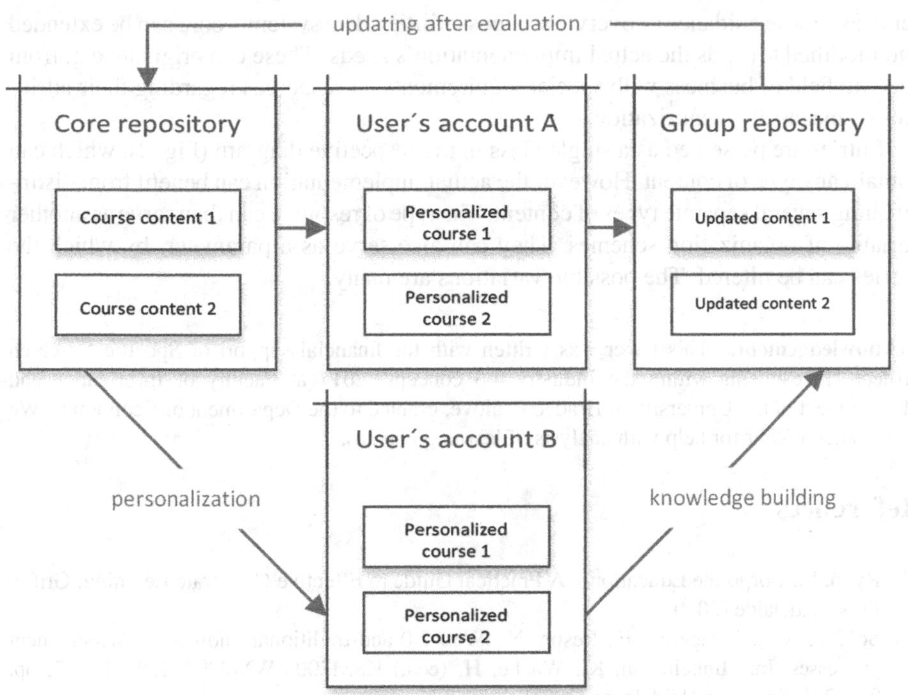

Fig. 2. The process of personalization leading to knowledge building

5 Conclusions and Discussion

The approaches and design proposals were demonstrated in the scope of business processes in corporate environment. The form of courses was selected as the representation of these processes. However as was already mentioned, the system´s proposal is applicable for knowledge management in various areas. The management of knowledge, information, work procedures etc. is important for every larger organization. Not only corporate organization - they can be just groups of people connected via the internet, who need to share some knowledge. Wikis are often used for this purpose and sometimes it is sufficient. However in situations where knowledge base has to be used often and efficiently, e.g. as a reference, the demands on organization and navigation are way bigger than wiki-based system can provide. Also in the case of frequently updated resources or resources with multidimensional categorization, the proposed design has major advantages over simple solutions such as wiki. Especially in larger organizations with many business processes.

As was already mentioned, the presented system´s proposal is not strictly delimited. It provides exact design proposals, grounded in both theory and research. Especially regarding usability aspects such as organization and knowledge-management aspects such as personalization. The final system proposal is mostly a conceptual solution rather than ready-to-use system. Therefore there are many areas, in which presented proposals serve as a loose guidance or overview of possibilities. The system´s core can be extended and modified towards the actual implementation´s needs. These can originate e.g. from specific field of business with special requirements on resources regarding their attributes or means of categorization.

Entries are presented as a single class in the respective diagram (Fig. 2), which can contain any type of content. However the actual implementation can benefit from distinguishing several separate types of content. The type of resource can then serve as another variable of organization schemes. They can also serve as a parameter, by which the entries can be filtered. The possible variations are many.

Acknowledgement. This paper was written with the financial support of Specific Research Project "Investments within the Industry 4.0 concept" 2017 at Faculty of Informatics and Management of the University of Hradec Kralove, granted to the Department of Economics. We thank Vaclav Zubr for help with analysis of learning courses.

References

1. Ryan, L.: Corporate Education—A Practical Guide to Effective Corporate Learning. Griffin Press, Adelaide (2010)
2. Scherp, A., Schwagereit, F., Ireson, N.: Web 2.0 and traditional knowledge management processes, In: Hinkelmann, K., Wache, H. (eds.) KSM2009-WM2009. LNI, P-145, pp. 222-231. Springer, Heidelberg (2009)
3. Ungaretti, A.S., Tillberg-Webb, H.K.: Assurance of learning: demonstrating the organizational impact of knowledge management and e-learning. Knowl. Manag. E-Learn. 41–60 (2011)
4. Peng, J., Jiang, D., Zhang, X.: Design and implement a knowledge management system to support web-based learning in higher education. Proc. Comput. Sci. **22**, 95–103 (2013). doi: 10.1016/j.procs.2013.09.085
5. Bartuskova, A., Krejcar, O.: Implementing knowledge and workflow management in learning management systems. In: Chiu, D.K.W., Marenzi, I., Nanni, U., Spaniol, M., Temperini, M. (eds.) ICWL 2016. LNCS, vol. 10013, pp. 20–26. Springer, Cham (2016). doi: 10.1007/978-3-319-47440-3_3
6. Rosenfeld, L., Morville, P., Arango, J.: Information Architecture: For the Web and Beyond, 4th edn. O'Reilly Media Inc, Sebastopol (2015)
7. Atkins, D., Brown, J.S., Hammond, A.L.: A review of the open educational resources (OER) movement: achievements, challenges, and new opportunities. William and Flora Hewlett Foundation (2007). http://www.hewlett.org/uploads/files/ReviewoftheOERMovement.pdf. Accessed Oct 2016
8. Rubin, B., Fernandes, R., Avgerinou, M.D., Moore, J.: The effect of learning management systems on student and faculty outcomes. Internet High. Educ. **13**(1–2), 82–83 (2010)
9. Ariadne Foundation. http://www.ariadne-eu.org/content/about
10. Multimedia Educational Resource for Learning and Online Teaching. http://www.merlot.org/merlot/index.htm

Dipolar Data Integration Through Univariate, Binary Classifiers

Leon Bobrowski(✉)

Faculty of Computer Science,
Bialystok University of Technology, Bialystok, Poland
l.bobrowski@pb.edu.pl

Abstract. Aggregation of large data sets is one of the current topics of exploratory analysis and pattern recognition. Integration of data sets is a useful and necessary step towards knowledge extraction from large data sets. The possibility of separable integration of multidimensional data sets by one dimensional binary classifiers is analyzed in the paper, as well as designing a layer of binary classifiers for separable aggregation. The optimization problem of separable layer designing is formulated. A dipolar strategy aimed at optimizing separable aggregation of large data sets is proposed in the presented paper.

Keywords: Data integration · Separability · Dipolar strategy · Univariate binary classifiers · Partially structured data

1 Introduction

Explored data sets may contain many high-dimensional feature vectors which represent objects (patients, events) [1]. We are considering a situation where a given data set can be divided into learning subsets in accordance with a particular patient's (objects) category. As an example, clinical learning subsets may contain such feature vectors which represent patients with only one disease. Separability of the learning sets means that each feature vector belongs to only one learning set [2].

Exploratory analysis is aimed at discovering interesting, useful patterns in a given data set. Data aggregation should result in decreasing the number of feature vectors. This means that some feature vectors are merged into a new vector. The decreasing of feature vectors dimensionality is also expected to be one of the results of data aggregation procedures.

We will consider the process of separable data aggregation [3]. The term *separable aggregation* means that only some feature vectors belonging to one learning set (to one category) can be merged. Merging those feature vectors which represent objects of different categories is not allowed in order to preserve data sets separability.

Pairs of feature vectors are referred to as *dipoles* in this paper. The term *clear dipole* refers to the case, where both of the feature vectors constituting a given dipole belong to the same category. Analogically, the term *mixed dipole* refers to the situation, where each of the feature vectors from a given pair belongs to a different category [3]. It has been proved that a layer composed of binary classifiers preserves the data sets

N.T. Nguyen et al. (Eds.): ICCCI 2017, Part I, LNAI 10448, pp. 73–82, 2017.
DOI: 10.1007/978-3-319-67074-4_8

separability during data transformation if and only if each mixed dipole is divided by at least one binary classifier from this layer [4].

The formal neuron is an example of a binary classifier. The output of the formal neuron is equal to one if and only if the weighted sum of input signals is greater or equal to some threshold. If this sum is less than the threshold, then the output is equal to zero.

Decision rule of the formal neuron may depend on many input signals. Designing a separable layer from formal neurons has been described in the earlier paper [4]. The procedure proposed in this paper has been based on multiple minimization of the convex and piecewise linear (*CPL*) criterion functions. The procedure od separable layer designing from formal neurons could be costly if neurons have many input signals. In the presented paper we are considering separable layer designing from such binary classifiers which have only one input signal. Such univariate binary classifiers can be also treated as a special type of formal neurons.

Univariate binary classifiers have also another important property. Such binary classifiers allow to deal with partially structured data when particular objects are represented by feature vectors of different dimensionality. Such partially structured data often occurs, for example, in deep learning tasks [6].

2 Partially Structured Learning Sets and Dipoles

Let us consider a set of m objects $O_j (j = 1, \ldots, m)$ which can be characterized by the numerical results $x_{j,i} (x_{j,i} \in R$ or $x_{ji} \in \{0, 1\})$ of possible measurements of n features $x_i (i = 1, \ldots, n)$.

We assume, that the collection $F(n) = \{x_1, \ldots, x_n\}$ of a large number n of features x_i used in the objects O_j characterization has been fixed a priori, and is constant. But we also assume, that not all features x_i of the object O_j have been measured and can not be used as the given object characterization by the n-dimensional feature vector $\mathbf{x}_j$:

$$(\forall j \in \{1, \ldots, m\})\ \mathbf{x}_j = [x_{j,1}, \ldots, x_{j,n}]^T \tag{1}$$

For each of m objects $O_j (j = 1, \ldots, m)$ the subset I_j contain the indices i of such features $x_i (i \in I_j)$ which have been not defined (not measured) for this object.

$$(\forall j \in \{1, \ldots, n\})\ (\forall i \in I_j)\ x_{j,i} \textit{ is not defined} \tag{2}$$

Often in practice, the undefined elements $x_{j,i}$ (2) are assumed to be equal to zero ($x_{j,i} = 0$). In such cases, *sparse* matrix (table) $\boldsymbol{X}$ is obtained [5]:

$$\boldsymbol{X} = [\mathbf{x}_1, \ldots, \mathbf{x}_m]^T \tag{3}$$

We assume that the data matrix $\boldsymbol{X}$ (3) has been divided into the K learning sets $C_k (k = 1, \ldots, K)$ in accordance with additional expert knowledge (*data labelling*):

$$C_k = \{\mathbf{x}_j : j \in J_k\} \tag{4}$$

where J_k are disjoined sets of the m_k indices j:

$$(\forall k' \neq k : k', k \in \{1, \ldots, K\}) J_{k'} \cap J_k = \emptyset \tag{5}$$

Definition 1: The learning sets $\mathbf{C}_k$ (4) are *partially structured* if they contain such feature vectors $\mathbf{x}_j = [x_{j,1},\ldots,x_{j,n}]^T$ (1) which have a part of such components $x_{j,i}$ which have been not defined (2).

Generally, such components $x_{j,i}$ of the feature vector $\mathbf{x}_j = [x_{j,1},\ldots,x_{j,n}]^T$ (1) which have not been defined should not be used in decision rules and can often be reduced. Reduced feature vectors $\mathbf{x}_j$ have different dimensionality $n_j (0 < n_j < n)$ The partially structured data sets are composed of such feature vectors $\mathbf{x}_j$ which have different dimensionality n_j. The partially structured data often occurs in practice, for example, in the deep learning tasks [5]. Special mathematical and computational techniques are needed for exploring these large data sets which are partially structured.

Definition 2: Two learning sets $\mathbf{C}_k$ and $\mathbf{C}_{k'}$ (4) are separable, if such elements $\mathbf{x}_j$ and $\mathbf{x}_{j'}$ which belong to different sets are not equal:

$$\textbf{\textit{if}}\,(k' \neq k),\ \textbf{\textit{then}}\,(\forall j' \in J_{k'})\ \textbf{\textit{and}}\,(\forall j \in J_k)\mathbf{x}_{j'} \neq \mathbf{x}_j \tag{6}$$

where the inequality $\mathbf{x}_{j'} \neq \mathbf{x}_j$ of the vectors $\mathbf{x}_{j'} = [x_{j',1}, \ldots, x_{j',n}]^T$ and $\mathbf{x}_j = [x_{j,1}, \ldots, x_{j,n}]^T$ means that there exists at least one feature $x_i (i \in I_{j'}$ and $i \in I_j)$ which was measured in both objects $O_{j'}$ and O_j and gave different measurement results $(x_{j'} \neq x_{j,i})$.

The assumption of the learning sets $\mathbf{C}_k$ (4) separability (6) is related to some constraints in the structure of the subsets I_j (2) of the features indices i. Only such partially structured data sets $\mathbf{C}_k$ (4) are considered in the paper in which the separability property (6) is fulfilled.

Definition 3: Two vectors $\mathbf{x}_j$ and $\mathbf{x}_{j'}$ which belong to the different learning sets $C_k(\mathbf{x}_j \in C_k)$ and $C_{k'}(\mathbf{x}_{j'} \in C_{k'})$ (4) constitute the *mixed* dipole $\{\mathbf{x}_j, \mathbf{x}_{j'}\}$.

Remark *1:* Two feature vectors $\mathbf{x}_j$ and $\mathbf{x}_{j'}$ which constitute the mixed dipole $\{\mathbf{x}_j, \mathbf{x}_{j'}\}$ have to be different $(\mathbf{x}_j \neq \mathbf{x}_{j'})$.

Definition 4: Two different vectors $\mathbf{x}_j$ and $\mathbf{x}_{j'}(\mathbf{x}_j \neq \mathbf{x}_{j'})$ which belong to the same learning set $C_k(\mathbf{x}_j \in C_k$ and $\mathbf{x}_{j'} \in C_k)$ (4) constitute the *clear* dipole $\{\mathbf{x}_j, \mathbf{x}_{j'}\}$.

3 Transformation of Learning Sets by Separable Layers of Univariate Binary Classifiers

Let us consider the layer of L univariate binary classifiers BC_l ($l = 1, \ldots, L$) with the decision rules $r_l = r(\theta_l, i(l); \mathbf{x})$ based on the feature vector $\mathbf{x} = [x_1,\ldots,x_n]^T$ ($\mathbf{x} \in R^n$):

$$(\forall l \in \{1,\ldots,L\}) \tag{7}$$

$$r_l = r(\theta_l, i(l); \mathbf{x}) = \begin{matrix} 1 & if & x_{i(l)} \geq \theta_l \\ 0 & if & x_{i(l)} < \theta_l \end{matrix}$$

where θ_l is the thresholds $q_l(q_l \in R^1)$ and *i(l)* is *index function* $i(l)(i(l) \in \{1,\ldots,n\})$ which links the *l*-th binary classifier BC_l to the *i(l)*-th component $x_{i(l)}$ of the feature vector $\mathbf{x} = [x_1,\ldots, x_n]^T$.

The complementary binary classifiers BC_l^c are also introduced by the below decision rules $r^c(\theta_l, i(l); \mathbf{x})$:

$$(\forall l \in \{1,\ldots,L\}) \tag{8}$$

$$r_l^c = r^c(\theta_l, i(l); \mathbf{x}) = \begin{matrix} 0 & if & x_{i(l)} \geq \theta_l \\ 1 & if & x_{i(l)} < \theta_l \end{matrix}$$

The complementary binary classifiers BC_l^c (8) are based on the same thresholds θ_l and the index function *i(l)* as the classifiers BC_l (7).

The layer of *L* binary classifiers BC_l (7) and BC_l^c (8) transforms each of the *m* input feature vectors $\mathbf{x}_j = [x_{j,1},\ldots,x_{j,n}]^T$ (1) into the output vector $\mathbf{r}_{j'}$ with *L* binary components $r_{j',l}$ defined by the Eq. (7) or (8):

$$(\forall j \in\{1,\ldots,m\})\ \mathbf{r}_{j'} = \left[r_{j',1},\ldots,r_{j'.L}\right]^T, \quad where\ (\forall l \in \{1,\ldots,L\}) r_{j',l} \in \{0,1\} \tag{9}$$

It is assumed here that the indices *j'* of the output vectors $\mathbf{r}_{j'}$ are some function $j' = j'(j)$ of the indices *j* of the feature vectors $\mathbf{x}_j$ (1). The layer of *L* binary classifiers results in data aggregation if more than one feature vector $\mathbf{x}_j$ (1) is transformed (1) into one output vectors $\mathbf{r}_{j'}$ [4].

The complementary binary classifiers BC_l^c (1) are introduced to avoid the situation where all components $r_{j,l}$ of the output vector $\mathbf{r}_{j'}$ (9) are equal to zero, i.e. $\forall l \in \{1,\ldots,L\}\ r_{j',l} = 0$ [4].

Definition 5: The layer of *L* binary classifiers is *separable* in respect to the learning sets C_k (4) if and only if elements $\mathbf{x}_j$ and $\mathbf{x}_{j'}$ of each *mixed* dipole $\{\mathbf{x}_j, \mathbf{x}_{j'}\}$ (*Definition* 2) are transformed (9) into different vectors $\mathbf{r}_j$ and $\mathbf{r}_{j'}(\mathbf{r}_j \neq \mathbf{r}_{j'})$.

Remark 2: The *l*-th binary classifier BC_l (7) and BC_l^c (8) of the layer divides the dipole $\{\mathbf{x}_j, \mathbf{x}_{j'}\}$ if and only if one of the two below inequalities is fulfilled:

$$\text{I.} \quad x_{j,i(l)} \geq \theta_l\ and\ x_{j',i(l)} < \theta_l \tag{10}$$

or

$$\text{II.} \qquad x_{j,i(l)} < \theta_l \; and \; x_{j',i(l)} \geq \theta_l \tag{11}$$

where the function $i(l)$ is defined in the decision rule (7) or (8), $\mathbf{x}_j = [x_{j,1},\ldots, x_{j,n}]^T$, and $\mathbf{x}_{j'} = [x_{j',1},\ldots, x_{j',n}]^T$.

In accordance with the decision rule (7) or (8), the inequalities (10) and (11) mean that only one vector $\mathbf{x}_j$ or $\mathbf{x}_{j'}$ from the mixed dipole $\{\mathbf{x}_j, \mathbf{x}_{j'}\}$ gives the output $r_l = 1$.

Remark 3: The layer of L binary classifiers BC_l (7) and BC_l^c (8) is *separable* in respect to the learning sets C_k (4) if and only if each *mixed* dipole $\{\mathbf{x}_j, \mathbf{x}_{j'}\}$ is divided in accordance with (10) or with (11) by at least one binary classifier of this layer.

4 Optimized Strategy of Separable Layer Designing from Univariate Binary Classifiers

In accordance with the *Remark* 3, the division of all the mixed dipoles $\{\mathbf{x}_j, \mathbf{x}_{j'}\}$ is a necessary and sufficient condition for the layer of L binary classifiers BC_l (7) and BC_l^c (8) separability. The optimization of the designing process is aimed at decreasing the number L of binary classifiers in the separable layer. This also leads to the decreasing of the transformed vectors $\mathbf{r}_j$ (9) dimensionality. In order to decrease the number of different vectors $\mathbf{r}_j$ (9) it is necessary to merge the greatest possible number of such feature vectors $\mathbf{x}_j$ which belong to the same learning set C_k (4).

The proposed strategy of the separable layer designing from binary classifiers BC_l (7) or BC_l^c (8) is based on evaluating particular values $\theta_{i,l}$ of the threshold θ_l (7) on the i-th axis (feature) x_i. The selected values $\theta_{i,l}$ of the threshold θ_i (7) have been located in the centers of such mixed dipoles $\{x_{j(l),i}, x_{j(l+1),i}\}$ on the i-th axis x_i which have the smallest, but greater than zero length $|x_{j(l),i} - x_{j(l+1),i}|$ (Fig. 1):

$$\theta_{i,l} = (x_{j(l),i} + x_{j(l+1),i})/2 \tag{12}$$

where $\mathbf{x}_{j(l)} = [x_{j(l),1},\ldots, x_{j(l),n}]^T$ and $\mathbf{x}_{j(k+1)} = [x_{j(l+1),1},\ldots,x_{j(l+1),n}]^T$, and $x_{j(l+1),i}\; x_{j(l),i} > 0$.

The example shown in Fig. 1 contains 8 objects O_j marked as "**o**" and 10 objects O_j marked as "**x**". Their objects O_j (j = 1, …, 18) could be represented by high dimensional feature vectors $\mathbf{x}_j$ (1) belonging to the learning sets C_1 and C_2 (4). All mixed dipoles $\{\mathbf{x}_j, \mathbf{x}_{j'}\}$ constituted by feature vectors $\mathbf{x}_j$ (1) have been divided based on the i-th feature x_i by three binary classifiers BC_l (l = 1, 2, 3) with the thresholds $\theta_{i,1}$, $\theta_{i,2}$, and $\theta_{i,3}$ adequately to the decision rule (7). The numbers m_l of mixed dipoles $\{\mathbf{x}_j, \mathbf{x}_{j'}\}$

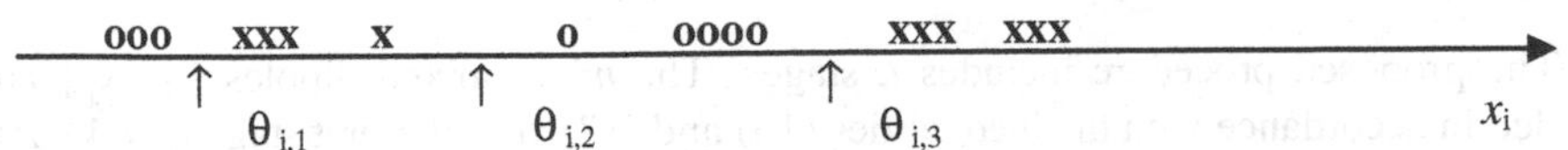

Fig. 1. Example of division of mixed dipoles $\{\mathbf{x}_j, \mathbf{x}_{j'}\}$ on the i-th axis x_i.

divided by particular classifier BC_l shown in the figure are equal to: $m_1 = 27$, $m_2 = 20$, and $m_3 = 48$.

The quality of the classifier BC_l (7) or BC_l^c (8) improves with increasing the number m_l of such mixed dipoles $\{\mathbf{x}_j, \mathbf{x}_{j'}\}$ which are divided by this classifier.

The optimal threshold $\theta_{i,l*}$ (11) located on the i-th axis x_i is characterized by the greatest number $m_{i,l*}$ of the divided mixed dipoles $\{\mathbf{x}_j, \mathbf{x}_{j'}\}$:

$$(\forall l \in \{1,\ldots,L\}) m_{i,l^*} \geq m_{i,l} \tag{13}$$

where $m_{i,l}$ is the number of the divided mixed dipoles $\{\mathbf{x}_j, \mathbf{x}_{j'}\}$ by the classifier $BC_{i,l}$ with the decision rule (7).

The optimal threshold $\theta_{i,l*}$ can be found for each feature x_i by using the inequalities (13). Based on these inequalities the optimal axis x_{i*} could also be identified. The optimal axis x_{i*} is characterized by globally the largest number $m_{i*,l*}$ of the divided mixed dipoles $\{\mathbf{x}_j, \mathbf{x}_{j'}\}$.

$$(\forall i \in \{1,\ldots,n\}) m_{i^*,l^*} \geq m_{i,l^*} \tag{14}$$

The globally optimal threshold $\theta_{i*,l*}$ (11) specified by the above inequalities allows to define the decision rule (7) of the optimal binary classifiers $BC_{i*,l*}$:

$$\textit{if } \mathrm{x}_{i^*} \geq \mathrm{q}_{i^*,l^*}, \textit{ then } r_{i^*,l^*} = 1 \textit{ else } r_{i^*,l^*} = 0 \tag{15}$$

where x_{i*} is the $i*$-th component of feature vector $\mathbf{x} = [\mathrm{x}_1,\ldots,\mathrm{x}_n]^T$.

The optimal binary classifier $BC_{i*,l*}$ (15) divides the mixed dipole $\{\mathbf{x}_j, \mathbf{x}_{j'}\}$ (*Definition* 2) if and only if the inequalities (10) or (11) are fulfilled with the $i*$-th components $\mathrm{x}_{j,\, i*}$ and $\mathrm{x}_{j',\, i*}$:

$$\text{I.} \quad \mathrm{x}_{j,i^*} \geq \theta_{i^*,l^*} \textit{ and } \mathrm{x}_{j',i^*} < \theta_{i^*,l^*} \tag{16}$$

Or

$$\text{II.} \quad \mathrm{x}_{j,i^*} < \theta_{i^*,l^*} \textit{ and } \mathrm{x}_{j',i^*} \geq \theta_{i^*,l^*} \tag{17}$$

where $\mathbf{x}_j = [\mathrm{x}_{j,1},\ldots, \mathrm{x}_{j,n}]^T$ and $\mathbf{x}_{j'} = [\mathrm{x}_{j',1},\ldots, \mathrm{x}_{j',n}]^T$.

The optimal binary classifier $BC_{i*,l*}$ (15) divides the maximal number $m_{i*,k*}$ (14) of mixed dipoles $\{\mathbf{x}_j, \mathbf{x}_{j'}\}$ on the basis of the i^*-th feature x_{i*}. Usually not all the mixed dipoles $\{\mathbf{x}_j, \mathbf{x}_{j'}\}$ are divided in accordance with the inequalities (16) or (17). The below procedure is aimed at division of all mixed dipoles $\{\mathbf{x}_j, \mathbf{x}_{j'}\}$.

Procedure of the separable layer designing

The proposed procedure includes L stages. The $m^{\mathrm{I}}_{i*,k*}$ mixed dipoles $\{\mathbf{x}_j, \mathbf{x}_{j'}\}$ are divided in accordance with the inequalities (16) and (17) after the first stage ($l = 1$). All these divided mixed dipoles $\{\mathbf{x}_j, \mathbf{x}_{j'}\}$ are removed after the first stage. The yet undivided mixed dipoles $\{\mathbf{x}_j, \mathbf{x}_{j'}\}$ are used in the second stage ($l = 2$). During the second stage the thresholds $\theta_{i,l}$ (12) can remain unchanged. Only the numbers $m^{\mathrm{II}}_{i,l}$ of the mixed dipoles

$\{\mathbf{x}_j, \mathbf{x}_{j'}\}$ which are divided during the second stage ($l = 2$) are recalculated on the basis of the inequalities (13) and (14).

The described scheme can be repeated in successive steps l ($l = 1,\ldots, L$) until all mixed dipoles $\{\mathbf{x}_j, \mathbf{x}_{j'}\}$ are divided (split). □

Theorem 1: If the learning sets C_k (4) are separable (6), then all the mixed dipoles $\{\mathbf{x}_j, \mathbf{x}_{j'}\}$ will be divided after a finite number L of the stages l by using the univariate splitting inequalities (16) and (17).

Proof: This theorem results directly from the described procedure of the separable layer designing which has been described earlier for the first stage ($l = 1$). The separable layer has been designed here from the univariate binary classifiers $BC_{i*,l*}$ (15).

In accordance with the *Definition* 1 of the separable sets, any two feature vectors $\mathbf{x}_j$ and $\mathbf{x}_{j'}$ which belong to different sets C_k and $C_{k'}$ (4) can-not be equal (6). The vectors $\mathbf{x}_j$ and $\mathbf{x}_{j'}$ are not equal (6) only if there exists at least one feature x_{i*} for which the inequality (16) or (17) is fulfilled. The inequalities (17) and (18) describe the univariate, mixed dipoles $\{x_{j,i*}, x_{j',i*}\}$ which are divided by the binary classifier $BC_{i*,l*}$ (15) based on the feature x_{i*}. At least one mixed dipole $\{\mathbf{x}_j, \mathbf{x}_{j'}\}$ is divided during each stage l of the described designing procedure.

The number of the mixed dipoles $\{\mathbf{x}_j, \mathbf{x}_{j'}\}$ created from the learning sets C_k (4) can be very large, but is finite. Therefore, after the finite number L of the stages l all the mixed dipoles $\{\mathbf{x}_j, \mathbf{x}_{j'}\}$ will be divided. □

Theorem 2: The layer of the L optimal binary classifier $BC_{i*,l*}$ (15) transforms K separable learning sets C_k (4) into the same number K of the separable sets $R_k = \{\mathbf{r}_j : j \in J_k\}$ composed of the transformed vectors $\mathbf{r}_j$ (9).

The proof of the similar Theorem has been provided in the work [4].

The number m_k of the transformed vectors $\mathbf{r}_j$ (9) in each set $R_k = \{\mathbf{r}_j : j \in J_k\}$ is equal to the number of the feature vectors $\mathbf{x}_j$ in the learning sets C_k (4), but not all transformed vectors $\mathbf{r}_j$ (9) in particular set R_k are different. Some different feature vectors $\mathbf{x}_j$ from the same learning set C_k (4) can be merged in the same vectors $\mathbf{r}_j$ (9). As a result, the numbers $m_{k'}$ of different vectors $\mathbf{r}_j$ (9) in some sets $R_k = \{\mathbf{r}_j : j \in J_k\}$ is expected to be smaller than the numbers m_k of feature vectors $\mathbf{x}_j$ in the learning sets C_k (4). The learning sets C_k (4) are expected to be reduced as the result of the feature vectors $\mathbf{x}_j$ transformation (9) by the separable layer of univariate binary classifiers [4].

5 Data Integration by Separable Layer of Univariate Binary Classifiers

The separable layer of L optimal binary classifier BC_l (15) transforms each of the m input feature vectors $\mathbf{x}_j = [x_{j,1},\ldots, x_{j,n}]^T$ (1) into the output vectors $\mathbf{r}_j = [r_{j,1},\ldots, r_{j,L}]^T$ (9) with the L binary components $r_{j,l}$. The parameter L is equal to the number of the stages l during the procedure of separable layer designing. The binary components $r_{j,i}$ of the output vectors $\mathbf{r}_j$ can be defined on the basis of the decision rule (7) of the optimal classifiers BC_l into the below manner:

$$(\forall j \in \{1,\ldots,m\}), (\forall i \in \{1,\ldots,n\}) (\forall l \in \{1,\ldots,L\}) \tag{19}$$

$$r_l(\theta_{i,l};\ x_{j,i}) = \begin{matrix} 1 & \text{if} & x_{j,i} \geq \theta_{i,l} \\ 0 & \text{if} & x_{j,i} < \theta_{i,l} \end{matrix}$$

Where $r_j(\theta_{i,l};\ x_{j,i})$ is the binary output of the classifier BC_l which depends on the i-th component $x_{j,i}$ of the feature vector $\mathbf{x}_j$ and on the threshold $\theta_{i,l}$.

In accordance with the optimal designing rules (15) and (16), there may exist up to L values $\theta_{i,l}$ of the threshold for the each axis (feature) x_i. We can assume without constraints that the threshold values $\theta_{i,l}$ were arranged in the increasing order:

$$(\forall i \in \{1,\ldots,n\}) (\forall l \in \{1,\ldots,L-1\}\theta_{i,l} < \theta_{i,l+1} \tag{20}$$

We can remark on the basis of the above assumptions that the separable layer of L optimal binary classifiers BC_l (19) integrates some feature vectors $\mathbf{x}_j$ from one of the learning sets $\boldsymbol{C}_k$ (4) into the below homogenous subsets $D_{j'}$ labelled by different output vectors $\mathbf{r}_{j'}$ of this layer ($\mathbf{r}_{j'} \neq \mathbf{r}_{j''}$ for $j' \neq j''$):

$$D_{j'} = \{\mathbf{x}_j : \mathbf{r}(\mathbf{w}, \boldsymbol{\theta}; \mathbf{x}_j) = \mathbf{r}_{j'}\} \tag{21}$$

where $\mathbf{r}(\boldsymbol{\theta};\ \mathbf{x}_j) = [r_1(\theta_{i(1),1};\ x_{j,i(1)}),\ r_2(\theta_{i(2),2};\ x_{j,i(2)}),\ldots,\ r_L(\theta_{i(L),L};\ x_{j,i(L)})]^T$ is the decision rule of the separable layer which is based on the decision rules $r_l(\theta_{i,l};\ x_{j,i(l)})$ (19) of particular classifiers BC_l.

The decision rule $\mathbf{r}(\boldsymbol{\theta};\ \mathbf{x}_j)$ of the layer depends on the vector $\boldsymbol{\theta} = [\theta_{i(1),1},\ldots,\ \theta_{i(L),L}]^T$ containing the thresholds $\theta_{i,l}$ of particular binary classifiers BC_l (7) or BC_l^c (8).

Let us assume that the thresholds $\theta_{i,l}$ have been ordered in non-decreasing manner ($\theta_{i,l} \leq \theta_{i,l+1}$). It can be seen on the basis of the above assumption that each subset $D_{j'}$ (21) contains such $m_{j'}$ feature vectors $\mathbf{x}_j = [x_{j,1},\ldots,\ x_{j,n}]^T$ (1) from the learning set C_k (4) which are defined by the below inequalities:

$$D_{j'} = \{\mathbf{x}_j \in C_k : (\forall i \in \{1,\ldots,n\}) (\forall l \in \{1,\ldots,L-1\}) \\ \theta_{i,l} \leq x_{j,i} < \theta_{i,l+1}\} \tag{22}$$

The inequalities mean that the subset $D_{j'}$ (21) has the shape of a cuboit in the n-dimensional feature space $R^n (\mathbf{x} \in R^n)$.

The homogenous data subset $D_{j'}$ (22) constitutes some pattern in the feature space if it contains a distinctively large number $m_{j'}$ of the feature vectors $\mathbf{x}_j$ from one learning set C_k (4). The proposed optimization strategy based on the Eqs. (14) and (15) is aimed at increasing the numbers $m_{j'}$ of the feature vectors $\mathbf{x}_j$ in selected subsets $D_{j'}$ (22). The result of such optimization strategy depends on the structure of a given data set $\boldsymbol{X}$ (3). It can be expected that in many cases the proposed technique of the data table $\boldsymbol{X}$ (3) integration could be very effective. In such cases the separable dipolar transformation of elements $\mathbf{x}_j$ of the learning data sets C_k (5) results in a low number of the different transformed vectors $\mathbf{r}_{j'}$ (9) and in numerous data subset $D_{j'}$ (22).

6 Concluding Remarks

The described method of the separable layer designing from univariate, binary classifiers BC_l (16) allows to reduce m feature vectors $\mathbf{x}_j$ (1) into $m'(m' < m)$ different, output vectors $\mathbf{r}_{j'}$ (21). Each transformed vectors $\mathbf{r}_{j'}$ (21) is composed of L binary components $r_{j', l}$ (9). Large, homogeneous subsets $D_{j'}$ (21) represented by the vectors $\mathbf{r}_{j'}$ (21) result in a high degree of data aggregation performed by the separable layer.

The separable layer of L binary classifiers reduces m vectors $\mathbf{x}_j$ (1) into $m'(m' < m)$ different vectors $\mathbf{r}_{j'}$ (21). The output vectors $\mathbf{r}_{j'}$ (9) from the layer can be further aggregated by next separable layer of binary classifiers. The hierarchical network of separable layers allows to transform all feature vectors $\mathbf{x}_j$ (1) from one learning set C_k (4) into one output vector $\mathbf{r}_k$ (21) of such network [4].

Multilayer hierarchical networks can be designed from univariate, binary classifiers on the basis of the dipolar separability technique described in the paper. This approach to hierarchical networks designing could be a complementary alternative to deep learning methods [6].

It is worth emphasizing another important application of separable layers. The transformation of the learning sets C_k (4) by the separable layer of L binary classifier BC_l (19) allows to replace partially structured feature vectors $\mathbf{x}_j$ of different dimensionality n_j by the well structured transformed vectors $\mathbf{r}_j$ (9) of the equal dimensionality L. Such components $x_{j,i}$ of feature vectors $\mathbf{x}_j$ which are not defined (2) can be treated as missing data [2]. In this context, the transformation of feature vectors $\mathbf{x}_j$ by the separable layer may be understood as a data supplements tool.

Extraction of numerous data subsets $D_{j'}$ (22) can also be used for the purpose of biclustering [7]. The biclusters $D_{j'}$ (22) obtained as a result of the separable data aggregation through dipolar univarite classifiers BC_l (19) can play a complementary role to collinear biclusters [8].

Acknowledgments. The presented study was supported by the grant S/WI/2/2013 from Bialystok University of Technology and funded from the resources for research by Polish Ministry of Science and Higher Education.

References

1. Hand, D., Smyth, P., Mannila, H.: Principles of Data Mining. MIT Press, Cambridge (2001)
2. Duda, O.R., Hart, P.E., Stork, D.G.: Pattern Classification. Wiley, New York (2001)
3. Bobrowski, L.: Data mining based on convex and piecewise linear criterion functions. Technical University Białystok (2005) (in Polish)
4. Bobrowski, L.: Piecewise-linear classifiers, formal neurons and separability of the learning sets. In: Proceedings of 13th International Conference on Pattern Recognition, ICPR 1996, 25–29 August 1996, Vienna, Austria, pp. 224–228 (1996)
5. Golub, G.H., Van Loan, C.F.: Matrix Computations, 4th edn. Johns Hopkins University Press, Baltimore (2013)
6. Arel, I., Rose, D.C., Karnowski, T.P.: Deep Machine learning–A new frontier in artificial intelligence research–A survey. IEEE Comput. Intell. Mag. **5**, 13–18 (2013)

7. Madeira, S.C., Oliveira, S.L.: Biclustering algorithms for biological data analysis: a survey. IEEE Trans. Comput. Biol. Bioinform. **1**(1), 24–45 (2004)
8. Bobrowski, L.: Biclustering based on collinear patterns. In: Rojas, I., Ortuño, F. (eds.) IWBBIO 2017. LNCS, vol. 10208, pp. 134–144. Springer, Cham (2017). doi:10.1007/978-3-319-56148-6_11

Intelligent Collective: The Role of Diversity and Collective Cardinality

Van Du Nguyen[1(✉)], Mercedes G. Merayo[2], and Ngoc Thanh Nguyen[1]

[1] Department of Information Systems, Faculty of Computer Science and Management, Wroclaw University of Science and Technology, Wrocław, Poland
{van.du.nguyen,Ngoc-Thanh.Nguyen}@pwr.edu.pl

[2] Department of Sistemas Informáticos y Computación, Universidad Complutense de Madrid, Madrid, Spain
mgmerayo@ucm.es

Abstract. Nowadays, there appears to be ample evidence that collectives can be intelligent if they satisfy *diversity, independence, decentralization,* and *aggregation.* Although many measures have been proposed to evaluate the quality of collective prediction, it seems that they may not adequately reflect the intelligence degree of a collective. It is due to the fact that they take into account either the accuracy of collective prediction; or the comparison between the capability of a collective to those of its members in solving a given problem. In this paper, we first introduce a new function that measures the intelligence degree of a collective. Following, we carry out simulation experiments to determine the impact of diversity on the intelligence degree of a collective by taking into account its cardinality. Our findings reveal that diversity plays a major role in leading a collective to be intelligent. Moreover, the simulation results also indicate a case in which the increase in the cardinality of a collective does not cause any significant increase in its intelligence degree.

Keywords: Wisdom of Crowds · Intelligent collective · Integration computing

1 Introduction

In recent years, research on Wisdom of Crowds and its applications have shown that a collective is superior to its individuals in solving a wide variety of difficult problems [1–4]. Intuitively speaking, this capability is due to the fact that a collective may have new information, new perspectives, etc. to a given problem that single individual does not possess. In [2] Surowiecki has put forward a hypothesis: *"A collective is more intelligent than single individuals"*. However, it can be argued that not all collectives of individuals can be considered intelligent. A collective can be intelligent if it satisfies: *diversity*, *independence*, *decentralization*, and *aggregation*. By diversity, we have in mind a variety of individual backgrounds or individual opinions on the problem that needs to be solved. Independence means that an individual opinion must be made independently of others in the collective [2, 3, 5]. Moreover, the existence of decentralization guarantees that members freely in making their opinions and thus ensures the existence of

N.T. Nguyen et al. (Eds.): ICCCI 2017, Part I, LNAI 10448, pp. 83–92, 2017.
DOI: 10.1007/978-3-319-67074-4_9

diversity in a collective. The last criterion, aggregation, is assumed as an appropriate method in which the individual opinions are combined.

In this paper, the term collective is considered as a set of predictions given by a number of members (such as humans or agent systems) on a given judgment or prediction problem. These predictions can be different from each other because collective members may have different backgrounds or different knowledge bases. On the basis of individual predictions, we need to determine a consistent one, called *Collective Prediction* that can be considered as representative of the collective as a whole. In general, it can be argued that the following conventional measures can be used to determine the intelligence degree of a collective: the difference between the collective prediction and the proper value [2, 6–9] (*Diff*); the number of times in which the collective prediction outperforms the individual ones [10, 11] (*WR*); the quotient of the individual errors and the collective error [11] (*CIQ*).

Let us consider the following situation: a number of people are asked for giving predictions on the outcome of a future event. Suppose that we have two collectives with the same cardinality. The predictions in the first collective are closer to the proper value than those in the second one. However, the collective predictions of these collectives reflect the proper value to the same degree. Intuitively speaking, the first collective is more intelligent that the second one. According to *Diff* measure, however, these collectives have the same intelligence degree. Conversely, according to *CIQ* measure, the second collective is more intelligent. Even though, in the case of using *WR*, two collectives will have the same intelligence degree if their collective predictions outperform all individual predictions regardless the difference between collective prediction and the proper value. From these limitations, in this paper, we will define a new function for measuring the intelligence degree of a collective that takes into account both collective prediction as well as individual predictions.

Additionally, diversity can be considered as the most important criterion of intelligent collectives. On the one hand, a collective involving diverse members may add new information, new perspectives to the problem that needs to be solved. On the other hand, the diversity of individual predictions has been proven to come in useful to avoid the phenomenon of the so-called correlated error [12–14]. Moreover, a factor that easily controls to enhance the diversity of a collective is to enlarge its cardinality [15]. Taken together, in this paper, we investigate the impact of diversity on the intelligence degree of a collective by taking into account its cardinality. To the best of our knowledge, these research problems have not been widely investigated in the literature.

The remainder of the paper has the following structure: Sect. 2 introduces some related works. The measure of the intelligence degree of a collective is presented in Sect. 3. Section 4 introduces the research model of the paper. Section 5 reports the simulation experiments and their evaluation. The final section includes some conclusions and future work.

2 Related Works

According to the statistical analysis from the game show *"Who Wants to be a Millionaire?"* and many other experiments [2], Surowiecki has put forward a hypothesis: *"A collective is more intelligent than single individuals"*. However, a raising question is whether all collectives are intelligent. A collective can be considered intelligent if it satisfies *diversity, independence, decentralization,* and *aggregation*.

In general, based on the difference between the collective prediction and the proper value [2, 6–9], we can state that a collective is more intelligent than another one if its collective prediction is closer to the proper value. This measure is widely used in the related works. For instance, in the experiments conducted by Galton in 1907 with 800 individuals on the problem of estimating the weight of an ox [6]. The collective guess was only 0.8% off from the actual weight of the ox. Similarly, in the problem of guessing the number of beans in a jar [2], the collective guess is only 2.5% off from the actual number of beans in the jar. Later, in [10, 11] the authors have introduced two additional measures. The first one, called *win ratio*, is based on the number of times in which the collective prediction outperforms the individual ones. Meanwhile, the second one is based on the quotient between the individual errors and the collective error. In general, these measures are mainly based on the capability of a collective in comparison with its members in solving a given problem.

However, these measures also have some limitations. In case two collective predictions reflect the proper value to the same degree, both of them will have the same intelligent degree (using *Diff* measure). Moreover, a collective involving widely dispersed predictions will be more intelligent than that involving consistent predictions (using *QIC* measure). Nevertheless, two collectives have the same intelligence degree if their collective predictions outperform all individual predictions without considering the difference between collective prediction and the proper value (using *WR* measure). If we consider the problem of predicting the future temperature of a region the experiments conducted by Wagner in [11], the collective prediction is only 0.89 (Celsius) off from the actual temperature (29.00). However, it still does not satisfy the predefined intelligent criterion (*CIQ > 10*). Conversely, in the experiments on predicting the number of jellybeans, the collective prediction is 44.37 while the actual quantity is 46. That is, the collective satisfies the intelligence criterion.

It can be said that diversity is one of the most important criteria for determining the intelligent of a collective. There are two main kinds of diversity: diversity in the composition of collective members and diversity of individual predictions on the given problem. The former kind of diversity is understood as the variety of individual backgrounds, knowledge bases, and so forth. Meanwhile, the latter kind corresponds to the difference among individual opinions on the same problem. To date, these kinds of diversity have been proven useful in leading to a collective to be intelligent [5, 16–21]. Moreover, a common factor that can easily control to enhance the diversity level of a collective is to enlarge its cardinality. In [22], the experiments conducted by Wagner and colleagues have revealed that a large collective has a positive impact on the collective prediction. Similarly, in [23] collectives of 18 members are better than those of 8 members in predicting the demand for 38 summer trips. In previous works [24, 25], based on Euclidean space, we have also confirmed that the large collective has a positive impact on the accuracy of collective prediction.

3 Intelligence of a Collective

Let U be a set of values representing the potential predictions of a given problem in the real world. By symbol $\prod(U)$ we denote the set of all non-empty finite subsets with repetitions of U. The collective of predictions provided by collective members on a specific problem, denoted by $X \in \prod(U)$, has the following form:

$$X = \{x_1, x_2, \ldots, x_n\} \quad (1)$$

where n represents the number of predictions in a collective.

Based on consensus choice, there are two most popular criteria used to determine the collective prediction of a collective. They are 1-Optimality and 2-Optimality (O_1 and O_2 respectively). In this paper, we used criterion O_2:

$$d^2(x^*, X) = min_{y \in U}\, d^2(y, X)$$

where x^* presents the collective prediction of collective X and $d^2(y, X) = \sum_{i=1}^{n} d^2(y, x_i)$.

Definition 1. *The intelligence degree of a collective is defined by a function Int as follows:*

$$Int: \prod(U) \to [0,\ 1] \quad (2)$$

where $\prod(U)$ *is set of non-empty subsets with repetition of universe U.*

In order to define a function that serves for measuring the intelligence degree of a collective, we consider the following assumptions.

Definition 2. *The function Int should satisfy the following conditions:*

1. $\forall X \in \prod(U)$:
   (a) $\forall x_i \in X$: If $(r = x_i) \wedge (r = x^*)$, then $Int(X) = 1$.
   (b) $\forall x_i \in X$: If $(d(r, x_i) = 1) \wedge (d(r, x^*) = 1)$, then $Int(X) = 0$.
2. $\forall X, Y \in \prod(U)$:

$$\text{If } (d(r, x^*) = d(r, y^*)) \wedge \left(\frac{d(r, X)}{n} \leq \frac{d(r, Y)}{m}\right), \text{ then } Int(X) \geq Int(Y).$$

where x^*, y^* *represent the collective predictions of collective X and Y respectively, r represents the proper value.*

The first condition deals with some special cases of the intelligence measure. Intuitively, it can be said that a collective achieves the maximal intelligence degree if its predictions are on the target. However, if individual predictions present the maximal difference from the proper value, then the intelligence degree of the collective will be minimal. The next condition presents the intelligence degree of a collective in comparison with another collective. If the individual predictions in collective X are closer to the proper value than those in collective Y and their collective predictions reflect the proper value to the same degree, then the intelligence degree of collective X is higher than that of collective Y.

Definition 3. *Function Int satisfying the conditions in Definition 2 has the following form:*

$$Int(X) = 1 - \left((1 - \alpha) \times \frac{d(r, X)}{n} + \alpha \times d(r, x^*) \right) \tag{3}$$

In [26], it was shown that: $d(r, x^*) \leq \frac{d(r, X)}{n}$. Therefore, the values of *Int* are bounded between 0 and 1. We consider the following numerical example of the intelligence degree.

Example 1. Given collectives $X = \{x_1, x_2\}$, $Y = \{y_1, y_2\}$ and $Z = \{z_1, z_2\}$ such that:

- $d(r, x_1) = d(r, x_2) = 1$ and $d(r, x^*) = 0$
- $d(r, y_1) = d(r, y_2) = 1$ and $d(r, y^*) = 1$
- $d(r, z_1) = d(r, z_2) = 0.5$ and $d(r, z^*) = 0$

The values of *Int* for these collectives with some extreme values of α are shown in Table 1.

Table 1. A numerical example of intelligence measure

	[α = 0.0]	[α = 0.5]	[α = 1.0]
Int(X)	0.0	0.5	1.0
Int(Y)	0.0	0.0	0.0
Int(Z)	0.5	0.75	1.0

According to Table 1, if function *Int* only takes into account the differences between the proper prediction and the individual predictions ([α = 0.0]), then collective X and Y have the same intelligence degree. It seems irrational because the collective prediction of collective X is better than that of collective Y. Similarly, in the case of [α = 1.0], both collective X and Z have the same intelligence degree. However, the individual predictions in collective Z are closer to the proper value than those in collective X. In the case of [α = 0.5], the intelligence degree of collective Z is highest in comparison with others.

4 Research Model

As aforementioned, the main concern of the paper is to investigate the impact of diversity and cardinality on the intelligence degree of a collective (see Fig. 1). The collective cardinality presents the size of a collective, whereas diversity is considered as the variety of individual predictions in a collective.

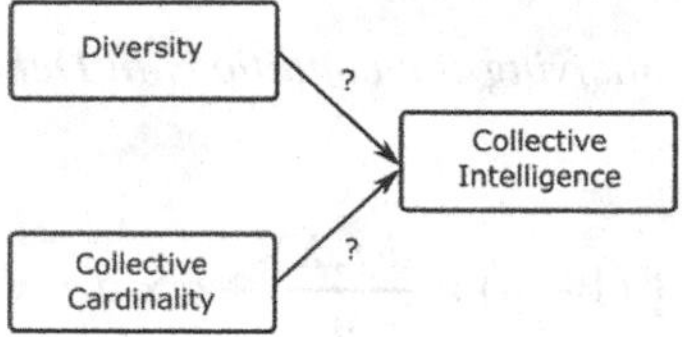

Fig. 1. Research model

In this work, we used the function defined in [8] to measure the diversity of a collective. This function takes into account the differences between individual predictions.

$$c(X) = \begin{cases} \frac{1}{n(1-n)} \sum_{i=1}^{n} d(x_i, X), & for\ n > 1 \\ 0, & otherwise \end{cases} \tag{4}$$

where $d(x_i, X) = \sum_{j=1}^{n} d(x_i, x_j)$ and $d(x_i, x_j)$ represents the difference between x_i and x_j. It can be said that a collective whose predictions are close together, will have the small diversity value.

5 Experimental Results and Their Evaluation

5.1 Simulation Design

In this section, we will present simulation experiments to determine how diversity and collective cardinality affect the intelligence degree of a collective. Suppose that the difference between the proper value and potential predictions in a collective does not exceed a predefined threshold ∂, that is, for all $x \in U$ we have that $d(r, x) \leq \partial$. In this paper, we use $\partial = 500$, and the proper value is 1000. For such assumption, U is a set of integers whose values range from 500 to 1500. We simulated collectives with different levels of diversity. The magnitude of each diversity level used here is 5 times smaller than the value of ∂. Specifically, the diversity levels will be D100, D200, D300, D400, and D500 corresponding to the values of diversity that belong to the intervals [0, 100), [100, 200), [200, 300), [300, 400), and [400, 500), respectively. We have no prior knowledge of choosing an appropriate value of cardinality. Thus, the collective cardinalities used are 9, 109, and 209.

5.2 The Impact of Diversity on the Intelligence of a Collective

The accuracy of the collective prediction cannot be less important than the average of the differences between individual predictions and collective prediction in the measure of the intelligence of a collective. Taking it into account in Eq. (2) we use $\alpha = 0.5$, $\alpha = 0.7$, and $\alpha = 0.9$. For each setting, we simulated 100 collectives with the same cardinality but with different diversity levels. Moreover, in Eq. (2), all differences between individual predictions and the proper value are normalized to [0, 1], therefore, in the next simulation we use the following formula for such normalization.

$$\forall x_i \in U : \delta(r, x_i) = \frac{d(r, x_i)}{\partial} \tag{5}$$

The simulation results are presented in Figs. 2, 3 and 4.

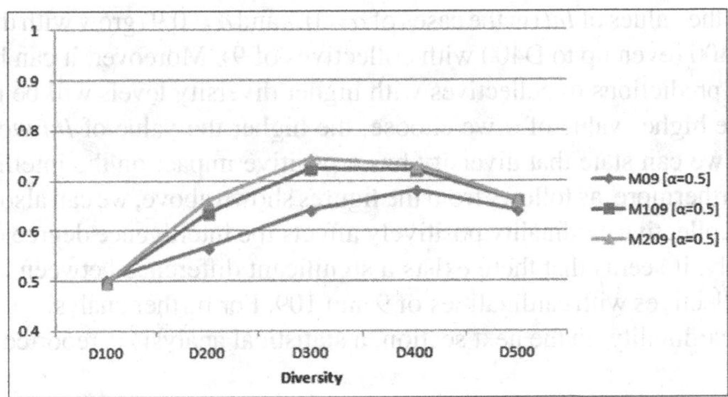

Fig. 2. Impact of diversity on the intelligence degree of a collective [α = 0.5]

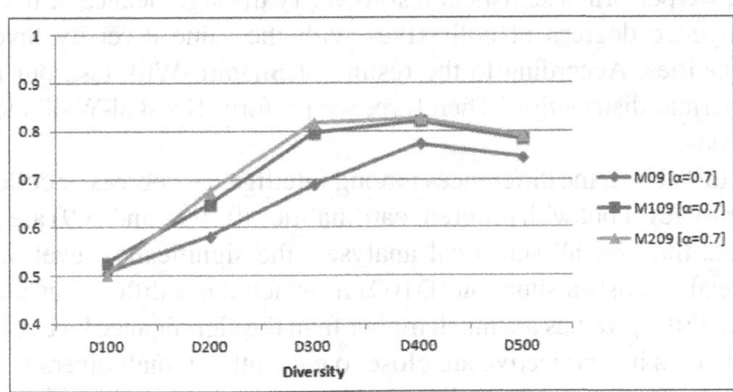

Fig. 3. Impact of diversity on the intelligence degree of a collective [α = 0.7]

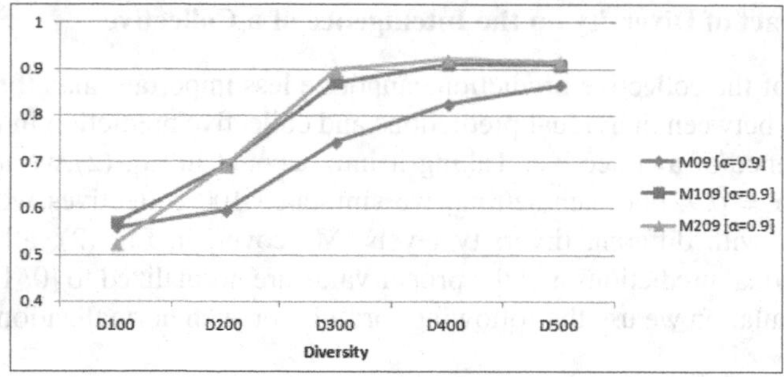

Fig. 4. Impact of diversity on the intelligence degree of a collective [α = 0.9]

Intuitively, the values of *Int* (in the case of α = 0.5) grow with the diversity level up to D300 (even up to D400 with collectives of 9).

Similarly, the values of *Int* (in the cases of $\alpha = 0.7$ and $\alpha = 0.9$) grow with the diversity level up to D400 (even up to D400 with collectives of 9). Moreover, it can be said that the collective predictions of collectives with higher diversity levels will be more accurate. Thus, the higher value of α we choose, the higher the value of *Int* we get. From these results, we can state that diversity has a positive impact on the intelligence of a collective. Furthermore, as follows from the figures shown above, we can also intuitively state that the collective cardinality positively affects the intelligence degree of a collective. Especially, it seems that there exists a significant difference between intelligence degrees of collectives with cardinalities of 9 and 109. For further analysis of the impact of collective cardinality, in the next section, a statistical analysis is reported

5.3 Statistical Analysis

In this section, we perform a statistical test to verify the significance of the differences between intelligence degrees of collectives with the same diversity level but with different cardinalities. According to the results of Shapiro-Wilk test, our data do not come from a normal distribution. Therefore, we perform Kruskal-Wallis test for such statistical analysis.

According to Table 2, the differences among intelligence degrees of collectives with the same diversity level but with different cardinalities (9, 109, and 209) are statistically significant. Note that, for all statistical analyses, the significance level 0.05 is used. However, there also exists a situation (D100) in which these differences are not statistically significant (the p values are much higher than the significance level). It is because when the predictions in a collective are close to each other (small diversity level), then the increase in the cardinality of a collective does not cause any increase in its intelligence degree.

Table 2. p values of statistical analysis for collectives with cardinalities 9, 109, 209

	D100	D200	D300	D400	D500
$\alpha = 0.5$	**0.993**	1.4134E-06	1.09699E-14	1.30906E-10	3.37784E-07
$\alpha = 0.7$	**0.744**	2.10282E-05	2.80036E-19	3.68219E-12	2.29235E-14
$\alpha = 0.9$	**0.479**	4.61576E-08	1.58988E-21	7.22501E-23	4.92926E-12

6 Conclusions and Future Works

This paper has introduced a new function that aims at measuring the so-called intelligence degree of a collective. This function is defined based on the differences between the collective prediction and the proper value as well as between individual predictions and the proper value. Subsequently, we have studied the impact of diversity on the intelligence degree of a collective taking into account its cardinality. The simulation results indicated that these factors play a major role in leading a collective to be intelligent. However, the simulation results also indicate the case in which the increase of the cardinality of a collective does not cause any increase in its intelligence degree. In the future work, we will extensively investigate the research problem to determine a proper value of α in the function measuring the intelligence degree of a collective. Also, the criteria proposed by Surowiecki [2] will be taken into account to build a model for intelligent collectives. Moreover, the problem of processing large collectives [27], and transferring knowledge among collective members [28] should also be considered.

Acknowledgement. This article is based upon work from COST Action KEYSTONE IC1302, supported by COST (European Cooperation in Science and Technology) and partially supported by the projects DArDOS (TIN2015-65845-C3-1-R (MINECO/FEDER)) and SICOMORo-CM (S2013/ICE-3006).

References

1. Clemen, R.T.: Combining forecasts: a review and annotated bibliography. Int. J. Forecast. **5**, 559–583 (1989)
2. Surowiecki, J.: The Wisdom of Crowds. Doubleday/Anchor, New York (2005)
3. Armstrong, J.S.: How to make better forecasts and decisions: avoid face-to-face meetings. Foresight Int. J. Appl. Forecast. **5**, 3–8 (2006)
4. Nielsen, M.: Reinventing Discovery: The New Era of Networked Science. Princeton University Press, Princeton (2011)
5. Page, S.E.: The Difference: How the Power of Diversity Creates Better Groups, Firms, Schools, and Societies. Princeton University Press, Princeton (2007)
6. Galton, F.: Vox populi (The wisdom of crowds). Nature **75**, 450–451 (1907)
7. Graefe, A., Armstrong, J.S.: Comparing face-to-face meetings, nominal groups, Delphi and prediction markets on an estimation task. Int. J. Forecast. **27**, 183–195 (2011)
8. Nguyen, N.T.: Advanced Methods for Inconsistent Knowledge Management. Springer, London (2008)
9. Nguyen, N.T., Sobecki, J.: Using consensus methods to construct adaptive interfaces in multimodal web-based systems. J. Univers. Access Inf. Soc. **2**(4), 342–358 (2003)

10. Kawamura, H., Ohuchi, A.: Evolutionary emergence of collective intelligence with artificial pheromone communication. In: 26th Annual Conference of the IEEE Industrial Electronics Society, vol. 4, pp. 2831–2836 (2000)
11. Wagner, C., Vinaimont, T.: Evaluating the wisdom of crowds. J. Comput. Inf. Syst. **11**, 724–732 (2010)
12. Lorge, I., Fox, D., Davitz, J., Brenner, M.: A survey of studies contrasting the quality of group performance and individual performance. Psychol. Bull. **55**, 337 (1958)
13. Gigone, D., Hastie, R.: Proper analysis of the accuracy of group judgments. Psychol. Bull. **121**, 149–167 (1997)
14. Larrick, R.P., Soll, J.B.: Intuitions about combining opinions: misappreciation of the averaging principle. Manag. Sci. **52**, 111–127 (2006)
15. Nguyen, V.D., Merayo, M.G.: Intelligent collective: some issues with collective cardinality. J. Inf. Telecommun. (2017). doi:10.1080/24751839.2017.1323702
16. Hong, L., Page, S.E.: Groups of diverse problem solvers can outperform groups of high-ability problem solvers. Proc. Natl. Acad. Sci. USA **101**, 16385–16389 (2004)
17. Armstrong, J.S.: Combining forecasts. In: Armstrong, J.S. (ed.) Principles of Forecasting, pp. 417–439. Springer, New York (2001)
18. Campbell, K., Mínguez-Vera, A.: Gender diversity in the boardroom and firm financial performance. J. Bus. Ethics **83**, 435–451 (2008)
19. Robert, L., Romero, D.M.: Crowd size, diversity and performance. In: Proceedings of the 33rd Annual ACM Conference on Human Factors in Computing Systems, pp. 1379–1382. ACM, Seoul, Republic of Korea (2015)
20. Kelley, T.L.: The applicability of the Spearman–Brown formula for the measurement of reliability. J. Educ. Psychol. **16**, 300–303 (1925)
21. Simons, A.M.: Many wrongs: the advantage of group navigation. Trends Ecol. Evol. **19**, 453–455 (2004)
22. Wagner, C., Suh, A.: The wisdom of crowds: impact of collective size and expertise transfer on collective performance. In: 2014 47th Hawaii International Conference on System Sciences, pp. 594–603 (2014)
23. Cui, R., Gallino, S., Moreno, A., Zhang, D.J.: The operational value of social media information. Available at SSRN (2015)
24. Nguyen, V.D., Nguyen, N.T.: A method for improving the quality of collective knowledge. In: Proceedings of ACIIDS, pp. 75–84 (2015)
25. Nguyen, V.D., Nguyen, N.T.: An influence analysis of the inconsistency degree on the quality of collective knowledge for objective case. In: Proceedings of ACIIDS, pp. 23–32 (2016)
26. Nguyen, N.T., Nguyen, V.D., Hwang, D.: An influence analysis of the number of members on the quality of knowledge in a collective. J. Intell. Fuzzy Syst. **32**, 1217–1228 (2017)
27. Vossen, G.: Big data as the new enabler in business and other intelligence. Vietnam J. Comput. Sci. **1**, 1–12 (2013)
28. Cao, S.T., Nguyen, L.A.: Query–subquery nets for Horn knowledge bases in first-order logic. J. Inf. Telecommun. **1**, 79–99 (2017)

RuQAR: Querying OWL 2 RL Ontologies with Rule Engines and Relational Databases

Jarosław Bąk(✉) and Michał Blinkiewicz(✉)

Institute of Control and Information Engineering, Poznan University of Technology, Piotrowo 3a, 60-965 Poznan, Poland
{jaroslaw.bak,michal.blinkiewicz}@put.poznan.pl

Abstract. We present RuQAR, a tool that supports the ABox reasoning as well as query answering with OWL 2 RL ontologies. RuQAR provides a non-naive method of transforming such ontologies into rules which can be executed by a forward chaining rule engine. Thus, query answering can be performed using functions available in a rule engine. Moreover, RuQAR supports a relational database access which extends reasoning scalability. We evaluate our tool using the LUBM benchmark ontology and data stored in relational databases. We describe our approach, RuQAR's implementation details as well as future research and development.

Keywords: Query answering · OWL 2 RL · Rule engine · Database access

1 Introduction and Motivation

The second version of the Web Ontology Language (OWL 2) offers three profiles providing significant advantages in different application scenarios. These profiles are: OWL 2 EL, OWL 2 RL and OWL 2 QL. All of them are defined as syntactic restrictions of OWL 2 [11] with different computational complexity. OWL 2 RL, which we are focused on, provides the implementation of polynomial time reasoning algorithms in a standard rule engine. Moreover, this profile has been designed to perform reasoning tasks in a forward chaining rule system by implementing a set of predefined rules. However, a naive implementation of OWL 2 RL reasoner is known to perform poorly with large ABoxes [7]. Furthermore, the official list[1] of OWL 2 reasoners supporting OWL 2 RL is limited. Moreover, there is a lack of tools that can generate rules for different rule engines.

Usually, a rule-based system processes data only in its working memory which is limited by available RAM space. According to a forward chaining mechanism (bottom-up evaluation), commonly used in reasoning tasks, a user gets conclusions as a set of inferred facts. In this set it is hard to find a fact or facts which the user is interested in. Thus, there is a need for executing a query in order

[1] http://www.w3.org/2001/sw/wiki/OWL/Implementations.

N.T. Nguyen et al. (Eds.): ICCCI 2017, Part I, LNAI 10448, pp. 93–102, 2017.
DOI: 10.1007/978-3-319-67074-4_10

to obtain the necessary results. This is a better way than looking through the working memory manually. Moreover, the forward chaining approach performs reasoning with all facts in the working memory. Therefore, some of the inferred facts are useless and many rules are fired unnecessarily. As a result the efficiency of the query answering process is decreased. One way of increasing the efficiency and scalability is to store data outside the working memory and load facts only when needed. Thus, the scalability and efficiency of reasoning as well as query answering will be increased.

According to the aforementioned issues we are motivated to provide an easy-to-use framework for performing the ABox reasoning with OWL 2 RL ontologies in any forward chaining rule engine. Moreover, we want to support efficient query answering with relational data that is semantically described by the use of mappings between an ontology and a database schema. In this paper we are focused on query answering with OWL 2 RL ontologies executed by forward chaining rule engines. However, presented approach can be applied to ontologies that are more expressible than OWL 2 RL. It is possible because we use HermiT[2] in order to execute the TBox reasoning (with the terminological part of an ontology) first. Then, we start the ABox reasoning (with the assertional part of the ontology). Thus, we can employ a rule-based engine in order to execute reasoning and query answering.

The main goal of this paper is to present database connectivity of the RuQAR (Rule-based Query Answering and Reasoning) framework in which query answering and reasoning can be performed using Drools[3] and Jess[4]. Moreover, we present RuQAR's evaluation using the LUBM ontology benchmark [6].

The remaining part is organized as follows. Section 2 discusses the related work. Section 3 presents the overview of the translation of an OWL 2 RL ontology into rules. Relational database access is presented in Sect. 4 whereas Sect. 5 describes the implementation details and experimental evaluation of RuQAR. Section 6 contains concluding remarks and the description of future work.

2 Related Work

A storage method for ABox as well as reasoning results in a relational database is described in [4]. The presented OwlOntDB system proposes a novel database-driven forward chaining method that executes scalable reasoning over OWL 2 RL ontologies with large ABoxes. However, OwlOntDB does not support query answering "on-the-fly". Addition of one fact requires to perform reasoning and materializing once again. Without this, an answer may be not complete or sound.

OWL 2 RL rule-based reasoners are presented in [12]. In this case Jess and Drools perform inferences with rules that directly represent the semantics of the OWL 2 RL Profile. As a result, these rules can be perceived as the naive

[2] http://www.hermit-reasoner.com/.
[3] http://www.jboss.org/drools.
[4] http://jessrules.com/.

ones. Moreover, in this approach a non-triple based representation of facts and patterns in rules is applied which makes it difficult to use in other applications.

In [9] another scalable OWL 2 RL reasoner is presented. In this case an inference engine is implemented within the Oracle database system. The proposed reasoner introduces novel techniques for parallel processing with special optimizations for computing *owl:sameAs* property. However, in this approach "on-the-fly" query answering is also not supported.

The most closely related work regarding ontology transformation is an approach employed in DLEJena [10]. Nevertheless, DLEJena is able to use only one reasoning tool (Jena in contrast to Jess and Drools in our case). Moreover, we employ slightly different translation approach. DLEJena uses template rules to produce instantiated rules whereas we provide a Java-based generation of rules. Such an approach do not produce redundant instantiated rules as in [10]. Furthermore, in DLEJena the entailment rules are created at runtime whereas RuQAR produces ABox rules ahead of the reasoning process.

3 Ontology Translation Method

When applying a rule engine to an ontology-based reasoning one needs to translate the ontology into rules and facts. In our previous work [2] we proposed an approach that splits such a reasoning into two successive processes: the TBox reasoning (which solves the concept subsumption problem) and the ABox reasoning (which solves the instance checking problem). Moreover, we provided a method of translating an OWL 2 based ontology into two sets: one of rules and one of facts. In this section we present the main overview of previously proposed approach which is necessary to understand the following sections. However, more details can be found in [1,2].

Since we focus on execution of rule-based reasoning with different rule engines we proposed the Abstract Syntax of Rules and Facts (ASRF). Rules and facts generated by our translation method are both expressed in ASRF first. Then, it is required to translate ASRF expressions into the native language of a chosen rule engine. However, RuQAR provides the translation into Jess and Drools out of the box.

The translation schema of an OWL 2 ontology into ASRF sets is presented in Fig. 1. It consists of the following steps:

1. An OWL 2 ontology is loaded into the HermiT engine with assumption that this ontology is consistent.
2. The TBox reasoning is executed by HermiT. As a result, a new classified version of the ontology (new TBox) is obtained.
3. The ontology is translated into rules and facts expressed in ASRF. However, in case when the ABox is empty, a set of facts is also empty.

According to the aforementioned schema, by having two ASRF sets we separate the TBox part (set of rules) from the ABox part (set of facts) of an ontology.

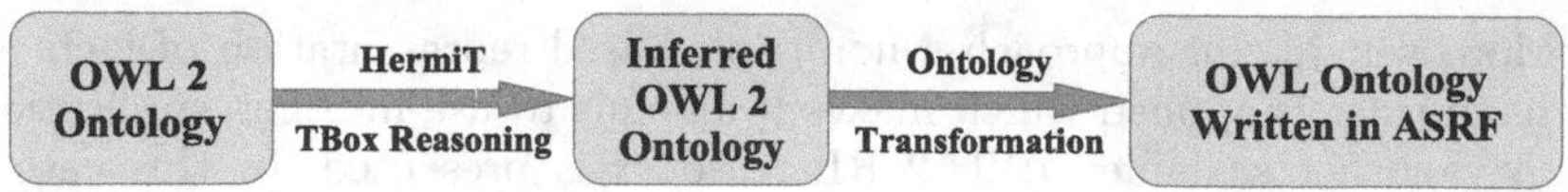

Fig. 1. Translation schema of an OWL 2 ontology into the ASRF syntax.

Thus, we are able to perform the ABox reasoning with a forward chaining rule engine after the translation of both ASRF sets into the engine's language.

After the classification performed by HermiT, our translation of an OWL 2 ontology into a set of ASRF rules is performed in the following way. For each supported OWL 2 RL/RDF rule and the corresponding OWL 2 RL axiom in the given ontology a rule that reflects the expression in this ontology is created. In other words, rather than transforming the semantics of OWL 2 RL into rules we create rules according to this semantics combined with a given ontology. For example, when an ObjectProperty *hasCousin* is defined as a SymmetricObjectProperty our method will generate a rule that follows the semantics of the property. As a result, when an instance of *hasCousin* occurs, a symmetric instance should be inferred (the following shortcuts are made: S for Subject, P for Predicate and O for Object):

$$\begin{aligned} &If \quad (Triple\ (S\ \ ?w)\,(P\ \ \text{“}hasCousin\text{”})\,(O\ \ ?z)) \\ &Then\ \ (Triple\ (S\ \ ?z)\ (P\ \ \text{“}hasCousin\text{”})\ (O\ \ ?w)) \end{aligned} \tag{1}$$

Therefore, rule (1) reflects the semantics of *prp-symp* rule from Table 5 in [11]. Such semantically equivalent rules containing a direct reference to a given ontology are created for each OWL 2 RL axiom that exist in this ontology. Each generated rule is an instantiated version of the corresponding OWL 2 RL/RDF rule for a particular TBox. As a result generated rules should be perceived as ontology instance related rules (i.e. instantiated rules or ABox rules). We call these rules ontology-dependant since they express the semantics of a particular TBox and are intended for the ABox reasoning (with facts). Thus, the rules can be directly applied in a forward chaining rule engine after the translation from ASRF to the engine's language. As a result, an execution of a reasoning with the assertional part is provided. Such an approach has a positive impact on the efficiency of reasoning since the semantics of TBox is directly reflected in the generated rules. Furthermore, the average number of conditions in the bodies of rules is smaller than in the corresponding OWL 2 RL/RDF rules, which consequently increases the efficiency of reasoning.

Current RuQAR implementation lacks support of some rules defined in OWL 2 specification [11]. We decided to use the simplest subset of OWL 2 RL/RDF rules which is easily implementable in any reasoning engine. Moreover, we excluded rules that are "constraint" rules (e.g. *cls-nothing2* from Table 6 in the OWL 2 RL Profile) and rules that are not used in the ABox reasoning (e.g. rules from Table 9 in [11]). Nevertheless, some rules need to be implemented, e.g. *cls-maxqc3* from Table 6 in [11].

Our translation method may provide more entailments during reasoning than those represented by OWL 2 RL/RDF rules. It results from the fact that we apply a DL-based reasoner and the TBox reasoning first. Nonetheless, it is determined by the expressivity of a given ontology. However, the application of our approach to an ontology that contains expressions beyond OWL 2 RL Profile will not provide the same entailments as derived by an appropriate DL-based reasoner. As a result, the reasoning with RuQAR will be sound but not complete. We observed such a case in our evaluation with LUBM ontologies where all results obtained with RuQAR were within entailments derived by Pellet[5]. This is a correct result since constructions used in LUBM ontologies are beyond the OWL 2 RL axioms.

4 Mapping Ontology Predicates to Relational Data

In order to enable semantic access to relational data, it is necessary to express relational concepts in terms of ontology concepts, that is to define mappings between a relational schema and ontology classes (concepts) and relations (roles). Given such mappings, one can transform relational data to RDF triples and process that copy in semantic applications. This method has an obvious drawback, such as maintaining synchronization. Another method is to create a data adapter based on query rewriting. Such adapters can rewrite SPARQL [5] query to SQL [3] query and execute it in RDBMS. Such a method could be fast in data retrieval, but without a reasoner, the full potential of ontology cannot be exploited. The third method is to generate semantic data from relational data "on-the-fly", on demand for the requesting application, and then process that data with a reasoner. We use such a method to fill a gap between the representation of relational data and the semantically described data.

A very important step in our mapping approach consists of linking data stored in a relational database to a knowledge base (an ontology). We accomplish this by creating mapping rules which contain SQL queries in their heads. These rules serve as mappings that are used to relate knowledge predicates and the corresponding database.

In our mapping method we assume that an ontology which will be used and translated into rules is properly constructed (i.e. the ontology is classified without inconsistencies). Then, it is required to define predicate-database mappings. Each predicate-database mapping is defined as a rule of the following form:

$$Ontology\_predicate \rightarrow SQL\_statement \tag{2}$$

A result of each SQL statement should return rows with one, two or three columns representing instances of a class, a property or a triple, respectively. The body of each mapping rule contains an ontology element which will be instantiated when an SQL query, that resides in the head, is executed. In other words,

[5] https://github.com/stardog-union/pellet.

every execution of an SQL query provides a set of RDF triples. We assume that every SQL query has one of the following permissible forms (query patterns):

$$SELECT\ Col_1\ FROM\ *\ WHERE\ (Col_1\ is\ not\ NULL); \tag{3}$$

$$SELECT\ Col_1,\ Col_2\ FROM\ *\ WHERE\ ((Col_1\ is\ not\ NULL)\ AND\ (Col_2\ is\ not\ NULL)); \tag{4}$$

$$SELECT\ Col_1,\ Col_2,\ Col_3\ FROM\ *\ WHERE\ ((Col_1\ is\ not\ NULL)\ AND\ (Col_2\ is\ not\ NULL))\ AND\ (Col_3\ is\ not\ NULL)); \tag{5}$$

where:

- COl_1, COL_2, COL_3 are the attributes (columns) that occur in the result of a query,
- $*$ is an SQL statement; it can contain SQL commands that are available in the SQL server - e.g. nested *Select* query or a table name,
- $(COL_x\ is\ not\ NULL)$ means NULL results are not allowed.

The default meaning of a query pattern is to provide an access to a relational database by obtaining results of the query execution. Pattern (3) can be used to obtain instances of a class; pattern (4) gathers instances of a property whereas pattern (5) aims at loading different kind of triples. Patterns are designed in a way it is easy to execute them with or without values. Execution of each pattern without values returns all mapped instances. Otherwise, only requested instances will be returned (if they occur in a database). In that way values can be perceived as constraints. For example, let assume that we have the following pattern for a property *worksFor*:

$$SELECT\ IDEmployee,\ IDCompany\ from\ Employee\ WHERE\ ((IDEmployee\ is\ not\ NULL)\ AND\ (IDcompany\ is\ not\ NULL)); \tag{6}$$

Pattern (6) tries to obtain instance(s) of relation *worksFor* between an employee's id and a company's id. If we do not know any value of used columns (*IDEmployee* or *IDCompany*) we can execute query which looks exactly the same as pattern (6). However, in case when values of variables are known we execute the following queries:

$$SELECT\ IDEmployee,\ IDCompany\ from\ Employee\ WHERE\ ((IDEmployee\ in\ (4,\ 5,\ 6))\ AND\ (IDcompany\ is\ not\ NULL)); \tag{7}$$

$$SELECT\ IDEmployee,\ IDCompany\ from\ Employee\ WHERE\ ((IDEmployee\ is\ not\ NULL)\ AND\ (IDcompany\ in\ (11, 13, 15))); \tag{8}$$

In that case *IDEmployee* values 4, 5, 6 and *IDCompany* values 11, 13, 15 are known. As a result we obtain instances of *worksFor* property that contain those values. However, we can execute query (9) that contains both values:

$$SELECT\ IDEmployee,\ IDCompany\ from\ Employee\ WHERE\ ((IDEmployee\ in\ (4))\ AND\ (IDcompany\ in\ (100))); \tag{9}$$

Each mapping rule that follows our method includes SQL pattern (query) in the head while the body contains an ontology predicate (class or property) or a triple. Every time, when an instance of a class/property/triple is required, we add a special trigger fact which activates a rule. When the rule fires, the corresponding SQL query is executed and results are added to a reasoning engine. As a result, query answering process fires rules with SQL queries only when there is a need for accessing data. It is important to note that queries that follow our patterns may combine data from different tables and we can execute complex queries, i.e. we can use nested *SELECT* statements (they should be inserted in * place).

5 Implementation and Experiments

RuQAR implements our approach of translating OWL 2 RL ontologies into sets expressed in the ASRF syntax. The database interface including our mapping method is also provided. Current version supports JDBC connectivity. The tool is developed in Java. RuQAR is able to execute ABox reasoning and query answering with two state-of-the-art rule engines: Drools and Jess. Moreover, RuQAR may be used as a library and can be employed in applications that require efficient ABox reasoning. The tool uses the OWL API [8] in order to load and process ontology files. We use Drools 5.5 and Jess 7.1. We employ MS SQL Server 2012 to store an ontology in a relational database.

RuQAR also supports automatic transformation of an ontology data into a relational database. However, the transformation is a very basic one. The ABox part of an ontology is transformed into a database with one table containing three columns: subject, predicate and object. Thus, it is easy to automatically generate mappings between an ontology and the corresponding database. The transformation and generation of mapping rules require only an access to a database server and an ontology. Then, both processes can be executed automatically.

We evaluated RuQAR's query answering feature using LUBM test ontology taken from the KAON2 website[6]. We used different datasets of each ontology (LUBM_0, ..., LUBM_4) where the higher number means bigger ABox set. Herein, we present results from the largest set because of the limited space. We performed the evaluation with the following engines: Jess, Drools and Pellet. Our tests were executed on a Windows 10 desktop machine with: i7-4820K CPU 3,7 GHz, Java 1.7 update 79 while the maximum heap space was set to 15 GB.

Our evaluation takes into account the execution of 14 LUBM queries in two cases: (i) TBox and ABox are stored in the main memory and (ii) TBox is stored in the main memory while ABox resides in a relational database. Evaluation schema for the first case (called IM case) was the following:

1. Perform the TBox reasoning with HermiT.
2. Transform the classified ontology into ASRF rules and facts.
3. Generate rules and facts for a rule engine.
4. Load rules and facts into a rule engine.

[6] KAON2 webpage: http://kaon2.semanticweb.org/.

5. Run reasoning.
6. Execute queries.

Evaluation schema for the second case (called DB case) was the following:

1. Store the ABox part of an ontology into a corresponding relational database.
2. Generate mapping rules for each class and property.
3. Perform the TBox reasoning with HermiT.
4. Transform the classified ontology into ASRF rules.
5. Generate rules for a rule engine.
6. Load both sets of rules into a rule engine.
7. Execute queries and perform reasoning.

Aforementioned cases were used for Jess and Drools. For the Pellet engine we loaded an ontology, performed the TBox and ABox reasoning separately, and then we executed 14 LUBM queries.

It is worth noting that in the DB case reasoning was performed only during the execution of queries. It means that this kind of execution should be perceived as top-down or goal-oriented reasoning. In other cases we had to perform reasoning first and then we were able to execute queries (without reasoning we would not be able to obtain complete results).

In each case we recorded: reasoning times, query answering times and counted the results. However, we executed tests using data stored in the working memory of an engine and with data stored in a relational database (only with Drools and Jess which follow our mapping method). Moreover, we validated the engines in order to prove that they produced identical results (an akin empirical approach was employed in [4,12]). LUBM queries were defined in: the Jess Language, the Drools Rule Language and in SPARQL (for the Pellet engine).

Figures 2 and 3 present results of our query answering evaluation. Each test was executed three times and average times are presented. Results are presented in milliseconds. Figure 2 shows IM evaluation. Times presented herein are without reasoning (TBox and ABox) since we wanted to show differences between query execution times. In each engine, combined (TBox+ABox) reasoning times were the following: 8,5 s in Drools, 10,6 s in Jess and 14,8 s in Pellet. In this case Drools performed reasoning and executed queries in the fastest way.

Figure 3 shows DB evaluation compared to the IM one. In order to make comparison adequate[7] we summarized reasoning times with query execution times in IM results. As we can see from the results Drools and Jess usually perform better when we use database as ABox storage, especially in comparison to Pellet. Queries 2, 8 and 9 require loading of huge number of triples. In queries 8 and 9 both engines load more than a half of all triples stored in a relational database. As a result, loading data from a database has strong impact on the efficiency. Nevertheless, our evaluation shows that querying data stored in a relational database using rule engine is possible and efficient. The important advantage comes from the fact that when using RuQAR with relational databases the answer for

[7] In DB case reasoning is performed during the execution of each query.

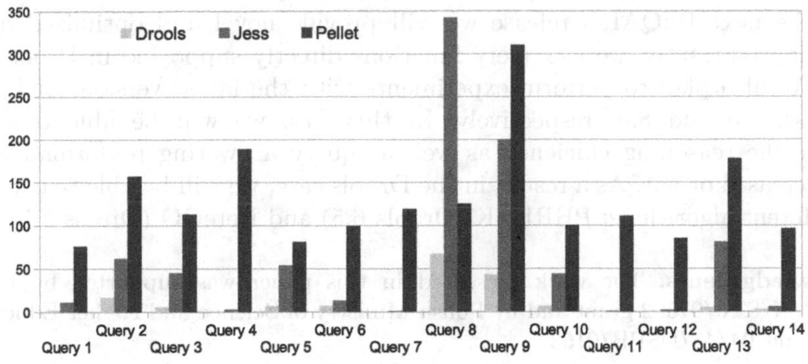

Fig. 2. LUBM queries executed in the working memory. Times in ms.

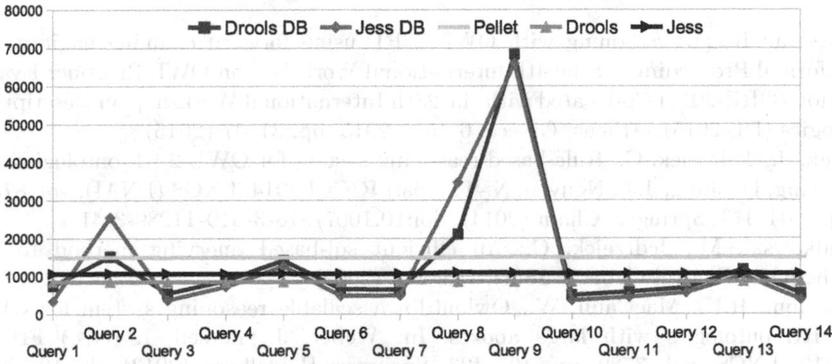

Fig. 3. LUBM queries executed with data stored in a relational database and comparison with the working memory tests. Times in ms.

a query is always up to date since queries are executed on the current state of a database ("on-the-fly"). In any other case when data change the whole reasoning process needs to be performed once again before any query can be executed. More information about RuQAR, ASRF and efficiency issues can be found at RuQAR's web page: http://etacar.put.poznan.pl/jaroslaw.bak/RuQAR.php.

6 Conclusions and Future Work

In this paper we presented a query answering method and a relational database connectivity implemented in the RuQAR framework that is aimed to be used with OWL 2 RL ontologies translated into a set of rules. Moreover, we described and performed an evaluation of RuQAR with Drools, Jess and Pellet. We compared them when executing queries in the working memory and with the use of a relational database. Our results show that it is better to use a rule engine when executing the ABox queries.

In the next RuQAR's release we will provide novel and optimized query processing (currently, we use query functions directly supported in Drools and Jess). We also plan to perform experiments with the latest versions of Drools and Jess, 6.5 and 8.0, respectively. In this case we will be able to check whether the reasoning efficiency as well as query answering performance has been increased or not. As a result, in the Drools case, we will be able to compare two different algorithms: PHREAK (Drools 6.5) and ReteOO (Drools 5.5).

Acknowledgments. The work presented in this paper was supported by UMO-2011/03/N/ST6/01602 grant and by Polish Ministry of Science and Higher Education under grant 04/45/DSPB/0163.

References

1. Bak, J.: Ruqar: reasoning with OWL 2 RL using forward chaining engines. In: Informal Proceedings of the 4th International Workshop on OWL Reasoner Evaluation (ORE-2015) Co-located with the 28th International Workshop on Description Logics (DL 2015), Athens, Greece, 6 June 2015, pp. 31–37 (2015)
2. Bak, J., Jedrzejek, C.: Rule-based reasoning system for OWL 2 RL ontologies. In: Hwang, D., Jung, J.J., Nguyen, N.-T. (eds.) ICCCI 2014. LNCS (LNAI), vol. 8733, pp. 404–413. Springer, Cham (2014). doi:10.1007/978-3-319-11289-3_41
3. Falkowski, M., Jedrzejek, C.: An efficient sql-based querying method to rdf schemata. Control Cybern. **38**(1), 193–213 (2009)
4. Faruqui, R.U., MacCaull, W.: Owlontdb: a scalable reasoning system for OWL 2 RL ontologies with large aboxes. In: Weber, J., Perseil, I. (eds.) FHIES 2012. LNCS, vol. 7789, pp. 105–123. Springer, Heidelberg (2013). doi:10.1007/978-3-642-39088-3_7
5. The W3C SPARQL Working Group. Sparql 1.1 overview (2013). http://www.w3.org/TR/sparql11-overview/
6. Guo, Y., Pan, Z., Heflin, J.: Lubm: a benchmark for owl knowledge base systems. Web Semant. **3**(2–3), 158–182 (2005)
7. Hogan, A., Decker, S.: On the ostensibly silent 'W' in OWL 2 RL. In: Polleres, A., Swift, T. (eds.) RR 2009. LNCS, vol. 5837, pp. 118–134. Springer, Heidelberg (2009). doi:10.1007/978-3-642-05082-4_9
8. Horridge, M., Bechhofer, S.: The owl api: a java api for working with owl 2 ontologies. In: OWLED (2009)
9. Kolovski, V., Wu, Z., Eadon, G.: Optimizing enterprise-scale OWL 2 RL reasoning in a relational database system (2010)
10. Meditskos, G., Bassiliades, N.: Dlejena: a practical forward-chaining owl 2 rl reasoner combining jena and pellet. J. Web Sem. **8**(1), 89–94 (2010)
11. Motik, B., Grau, B.C., Horrocks, I., Wu, Z., Fokoue, A., Lutz, C.: OWL 2 Web Ontology Language Profiles, 2nd edn. W3C Recommention, Cambridge (2012)
12. O'Connor, M.J., Das, A.: A pair of owl 2 rl reasoners. In: Klinov, P., Horridge, M. (eds.) CEUR Workshop Proceedings of OWLED, vol. 849. CEUR-WS.org (2012)

The Efficiency Analysis of the Multi-level Consensus Determination Method

Adrianna Kozierkiewicz-Hetmańska(✉) and Mateusz Sitarczyk

Faculty of Computer Science and Management,
Wroclaw University of Science and Technology,
Wybrzeze Wyspianskiego 27, 50-370 Wroclaw, Poland
adrianna.kozierkiewicz@pwr.edu.pl, 203383@student.pwr.edu.pl

Abstract. The task of processing large sets of data which are stored in distributed sources is still a big problem. The determination of a one, consistent version of data could be very time- and cost-consuming. Therefore, the balance between the time of execution and the quality of the integration results is needed. This paper is devoted to a multi-level approach to data integration using the Consensus Theory. The experimental verification of multi-level integration methods has proved that the division of integration task into smaller subproblems gives similar results as the one-level approach, but improves a time performance.

1 Introduction

Nowadays, data is often used in making decisions and frequently originates from a selection of autonomous sources. Moreover, this data may be incomplete and inconsistent, therefore, their integration into one, consistent and reliable version is a complex task, which can be very expensive and time-consuming. The solution of the mentioned problems can be dividing an initial set of input data into smaller, more manageable groups. Next, for each unit the solution is designated independently. The final, consistent outcome is a result of merging of partial results.

Due to the fact, that for data integration we apply the widely accepted Consensus Theory [19], in this paper we consider consensus determination problem. We refer to results of the aforementioned data integration by "a consensus", which can be designated in two ways: by using a one- or a multi-level approach.

The multi-level consensus determination involves two main problems. The first one concerns how input data should be divided into smaller groups in order to acquire the best final consensus. This problem has been primary solved in [12]. The previous researches have showed that data should be divided into highly divergent subsets to achieve the best quality of the final outcome. The second problem focuses on an influence of number of levels on the effectiveness of the whole process. To the best of our knowledge, this problem has not been widely investigated. Therefore, the main contribution of this paper is examining that adding an additional steps in multi-level consensus determination procedure change the quality of the final consensus and improve the execution time.

N.T. Nguyen et al. (Eds.): ICCCI 2017, Part I, LNAI 10448, pp. 103–112, 2017.
DOI: 10.1007/978-3-319-67074-4_11

The article is structured as follows. In the next Section the short overview about previous researches are described. In Sect. 3 authors present the introduction to the Consensus Theory and basic notions used in the rest of the paper. Section 4 contains the general idea of a multi-level consensus determination approach. In Sect. 5 the results of experimental verification with proper analysis is presented. Section 5 concludes this paper.

2 Related Works

Solving the consensus problem is not new and many authors in theirs publications bring it up. The general problem of choosing a theory can be formulated as follows: *for a given set X being a subset of a universe U the choice concerns on a selection of a subset of X*. Choosing the subset of X is determined by some criteria. Barthelemy and Monjardet [5] researched two classes of problems related with consensus theory:

- problems, in which a certain and a hidden structure is searched
- problems, in which inconsistent data related to the same subject is unified.

In this paper we consider the second class of problems. We also know four different approaches of solving the consensus problem: axiomatic, constructive, optimisation and boolean reasoning.

In the axiomatic approach a set of axioms has been defined to specify the conditions, which should be fulfilled by consensus functions. In accordance with solving problem the form of axioms can be different. In the literature [17,19] 10 postulates for consensus choice functions are presented: reliability, unanimity, simplification, quasi-unanimity, consistency, Condorcet consistency, general consistency, proportion, 1-optimality, 2-optimality.

In the constructive approach to solving consensus problems, definitions of a microstructure and a macrostructure of a universe U of objects are required. A microstructure of U is defined as a structure of its elements and a macrostructure as a relation between elements of U. In the literature the following microstructures have been investigated: linear orders [1,16]; semilattices [3]; n-tree [2,8]; ordered partitions and coverings [7,19]; non-ordered partitions [4]; weak hierarchies [15] and ontologies [11,13]. In the optimisation approach solving consensus problems is based on some optimality rules. There are three classes of these rules: global optimality rules, Condorcet's optimality rules and maximal similarity rules.

In the last approach, Boolean reasoning, a consensus problem is looked upon as an optimization problem. This problem is coded as Boolean formula. The first part of those formulas determines a solution of the problem [6,20].

The multilevel solution of consensus problems for the first time was mentioned in [18]. In previous works [10,14] one-level and two-level consensuses satisfying the 1-optimality and 2-optimality criterion have been compared. For assumed macro- and microstructure the mean error for a different number of classes k and vector lengths n has been examined. For the 1-optimality criterion researches have demonstrated that the mean of error for a one-level consensus is smaller by 1,2% than for a two-level consensus. For the 2-optimality criterion achieved results showed that two-level algorithm has given results worse by 5% and the one-level method worse by 1% in comparison to optimal solution.

In the article [12] authors have checked how coherence of vectors in classes, using Fleiss' kappa value, has affected on the consensus quality in the two-level consensus problem. Authors demonstrated, that a higher variety between vectors in one class, the better quality of the final consensus can be achieved. So far, the consensus problems for more than two levels have not been investigated in details. Therefore, this paper is an extension of the previous research.

3 Basic Notions

Let U be a finite, nonempty set of a universe of objects. Each object could reflect the potential elements of a knowledge referring to a certain world. By the symbol 2^U we denote a powerset of U, that is the set of all subsets with repetitions of U. The symbol $\Pi_b(U)$ denotes the set of all b-element subset (with repetitions) of the set U for $b \in N$. Thus $\Pi(U) = \bigcup_{b \in N} \Pi_b(U)$ is the set of all nonempty subsets with repetitions of the universe U. Each X which belongs to $\Pi(U)$ is called a knowledge profile (a profile).

Definition 1. *The macrostructure of the set U is a distance function $\delta : U \times U \rightarrow [0, 1]$ which satisfies the following conditions ([18]):*

1. $\forall_{v,u \in U}, \delta(v, u) = 0 \Leftrightarrow v = u$
2. $\forall_{v,u \in U}, \delta(v, u) = \delta(u, v)$

Definition 2. *For an assumed distance space (U, δ), the consensus choice problem requires establishing the consensus choice function. By a consensus choice function in space (U, δ) we call a function:*

$$C : \Pi(U) \rightarrow 2^U \tag{1}$$

By $C(X)$ we denote the representation of $X \in \Pi(U)$. By such $c \in C(X)$ we call a consensus of a profile X.

In [18,19] authors present some postulates for consensus choice functions. The *1-optimality* and *2-optimality* criterions play the important role in solving the consensus choice problem. Let us assume the following notions to formally define these postulates:

- $\delta^1(x, X) = \sum_{y \in X} \delta(x, y)$
- $\delta^2(x, X) = \sum_{y \in X} (\delta(x, y))^2$

Definition 3. *For a profile $X \in \Pi(U)$ a consensus choice function C satisfies the postulate of:*

- 1-optimality *iff* $(x \in C(X) \Rightarrow (\delta^1(x, X) = \min_{y \in U} \delta^1(y, X))$,
- 2-optimality *iff* $(x \in C(X) \Rightarrow (\delta^2(x, X) = \min_{y \in U} \delta^2(y, X))$.

The *1-optimality* postulate requires the consensus to be as near as possible to elements of the profile and could be recognised as the best representation of the profile. The *2-optimality* criterion can be referred as the most 'fair' consensus.

4 Multi-level Consesnsus Determination Approach

As it was mentioned in previous section, the problem of processing of big set of data is often very difficult or even impossible to complete in reasonable time. Therefore, the problem of consensus determination can be divided into smaller problems which are easier to solve. The general idea of two-level approach to the consensus choice requires to conduct an initial division of the profile into k classes. For each class the consensus is determined using an ordinary (one-level) algorithm. The final consensus is appointed as a solution obtained in the previous stage. Obviously, the procedure can be divided into even more stages which outputs serve as input of subsequent steps and then we say about multi-level consensus determination method. The general idea is presented in the Fig. 1. It requires establishing some auxiliary elements like: distance space (U, δ), the clustering method or one-level consensus determination method.

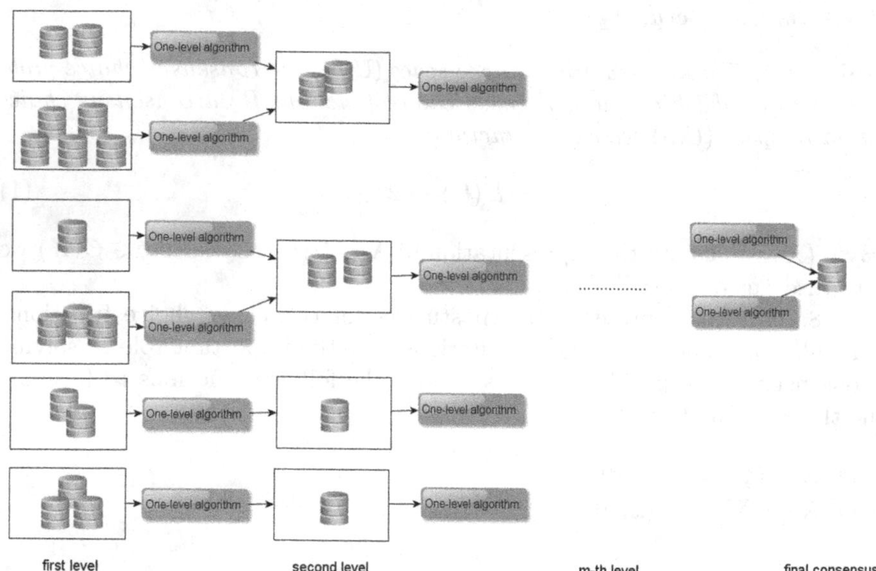

Fig. 1. General idea of multi-level consensus determination method

In this paper we assume that the profile X consists of n binary vectors of the length equal to N. Formally, we can define distance space as: $U = \{u_1, u_2, ...\}$ where elements of the universe are binary vectors and $\delta(w, v) = \sum_{j=1}^{N} |w^j - v^j|$ for such $w, v \in U$ that $w = (w^1, w^2, ..., w^N), v = (v^1, v^2, ..., v^N)$, $v^q, w^q \in \{0, 1\}$, $q \in \{1, ..., N\}$. The profile is defined as: $X = \{a_1, a_2, ..., a_n\} \in \Pi(U)$, where: $a_i = (a_{i1}, a_{i2}, ..., a_{iN}), i \in \{1, ..., n\}$.

This paper is devoted to examining how the number of levels influence on the algorithm's efficiency. For simplicity of calculations, we assume that the profile X is divided into k classes in the random way. In our previous work [9] we have proved that if the cardinality of profile is the odd number it is the enough condition to determine a reliable consensus. Thus, in this work, the set following numbers of classes on each level is presented in Table 1.

Table 1. The assumed number of classes in each level

Levels number	Level I	Level II	Level III	Level IV	Level V
1	1	-	-	-	-
2	13	1	-	-	-
3	91	13	1	-	-
4	455	91	13	1	-
5	1365	455	91	13	1

For the assumed distance space (U, δ) the one-level method of determining the consensus satisfying 1-optimality is conducted in the following steps [18]:

Algorithm 1. 1-optimality consensus determination method

Require: $X = \{a_1, a_2, ..., a_n\}$

1: Set $j = 1$;
2: Set $f_j = 0$, $i = 1$;
3: IF $a_{ij} = 1$ THEN $f_j + +$;
4: $i + +$;
5: IF $i \leq n$ GOTO 3;
6: $j + +$;
7: IF $j \leq N$ GOTO 2;
8: Set $j = 1$;
9: IF $f_j \geq \frac{n}{2}$ THEN $x_j^* = 1$ ELSE $x_j^* = 0$
10: $j + +$;
11: IF $j \leq N$ GOTO 10;

To determine the 2-optimality consensus we use the heuristic algorithm presented below [18]:

Algorithm 2. 2-optimality consensus determination method

```
Require: X = {a_1, a_2, ..., a_n}
Random x, set md = δ^2(x, X);
Set j = 1;
x_j = x_j ⊕ 1;
IF δ^2(x, X) < md THEN md = δ^2(x, X) ELSE x_j = x_j ⊕ 1;
j + +;
IF j ≤ N GOTO 3 ;
```

5 Experimental Evaluation of One- and Multi-level Binary Vector Integration

The experiment was conducted using a dedicated environment. As in other researches [10,14], in this experiment we also used vector length $N = 10$, however number of classes was different for each sample as it was mentioned in previous Section (Table 1). Apart from the distance value we also simulated a multithread multilevel integration (spreading each class on different thread or machine), and studied time of performance. All analysis are made for a significance level $\alpha = 0.01$ and for the profile size of $n = 15015$ vectors.

At the beginning we analyse normalised consensus distance from the profile received from the algorithm of consensus choice for function which satisfying the 1-optimality criterion. The measured distance can be formally defined as: $\frac{\sum_{y \in X} \delta(x,y)}{n}$. Before selecting a proper test we analysed the distribution of obtained data by using Shapiro-Wilk test. Because all $p - values$ are greater than $\alpha = 0.01$, we cannot reject the null hypothesis and claim that the samples come from a normal distribution.

Results of multiple-sample test suggest, that samples for each integration level have different variances, therefore to the further analysis we use Welch and Brown-Forsythe one-way ANOVA test for independent samples. The value of statistic test is equal 42.995 and $p - value < 0.000001$ is less than α, therefore we accept the alternative hypothesis and claim that samples are different.

In the next step we analyse difference of samples' average in pairs by using Tamhane's T2 post-hoc test. Results show, that the algorithm determining the two-level consensus is worse than the algorithm determining one-level consensus by 0.11%. The three-level consensus is worse than the one-level consensus by 0.22% and 0.11% than the two-level consensus. The four-level consensus is worse than the one-level consensus by 0.29% and 0.06% than the three-level consensus. The five-level consensus is worse than the one-level consensus by 0.31% and 0.02% than the four-level consensus. Differences between samples' averages are also illustrated on the left side of the Fig. 2.

In the similar way we conducted tests for algorithm of consensus choice for function which satisfying the 2-optimality criterion. The average distance is defined as: $\frac{\sum_{y \in X} \delta^2(x,y)}{n}$. We also compare achieved results from heuristic with the optimal solution achieved using the brute-force algorithm. The results of

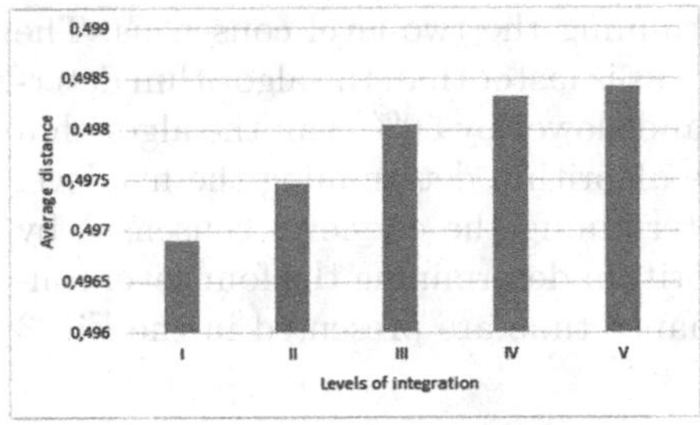

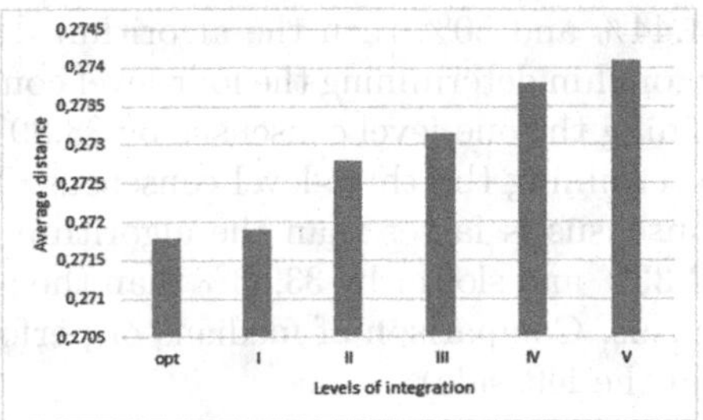

Fig. 2. Average distance between consensus (satisfying 1- and 2 - optimality postulate) and elements of profile

Shapiro-Wilk test confirms the hypothesis that sample come from a normal distribution. Welch and Brown-Forsythe one-way ANOVA test for independent samples points out that at least one sample is different (the statistic value test: 85.868, $p - value < \alpha$).

In the last step for these samples, we analyse difference of samples' average in pairs by using Tamhane's T2 post-hoc test. Conducted computations show that the algorithm determining the one-level consensus is worse than the optimal algorithm by 0.04%. The algorithm determining the two-level consensus is worse than the algorithm determining one-level consensus by 0.32% and 0.37% than the optimal algorithm. The algorithm determining the three-level consensus is worse than the algorithm determining one-level consensus by 0.45%, 0.12% than the algorithm determining two-level consensus and 0.5% than the optimal algorithm. The algorithm determining the four-level consensus is worse than the algorithm determining one-level consensus by 0.69%, 0.24% than the algorithm determining three-level consensus and 0.74% than the optimal algorithm. The algorithm determining the five-level consensus is worse than the algorithm determining one-level consensus by less than 0.8%, 0.11% than the algorithm determining four-level consensus and 0.84% than the optimal algorithm. Samples' averages are shown on the right side of the Fig. 2.

In the second stage of our researches we analyse performance time of determining 1-optimality and 2-optimality consensuses. The $p-value$ of Shapiro-Wilk test for some samples are smaller then 0.01 therefore, some samples do not come from a normal distribution. Next, we conduct two Kruskal-Wallis tests for 1-optimality and 2-optimality criterion, respectively. For both group of samples $p - values$ are significantly smaller than α, thus the execution times of the multi-level algorithm are different for different numbers of levels.

To compare samples we compute the median for each and analyse differences of them in pairs. Initially, we examine methods where determining consensuses satisfied 1-optimality criterion. Achieved results show, that the algorithm determining the two-level consensus is faster than the algorithm determining the one-level consensus by 88.89%. The algorithm determining the three-level consensus is faster than the algorithm determining the one-level consensus by

94.44% and 50% then the algorithm determining the two-level consensus. The algorithm determining the four-level consensus is faster than the algorithm determining the one-level consensus by 88.89% and slower by 50% than the algorithm determining the three-level consensus. The algorithm determining the five-level consensus is faster than the algorithm determining the one-level consensus by 83.33% and slower by 33.33% than the algorithm determining the four-level consensus. Comparison of medians of performance time are presented in the Fig. 3 (on the left side).

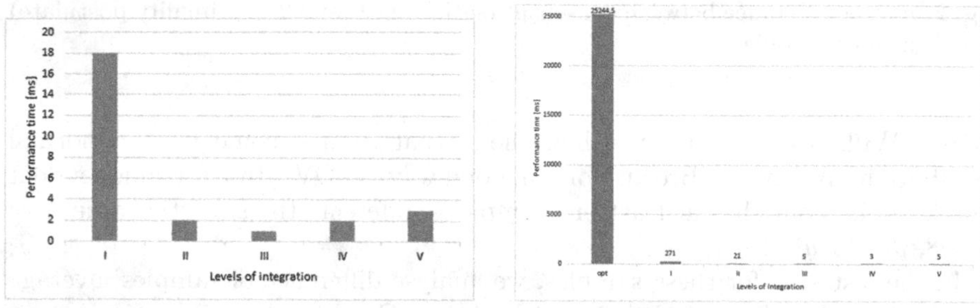

Fig. 3. Medians of performance time of integration for 1-optimality and 2-optimality criterion

Corollary. The last thing, which we analysed, is performance time of the algorithm of consensus choice for the function which satisfying the 2-optimality criterion. Achieved results show that the algorithm determining the one-level consensus is faster than the optimal algorithm by 98.93%. The algorithm determining the two-level consensus is faster than the algorithm determining the one-level consensus by 92.25% and 99.92% than the optimal algorithm. The algorithm determining the three-level consensus is faster than the algorithm determining the one-level consensus by 98.15%, 76.19% than the algorithm determining the two-level consensus and 99.98% than the optimal algorithm. The algorithm determining the four-level consensus is faster than the algorithm determining the one-level consensus by 98.89%, 40% than the algorithm determining the three-level consensus and 99.99% than the optimal algorithm. The algorithm determining the five-level consensus is faster than the algorithm determining the one-level consensus by 98.15%, 99.98% than the optimal algorithm and slower by 40% than the algorithm determining the four-level consensus. Huge differences of performance time between the brute-force and the heuristic algorithm are illustrated in Fig. 3 (on the right side).

6 Future Works and Summary

In this paper the comparison of multilevel consensus algorithms fulfilling the 1-optimality and 2-optimality criterion has been presented. The multi-level consensus determination methods have been tested based on two factors: the quality of the final consensus and algorithms performance times. Our researches have showed, that adding additional level of consensus determination algorithm has caused growth of the average distance between consensus and elements of profile for both: 1-optimality and 2-optimality criterion. In comparison to the one-level consensus determination method, examined distance has been worse less than 0.5% and less 1% for 1-optimality and 2-optimality criterion, respectively.

The analysis of performance times has demonstrated that adding additional level speeds up algorithm, however, only to some extent. When the threshold point is achieved adding a next level to the algorithm has increased the performance time. From the execution time point of view, the consensus determination algorithm is the most efficient for three and four levels, in respect of 1- and 2-optimality criterion. The three-level algorithm determining consensus satisfying 1-optimality criterion is 98.15% faster than the one-level algorithm. The algorithm dedicated for the determination of the 2-optimal consensus is about 99.

In the future more research concerning a multilevel consensus is planned. We intend to find the best number of algorithm's levels depending on number of vectors n, number of classes k and variety of vectors in one class, to achieve the best performance time and the quality in the same time with an acceptable error. Moreover, the algorithm determining 2-optimality consensus is heuristic, therefore authors want to study and develop another algorithm which will give better consensus quality for 2-optimality criterion. Additionally, similar analysis for different micro- and macrostructures, like tuples, are planned.

Acknowledgment. This article is based upon work from COST Action KEYSTONE IC1302, supported by COST (European Cooperation in Science and Technology).

References

1. Arrow, K.J.: Social Choice and Individual Values. Wiley, New York (1963)
2. Barthelemy, J.P.: Thresholded consensus for n-trees. J. Classif. **5**, 229–236 (1988)
3. Barthelemy, J.P., Janowitz, M.F.: A formal theory of consensus. SIAM J. Discrete Math. **4**, 305–322 (1991)
4. Barthelemy, J.P., Leclerc, B.: The median procedure for partitions. In: DIMACS Series in Discrete Mathematics and Theoretical Computer Science, vol. 19, pp. 3–33 (1995)
5. Barthelemy, J.P., Monjardet, B.: The median procedure in cluster analysis and social choice theory. Math. Soc. Sci. **1**, 235–267 (1981)
6. Brown, F.N.: Boolean Reasoning. Kluwer Academic Publisher, Dordrecht (1990)
7. Daniłowicz, C., Nguyen, N.T.: Consensus-based partition in the space of ordered partitions. Pattern Recogn. **21**, 269–273 (1988)
8. Day, W.H.E.: Consensus methods as tools for data analysis. In: Bock, H.H. (ed.) Classification and Related Methods for Data Analysis, pp. 312–324. North-Holland, Amsterdam (1988)
9. Kozierkiewicz-Hetmańska, A.: Analysis of susceptibility to the consensus for a few representations of collective knowledge. Int. J. Softw. Eng. Knowl. Eng. **24**(5), 759–775 (2014)

10. Kozierkiewicz-Hetmańska, A.: Comparison of one-level and two-level consensuses satisfying the 2-optimality criterion. In: Nguyen, N.-T., Hoang, K., Jędrzejowicz, P. (eds.) ICCCI 2012. LNCS, vol. 7653, pp. 1–10. Springer, Heidelberg (2012). doi:10.1007/978-3-642-34630-9_1
11. Kozierkiewicz-Hetmańska, A., Pietranik, M.: The knowledge increase estimation framework for ontology integration on the concept level. J. Intell. Fuzzy Syst. **32**(2), 1161–1172 (2017)
12. Kozierkiewicz-Hetmańska A., Pietranik, M.: Assessing the quality of a consensus determined using a multi-level approach, In: Proceedings of the 2017 IEEE International Conference on INnovations in Intelligent SysTems and Applications (2017, to appear)
13. Kozierkiewicz-Hetmańska, A., Pietranik, M., Hnatkowska, B.: The knowledge increase estimation framework for ontology integration on the instance level. In: Nguyen, N.T., Tojo, S., Nguyen, L.M., Trawiński, B. (eds.) ACIIDS 2017. LNCS, vol. 10191, pp. 3–12. Springer, Cham (2017). doi:10.1007/978-3-319-54472-4_1
14. Kozierkiewicz-Hetmańska, A., Nguyen, N.T.: A comparison analysis of consensus determining using one and two-level methods. In: Advances in Knowledge-Based and Intelligent Information and Engineering Systems, pp. 159–168 (2012)
15. McMorris, F.R., Powers, R.C.: The median function on weak hierarchies. In: DIMACS Series in Discrete Mathematics and Theoretical Computer Science, vol. 37, pp. 265–269 (1997)
16. Mirkin, B.G.: Problems of group choice, Nauka Moscow (1974)
17. Nguyen, N.T.: Using distance functions to solve representations choice problems. Fundamenta Informaticae **48**, 295–314 (2001)
18. Nguyen, N.T.: Consensus Choice Methods and their Application to Solving Conflicts in Distributed Systems, Wroclaw University of Technology Press (2002). (in Polish)
19. Nguyen, N.T.: Advanced Methods for Inconsistent Knowledge Management. Springer-Verlag, London (2008)
20. Pawlak, Z.: Rough Sets-Theoretical Aspects of Reasoning About Data. Kluwer Academic Publisher, Dordrecht (1991)

Collective Intelligence Supporting Trading Decisions on FOREX Market

Jerzy Korczak, Marcin Hernes(✉), and Maciej Bac

Wrocław University of Economics, Wrocław, Poland
{jerzy.korczak,marcin.hernes,maciej.bac}@ue.wroc.pl

Abstract. The aim of the paper is to present an approach to support decision-making on financial markets using an idea of collective intelligence implemented as a multi-agent system, called A-Trader. A-Trader is integrated with the Meta Trader system which provides online data, including ticks of any securities, goods or currency pairs. Many of the implemented agents apply AI methods, communicate their trading advices to the supervisor that integrates all information and suggests the trading decision. The first part of the paper presents the architecture and functionalities of A-Trader. The structure and functionality of the agents and approach to the building of the trading strategies are detailed. The last section describes the results of the performance evaluation of selected trading strategies on FOREX.

Keywords: Multi-agent systems · Supporting trading decision-making · FOREX market

1 Introduction

Trading decision making is usually supported by methods based on financial mathematics, statistics, economics or artificial intelligence. These methods are often implemented as software agents in trading systems. Many of these systems operate on the FOREX market (Foreign Exchange Market), where currencies are traded against one another in pairs, for instance EUR/USD, USD/PLN. In the paper, the main focus will be upon High Frequency (HF) quotes and online algorithmic trading.

In comparison with traditional trading, High Frequency Trading (HFT) puts strong emphasis on the price formation process, short-term positions, high speed, and sophisticated algorithms based on efficient and robust indicators and modern IT. High frequency traders are taking decisions on the basis of real-time quotes changes in order to achieve satisfactory rate of return.

Generally speaking, to support traders, the systems must provide as soon as possible advice on what position should be taken: buy, sell, or hold. In a trading system the advice might be computed by one or many algorithms, in terms of software implementation, by one or collection of agents.

One of the first multi-agent systems in the domain of finance was proposed by [13]. This system was built of several agents applying the specialized financial models. Bohm and Wenzelburger [1] describe an evaluation of the shares portfolio optimization strategy by three agents: rational agent, interference agent and technical analysis agent.

N.T. Nguyen et al. (Eds.): ICCCI 2017, Part I, LNAI 10448, pp. 113–122, 2017.
DOI: 10.1007/978-3-319-67074-4_12

The work [14] presents a multi-agent system, where the agents' intelligence is supported by the Fuzzy Expert System. Works [2, 5, 15] describe a system where two groups of agents apply the methods of fundamental and technical analysis trying to shape market dynamics. Sher [12] uses neutral networks and neuro fuzzy computing for taking into account the geometrical patterns of the financial data when making predictions and trades. Khosravi et al. [7] present an agent that uses multiple behavioral techniques to make bidding decisions fighting with market uncertainty.

The above cited solutions did not fully satisfy the user expectations. They demonstrated low performance on HFT and costly maintenance.

The aim of this paper is to present an approach to model and advise investors on the FOREX market using an idea of collective intelligence implemented as a multi-agent system, called A-Trader. A-Trader is a kind of trading signal generators that advise recommended open and closed positions for online FOREX traders. The trading opportunities are provided by consensual advice, generated by multiple software agents that use technical and fundamental analysis as well as behavioral sentiments. Software agents in the system generate buy-sell decisions with the use of diversified methods. The performance of the agents is constantly evaluated to select the agents giving the best advice on the current market situation. In other words, the agents' highest performing decisions constitute the basis of the trader's buy-sell transactions. Trading agents in the system form the collective investment strategies. The strategies are constantly evaluated, and those with the highest evaluation can be chosen to advise the trader. The return on investment, usually taken as a key measure, is not considered as the only evaluation criterion. The other aspects having influence on the effectiveness of the investment strategies, such as investment risk [6] or transaction costs should also be taken into consideration. To evaluate the system's performance, the A-Trader uses a multi-criteria function that calculates several profit and risk measures, such as Sharpe ratio, number of profitable and unprofitable transactions, the average coefficient of volatility.

The structure of the paper is as follows. The first part of the article presents the architecture and functionalities of the system. Next, the structure and functionality of agents and the building of the trading strategies are discussed. The final section describes the research experiment and the results of the performance evaluation of selected trading strategies.

2 Architecture and Functionalities of A-Trader

In general, a multi-agent system is composed of the agents which are capable of generating independent decisions. It should be noted that decisions can be consistent or contradictory, e.g. two independent agents may generate buy and sell decision at the same time.

The architecture and the operational idea is presented in Fig. 1.

The architecture has been already defined in our previous publications [8, 9]. The key agent of the system is the Supervisor (S). Its goal is to generate profitable trading advice that reduces the investment risk. The Supervisor coordinates the computing Basic and Intelligent agents, and provides the final advice to the trader. It is charged to

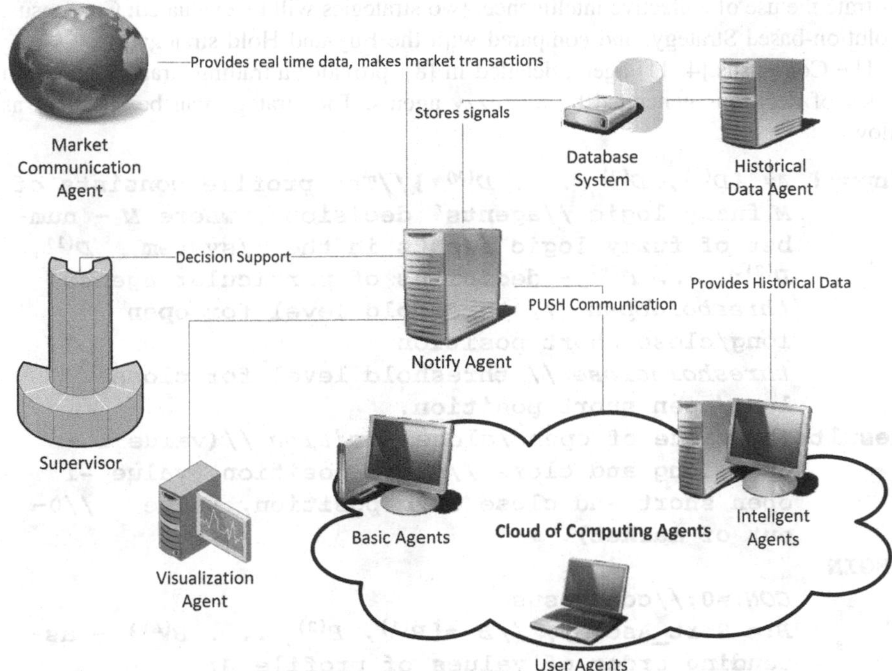

Fig. 1. A-Trader system architecture.

resolve conflicts and to assess the effectiveness in investing and risk. The central component of the architecture is the Notification Agent (NA) that receives the quotations, distributes messages and data to various agents, and controls the system operation.

The communication of the system with the external environment is ensured by the Market Communication Agents (MCA). These agents deliver news from financial markets and quotations of the available securities. They also transmit open and close position orders.

Visualization of the results of the agents is an important issue in verifying the correctness of agents operations. The task is carried out by the User Communication Agents (UCA). The Communication Agent allows the trader to communicate its own recommendations to the Intelligent Agents. It is possible to modify the parameters of a selected agent or to suggest the Supervisor which mechanisms are supposed to influence investment decisions, and to what extent.

3 Building Trading Strategies

The trading strategies of A-Trader indicate when to close/open a long or short position, taking into consideration signals generated by agents. These strategies apply more sophisticated algorithms than the typical technical analysis functions [10]. In order to illustrate the use of collective intelligence, two strategies will be evaluated: Consensus, Evolution-based Strategy, and compared with the Buy-and-Hold strategy.

The Consensus [4, 11] agent, detailed in [8], provides a trading strategy based on the set of decisions generated by all fuzzy agents. The strategy can be specified as follows:

```
Input: A= {D(1), D(2), .... D(M) }//The profile consists of
       M fuzzy logic //agents' decisions, where M - num-
       ber of fuzzy logic agents in the //system, D(1),
       D(2), .... D(M) - decisions of particular agents
       thresholdopen  // threshold level for open
       long/close short position
       thresholdclose // threshold level for close
       long/open short position.
Result:The value of open /close position //(value 1 -
       open long and close //short position, value -1
       open short and close long position, value   //0-
       out of market).
BEGIN
       CON:=0;//consensus
       B:= Sort_Asc(A); //B ={B(1), B(2), .... B(M)} - as-
       cending order of values of profile A;
       i:= Floor((M+1)/2); //i - auxiliary variable
       j:= Ceil((M+2)/2); //j - auxiliary variable
       Set CON as any value from interval [B(i), B(j)];
       position=0;
If CON >= thresholdopen then position = 1;
If CON <= thresholdclose then position = -1;
END.
```

The Evolution-based Strategy is inspired by the works of [3]. This strategy defines the best thresholds for open/close positions on the basis of technical analysis, fundamental analysis, and behavior-based agents. The evolutionary algorithm determines which agents should be taken into account in decision making. It also finds out how important the advice of a given agent is. The algorithm searches the space of agent decisions and sets the weighting of their importance. Genotype (Fig. 2) contains the weightings for every agent separately for opening/closing short/long position.

Except for weightings, every advising agent also has "compulsory" parameters which mean that the signal value of the agent has to be positive, negative or "don't care". It also sets Take Profits value, Stop Loss and Trailing Stops for long and short positions separately. The result of the evolutionary algorithm is a phenotype that can be

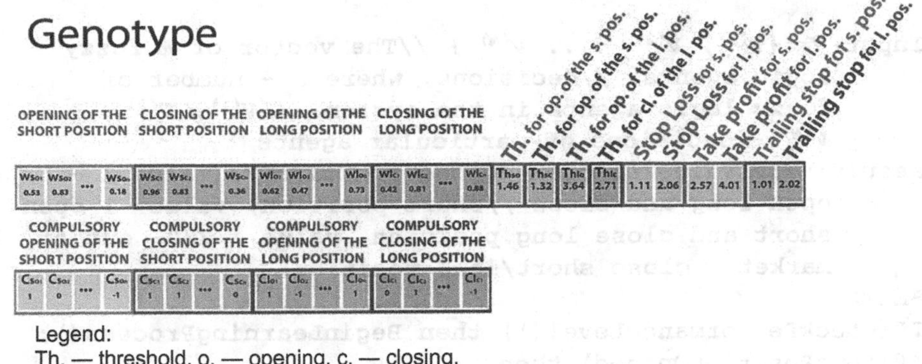

Fig. 2. Genotype in the evolution-based strategy.

interpreted as a set of decision rules. An example of the set of rules to open short position for the agent at time T_0 can be defined as follows:

$$\begin{gathered}(A_1T_0 * w_{so_1} + \cdots + A_nT_0 * w_{so_n} > Th_{so}) \\ \wedge \\ ((A_1T_0 * C_{so_1} > 0)\bigvee(C_{so_1} = 0)) \\ \wedge \\ \vdots \\ \wedge \\ ((A_nT_0 * C_{so_n} > 0)\bigvee(C_{so_n} = 0))\end{gathered} \tag{1}$$

where:

A_nT_0 – value of *Agent n* signal in time T_0,
w_{so_n} – weighting for opening short position for *Agent n*,
Th_{so} – threshold for opening short position,
C_{so_n} – compulsory parameter for opening short position for *Agent n*.

The conditions for opening/closing short/long position consist of two parts. In the first part the algorithm checks if a threshold is reached. To check the threshold, the algorithm multiplies values of signals of each agent by the corresponding weightings, then sums up all the results. If the sum is over the opening short position threshold, the first part of the condition is fulfilled. In the second part of the condition, the algorithm checks if all compulsory rules are fulfilled. The strategy can be specified as follows:

```
Input: V= {V(1), V(2), .... V(M) } //The vector of M fuzzy
      logic agents' //decisions, where M - number of
      fuzzy logic agents in the system, //V(1), V(2), ....
      V(M) - decisions of particular agents
Result: The value of open /close position //(value 1 -
      open long and close //short position, value -1 open
      short and close long position, value   //0- out of
      market - close short/long position).
BEGIN
If(CheckPerformanceLevel()) then BeginLearningProcess();
If(LongPositionOpened) then
If(CheckClosingLongPositionCondition(V)) then
  CloseLongPosition();
  Return -1;
If(ShortPositionOpened) then
If(CheckClosingShortPositionCondition(V))then
  CloseShortPosition();
  Return 1;
If(CheckOpeningLongPositionCondition(V))then
  OpenLongPosition();
  Return 1;
If(CheckOpeningShortPositionCondition(V)) then
  OpenShortPosition();
  Return -1;
  Return 0;
END.
```

A-Trader built-in strategies can be reused and extended. The trader may easily add a new agent, a new source of information, by filling out a generic pattern of the agent structure. More details about the implemented trading strategies can be found in [8, 9].

4 Research Experiment

The strategies' performance analysis is carried out for data within the M1 range of quotations from the FOREX market. The following assumptions were made for the purpose of this analysis, namely:

1. USD/PLN quotations were selected from randomly chosen periods, notably:
   - 12-11-2015, 0:00 am to 12-11-2015, 23:59 pm,
   - 18-11-2015, 0:00 am to 19-11-2015, 03:00 pm,
   - 20-11-2015, 0:00 am to 20-11-2015, 23:59 pm,
2. At the verification, the trading signals (for open long/close short position equals 1, close long/open short position equals −1) are generated by the strategies Consensus,

Evolution-based Strategy. The Buy and Hold (B&H) strategy is used as a benchmark.

3. It was assumed that the unit of performance analysis ratios (absolute ratios) is pips (a change in price of one "point" in FOREX trading is referred to as a pip).
4. The transaction costs are directly proportional to the number of transactions.
5. The capital management – it was assumed that in each transaction the investor engages 100% of the capital held. The capital management strategy may be determined differently by the trader.
6. The performance analysis was performed with the use of the following measures (ratios):
   - rate of return (ratio x_1),
   - the number of transactions,
   - gross profit (ratio x_2),
   - gross loss (ratio x_3),
   - the number of profitable transactions (ratio x_4),
   - the number of profitable transactions in a row (ratio x_5),
   - the number of unprofitable transactions in a row (ratio x_6),
   - Sharpe ratio (ratio x_7),
   - the average coefficient of volatility (ratio x_8),
   - the average rate of return per transaction (ratio x_9), counted as the quotient of the rate of return and the number of transactions.
7. To compare the agents' performance, the following evaluation function was elaborated:

$$y = (a_1x_1 + a_2x_2 + a_3(1 - x_3) + a_4x_4 + a_5x_5 + a_6(1 - x_6) + a_7x_7 + a_8(1 - x_8) + a_9x_9) \quad (2)$$

where x_i denotes the normalized values of ratios mentioned in item 6 from x_1 to x_9. It was adopted in the test that coefficients a_1 to a_9 = 1/9. It should be mentioned that these coefficients may be modified with the use of, for instance, an evolution-based method, or determined by the trader in accordance with his or her preferences.

Table 1 presents results of performance analysis. The analysis of the ratios by the traders is very difficult and taking decisions in real-time is limited. However, the results of the experiment allow us to rank strategies in the given periods. In the first period the Consensus was the best one. In the second period the best was the Evolution-based Strategy. Ranking in the third period is similar to the second period. The Consensus strategy was ranked highest most often (2 out of 3 periods), although the rate of return of this strategy was not always the highest. However the Consensus strategy characterizes the low level of risk measures and that is the reason for highest ranking. The Evolution-based strategy generated the high level of average rate of return per transaction. The results Buy and Hold strategy were added in the Table 1 as a trading benchmark.

It should be noted that in other systems (e.g. in the MetaTrader, XTRADE, Trade Chimp), evaluation is, in most cases, performed "manually" by the trader. This is very

Table 1. Trading performance

Ratio	Consensus			Evolution-based strategy			B & H		
	Period 1	Period 2	Period 3	Period 1	Period 2	Period 3	Period 1	Period 2	Period 3
Rate of return [Pips]	680	178	243	432	198	202	−41	173	198
Number of transactions	19	37	32	5	4	6	1	1	1
Gross profit [Pips]	431	175	164	231	211	129	0	173	198
Gross loss [Pips]	177	77	97	128	178	117	−41	0	0
Number of profitable transactions	12	16	17	3	2	3	0	1	1
Number of profitable consecutive transactions	5	6	4	2	1	2	0	1	1
Number of unprofitable consecutive transactions	2	2	3	1	1	1	1	0	0
Sharpe ratio	1.08	1.77	2.80	0.92	1.15	2.09	0	0	0
Average coefficient of volatility	0.62	0.11	1.03	2.26	1.28	1.17	0	0	0
Average rate of return per transaction	35.79	4.81	7.59	86.40	49.50	33.66	−41	173	198
Value of evaluation function (y)	**0.58**	**0.28**	**0.47**	**0.42**	**0.39**	**0.36**	**0.08**	**0.26**	**0.24**

time-consumption process, therefore taking decision in near-real time is rather impossible. Besides, these systems offer only the functions to compute the basic financial ratios (rate of return, number of transactions, highest profit, etc.). A-Trader system, instead, offers also additional ratios, e.g. risk measures and the value of global evaluation function (see Table 1).

The Supervisor agent uses the evaluation function in order to the performance evaluation of particular strategies for opening/closing positions in the system. This process is performing automatically, in near real-time. Supervisor agent also advises the investor to take final decisions on the basis of decisions generated by the strategy with the highest level of performance. In order to considering user preference concerning the criterion of the importance of particular evaluation ratios, he/she can change a_i and x_i parameters of the evaluation function. The transaction costs are also considered in evaluation. It is computed as a relation between the number of transactions and the average rate of return from the transaction. However, this simple principle cannot be adopted more generally, because a large number of transactions have an impact in reducing the strategy's efficiency level, especially for transactions with a high rate of return.

5 Conclusions

The idea of collective intelligence, implemented in as the multi-agent system, has been positively validated on high frequency financial time series. The collection agents, shown in the two illustrated trading strategies, has provided much better investment decisions than all individual agents. However, there is no one strategy which definitely dominates over all the others.

The use of performance evaluation function allowed the automatic setting of the best strategy in a time close to real-time, and this has, in turn, a positive influence on investment effectiveness. The numerous experiments on the financial time series allow us to draw the conclusion that the level of performance of any particular strategy changes, depending on the current situation on the FOREX market.

The results of the study are promising, however more experiments are needed to deeper conclusions. Therefore the scope of the tests should be considerably extended to get more general outcomes.

In the near future, this research will be continued on the development of new forecasting strategies, issues of agent synergy and cooperation, as well as more trader-oriented methods of performance evaluation.

References

1. Bohm, V., Wenzelburger, J.: On the performance of efficient portfolios. J. Econ. Dyn. Control **29**(4), 721–740 (2005)
2. Chiarella, C., Dieci, R., Gardini, L.: Asset price and wealth dynamics in a financial market with heterogeneous agents. J. Econ. Dyn. Control **30**(9–10), 1755–1786 (2006)
3. Eiben, A.E., Smith, J.E.: Introduction to Evolutionary Computing. Springer, Berlin (2003)
4. Hernes, M., Sobieska-Karpińska, J.: Application of the consensus method in a multiagent financial decision support system. Inf. Syst. e-Bus. Manag. **14**(1), 167–185 (2016)
5. Ivanović, M., Vidaković, M., Budimac, Z., Mitrović, D.: A scalable distributed architecture for client and server-side software agents. Vietnam J. Comput. Sci. **4**(2), 127–137 (2017)
6. Jajuga, K., Jajuga, T.: Inwestycje: Instrumenty Finansowe, Ryzyko Finansowe, Inżynieria Finansowa. PWN, Warszawa (2000)
7. Khosravi, H., Shiri, M.E., Khosravi, H., Iranmanesh, E., Davoodi, A.: TACtic-a multi behavioral agent for trading agent competition. In: Sarbazi-Azad, H., Parhami, B., Miremadi, S.-G., Hessabi, S. (eds.) CSICC 2008. CCIS, vol. 6, pp. 811–815. Springer, Heidelberg (2008). doi:10.1007/978-3-540-89985-3_109
8. Korczak, J., Hernes, M., Bac M.: Risk avoiding strategy in multi-agent trading system. In: Proceedings of Federated Conference Computer Science and Information Systems (FedCSIS), Kraków, pp. 1131–1138 (2013)
9. Korczak, J., Hernes, M., Bac, M.: Fuzzy logic as agents' knowledge representation in A-Trader system. In: Ziemba, E. (ed.) Information Technology for Management. LNBIP, vol. 243, pp. 109–124. Springer, Cham (2016). doi:10.1007/978-3-319-30528-8_7
10. Lento, C.: A combined signal approach to technical analysis on the S&P 500. J. Bus. Econ. Res. **6**(8), 41–51 (2008)
11. Nguyen, N.T.: Using consensus methodology in processing inconsistency of knowledge. In: Last, M., et al. (eds.) Advances in Web Intelligence and Data Mining. SCI, pp. 161–170. Springer, Heidelberg (2006). doi:10.1007/3-540-33880-2_17

12. Sher, G.I.: Forex trading using geometry sensitive neural networks. In: Soule, T. (ed.) Proceedings of the 14th Annual Conference Companion on Genetic and Evolutionary Computation (GECCO 2012), pp. 1533–1534. ACM, New York (2012)
13. Sycara, K.P., Decker, K., Zeng, D.: Intelligent agents in portfolio management. In: Jennings, N., Wooldridge, M. (eds.) Agent Technology, pp. 267–282. Springer, New York (2002)
14. Tatikunta, R., Rahimi, S., Shrestha, P., Bjursel, J.: TrAgent: a multi-agent system for stock exchange. In: Proceedings of the 2006 IEEE/WIC/ACM International Conference on Web Intelligence and Intelligent Agent Technology (WI-IATW 2006), pp. 505–509. IEEE Computer Society, Washington, DC (2006)
15. Westerhoff, F.H.: Multiasset market dynamics. Macroecon. Dyn. **8**(2011), 596–616 (2011)

Social Networks and Recommender Systems

Testing the Acceptability of Social Support Agents in Online Communities

Lenin Medeiros(✉) and Tibor Bosse

Behavioural Informatics Group, Vrije Universiteit Amsterdam, De Boelelaan 1081, 1081 HV Amsterdam, Netherlands
{l.medeiros,t.bosse}@vu.nl

Abstract. This paper describes the first steps towards development and evaluation of an 'artificial friend', i.e., an intelligent agent that provides support via text messages in social media in order to alleviate the stress that users experience as a result of everyday problems. The agent consists of three main components: (1) a module that processes text messages based on text mining and classifies them into categories of problems, (2) a module that selects appropriate support strategies based on a validated psychological model of emotion regulation, and (3) a module that generates appropriate responses based on the output of the first two modules. The application has been tested in a pilot study involving 33 participants that were asked to interact with different variants of the agent via the social network Telegram. The results provide hints that the agent is appreciated over a baseline version that generates random support messages, but also point at some possibilities to further improve the agent.

Keywords: Social media · Empathic agents · Chatbots · Pilot study · Text mining · Emotion regulation

1 Introduction

Situations such as work deadlines, non-serious health issues, flight delays, broken relationships, loss of family members, etc. can be categorized as 'everyday problems', which many of us experience from time to time. They are known to be important sources of stress [1]. To help people to cope with everyday stress, *peer support* seems to be a promising means [5,8,13,16]. In our current society, one of the quickest and most frequently used approaches to provide peer support is to use online social networks like Twitter or Facebook [17], since this type of support only requires sending a short text message at appropriate moments. Indeed, as concluded in [14], sharing problems and showing affection are among the most common reasons why people use social media.

In spite of its strong potential, helpful peer support is not always available for users of social media, for the simple reason that some people have fewer friends than other. Moreover, even people who have many friends do not always want to

N.T. Nguyen et al. (Eds.): ICCCI 2017, Part I, LNAI 10448, pp. 125–136, 2017.
DOI: 10.1007/978-3-319-67074-4_13

share their problems online, particularly when their problems are very personal. Besides that, research has shown that people who deliver peer support are more vulnerable to developing stress-related complaints themselves [12].

To deal with the issues listed above, this paper is part of a project that explores the possibilities of *computer-generated peer support* via online social networks. More specifically, we introduce the concept of 'artificial friends' that have the ability to analyze text messages that people share via online social networks, and generate appropriate responses to these messages with the aim of helping them deal with their 'everyday problems'. The main source of inspiration for this vision is a number of promising recent initiatives in developing artificial agents that support human beings in similar domains [4,18]. We envision our system as an intelligent software agent or chatbot (possibly, but not necessarily embodied in the form of an avatar), which has the ability to analyze messages posted in social media, understand which messages potentially seek for peer support, and generate personalized response messages in order to reduce the user's experience of stress.

This paper presents a first prototype of such a support agent and aims to test how users would react to it. It builds upon a previous study in which the requirements for the system were elicited [11]. The current paper describes the algorithms used to make the agent process incoming messages and generate appropriate responses, and describes a preliminary evaluation of the application that was performed via a pilot study with 33 participants.

The remainder of this paper is structured as follows. Section 2 describes the background of the current research, including the results of a requirements analysis that was performed to obtain more insight in how the proposed support agent should behave. After that, Sect. 3 presents the support agent itself. Section 4 describes a pilot study that was conducted in order to evaluate the usability of the support agent, and Sect. 5 discusses the results of this study. Finally, Sect. 6 concludes the paper with a discussion.

2 Background

2.1 Related Work

Peer support, as defined by Kim et al. [10], has been considered as an effective method for promoting health, in particular in situations where people have to cope with stress [2,3,8]. Recently, the concept of peer support has been introduced within Artificial Intelligence, resulting in various types of intelligent virtual agents that provide support, varying from virtual depression therapists [4] to virtual buddies that help victims of cyberbullying [18]. Our project builds upon this research by bringing the idea of supportive virtual agents into the world of social media, resulting in a supportive agent (a chatbot) to help coping with 'everyday stress'.

Several other studies exist that involve chatbots. For instance, Gianvecchio et al. propose an approach to distinguish humans from chatbots [6], and Holgraves & Han use a chatbot as a tool to study online conversational behavior

[9]. A chatbot in the form of a virtual guide for tourists in heritage tours is put forward in [15]. This work can be seen as complimentary to our research since our long-term goal is to come up with a chatbot that acts as a human user providing social support (and not only practical information) to others to make them feel better understood.

2.2 Requirements Analysis

Before starting to develop the support agent as described in the introduction, we were interested in: (1) identifying the most common types of stressful situations shared by people via social networks and (2) determining the strategies used by users to support stressed friends in these situations.

Table 1. Types of shared stressful situations and respective examples.

Type	Example
Relationship	"I miss my boyfriend"
Work	"I'm working too much"
Death	"My aunt passed away"
Financial	"I have so many bills to pay"
Disease	"I'm ill"
Exams	"I didn't pass my exam"
Other	"My dog is crying"

To accomplish this, we performed a number of tasks (see [11] for details). First, we collected data provided by participants through an anonymous survey in order to check what kinds of stressful situations they had shared with their friends via social networks. This resulted in 7 categories, which are shown in Table 1, along with an illustrative example for each type of stressful situation.

Next, we also asked the participants what kind of support they received in response to sharing their problems. The results were clustered according to 6 types of 'support strategies', which are based on Gross' theory on emotion regulation [7]. This is a commonly accepted theory in the literature, which has been proposed to be applicable to self-regulation as well as interpersonal emotion regulation [19]. The following six strategies are used: *situation selection (s.s.)* (e.g., avoiding undesired stressful situations), *situation modification (s.m.)* (e.g., changing a given stressful situation), *attentional deployment (a.d.)* (e.g., stopping thinking about a situation), *cognitive change (c.c.)* (e.g., looking at the 'bright side' of a situation) and *general emotional support (g.e.s.)* (e.g., simply showing empathy, love, trust and/or caring). Note that the first five strategies are taken directly from Gross. The sixth strategy was added because the data indicated that people sometimes use a more general type of emotional support.

Table 2. Frequency table for all types of support identified. The agent uses these data to select its support strategy.

	g.e.s	c.c	a.d	s.s	s.m
Relationship	29%	21%	18%	14%	18%
Work	17%	44%	17%	0%	22%
Death	53%	27%	0%	0%	20%
Financial	40%	40%	0%	0%	20%
Disease	11%	56%	22%	0%	11%
Exams	0%	25%	0%	0%	75%
Other	56%	22%	0%	11%	11%

Based on this analysis, it was possible to relate the various types of problems shared via social networks to the various types of support given for these problems. The results are shown in Table 2, and will be used in Sect. 4 to develop our support agent. The most important requirements resulting from this phase are the following:

1. The most typical stressful situation shared by people via their social networks concerns their own relationships.
2. In case people share stressful situations about relationships or death, the most frequently used support strategy is general emotional support.
3. In case people share stressful situations about work or diseases, the most frequently used support strategy is cognitive change.

3 Support Agent

A prototype of the support agent was developed using Python, in the form of a bot for the Telegram Messenger App, which provides a public API. We also used MongoDB to manage the data generated by our bot. In the pilot study, various versions of the bot have been used for different experimental conditions (this will be explained in Sect. 5), but in the current section we only discuss the specification concerning the complete version of the bot.

The algorithm behind our agent is based on the results obtained from the study reported in Sect. 2 (see Table 2). The high level workflow of the application is described in the following (note that the respective code is available on GitHub via https://github.com/leninmedeiros/Stress-Support-Bot and see also Fig. 1):

1. A given user sends a message to the bot (we are assuming that any message sent by a given user is a description a of stressful situation);
2. To process the incoming message, the bot will first identify the type of stressful situation shared by the user. To this end, it uses sets of key words (for the different categories shown in Table 2). Such bags of terms (also available in the GitHub repository) were designed based on the data obtained from the

Algorithm 1. Processing incoming messages from a stressed user.

```
function ProcessIncomingMessages(m1)
    situation ← ClassifyTheStressfulSituation(m1)
    strategy ← SelectStrategy(situation)
    m2 ← ConstructResponse(situation, strategy)
    return m2

function ConstructResponse(situation, strategy)
    response ← SelectTemplate(strategy)
    response ← SetTemplateForSituation(situation, response)
    return response
```

requirements analysis and the most common synonyms of these words. The current version can only deal with English words;

3. After classifying the type of the stressful situation, the bot will select the proper support strategy. This decision is made based on the data obtained previously as well. For example, as shown in Table 2), for 29% of the cases when people share problems about a relationship, the support strategy used is 'general emotional support'. These percentages are used as probabilities in our application in order to select a support strategy;
4. Finally, after having both the problem and the support strategy identified, the bot will send a support message back to the user. To construct a message, first a template message is randomly selected from a list of templates that match the support strategy (again, these templates were developed based on the data collected previously). After that, the template is filled in with the appropriate terms to refer to the stressful situation.

4 Pilot Study

In a first attempt to evaluate the usefulness of the developed support agent, a pilot study was designed in which participants were asked to interact with (different variants of) the bot and provide feedback by means of a questionnaire. The details of this study are discussed below.

4.1 Participants

A sample of 33 participants was recruited (19 male and 14 female; age between 21 and 30 years). All participants were friends and acquaintances of the authors, and most of them had an academic background.

4.2 Agent Variants

Before the start of the pilot study, four variants of the support agent were created (1 bot per variant was launched). Bots 1 and 2 used our algorithm to generate

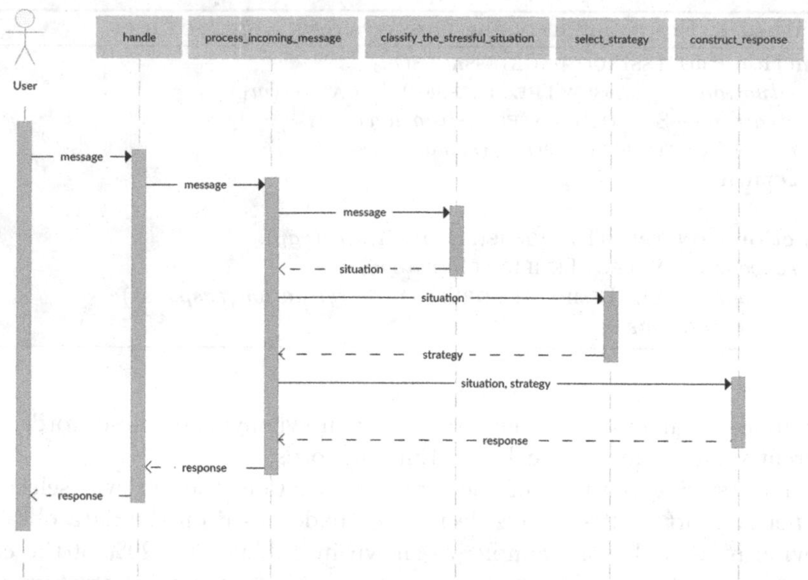

Fig. 1. Structure of the developed bot represented by a sequence diagram. This diagram shows the most important functions of the system with their respective inputs and outputs.

a support message, whereas bots 3 and 4 selected a random support message. Moreover, bots 1 and 3 explicitly mentioned the name of the stressful situations within the support message, whilst bots 2 and 4 did not.

To illustrate the difference between the four variants, assume a user would enter the message 'My cat died'. For this sentence, bot 1 and 2 would use the algorithm described in the previous section to select a suitable response strategy, which would most likely be 'general emotional support'. Instead, bot 3 and 4 would randomly select a strategy. Furthermore, bot 1 and 3 would use the term 'death' (or 'tragedy') in its response message (assuming that the incoming message was classified correctly), whereas bot 2 and 4 would not. So, examples of response messages that could be generated by the four different bots are as follows:

1. "I'm really sorry to hear about this tragedy, but keep in mind that there are good people around you that care about you!"
2. "I'm really sorry to hear about this situation, but keep in mind that there are good people around you that care about you!"
3. "I'm really sorry to hear about this tragedy, but perhaps you should just ignore it and focus on positive things!"
4. "I'm really sorry to hear about this situation, but perhaps you should just ignore it and focus on positive things!"

These four variants were created because we were interested in whether mentioning the name of the identified stressful situation has an effect on how users perceive the bot and, also, whether our algorithm for selecting a support strategy has an effect, compared to randomly selecting a strategy.

4.3 Design

To make effective use of the (relatively low) number of participants, a mixed design was used. To this end, participants were randomly allocated to one of four conditions: (A) first bot 1, then bot 3; (B) first bot 3, then bot 1; (C) first bot 2, then bot 4 and (D) first bot 4, then bot 2.

Hence, each participant was asked to interact with two variants of the bots (randomly selected). For instance, participants in group A first interacted with bot 1, and after that they interacted with bot 3. This design allowed us to compare bot 1 and 3 separately (by using data from condition A and B in a within subjects comparison), as well as to compare bot 2 and 4 (by using data from condition C and D). In this way, the effect on the strategy selection algorithm could be investigated. By keeping the amount of participants per group balanced, an ordering effect was avoided. Moreover, by comparing data from condition A and B with data from condition C and D (in a between subjects comparison), the effect of mentioning the name of the stressful situation could be studied.

4.4 Task and Procedure

Participants were invited by e-mail to take part in our study. Upon accepting the invitation, they were asked to open a URL, which displayed a page with the following instructions:

1. If you already use the Telegram Messenger and have an account, please go to step 3;
2. Download the Telegram Messenger for Android, iOS, etc. You can easily check how to do this by going to https://telegram.org/;
3. After having the Telegram Messenger app running in one of your devices, search for the contact 'stress support one bot';
4. Think about 10 stressful situations that you are facing, have faced in the past, or you believe anyone else could face (like "my aunt died", "I fought with my girlfriend", "I have so many bills to pay", "I'm ill", "my pet needs to go to a vet", "I have lots of exams next week", etc.). Put these situations in 10 different messages and send them to the bot (type a given situation, press enter, wait for the bot's response and continue);
5. After finishing step 4, search for the contact 'stress support two bot';
6. Repeat step 4 but in this case you have to send the messages to the bot 'stress support two bot';
7. We will not share any information about you and the entire procedure of talking to the bots. Taking part in the experiment usually lasts no more than 20 min. After that, you can start answering our survey.

The survey was developed using Qualtrics[1], and consisted of the following four statements (named S1–S4), for which users had to indicate to what extent they agreed on a Likert-scale from 1 (strongly disagree) to 7 (strongly agree):

1. It felt like the bot understood the problems I shared (S1)
2. The support messages sent by the bot were appropriate (S2)
3. The support messages sent by the bot were helpful (S3)
4. The support messages sent by the bot could have been sent by a real human as well (S4)

4.5 Variables

The participants' ratings for the four statements presented in the previous section were the main dependent variables of the study. The type of bot (as explained previously) was the independent variable. We hypothesized that using the developed algorithm to select the support strategy would make the bot's behavior appear more human-like. Hence we formulated the following hypothesis:

Hypothesis 1. *The ratings for statement S4 will be significantly higher for bot 1 than for bot 3, and higher for bot 2 than for bot 4.*

In addition, we expected that including the name of the stressful situation would enhance people's feeling that the bot understands their problems. For this reason we formulated another hypothesis:

Hypothesis 2. *The ratings for statement S1 will be significantly higher for bot 1 and 3 than for bot 2 and 4.*

Statement S2 and S3 were included to get a general idea about the usefulness of the support agent, and the possibilities to apply it in practice. However, no explicit hypotheses were formulated regarding these statements.

Finally, to gain more insight in the agent's ability to classify the incoming messages into the different categories (i.e., independent of the user's perception of the agents), the agent's *accuracy* is investigated. This accuracy is defined as the number of correctly classified messages divided by the total number of messages.

5 Results

Paired t-tests were performed to check if there were any significant differences between the ratings given for the different conditions mentioned in Sect. 4. In particular, the ratings given for bot 1 were compared to those given for bot 3, and the ratings given for bot 2 were compared to those given for bot 4.

[1] https://www.qualtrics.com/.

When comparing bot 1 to bot 3, the average ratings for the four statements were: S1: 3.7 vs. 3.8 ($p = 0.39$), S2: 4.8 vs. 4.3 ($p = 0.07$), S3: 4.0 vs. 4.0 ($p = 0.5$), and S4: 4.8 vs. 4.6 ($p = 0.17$). Hence, we have not found any conclusive evidence for a difference between those two bots for any of the four characterstics that were tested, although for statement S2 bot 1 received substantially higher ratings.

When comparing bot 2 to bot 4, the average ratings for the four statements were: S1: 3.3 vs. 3.2 ($p = 0.35$), S2: 4.1 vs. 4.1 ($p = 0.50$), S3: 4.1 vs. 4.0 ($p = 0.34$), and S4: 4.3 vs. 3.9 ($p < 0.05$). From this, it may be concluded that participants rated the support messages generated by bot 2 (using our model) as more likely to be sent by a real human (= statement S4) than the messages generated by bot 4 (which were generated randomly).

Consequently, we can conclude that Hypothesis 1 is partly fulfilled: the ratings for statement S4 were significantly higher for bot 2 than for bot 4, but were not significantly higher for bot 1 than for bot 3.

Next, in order to isolate the effect of stating or not stating the name of the stressful situation in the support messages, we compared the ratings for bots 1 and 3 (as one group) with the ratings for bots 2 and 4 (as another group), using unpaired t-tests. In this case, the respective p-values found for the four statements were: 0.08, 0.11, 0.45 and 0.08. Unfortunately, these values do not allow us to conclude that there are any statistical differences in the ratings for the bots that mentioned the stressful situation explicitly compared to the ones that did not. Hence, also Hypothesis 2 cannot be confirmed. Nevertheless, although not significant, the ratings for statement S1 were higher for bot 1 and 3 than for bot 2 and 4 ($p = 0.08$). This is still an encouraging finding, which provides opportunities for further exploring this research direction in the future.

Additionally, we (manually) calculated the percentage of the messages for which the bot correctly classified its type (i.e., relationship, work, etc.). This turned out to be the case in 81.4% of the cases. Given the fact that we did not put much effort yet into optimizing the classification module (using a simple bag-of-words approach), we believe that this is another promising result.

Finally, the participants provided us some general impressions and feedbacks about the idea to have an 'artificial friend' in their social network. Unanimously, they pointed out that being able to have an elaborated dialogue with the users is crucial for the bot. Some of them believed that the size and the type of the support messages should depend on the size of the messages sent by the users. Most of them would use such a bot if they were also seeking for practical advices (for instance, a stranded passenger in an airport gate who got upset and wants to calm down as well as to find a gate).

6 Discussion

This paper describes the first steps towards the development and evaluation of a so-called 'artificial friend', i.e., an intelligent agent that provides support via text messages in social media in order to alleviate the stress that users experience as a result of 'everyday problems' [1]. The assumption underlying this research is that such an agent can help reducing the stress people experience in these situations by generating tailored response messages, and that this is particularly helpful in cases where users do not receive comforting responses from their human peers.

The presented agent consists of three main components, namely (1) a module that processes text messages based on text mining and classifies them into categories of problems, (2) a module that selects appropriate support strategies based on a validated psychological model of emotion regulation, and (3) a module that generates appropriate responses based on the output of the first two modules.

The application has been tested in a pilot study involving 33 participants that were asked to interact with different variants of the agent via the social network Telegram, and provide feedback by means of an online questionnaire. An analysis of the results leads to a number of conclusions. First, participants were generally positive about the application, in the sense that they provided average to above-average ratings for statements about the abilities of the bot. Second, the module that classifies incoming messages was evaluated positively, in the sense that it performs with an average accuracy of 81.4%. Third, the application's ability to select appropriate support strategies was deemed promising, as the hypothesis that this ability has a positive impact on users' experience (Hypothesis 1) was partly confirmed. Fourth, regarding the application's ability to name the experienced stressful situation explicitly, no conclusive effect on user experience was found (Hypothesis 2), but due to the relatively low power of the study it is advisable to investigate this issue in more detail in the future. Also, it is useful to emphasize that even the most minimal baseline condition used in the pilot study (bot 4) already possesses some intelligence, as the messages generated by this bot are constructed on the basis of insights from the earlier investigation.

Additionally, it should be noted that the accuracy of 81.4% in classifying stressful situations, although promising, is still too low to be of much use in a real world application. However, improving this accuracy will probably be not too difficult by using more sophisticated machine learning techniques (e.g., using a Vector Space Model to create a representation of the text in the messages). While doing this, also additional categories could be introduced, in order to reduce the percentage of messages that end up in the 'other' category. Along similar lines, a future version of the application might use of context information such as a history of the problems shared before. Such improvements will probably enhance users' experience of the bot, hence increasing the chances to find a significant effect on the independent variables.

Finally, in future work we intend to explore various possibilities to extend the functionality of the support agent. For example, instead of only addressing simple interactions consisting of one incoming message and one response message, it would be interesting to tackle more complex types of interaction. This could eventually result in entire human-agent conversations, as described in [18]. Another idea for follow-up research is to place the bot within an online group of people that are mutually helping each other to cope with stress. This would extend its scope from one-to-one settings to group settings, thereby further broadening the potential impact of this promising type of technology.

Acknowledgments. Lenin Medeiros' stay at Vrije Universtiteit Amsterdam was funded by the Brazilian Science without Borders program. This work was realized with the support from CNPq, National Council for Scientific and Technological Development - Brazil, through a scholarship with reference number 235134/2014-7.

References

1. Burks, N., Martin, B.: Everyday problems and life change events: Ongoing versus acute sources of stress. J. Hum. Stress **11**(1), 27–35 (1985)
2. Cobb, S.: Social support as a moderator of life stress. Psychosom. Med. **38**(5), 300–314 (1976)
3. Sheldon Cohen and Thomas: A Wills. Stress, social support, and the buffering hypothesis. Psychol. Bull. **98**(2), 310 (1985)
4. DeVault, D., Artstein, R., Benn, G., Dey, T., Fast, E., Gainer, A., Georgila, K., Gratch, J., Hartholt, A., Lhommet, M. et al.: Simsensei kiosk: a virtual human interviewer for healthcare decision support. In: Proceedings of AAMAS 2014, pp. 1061–1068. IFAAMAS (2014)
5. Eysenbach, G., Powell, J., Englesakis, M., Rizo, C., Stern, A.: Health related virtual communities and electronic support groups: systematic review of the effects of online peer to peer interactions. BMJ **328** (2004)
6. Gianvecchio, S., Xie, M., Wu, Z., Wang, H.: Measurement and classification of humans and bots in internet chat. In: USENIX Security Symposium, pp. 155–170 (2008)
7. James, J.: Gross. Emotion regulation: Affective, cognitive, and social consequences. Psychophysiology **39**(3), 281–291 (2002)
8. Heaney, C.A., Israel, B.A.: Social networks and social support. Health Behav. Health Educ. Theory Res. Pract. **4**, 189–210 (2008)
9. Holtgraves, T., Han, T.-L.: A procedure for studying online conversational processing using a chat bot. Behav. Res. Methods **39**(1), 156–163 (2007)
10. Kim, H.S., Sherman, D.K., Taylor, S.E.: Culture and social support. Am. Psychol. **63**(6), 518 (2008)
11. Medeiros, L., Bosse, T.: Empirical analysis of social support provided via social media. In: Spiro, E., Ahn, Y.-Y. (eds.) SocInfo 2016. LNCS, vol. 10047, pp. 439–453. Springer, Cham (2016). doi:10.1007/978-3-319-47874-6_30
12. Medeiros, L., Sikkes, R., Treur, J.: Modelling a mutual support network for coping with stress. In: Nguyen, N.-T., Manolopoulos, Y., Iliadis, L., Trawiński, B. (eds.) ICCCI 2016. LNCS, vol. 9875, pp. 64–77. Springer, Cham (2016). doi:10.1007/978-3-319-45243-2_6
13. O'Dea, B., Campbell, A.: Healthy connections: online social networks and their potential for peer support. In: Health Informatics: The Transformative Power of Innovation - Selected Papers from the 19th Australian National Health Informatics Conference, HIC 2011, Brisbane, Australia, 1–4 August 2011, pp. 133–140. IOS Press (2011)
14. Quan-Haase, A., Young, A.L.: Uses and gratifications of social media: a comparison of facebook and instant messaging. Bull. Sci. Technol. Soc. **30**(5), 350–361 (2010)

15. Santangelo, A., Augello, A., Gentile, A., Pilato, G., Gaglio, S.: A chat-bot based multimodal virtual guide for cultural heritage tours. In: PSC, pp. 114–120 (2006)
16. Takahashi, Y., Uchida, C., Miyaki, K., Sakai, M., Shimbo, T., Nakayama, T.: Potential benefits and harms of a peer support social network service on the internet for people with depressive tendencies: qualitative content analysis and social network analysis. J. Med. Internet Res. **11**(3) (2009)
17. Breda, W., Treur, J., Wissen, A.: Analysis and support of lifestyle via emotions using social media. In: Aberer, K., Flache, A., Jager, W., Liu, L., Tang, J., Guéret, C. (eds.) SocInfo 2012. LNCS, vol. 7710, pp. 275–291. Springer, Heidelberg (2012). doi:10.1007/978-3-642-35386-4_21
18. Van der Zwaan, J.M., Dignum, V., Jonker, C.M.: A conversation model enabling intelligent agents to give emotional support. In: Ding, W., Jiang, H., Ali, M., Li, M. (eds.) Modern Advances in Intelligent Systems and Tools. SCI, vol. 431, pp. 47–52. Springer, Heidelberg (2012). doi:10.1007/978-3-642-30732-4_6
19. Williams, M.: Building genuine trust through interpersonal emotion management: a threat regulation model of trust and collaboration across boundaries. Acad. Manag. Rev. **32**(2), 595–621 (2007)

Enhancing New User Cold-Start Based on Decision Trees Active Learning by Using Past Warm-Users Predictions

Manuel Pozo[1,2(✉)], Raja Chiky[1(✉)], Farid Meziane[3(✉)], and Elisabeth Métais[4(✉)]

[1] Institut Supérieur d'Eléctronique de Paris, LISITE Lab, 28, rue Notre-Dame-des-Champs, 75006 Paris, France
{manuel.pozo,raja.chiky}@isep.fr

[2] Blackpills, Paris, France
mp@blackpills.com

[3] Informatics Research Institute, University of Salford, Salford M5 4WT, UK
f.meziane@salford.ac.uk

[4] CEDRIC Lab, CNAM, 292 rue Saint Martin, Paris, France
metais@cnam.fr

Abstract. The cold-start is the situation in which the recommender system has no or not enough information about the (new) users/items, i.e. their ratings/feedback; hence, the recommendations are not accurate. Active learning techniques for recommender systems propose to interact with new users by asking them to rate sequentially a few items while the system tries to detect her preferences. This bootstraps recommender systems and alleviate the new user cold-start. Compared to current state of the art, the presented approach takes into account the users' ratings predictions in addition to the available users' ratings. The experimentation shows that our approach achieves better performance in terms of precision and limits the number of questions asked to the users.

Keywords: Active learning for recommender systems · Cold-start problem · New users problem · Decision trees

1 Introduction

The new user cold start is the situation where a recommender system cannot generate personalized recommendations for a new user because it has not learnt yet his preferences. This issue is commonly encountered in collaborative filtering recommendations as they rely mainly on the users' feedback to predict future users' interests [1]. In addition, new users start evaluating the system from their first usage [2]. This is a challenge for both academia and industry because the recommendations' accuracy is directly related to the users' satisfaction and fidelity [3].

The techniques used to alleviate the new user cold start can be categorized into passive learning and active learning. Passive collaborative filtering techniques [1] learn from sporadic users' ratings; hence learning new users preferences is slow [4]. Active techniques interact with the new user in order to retrieve

N.T. Nguyen et al. (Eds.): ICCCI 2017, Part I, LNAI 10448, pp. 137–147, 2017.
DOI: 10.1007/978-3-319-67074-4_14

a bunch of ratings that allow them to learn the user's preferences. We focus on active learning techniques for collaborative filtering because they quickly and accurately bootstrap the generation of recommendations for users. In addition, collaborative filtering only requires users preferences; analyzing users' ratings from (new) users can achieve better recommendation in cold-start than exploiting other users' attributes (e.g. age, genre) [5].

A naive active learning approach is to question users about their interests and get their answers [6]. Such questions may include: 'Do you like this movie?' with possible answers such as: 'Yes, I do'; 'No, I do not'; 'I have not seen it'. In this context the questions are items and the answers are the users' preferences to these items. However, users are not willing to answer many questions [3,7]. As a consequence, the main goal is to present short (a maximum of 5–7 questions [2]), but very informative questionnaires. Active learning creates personalized questionnaires which leads to a progressive understanding of the user's preferences. In fact, the personalization of the questionnaires is close to a recommender system concept, although the latter seeks the items the user likes and the former seeks the items the user recognizes.

Our contribution relies on an active learning technique based on decision trees that exploits both available users' ratings and warm users' ratings predictions in order to improve the questionnaire. The experimentation shows that our approach enhances previously suggested ones in terms of accuracy and in a smaller number of questions.

This paper is organized as follows: Sect. 2 presents the state of the art for active learning using decision trees techniques. Section 3 presents our contribution to enhance active learning based on past warm users' rating predictions. Section 4 shows the experimentations performed and the results of our approach. Finally we conclude and present our future works in Sect. 5.

2 Related Work

In this research, we focus on active learning techniques in the domain of collaborative filtering recommendations, particularly those using decision trees because: (1) the sequential question paradigm allows a personalization of the questionnaire, and (2) they aim to well profile a new user by posing as less questions as possible. Other techniques, such as Entropy0, Logarithmic Popularity Entropy (LPE) and Harmonic Entropy Logarithmic Frequency (HELF), are not discussed [6]. Specifically, we do not mention passive learning techniques [8–10], content-based techniques [11] or other hybrid techniques [12–14]. For more information about these approaches, the readers can refer to [3,4,7].

Recent researches focus on user partitioning techniques that allow to group users of similar tastes into clusters or nodes, and then find out to which group the new user belongs to. In [15] authors use clustering techniques to find the correct users neighbors that match the new user with other users' profiles. This makes it easier to generate recommendations in cold-start. In [2] the authors use non supervised ternary decision trees to model the questionnaire. The decision trees are built off-line to be completely available for new users that receive the questions sequentially. To move to a new question they answer the current one by clicking on one of the three possible answers ('like', 'hate' and 'unknown'). The users' answers lead to a different child node of the trees. This creates a personalized tree path that depends on the past users' answers. On the other hand, this technique uses a collaborative filtering approach to choose questions. Using available users' ratings, they seek the best discriminative item in order to split the population of users into three nodes (users who liked, those who hated and those who do not know this item). The best item is the one which minimizes a statistical error within the users' ratings of the node.

In [16] authors suggest to apply matrix factorization while building every node of a decision tree, yet this is computationally expensive. In [17] the authors proposed to "learn" the active learning technique. They assumed that warm users can be thought of as new users from whom some ratings are known. Thus, this is seen as a supervised decision trees which internally reduces the accuracy of the technique by picking the best discriminative items. Moreover, they split the tree nodes into six, a 1–5 natural scale rating and an unknown node.

The approach in this paper uses decision trees as in [2,17], and it picks better discriminative items and hence better bootstraps the new users' preferences.

3 Contribution

Supervised and non supervised active learning decision trees aim to find out the best discriminative items for every node of the tree in order to better capture the new users preferences.

Formally, let R be the available ratings. The rating of a user u in an item i is defined by $r_{u,i} \in R$. In addition, let t be a node in the decision trees. We define U_t, I_t, and R_t as the set of users, items and ratings currently in the node t. Furthermore, $R_t(u)$ and $R_t(i)$ are ratings of the user u and item i in the node t. Given the current node t, these techniques iterate over all candidate items $i \in I_t$ by analyzing users' ratings on i. The users populations are then grouped into users' who rated item i and users who did not. Typically, the latter is more populated due to the sparse nature of the available dataset. Furthermore, the users who rated item i can be grouped into further categorizations, e.g. users who liked/hated, or who rated '1, 2, 3, 4, 5'. Then, the population of users in these nodes, and their ratings, are used to evaluate the performance of choosing i as one discriminative item of R_t.

Our contribution exploits the predictions over the existing R. Thus, we define P as the predicted set of R, so that for each $r_{u,i} \in R$ there is a prediction $p_{u,i} \in P$. The set P is computed by using collaborative filtering techniques, e.g. matrix factorization. Highlight that the number of users, items, and entries in R and P are the same. Finally, P_t is the set of predictions currently in the node t, and $P_t(u)$ and $P_t(i)$ are the set of users and items predictions in the node t.

Current decision trees techniques exploit only the available ratings in R in order to (1) find the discriminative items, (2) split the users' population, and (3) compute predictions over the candidate items. These techniques use a simple item prediction method based on the "item rating average" in order to evaluate a prediction accuracy and to compute prediction labels for candidate items. Note that this technique is fast and accurate in large datasets, which allows a quicker generation of predictions from the available ratings in R_t. On the other hand, using more accurate prediction techniques is possible but (1) it can be very expensive and time consuming to do it for every node of the tree, and (2) the predictions needed in decision trees are item-oriented regardless of the user (the same prediction value to any user).

We propose to change this paradigm by using more accurate predictions over the available ratings R. The main idea is to introduce the prediction P as a new source. Hence, R and P are available from the root node of the tree. Then, when the node is split into child nodes, R_t is split into $R_{t-child}$. As long as we want to preserve that for every rating $r_{u,i} \in R_t$ there is an associated prediction $p_{u,i} \in P_t$, for every node t, we split P_t into $P_{t-child}$ as well. In addition, we propose to use the available ratings in R only to split the users population, and P to find out the best discriminative items to enhance the prediction label of candidate items.

This makes sense since finding discriminative items and label predictions are associated with computing an error. As long as P is built by using more accurate methods than the "item rating average", this error is minimized efficiently. We propose using efficient algorithms, such as matrix factorization [18]. The main drawback of using matrix factorization is that it computes different item predictions for different users. The decision trees require a unique item prediction value to be applied to any user. In [2,17] the authors use the "item rating average" within R_t. We suggest using a similar method, with a major difference that is computing the "item prediction average", which is indeed the average of the predictions within P_t.

In fact, collaborative filtering methods are very accurate for recommending items to users by replicating the users' rating behavior. As a consequence, they are good as well in guessing the average prediction of users, items, and in general the average rating value of the dataset. Figure 3 supports this statement. In addition, this is true as well for the "item prediction average". Figure 2 develops this by considering different group of users split by quantity of ratings. We observe that "item prediction average" based on matrix factorization predictions (MF-Avg) are close to the item rating average predictions (Item-Avg), while as normal the matrix factorization (MF) outperforms these predictions.

3.1 Apply Warm Predictions to Decision Trees Algorithms

The difference between supervised and non-supervised approaches is that the former considers that some users' ratings can be used to validate the technique. As a consequence, these ratings can be used as a validation set to evaluate the accuracy of the tree node. On the contrary, since non supervised techniques do not have any validation, they compute a statistical error based on the available ratings in the node. Nevertheless, in both approaches a validation is not possible in the 'unknown' nodes, since by definition, there is no rating label for these

Statistic	Movielens	MF
1st Quartile	3.00	3.13
Median	4.00	3.58
Mean	3.51	3.51
3rd Quartile	4.00	3.96

Fig. 1. Statistics for available ratings and matrix factorization (MF) predictions of Movielens 10M [19].

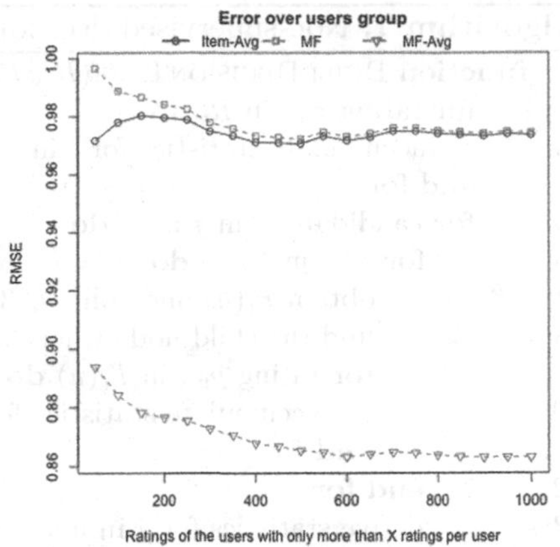

Fig. 2. Prediction techniques and average comparisons regarding the RMSE for Movielens 10M [19].

users to this item. As a consequence, a statistical error is mandatory in this case. Our approach uses similar statistics as [2] (Fig. 1).

Non Supervised Approach. In [2], the authors define a set of statistics and an internal error using these statistics to find out the best discriminative item. In this approach, the best item is the one which reduces this error. In addition, as long as the tree nodes contain many ratings, they use the item rating average method to compute item label predictions for items.

In our approach we use the same statistics to compute the same error, with two major differences. First, the available ratings are only used to split the population of users. As a consequence, the statistics and the items predictions are computed by using the proposed set of predictions P. Second, once a discriminative item is chosen in a parent node it does not pass to the child nodes. This is done for two reasons: (1) to avoid to choose the same item, and hence, to avoid to pose twice the same question to the same user, and (2) to delete the influence of the items' ratings in the child nodes. In fact, one can avoid choosing an item without deleting their ratings as done in [2]. This approach is described in Algorithm 1.

Supervised Approach. In [17], the authors suggested using warm-users as cold-users from whom some interests are known. This assumption allows to create a supervised decision trees where some labels are known for validation purposes.

We suggest again to use the predictions P over the available ratings in R in order to enhance this technique. We make use of R to split the users' population, meanwhile P is used to (1) validate the approach, and (2) obtain items label for

Algorithm 1. Non-supervised decision tree algorithm

```
function BuildDecisionTree(R_t, P_t, currentTreeLevel)
    for rating r_{u,i} in R_t do
        accumulate statistics for i in node t using p_{u,i}
    end for
    for candidate item j in I_t do
        for r_{u,j} in R_t(j) do
            obtain P_t(u) and split U_t 3 child nodes based on j
            find the child node where u has moved into
            for rating p_{u,i} in P_t(u) do
                accumulate statistics for i in node t - child using p_{u,i}
            end for
        end for
        derive statistics for j in node tU from the tL and tD statistics
        candidate error: e_t(j) = e_{tL}(j) + e_{tD}(j) + e_{tU}(j)
    end for
    candidate item i* = argmin_i
    compute p_{i*} by using item prediction average
    if currentTreeLevel ¡ maxTreeLevel then
        create 3 child nodes U_{t-child} based on i* ratings
        for child in child nodes do
            exclude i* from R_{t-child}
            BuildDecisionTree( R_{t-child}, P_{t-child}, currentTreeLevel +1 )
        end for
    end if
    return i*
end function
```

the chosen discriminative items. The validation requires the items predictions, which in [17] is given by the item rating average within the child node. As long as P contains the predicted values $p_{u,i}$, this validation is more accurate. In addition, the item prediction average over P is also used to obtain a prediction, and we split the nodes into 3 child nodes ('like', 'dislike', 'unknown') rather than 6. Algorithm 2 shows this approach.

3.2 Complexity of the algorithm

The complexity of our approaches for non-supervised decision trees and supervised decision trees is very similar to [2,17]. In fact, these algorithms follow a similar procedure. The complexity of splitting the users in node t is $O(\sum_{u \in U_t} |R_t(u)|^2)$, and thus, for all the nodes in the same level we use $O(\sum_{u \in U} |R(u)|^2)$. As a consequence, the complexity to build a tree of N questions is $O(N \sum_{u \in U} |R(u)|^2)$. In fact, adding the prediction set P does not affect the complexity of the algorithms, although, the memory footprint of the approaches may vary according to their implementations. Considering that rating and prediction datasets are coded equally, our approach consumes double of the memory size to store the set P.

Algorithm 2. Supervised decision tree algorithm

```
function BuildDecisionTree(U_t, R_{t-train}, R_{t-validation}, P_t, currentTreeLevel)
    for user u ∈ U_t do
        compute RMSE_u^1 on R_{t-validation}(u) and P_t(u)
    end for
    for candidate item j from R_{t-train} do
        split U_t 3 child nodes based on j
        for user u ∈ U_t do
            find the child node where u has moved into
            compute RMSE_u^2 on R_{t-validation}(u) and P_t(u)
            Δ_{u,i} = RMSE_u^1 − RMSE_u^2
        end for
    end for
    δ = aggregate all Δ_{u,i}; and pick candidate item i* = argmax_i δ_i
    compute p_{i*} by using item prediction average
    if currentTreeLevel < maxTreeLevel and Δ_{i*} ≥ 0 then
        create 3 child nodes U_{t-child} based on based on i* ratings
        for child in child nodes do
            exclude i* from R_{t-child}
            BuildDecisionTree( U_{t-child}, R_{t-child-train}, R_{t-child-validation},
        P_{t-child}, currentTreeLevel +1 )
        end for
    end if
    return i*
end function
```

4 Experimentation

The goal of our experimentation is two-folds (i) to present the behaviour of current techniques in smaller datasets and (ii) to show the performance of our presented approach. Recent techniques have presented their results using Netflix dataset. However, this dataset is no longer available for research. Hence, we use the Movielens 10M dataset [19], which contains 71567 users, 10681 items and 10 million ratings. Since our approach considers external techniques prediction as a new source, in order to build our decision trees we use matrix factorization [18] due to its accuracy. We compare our approach in non supervised decision trees, as in [2], and in supervised decision trees, as in [17].

In order to compare the approaches we use the RMSE metric oriented to users, which measures the squared difference between the real ratings and the predicted ratings:

$$RMSE_u = \sqrt{\frac{1}{N}\sum (r_{u,i} - p_i)^2} \tag{1}$$

where N is the number of ratings of the user u, p_i is the predicted label value of the candidate item in the question node and $r_{u,i}$ is the real rating of the user u for the item i. Hence, the evaluation of the error in one question is the average of the users error in this question number. As a consequence, for this metric the lower is the better.

The experimentation carried out in [2] splits the datasets into 90% training set, D_{train} and 10% test set, D_{test}. However, this is not a real cold-start context since the same user may appear in both training and test set. We suggest a real cold-start situation. We split the set of users in the datasets into 90% training set, U_{train} and 10% test set, U_{test}. Hence, the users in the training set help to build the decision trees and the users in the test set are considered as new user to evaluate the performance of the approach.

The process we have followed to run this experimentation is as follows. First we split the dataset into U_{train} and U_{test}. Second, we compute the collaborative filtering algorithms over ratings R in U_{train} and we extract the associated predictions P. Third, we train the approach of Golbandi by using U_{train}. Our approach is trained by using both ratings in training set R and the prediction of the training set P. Finally, the performance of the decision trees is evaluated by using the test set U_{test}. The users in this set are used to answer the questions. If the item is known, we compute the RMSE associated to this answer and this question. Then, the user answer a new question. At the end, we compute the average of the accumulated nodes RMSE.

Knowing that the experimentation may depend on the split of the dataset, we run it 50 times and then used the mean value of the RMSE. We use this process to evaluate the performance for the MovieLens 1M and MovieLens 10M. Figure 3(a) shows the results (the mean values and tendency curves) of this experimentation for both MovieLens datasets, where 'Golbandi' is the approach used in [2]. On the one hand, our approach achieves a much lower error in less number of questions. This matches with the needs of active learning; short but very informative questionnaires. This is possible due to the higher accuracy of the matrix factorization. On the the other hand, all the approaches tend to converge into a pseudo-asymptotic behavior. This is due to the fact that nodes in the bottom of the tree (nodes in 8th question) are less populated by users and thus predictions and profiles are less accurate. This particularity is not shown in [2,17] due to their very large dataset.

We perform a similar experiementation to compare our approach to [17]. This time the authors use a 4-fold set to evaluate their dataset: D_{train} set which is split into U_{train} and $U_{validation}$ sets, and D_{test} set which is split into U_{test} and U_{answer} sets. U_{train} and $U_{validation}$ are to train and validate the evolution of the algorithm. U_{test} and U_{answer} are used to evaluate the performance of the tree at the question q and to answer to that questions. As long as the validation phase aims to optimize the RMSE, the accuracy prediction of the matrix factorization enhances this metric. This yields to better questions and hence the accuracy of the decision tree is enhanced as well. Figures 3(b) shows better results than [17] as well. Further analysis are not described in this paper due to a lack of space.

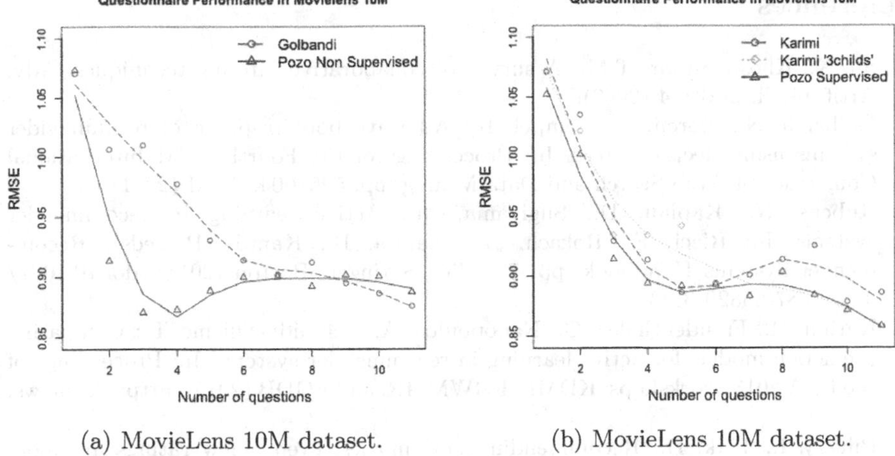

(a) MovieLens 10M dataset.

(b) MovieLens 10M dataset.

Fig. 3. Questionnaire performance in RMSE.

5 Conclusions and Future Work

The personalization of the active learning technique is crucial to better learn the new users preferences and decision trees are interesting techniques to model questionnaires. Indeed, decision trees can predict the items that new users have already used, although we consider that recent approaches do not correctly exploit the "prediction" inside the decision trees since they use very simple approaches (e.g. item rating average) to make it tractable.

The main idea of our contribution is to train an accurate collaborative filtering techniques with a ratings dataset to generate a prediction dataset. Then, both ratings and predictions dataset are used inside the decision trees. The former properly split the users' population while building the tree. The latter enhances the seek of the best discriminative items (questions) and better predict the associated labels. We have tested this approach in non supervised decision trees and supervised decision trees. The experimentation shows that our approach find better questions to present to users in order to better understand his preferences.

Our future work focuses on (1) detecting the new users preferences directly on the fly, and (2) using new techniques to exploit the information coming from questionnaires. We especially believe that the time (new) users spent to answer a question is very significant for the answer itself. Thus, we focus on "time-aware" recommendation techniques and decision trees to retrieve and exploit not only the users' answers but also the users' behavior.

Acknowledgments. This work has been supported by FIORA project, and funded by "DGCIS" and "Conseil Regional de l'Île de France".

References

1. Su, X., Khoshgoftaar, T.M.: A survey of collaborative filtering techniques. Adv. Artif. Intell. **2009**, 4 (2009)
2. Golbandi, N., Koren, Y., Lempel, R.: Adaptive bootstrapping of recommender systems using decision trees. In: Proceedings of the Fourth ACM International Conference on Web Search and Data Mining, pp. 595–604. ACM (2011)
3. Rubens, N., Kaplan, D., Sugiyama, M.: Active learning in recommender systems. In: Ricci, F., Rokach, L., Shapira, B., Kantor, P. (eds.) Recommender Systems Handbook, pp. 735–767. Springer, Boston (2011). doi:10.1007/978-0-387-85820-3_23
4. Karimi, R., Freudenthaler, C., Nanopoulos, A., Schmidt-Thieme, L.: Comparing prediction models for active learning in recommender systems. In: Proceedings of the LWA 2015 Workshops: KDML, FGWM, IR, and FGDB (2015). http://ceur-ws.org
5. Pilászy, I., Tikk, D.: Recommending new movies: even a few ratings are more valuable than metadata. In: Proceedings of the Third ACM Conference on Recommender Systems, pp. 93–100. ACM (2009)
6. Rashid, A.M., Albert, I., Cosley, D., Lam, S.K., McNee, S.M., Konstan, J.A., Riedl, J.: Getting to know you: learning new user preferences in recommender systems. In: Proceedings of the 7th International Conference on Intelligent User Interfaces, pp. 127–134. ACM (2002)
7. Elahi, M., Ricci, F., Rubens, N.: Active learning in collaborative filtering recommender systems. In: Hepp, M., Hoffner, Y. (eds.) EC-Web 2014. LNBIP, vol. 188, pp. 113–124. Springer, Cham (2014). doi:10.1007/978-3-319-10491-1_12
8. Hofmann, T.: Collaborative filtering via Gaussian probabilistic latent semantic analysis. In: Proceedings of the 26th Annual International ACM SIGIR Conference on Research and Development in Information Retrieval, pp. 259–266. ACM (2003)
9. Lemire, D., Maclachlan, A.: Slope one predictors for online rating-based collaborative filtering. In: SDM, vol. 5, SIAM 1–5 (2005)
10. Koren, Y., Bell, R., Volinsky, C.: Matrix factorization techniques for recommender systems. Computer **42**(8), 30–37 (2009)
11. Peis, E., del Castillo, J.M., Delgado-López, J.: Semantic recommender systems. Analysis of the state of the topic. Hipertext.net **6**, 1–5 (2008)
12. Ziegler, C.N., Lausen, G., Schmidt-Thieme, L.: Taxonomy-driven computation of product recommendations. In: Proceedings of the Thirteenth ACM International Conference on Information and Knowledge Management, pp. 406–415. ACM (2004)
13. Vozalis, M.G., Margaritis, K.G.: Using SVD and demographic data for the enhancement of generalized collaborative filtering. Inf. Sci. **177**(15), 3017–3037 (2007)
14. Barjasteh, I., Forsati, R., Masrour, F., Esfahanian, A.H., Radha, H.: Cold-start item and user recommendation with decoupled completion and transduction. In: Proceedings of the 9th ACM Conference on Recommender Systems, pp. 91–98. ACM (2015)
15. Rashid, A.M., Karypis, G., Riedl, J.: Learning preferences of new users in recommender systems: an information theoretic approach. ACM SIGKDD Explor. Newsl. **10**(2), 90–100 (2008)
16. Zhou, K., Yang, S.H., Zha, H.: Functional matrix factorizations for cold-start recommendation. In: Proceedings of the 34th International ACM SIGIR Conference on Research and Development in Information Retrieval, pp. 315–324. ACM (2011)

17. Karimi, R., Nanopoulos, A., Schmidt-Thieme, L.: A supervised active learning framework for recommender systems based on decision trees. User Model. User Adapt. Interact. **25**(1), 39–64 (2015)
18. Zhou, Y., Wilkinson, D., Schreiber, R., Pan, R.: Large-scale parallel collaborative filtering for the Netflix prize. In: Fleischer, R., Xu, J. (eds.) AAIM 2008. LNCS, vol. 5034, pp. 337–348. Springer, Heidelberg (2008). doi:10.1007/978-3-540-68880-8_32
19. Harper, F.M., Konstan, J.A.: The movielens datasets: history and context. ACM Trans. Interact. Intell. Syst. **5**(4), 19:1–19:19 (2015)

An Efficient Parallel Method for Performing Concurrent Operations on Social Networks

Phuong-Hanh Du, Hai-Dang Pham, and Ngoc-Hoa Nguyen(✉)

Department of Information Systems,
VNU University of Engineering and Technology, Hanoi, Vietnam
{hanhdp,dangph,hoa.nguyen}@vnu.edu.vn

Abstract. This paper presents our approach to optimize the concurrent operations on a large-scale social network. Here, we focus on the directed, unweighted relationships among members in a social network. It can then be illustrated as a directed, unweighted graph. With such a large-scale dynamic social network, we face the problem of having concurrent operations from adding or removing edges dynamically while one may ask to determine the relationship between two members. To solve this challenge, we propose an efficient parallel method based on (i) utilizing an appropriate data structure, (ii) optimizing the updating actions and (iii) improving the performance of query processing by both reducing the searching space and computing in multi-threaded parallel. Our method was validated by the datasets from SigMod Contest 2016 and SNAP DataSet Collections with the good experimental results compared to other solutions.

Keywords: Bi-directional BFS search · Concurrent operations on social networks · Multi-threaded parallel computing

1 Introduction

Today, social networks play a significant role in our networked society. Facebook, Twitter, WhatApp, etc. can be seen as the famous examples applied more commonly in our modern life. On social networks model, graph theory has been considered as a proper methodology. Furthermore, a member is generally modeled by a vertex, and the direct relationship between two members is represented by an edge. Actually, there are three remarks in this kind of graph we should consider: (i) the number of vertices and edges are enormous; (ii) the graph is dynamic due of the relationship changes among members and new member registered; and (iii) the shortest distance (SD) query is mostly performed in order to find the way to establish the connection between any two members. Thus, the SD query allows to analyze the influence of a user to the community [8]; to identify the closeness between two users; to find more related users or contents by using the socially-sensitive search [1]. Although the SD problem is frequently trivial, the task to answer optimal path queries is a considerable challenge in

N.T. Nguyen et al. (Eds.): ICCCI 2017, Part I, LNAI 10448, pp. 148–159, 2017.
DOI: 10.1007/978-3-319-67074-4_15

the context of having large-scale and quickly changing/elastic social network in reality [5].

In this paper, we demonstrate a method to enhance the performance of parallelizing the concurrent operations on large-scale and elastic social network. To gain this purpose, we propose an appropriate data structure for modeling the network, following a strategy to parallelize the updating operations and finally the sufficient approach to perform the computing queries (such SD) in parallel.

The rest of this paper is organized as follows. Section 2 presents preliminaries and related works. Section 3 details our efficient method for improving the performance of both updating and computing operations. In Sect. 4, we summary our experiment to verify and benchmark our approach. Finally, the last section provides some conclusions and future works.

2 Problem Formulation and Related Works

2.1 Data and Operation Model

As mentioned above, in this article, a social networks can then be represented as a graph $G(V, E)$ where V is the set of all members (and called vertices) and $E = (v_i, v_j)|v_i, v_j \in V$ represents the set of all directed relationships (called edges) (v_i and v_j are connected with a single unweighted link). The total number of edges to (incoming) and from (outgoing) a vertex v_i is called the degree of v_i and is represented as $deg(v_i)$.

For the vertex, it is conveniently represented by a number. That leads to encode the $|V|$ vertices from 0 to $|V| - 1$. For the graph edges, there are three main structure types: (i) edge lists, (ii) adjacency matrices and (iii) adjacency lists. In the large scale graph, the appropriate way to represent the large-scale edges of the graph is the adjacency list structure [2].

For the social networks modeled, the fundamental relationship operations can be "*Read*" and "*Write*". Once writing on a such network, an edge is simply added or deleted, whereas a *Read* on it is traversing its vertices. A graph traversal is also considered as a query on the graph. In a traversal, we are performing whether a Depth-First Search or a Breadth-First Search (BFS) method [7]. From that, three operation types will be specified and described in more details as follows:

- Add an edge ['A' u v]: Modify the current network by adding another relation (edge) from the member u to v. In the case of the input of the original graph, if the edge already exists, the graph remains unchanged. If one (or both) of the specified endpoints of the new edge does not exist in the graph, it should be added.
- Delete an edge ['D' u v]: Perform the removing of relation between (u v) from the current network. If the specified relation does not exist in the network, the later should remain unchanged.
- Compute the SD ['Q' u v]: Perform and return the SD from u to v in the current network. If there is no relation between these members or if neither of

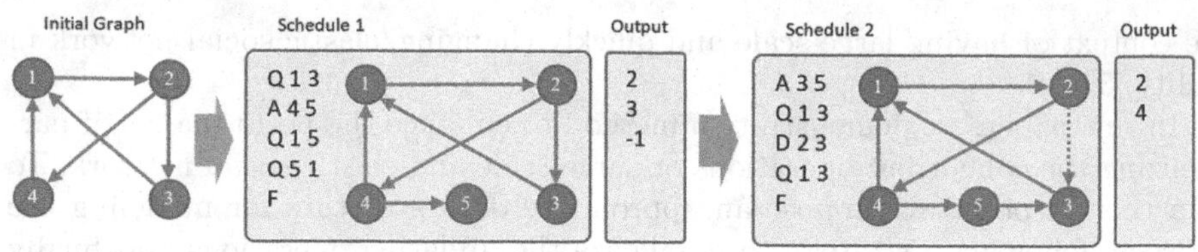

Fig. 1. Concurrent operations on graph

them exists in the network, the answer should be −1. The distance between any member and itself is always 0.

After that, the concurrent operations on G could be shown by a schedule of operations being susceptible updates and SD queries. For efficiently controlling the concurrency, we keep this schedule as a sequence of operations by their order of acceptance at the system. It is clear that if we assume the consistency of graph when performing each action this schedule, the final graph state will be consistent. The Fig. 1 indicates some example schedules of concurrent operations on a graph.

2.2 Related Works

To perform the actions on the graph, a substantial number of tools and libraries can feasibly be used to address this problem. NetworkX [12] and SNAP C++ library [4] are two popular libraries that can be shown here. However, their implementation is not optimal due of general purpose requirements: the SD is calculated only in sequence (performed by only one CPU core) and the directional selection is based on the number of enqueued vertices only.

For the large-scale graphs, the recent study of GraphLab in [3], PowerGraph in [10], GraphX in [11] are dedicated for processing the large graph in both distributed and parallel computation. These systems are efficient for general purposes in case having a dominant computing platform such as clusters and supercomputers [3]. Nevertheless, they are not adequate for the SD computation for the dynamic graph in the context of medium computing platforms.

The similar works for managing large-scale directed, dynamic graphs should describe five final teams participated the SigMod Programming Contest 2016 (SPC16) [13]. We will summary their main ideas here:

1. **H_minor_free**: this team was the Overall Winner in the SPC16. Their main ideas are based on considering the state of each edge: ALIVE for passable edges; DEAD for impassable edges; and UNKNOWN for the edges modified in the batch. For each batch, they experience 3-step process: (i) updating operations add UNKNOWN to the edges; (ii) utilizing bi-directional BFS (bBFS) processes SD queries using one of OpenMP threads; and (iii) updating edge states to ALIVE or DEAD after bBFS [14].

2. **uoa_team**: this group used the approach of multiversioning data structures for updating operations and using the heuristics to optimize multi-threading bBFS. They also used the *threadpool*11 and *concurrentqueue* libraries in order to enhance the performance of bBFS.
3. **akgroup**: the strategic solution of this team is based on the adjacency lists for both incoming and outgoing vertices. The updating operations have to be committed before performing the parallel heuristic bBFS [2].
4. **gStreamPKU**: the solution is based on (i) reducing the number of basic operations per query with Bit Compression and Optimizing program's spatial locality; and (ii) building parallelism Delta Graph (updating operations) to support fully concurrent bBFS query execution within a batch.
5. **while1**: this team used the idea of "transaction edge list" for performing the updating operations. All node and edge lists are duplicated for both NUMA nodes to ensure memory locality for processing parallel bBFS.

The last three teams got the third prize for that Contest.

3 Method for Performing Concurrent Operations

To process the concurrent operations on such social network represented by a directed, unweighted graph G, we suppose that all n concurrent operations have been executed by order of each operation $Op[i]$. This approach is strictly a serial schedule for concurrent operations. Thus, it is clear that G will be consistent after performed all concurrent operations. That schedule will be ingested by the Algorithm 1 illustrated as following:

Algorithm 1. Perform the concurrent operations

Input: Graph G and Op is the schedule of n operations (a,u,v) on the graph
Output: G committed all updates and list of SDs for all queries

```
for t = 0; t < n; t++ do
    (a,u,v) = Op[t];
    if a = 'Q' then
        Queries.push_back(t,u,v);//push back the tuple (t,u,v) into the Queries vector
    else
        Updates.push_back(t,a,u,v);//push back the tuple (t,a,u,v) into the Updates vector
    end if
end for
Make temporally edges for the updating operations; //Algorithm 2
Perform in parallel all queries;//Algorithm 4
Commit the updating operations;//Algorithm 3
```

In this algorithm, each tuple in both Queries and Updates vector has the timestamp t specified by order of operation in the concurrent list. We will use this parameter for checking whether an edge can be used or not when computing the SD queries.

3.1 Organizing the Data Structure

For managing the big data, the consecutive item list seems to be the best way to allow having the highest cache hit rate [2]. Moreover, in this research, we mainly examine large social networks having the number of members smaller than one billion (not Facebook case). Therefore, the member identification is represented by a 30-bit integer, and if we allocate 4-byte integer for a vertex, the two last bits for the state of edge can then be used.

To make in profit of the multi-core, multi-chip computational power, we use the idea of H_minor_free team [14] to model an edge: each edge can be one the three state: ALIVE, DEAD and UNKNOWN for executing concurrent operations. Therefore, the concurrent operations in the Fig. 1 can be illustrated by the Fig. 2.

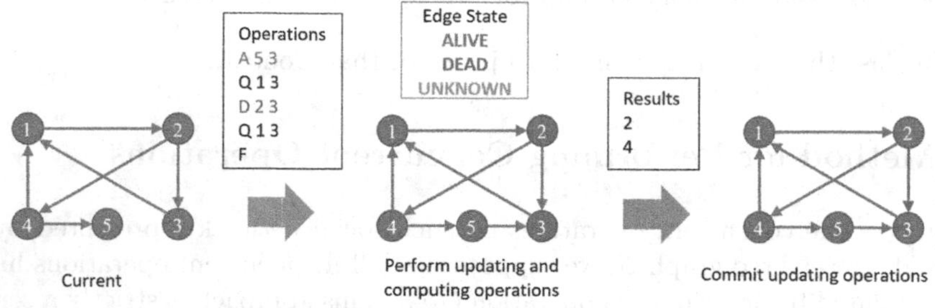

Fig. 2. Execution of updating operations

From the above idea, the graph data is represented by the adjacency lists. All incoming/outgoing nodes of a node u are stored in a sorted vector $incomingEdges[u]/outgoingEdges[u]$. Each item v of these vectors has the first 30-bits for the vertex; the last 2-bits for the state of edge (u, v). By using both incoming and outgoing edge lists, we can quickly explore the graph by both directions. That allows computing the SD queries by the bBFS algorithm [6].

3.2 Optimizing the Updating Operations

All updating operations will be firstly performed by adding/modifying an edge having the UNKNOWN state.

Algorithm 2. Make edges for the updating operations

```
Input: Updates: schedule of m updating operations (t,a,u,v); Graph G;
Output: G, incomingSum, outgoingSum modified by Updates
for each (t,a,u,v) in Updates do
    if a == 'A' then
        InsertNode(outgoingEdges[u],v); InsertNode(incomingEdges[v],u);
    else
        RemoveNode(outgoingEdges[u],v); RemoveNode(incomingEdges[v],u);
    end if
end for
```

The two global lists *incomingSum*[v] and *outgoingSum*[v] are used to store the all incoming and outgoing vertices from every vertex of *incomingEdges*[v] and *outgoingEdges*[v] respectively. Theses parameters will be used for computing the SD queries.

Algorithm 3. Commit the updating operations

```
Input: Updates: schedule of m updating operations (t,a,u,v); Graph G;
Output: G committed by all m updating operations
for each (t,a,u,v) in Updates do
    if a == 'A' then
        CommitAdd(outgoingEdges[u],v); CommitAdd(incomingEdges[v],u);
        outgoingSum[u] += outgoingEdge[v].size();
        incomingSum[v] += incomingEdges[u].size();
    else
        CommitDelete(outgoingEdges[u],v); CommitDelete(incomingEdges[v],u);
        outgoingSum[u] -= outgoingEdge[v].size();
        incomingSum[v] -= incomingEdges[u].size();
    end if
end for
```

Once all queries have been executed, we will commit all updating operations by modifying the state of updating edges to the right state by the Algorithm 3. Note that the operation *CommitAdd* of a vertex v into the sorted vector V_n will change the state of v (for the edge (n,v)) to ALIVE. Similarly, applying for the operation *CommitDelete*, the state of v for the edge (n,v) will be DEAD.

3.3 Optimizing the Query Processing

For the query processing, we interest firstly to determine the shortest relationship distance between two members of the social network. Since this is modeled by a directed and unweighted graph, the parallel BFS algorithm is adequate to answer that distance [6,9]. Another point should be emphasized here that all the capability of modern multi-core and multi-thread CPU have to be exploited. To do this, our strategy has to offer in parallel processing for all consecutive queries, which will be discussed more details as following.

i. Algorithm of shortest distance computation

For bBFS, the incoming edges and outgoing edges should be used. Two bitmap arrays (called *incomingMaps*/*outgoingMaps*) are also used for remarking traveled incoming/outgoing nodes. To reduce the search space, the strategy of predicting deeply on the graph has been proposed: following only the direction that has the smaller sum of next level enqueued vertices and their children [2].

Our strategy tuning the bBFS algorithm can be explained in the Algorithm 4. The *testBit*(v, *map*) function returns the value of bit at the position v in the buffer *map*. Meanwhile, *setBit*(v, *map*) function will set the bit at the position v in *map* to 1.

Algorithm 4. Compute SD (u,v) by the bi-directional BFS

Input: (t, u, v): containing the timestamp t and two vertices u and v; *incomingMap*, *outgoingMap*: global arrays for marking the traveled vertices of each worker; *incomingQueues*, *outgoingQueues*: global arrays for enqueueing the potential vertices to travel of each worker; *incomingSum*, *outgoingSum* and G

Output: the shortest distance from u to v

```
if u == v then
    Return 0;
end if
if (incomingEdges[v].size() == 0) || (outgoingEdges[u].size() == 0) then
    Return −1; //No path between u and v
end if
Get inMap and outMap from global incomingMaps and outgoingMaps for this thread
Get inQueue and outQueue from global incomingQueues and outgoingQueues for this thread
inQueue ← v; //initiate in-queue
setBit(v, inMap); //indicate v was visited
outQueue ← u; //initiate out-queue
setBit(u, outMap); // indicate u was visited
inCost = 0, outCost = 0; //distance to/from u/v
outSize = outgoingSum[u]; inSize = incomingSum[v];
while (outSize > 0)&&(inSize > 0) do
    if (outSize < inSize) then //following the outQueue
        outSize = 0; outCost+ = 1;
        for each vertex e in outQueue do
            for each vertex n in outgoingEdges[e] do
                if !testBit(n, outMap) then
                    state = GetState(n);
                    if (state==ALIVE)||(state&UNKNOWN && IsEdgeAlive(e,n,state,t)) then
                        if testBit(n, inMap) then
                            Clear all bits of inMap and outMap;
                            Return inCost + outCost;
                        end if
                        //ifnot, we push this vertex to outQueue
                        outQueue ← n; setBit(n, outMap);
                    end if
                end if
            end for
            outSize+ = outgoingSum[e];
        end for
    else//following the inQueue
        inSize = 0; inCost+ = 1;
        for each vertex e in inQueue do
            for each vertex n in incomingEdges[e] do
                if !testBit(n, inMap) then
                    state = GetState(n);
                    if (state==ALIVE)||(state&UNKNOWN && IsEdgeAlive(n,e,state,t)) then
                        if testBit(n, outMap) then
                            Clear all bits of inMap and outMap;
                            Return inCost + outCost;
                        end if
                        //ifnot, we push this vertex to inQueue
                        inQueue ← n; setBit(n, inMap);
                    end if
                end if
            end for
            inSize+ = incomingSum[e];
        end for
    end if
end while
Clear all bits of inMap and outMap;
Return −1;
```

For a UNKNOWN edge, we use the vector of updating operations for deciding use this edge or not. The idea of the $IsEdgeAlive(u, v, t, state)$ function has to answer this problem and illustrated as the following algorithm:

Algorithm 5. Verify if an edge (u, v) is ALIVE at the timestamp t

Input: (u,v,state,t): edge (u, v) with *state* at the query moment t; *Updates*: all updating operations
Output: TRUE if (u, v) is ALIVE at t; FALSE if not

```
i = lower_bound(Updates, (u, v, t, 0));
if i == Updates.begin() then
    Return (state&1) == 0;
end if
(lu,lv,la,lt) = Updates[i-1];
if (u == lu && v == lv) then
    Return (la=='A');
end if
Return (state&1) == 0;
```

In this algorithm, *lower_bound* is a function implemented in the C++ libraries. It finds (by using the binary search) and returns an iterator pointing to the first element in *Updates* that is not less than the tuple $(u, v, t, 0)$.

ii. Query parallel processing
Our solution for processing the consecutive queries is:

- Global incoming/outgoing queues are employed for each call of searching the SD. Then, each searching thread will use only proper in/out queues determined by an interval of global in/out queues.
- Each searching thread will also own the appropriate in/out map slots calculated from the global in/out maps. Since the searching process finishes, these in/out slots will be cleared for the next search.
- Cilkplus is used for performing queries in parallel because this method seems to be the most efficient one and achieve the outstanding performance in our solution [2].

Thus, the parallelization of consecutive queries is illustrated as the following algorithm:

Algorithm 6. Execute all consecutive queries on the graph G

Input: *Queries* having all SD queries; Graph G
Output: vector of results

```
//Perform in parallel the queries by Cilkplus method
for i = 0; i < Queries.size(); i++ do
    (u, v, t) ← Queries[i];
    distances[i] = shortest_distance(u, v, t);
end for
```

---

4 Experiments and Evaluation

Based on the proposed method, we built and implemented our solution in C++ language. The experiments were performed in the evaluation machine having 2 x Intel(R) Xeon(R) CPU E5-2697 v4 @ 2.30GHz (45 MB Cache, 18-cores per CPU), 128 GB for the main memory, CentOS Linux release 7.2.1511, gcc 6.3.0. This computing system was configured with maximum 36-threads in parallel (disable hyperthreading).

To validate our method for optimizing the concurrent operations on large-scale dynamic directed graph, two datasets from the Stanford Large Network Dataset Collection [15] and one from SigMod Programming Contest [13] are selected to evaluate the results. Here are some statistics of these graphs:

Table 1. Graph collection statistics

Parameter	SigMod dataset	Pokec dataset	LiveJournal dataset
Edges	1574074	68993773	30622564
Nodes	3232855	4847571	1632803
Diameter (longest shortest path)	9	16	11

Particularly, LiveJournal is a free on-line community with almost 10 million members. In addition, Pokec is the most popular on-line social network in Slovakia that connects more than 1.6 million people [15].

For testing workloads, we built a tool in order to generate the workloads by using the testing protocol of SigMod [13]. For each graph, we produced two experiment sets of workloads and results. The first one concentrates on the query operations. This experiment workload composes 1,000,000 operations involving queries/insertions/deletions ratios 80/10/10 respectively. We denote *8-1-1* for this set. In the second experiment set, we interest to the real situation which updating operations play an integral part in social networks. The set contains 1,000,000 operations as well, but differs from the respective proportion of queries/insertions/deletions of 50/40/10. *5-4-1* is denoted for this kind of experiment set.

To measure the time performance and the parallel efficiency, we run our solution (called **akGroupPlus**) and other solutions detailed in the Sect. 2 (**akgroup** - our old solution; **H_minor_free** - the Overall Winner) with distinct numbers of thread for each dataset. To be specific, all solution was tested with 1, 2, 4, 8, 16, 24, and 32 threads (since we have maximum 36-threads in our computing system). Every test was executed ten times for each dataset per each number of thread and gave the average result. The final results are illustrated in the figures below.

From the Fig. 3 that the execution time of the *akGroup* is the longest and its trend is obviously upward relative with the increase of the number of thread. Meanwhile, *akGroupPlus* and *H_minor_free* have the execution time reduced when performing in more threads. Additionally, our solution does not get the first rank when running on the only single thread, it, however, come on top from two threads onwards.

Concerning Pokec dataset, as it is demonstrated in Fig. 4, the running time of all solutions decrease considerably when the number of thread rise. In the Pokec 8-1-1, at one and two thread cases, *akGroup* team take the first place. Moreover, it can be seen notably that from four threads and more, our solution, *akGroupPlus*, comes first. Likewise, the similar pattern is showed in all solutions in the Pokec 5-4-1: akGroup takes the first place in case of performing no-parallel; other cases, *akGroupPlus* is the best one.

Regarding LiveJournal, similar to Pokec dataset, there is a gradual decline in the running time of three solutions with the rising number of threads. *akGroup* stands in

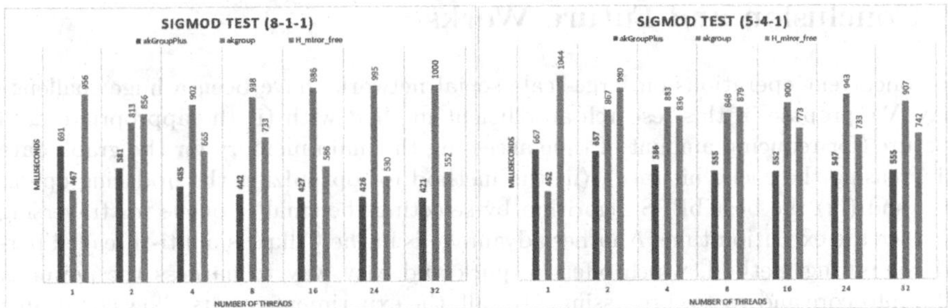

Fig. 3. Sigmod dataset test results

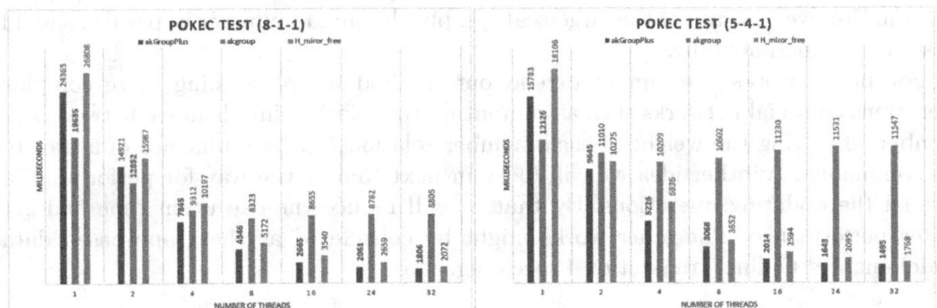

Fig. 4. Pokec dataset test results

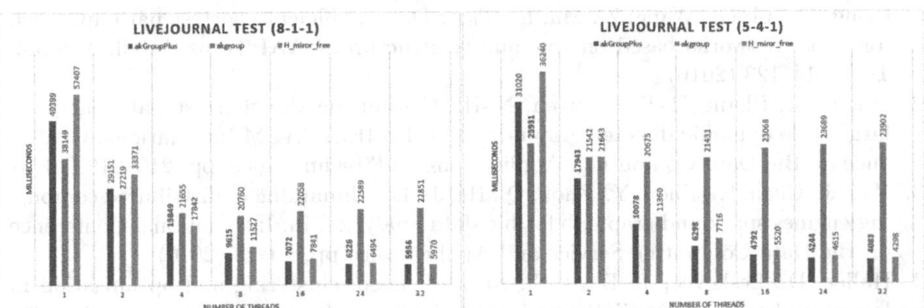

Fig. 5. LiveJournal dataset test results

first place at 1 and 2 threads for the 8-1-1 set and at 1-thread for the 5-4-1 set; the other cases, the best execution time is privileged to *akGroupPlus*.

In all datasets, the execution time does not decrease linearly when we increase the thread number. The reason here is firstly due to the concurrency among concurrent operations. Secondly, the cache locality for the graph data structure causes also the execution time: the more parallel thread number, the higher the miss rate. Thus, in case of the large-scale graph, the thread number in parallel should be selected experimentally.

5 Conclusion and Future Works

The concurrent operations on large-scale social networks have been a huge challenge today. We propose in this research an efficient method with (i) the appropriate data structure (for reducing amount of time accessing the main memory for the graph data by increasing the cache hit rate), (ii) the method for optimizing the updating operations, and (iii) the best bBFS algorithm by selecting the smaller queue for traversing to reduce the execution time. Another advantage is in the Cilkplus multi-threaded parallel computing method, which offers a quick and easy way to harness the power of both multi-core and vector processing. Overall, the experiment results allow confirming that *akGroupPlus* is the most efficient method and obtain the outstanding performance in comparison with other approaches for executing concurrent operations in parallel. The time execution is often reduced proportionally with the number of real parallel threads. However, in case of the large-scale graph, the thread number in parallel should be selected experimentally.

For future works, we aim to extend our method for performing more complex operations on social networks such as computing the relationship distance between two members involving the weight of each member relationship, the influence of a user to the community. Another idea we will focus in next time is the way for performing in parallel the updating operations. By that, it will reduce the execution time and get better performance. The other works might be considered as the graph-based data model enabled Online Transaction Processing.

References

1. Gong, M., Li, G., Wang, Z., Ma, L., Tian, D.: An efficient shortest path approach for social networks based on community structure. CAAI Trans. Intell. Technol. **1**(1), 114–123 (2016)
2. Du, P.-H., Pham, H.-D., Nguyen, N.-H.: Optimizing the shortest path query on large-scale dynamic directed graph. In: The 3rd IEEE/ACM International Conference on Big Data Computing, Applications and Technologies, pp. 210–216 (2016)
3. Wei, J., Chen, K., Zhou, Y., Zhou, Q., He, J.: Benchmarking of distributed computing engines spark and graphlab for big data analytics. In: International Conference on Big Data Computing Service and Applications, pp. 10–13 (2016)
4. Hallac, D., Leskovec, J., Boyd, S.: Network lasso: clustering and optimization in large graphs. In: ACM SIGKDD International Conference on KDD, pp. 387–396 (2015)
5. U, L.H., Zhao, H.J., Yiu, M.L., Li, Y., Gong, Z.: Towards online shortest path computation. IEEE Trans. Knowl. Data Eng. **26**(4), 1012–1025 (2014)
6. Chakaravarthy, V.T., Checconi, F., Petrini, F., Sabharwal, Y.: Scalable single source shortest path algorithms for massively parallel systems. In: IEEE 28th International Parallel and Distributed Processing Symposium, pp. 889–901 (2014)
7. Mondal, J., Deshpande, A.: Managing large dynamic graphs efficiently. In: Proceedings of the ACM SIGMOD 2012, pp. 145–156 (2012)
8. Yahia, S.A., Benedikt, M., Lakshmanan, L., Stoyanovich, J.: Efficient network aware search in collaborative tagging sites. Proc. VLDB Endow. **1**(1), 710–721 (2008)

9. Leiserson, C.E., Schardl, T.B.: A work-efficient parallel breadth-first search algorithm (or how to cope with the nondeterminism of reducers). In: Proceedings of the Twenty-Second Annual ACM Symposium on Parallelism in Algorithms and Architectures, pp. 303–314 (2010)
10. Gonzalez, J.E., Low, Y., Gu, H., Bickson, D., Guestrin, C.: PowerGraph: distributed graph-parallel computation on natural graphs. In: 10th USENIX Symposium on Operating Systems Design and Implementation, pp. 17–30 (2012)
11. Gonzalez, J.E., Xin, R.S., Dave, A., Crankshaw, D., Franklin, M.J., Stoica, I.: GraphX: graph processing in a distributed dataflow framework. In: 11th USENIX Conference on Operating Systems Design and Implementation, pp. 599–613 (2014)
12. Hagberg, A.A., Schult, D.A., Swar, P.J.: Exploring network structure, dynamics, and function using NetworkX. In: Proceedings of the 7th Python in Science Conference, pp. 11–15 (2008)
13. The ACM SIGMOD Programming Contest 2016: http://dsg.uwaterloo.ca/sigmod16contest/. Accessed 15 May 2017
14. H_minor_free: http://dsg.uwaterloo.ca/sigmod16contest/downloads/H_minor_free-poster.pdf. Accessed 15 May 2017
15. Stanford Large Network Dataset Collection: https://snap.stanford.edu/data/index.html. Accessed 15 May 2017

Simulating Collective Evacuations with Social Elements

Daniel Formolo(✉) and C. Natalie van der Wal

Department of Computer Science, Vrije Universiteit Amsterdam, Amsterdam, The Netherlands
{d.formolo,c.n.vander.wal}@vu.nl

Abstract. This work proposes an agent-based evacuation model that incorporates social aspects in the behaviour of the agents and validates it on a benchmark. It aims to fill the gap in this research field with mainly evacuation models without psychological and social factors such as group decision making and other social interactions. The model was compared with the previous model, its new social features were analysed and the model was validated. With the inclusion of social aspects, new patterns emerge organically from the behaviour of each agent as showed in the experiments. Notably, people travelling in groups instead of alone seem to reduce evacuation time and helping behaviour is not too costly for the evacuation time as expected. The model was validated with data from a real scenario and demonstrates acceptable results and the potential to be used in predicting real emergency scenarios. This model will be used by emergency management professionals in emergency prevention.

Keywords: Social contagion · Agents · Model · Evacuation · Simulation

1 Introduction

Preventing incidents in crowded environments involves many aspects such as investigating the effects of environmental factors like the number of doors, their positions, the location of other objects and escape areas. All these factors have a big impact on evacuation time and consequently influence the number of injuries and casualties. There are many details to be considered, some of them are predefined by the design of the building itself, such as stairs, width of exits and existing pillars. Other environmental aspects are organised only hours before an event starts, such as areas for queuing, gates and signs to direct the flow of people. Moreover, there are psychological and social factors that have an effect on the evacuation process as well, such as age, gender, language, being in a rush. To avoid dangerous situations, organisers follow security protocols and make scenario simulations. Simulations are good to indicate security lacks and unpredicted situations. Most of the evacuation simulation tools consider physical characteristics of the environment, ignoring behaviour and personality of people. Some of them consider queues formation, jamming, clogging, fluid movement of crowds and following behaviour [22].

Evacuation simulation models could become more precise by incorporating realistic human behaviours, as currently they do not. Most evacuation simulation models do not incorporate psychological and social factors. Observations of actual emergencies show that people tend to be slow to respond to evacuation alarms (taking up to 10 min) and

N.T. Nguyen et al. (Eds.): ICCCI 2017, Part I, LNAI 10448, pp. 160–171, 2017.
DOI: 10.1007/978-3-319-67074-4_16

take the familiar route out instead of the nearest exit [1, 7]. Most models simulate people like 'robots' taking rational decisions to reach a safe place, avoiding obstacles and suffering the influence of the environment in their speed to escape from a dangerous place. Including psychological and social factors in evacuation simulations could make these models more realistic and better in their predictions to ultimately save more lives.

As part of the EU Horizon 2020 project IMPACT[1], this work will propose and validate an evacuation simulation incorporating social factors, named the IMPACT model. It is a refinement and extension of an initial version of the model proposed in [18]. New features added are: helping behaviour, groups, age and gender. The following social features were refined: familiarity, response time and social contagion. The rest of the paper is organised as follows. Section 2 starts with a short literature review. Next, Sect. 3 introduces the conceptual and formal model with the new features. This is followed by reports of the simulation experiments in Sect. 4 and a summary and discussion in Sect. 5.

2 Related Work

There are many different approaches for computer models of crowd evacuation simulations. Zheng and colleagues describe seven approaches for computer evacuation models: (1) cellular automata, (2) lattice gas, (3) social force, (4) fluid dynamics, (5) agent-based, (6) game theory, (7) animal experiments [22]. Each one of these approaches combines physical aspects with statistics measurements of evacuation flows in diverse types of environments. According to Templeton and colleagues, current crowd simulations don't include psychological factors and therefore cannot accurately simulate large collective behaviour that has been found in extensive empirical research on crowd events [16]. On the other hand, Santos and Aguirre have reviewed the integration of social and psychological factors incorporated in evacuation simulation models [13]. They describe how social dimensions are incorporated in three evacuation simulation models: FIRESCAP [4], EXODUS[2] [6] and multi-agent simulation for crisis management (MASCM [8]).

MASCM includes social interaction in the way of evacuation leaders. For example, evacuation leaders can communicate 'please follow me' and start to walk along the evacuation route or find an evacuee at the distance or wait for the evacuee to approach. Even though leadership is modelled fairly accurate, there is no possibility of simulating the set of group decision-making processes involved in selecting a leader when there is no trained professional/evacuation leader present. Yet in evacuation situations, there are often no official leaders. Also, evacuees in MASCM go to the nearest group, but research suggests that such an action typically involves social factors including the relationship between the evacuees from the start of the evacuation. EXODUS includes 22 social psychological attributes and characteristics for each agent, including age, name, sex, breathing rate, running speed, dead/alive, familiarity with building, agility and patience. Yet, agents in EXODUS cannot have social micro-level interactions that would create

[1] http://www.impact-csa.eu/.
[2] http://fseg.gre.ac.uk/index.html.

a collective definition of the situation for groups and collective interactions. Other models that Santos and Aguirre reviewed do not model social dimensions, such as group decision making, but focus more on the physical constraints and factors such as walking speed, walkways, stairways etcetera to find the optimal flow of the evacuation process. This work aims to fill this gap in the field of evacuation simulations, by proposing a model that includes social interactions and collective decision making.

3 IMPACT Model

The evacuation dynamics were modelled using an agent-based model with the beliefs-desires-intentions and network-oriented modelling approaches [11, 17]. The first version of the model included physical, psychological and social aspects. Each agent in the simulation has his own characteristics and his behaviour is influenced by environment as well as by other agents around him. Table 1 describes the internal characteristics of each agent and the external influences on them as the initial version [18].

Table 1. Individual agent characteristics in the initial version of the model.

Characteristic	Description
Familiarity	When an agent is familiar, he chooses the nearest exit otherwise, he always evacuates via the main exit
Compliance	The compliance level of each agent can vary between 0 and 1 and has an effect on the agent's desires. In the initial version, the setting was fixed to 0.5 for each agent
Fear	Internal state influenced by external factors and other internal states
Belief of danger	The level of this state directly influences the decision to evacuate or not. It is a combination of fear and external stimuli
Desire walk randomly or evacuate	Depending on the level of Fear and Belief of danger, an agent can have the desire to walk randomly in the environment or to escape
Intention of walk randomly/evacuate	Based on the desires, the agent creates an intention (decision) to perform a certain action with a certain intensity
Speed	Maximum speed of the agent, depending on intensities of the intention. It varies from walking to running. Range: (1.42 m/s, 4.26 m/s)
Express emotion	Expression of fear to other agents within observation distance
Vision radius	Scope of vision of an agent
Fall	Each agent can fall, spending some time before standing up and continuing its path. The chance of falling is based on the number of agents around and the agent's own speed. The more people around the agent, the faster the agent runs and the higher the chance to fall
Observation event (external factor)	Agents who observe dangerous events change their beliefs and immediately start to evacuate
Observation alarm (external factor)	When the alarm sounds, there is a 50% chance for each agent to start to evacuate. (Representing risk-taking, as not all passengers react quickly to a fire alarm [7, 10])
Observation fear (external factor)	Agents can observe other's fear and belief expressions and decide to evacuate without seeing the danger him/herself

The model was implemented in the Netlogo multi-agent language [20]. Figure 1 gives an overview of the conceptual model, showing four agent modules. Individual characteristics influence all other modules. the perceived external stimuli from others (people-people interaction) and from environment (people environment interaction) affects directly the actions in such a way that the patterns simulations emerge organically from the behaviour of each agent and events in the environment.

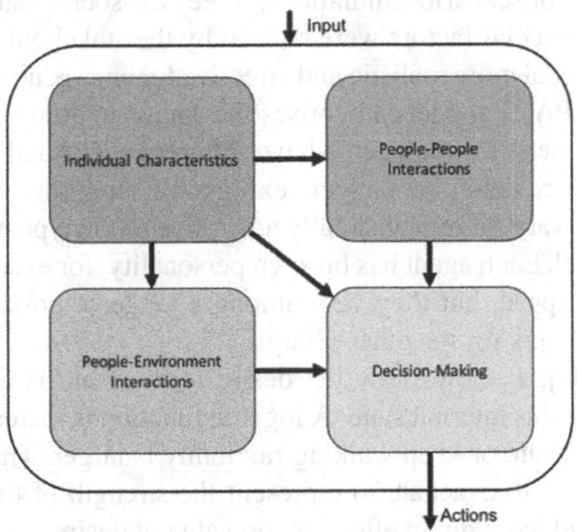

Fig. 1. Agent modules of the IMPACT model.

Comparing the current IMPACT model to the initial one, some characteristics were updated regarding speed, falls and compliance. For all of them, the updates are based on psychological and statistical analyses as described below.

- **Speed:** The walking speeds varied for each demographic group (children, adult males, adult females, elderly males, elderly females) and were based on the observational work of Willis et al. [21], ranging from 1.12 m/s to 1.58 m/s. We calculated running speeds by multiplying the walking speed for each demographic group by three – to account for the luggage, belongings, and clothes that people wear while travelling – to yield speeds between 3.36 m/s and 4.75 m/s. Moreover, a crowd congestion factor was added that reduces the speed according to the number of agents within the same square metre: ≤ 4 people (no speed reduction), 5 people (62.5% reduction), 6 people (75%), 7 people (82.5%), 8 people (95%). These speed adjustments were based on research by Still [15], where 8 is the maximum number of people per square metre and 4 the number of people from which speed reduces.
- **Falls:** The number of falls in the initial model seemed unrealistically high during structured simulations. So, we manually tuned the value to a more realistic level by visually inspecting the movement patterns during many different settings. This resulted in a new rule: if there are more than 4 people in the same square metre of the agent and if he is running more than 3 m/s, then there is a 0.5% chance of a fall for each new movement.

- **Compliance:** In the current version, the probability of compliance is based on data from Reininger et al.'s [12] study of gender differences in hurricane evacuation, modified for different age groups using data from Soto et al.'s [14] personality study. The model has 6 compliance values according to the category of the agent: male or female, and child, adult, or elderly. The precise levels can be found in Sect. 2 of [19].

Besides the improvements in speed, falls and compliance that have substantial changes to the outputs of scenario simulations, three new social features were added to the model. These new social factors were chosen by the stakeholders of the IMPACT project to make the model more realistic and effective for emergency prevention/predictions. The current IMPACT model categorises the agents in groups of age and gender. Based on these characteristics the agents have different speed and compliance levels, consequently their fear, belief of danger, express of emotions and desire to walk randomly or evacuate vary more realistically according to the type of agent and in relation to the initial model. Each agent has his own personality, for example, not all female elderly has the same speed, but they vary among a range of possible speeds for that category. The same occurs for the other groups.

As an example, Eq. 1 shows how the desire to evacuate is calculated and how compliance influences this internal state. A logistic function is thereafter used to decide if the intention to evacuate or keep walking randomly is larger. Then, this outcome is multiplied by the desire to evacuate to represent the strength of the intention. In this way, compliance Level has a direct effect on the value of desire to evacuate. The result is a number between (0, 1), whereby 0 means a minimal intention and 1 a maximal intention to evacuate. Each decision and internal state in the model has a formal rule, for lack of space, only the rule to calculate the desire to evacuate is shown below. All formal rules can be found in [19].

$$\begin{aligned}&\text{desire\_evacuate(t)} = \\ &\text{desire\_evacuate(t)} + \eta \cdot \\ &\Big(\Big(compliance \cdot \Big(\max\Big(\omega_{amplifyingevacuation} \cdot \text{belief\_dangerous(t)}, \omega_{amplifyingevacuation} \cdot \\ &fear(t), \omega_{amplifyingevacuation}\Big)\Big)\Big) - \text{desire\_evacuate(t)}\Big) \cdot \Delta t.\end{aligned} \tag{1}$$

Whereby,

ω = *predefined weights of the tuned model.*
η = *speed factor, defines how smooth or abrupt changes in the calculations happen.*
max = *function that returns the maximum value among the parameters.*

The current IMPACT model also includes group formations. Currently, it supports the most frequent types of groups with 2, 3 and 4 people. Agents in a group always move together with the same speed. The group speed is 40% of the difference between the minimum and maximum speed of group's members, assuming that the slower people in the group can be 'pulled' along faster or children can be carried, but not faster than the average speed of the group. Children are always part of a group, while adults and elderly can both travel alone as in groups. One of the group members is always the leader, the others follow him. However, the leader is constantly influenced by other members of the group since they are close to each other, so if other members express a high level of fear, the leader will increase his level of fear in a re-feed cycle until a stable value is reached that could be enough to make the leader decide to evacuate.

Besides that, helping behaviour was included. An agent can decide to help other people that felt on the floor. That means, the agent stops in front of the fallen agent until it stands up and after that, each agent follows his own path. The decision to help another agent is a statistical combination of gender, age and group identity of the helpers and fallers. For example, if the agent belongs to a group, there is a very high chance that his group members help him [2]. Men are most likely to help others and women, children, and older adults are most likely to receive help [3].

4 Simulation Results

Experimental Design. Simulation experiments with different factors and levels were designed to answer the research questions of this work: (1) Model comparison: does the model have significant differences after the improvements compared to the initial version? (2) Model validation: does the model correspond to reality and in how far? (3) Model simulations: what are the effects of the new features: helping and group formation on the evacuation time? A square (20 × 20 m) layout of a building with two exits (4 metres wide) was chosen to represent a general building layout. All environmental and personal factors such as width of the doors and level of compliance were kept constant among simulations. Only the factors and levels stated in each experimental setup were systematically varied. After inspecting the averages and variances in evacuation time of 100 simulations of a scenario with the most variability, Eq. 2 was used to find the minimum number of repetitions (56) to guarantee that the error in the outcome results are within 5% of the maximum error with 95% of confidence. In total, 60 repetitions of each variation were run and the results represent the average of these runs.

$$n \geq [100 \cdot Z \cdot s/r \cdot \bar{x}]^2 = 56.61599 \rightarrow 60 \text{ samples} \tag{2}$$

Whereby,
Z = *confidence interval of 95%*; s = *standard deviation, 53.4287*
r = *maximum error of 5%*; $\bar{x}$ = *evacuation time average of 100 samples*

Simulation Results Comparing the Models. This experiment was conducted according to Table 2 comparing the influence of each factor of the initial model [15] with the current IMPACT model. Figure 2 compares the average evacuation time of the two models, varying familiarity and social contagion as described in Table 2. For all cases the current IMPACT model presented higher evacuation times than the old model. Even comparing social contagion of the models with themselves, it is clear that the influence of social contagion on the current IMPACT model, was not visible in the old one. In relation to the number of falls, there was a drastically reduction for reasonable values. For example, in a simulation with 800 people the initial model has 368 falls on average, while the current IMPACT model has 16. For both, when the social contagion is activated it results in the duplication of falls, and no significant variability was observed in the average response time to evacuate after the incident started.

Table 2. Factors and levels in simulation experiment social contagion and familiarity.

Level	Factor				
	Crowd density	Familiarity	Social contagion	Environment type	Model
1	Low: 2 people/m^2	0%	On	Square room, 2 doors	Initial model [15]
2	Medium: 4 people/m^2	50%	Off		Current model
3	High: 8 people/m^2	100%			

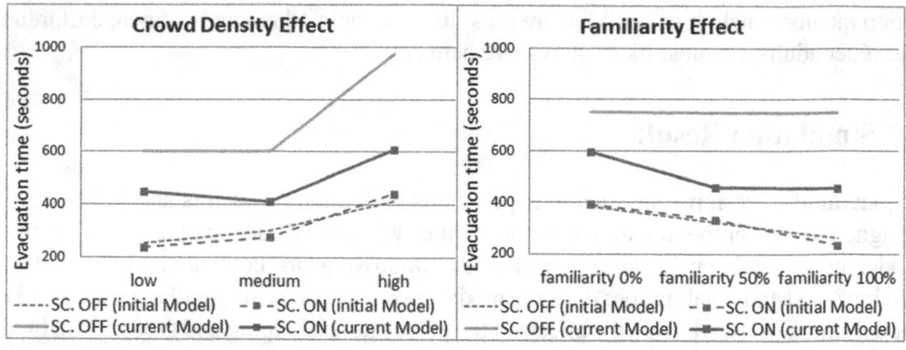

Fig. 2. Average evacuation time of the initial and current IMPACT models.

Simulation Results Exodus Benchmark. The project Exodus [6] was selected as benchmark of the model. It is a traditional model accepted by the specialists in this area [9]. The environment selected is called SGVDS1. It is a ship: a complex environment composed of 3 floors divided into sectors, with many escape route possibilities to the 4 exit areas. Figure 3 shows the floor plan of the ship which was imported to the simulator. An experiment was conducted comparing three versions of the current IMPACT model with the benchmark as the baseline model, see Table 3. The current IMPACT model covers more aspects than those required by the benchmark protocol. In order to make a fair validation some of the current IMPACT model's variables were fixed:

- Familiarity: It was assumed that everybody was not familiar with the environment.
- Relationship: It was assumed that no one has relationships with other passengers.
- Social contagion: Considered depending on experimental condition, see Table 3.
- The passenger`s speed in experiment 1 follow the patterns indicated in [5]. In experiment conditions 2 and 3, the speed was calculated by the current IMPACT model.
- Groups and help: Not considered in any experimental condition.

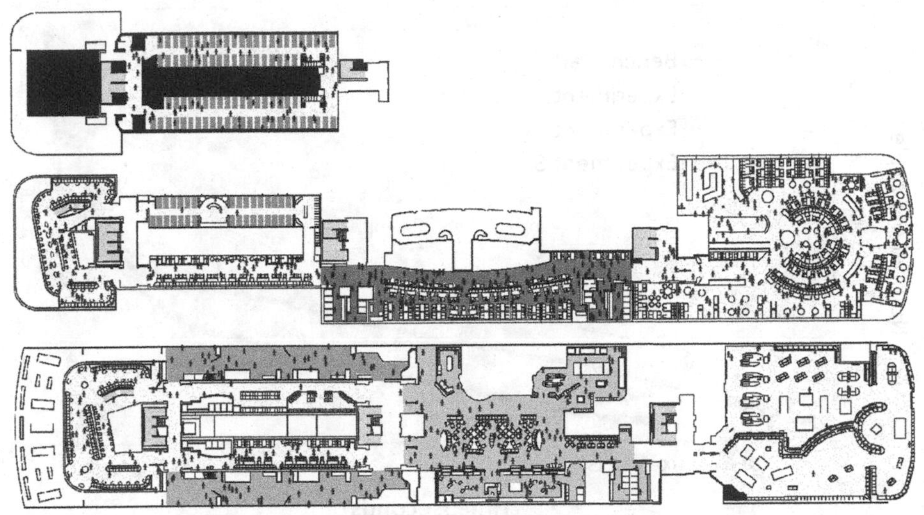

Fig. 3. Scenario of the software simulation.

The validation considers The Final Evacuation Time in Seconds (FET); the percentage difference between the predicted and Total Assembly Time (TAT); and the curve differences between the predicted and expected arrivals to the assembly areas (exits). This last measurement is calculated based on Euclidean Relative Difference (ERD), Euclidean Projection Coefficient (EPC) and Secant Cosine (SC).

Table 3. Results over validation protocol for the overall arrival times.

Condition:	Benchmark	Experimental Condition 1	Experimental Condition 2	Experimental Condition 3
Explanation:	Exodus SGVDS 1	No social contagion, response time and speed taken from the benchmark	No social contagion, response times and speed calculated by the model	Social contagion ON, response times and Speed calculated by the model
FET:	585	498.6	543.4	516.6
TAT:	0	14.77	7.11	11.69
ERD:	0	0.568171	0.575657	0.565754
EPC:	0	0.724621	0.731295	0.731634
SC:	0	0.522105	0.423135	0.451471

In [5] it is stated that a 'good' TAT, should be below 40%. For all conditions this is true, indicating that the TAT in the experimental conditions are all 'good' and below 40%. For ERD, all experimental conditions are over, but close to, the expected boundary that is ≤ 0.45, while for EPC, the results do stay within the expected boundaries of $0.6 \leq \text{EPC} \leq 1.4$. For SC the values are below the expected boundary ≥ 0.6, and again close to the acceptance threshold. Figure 4 shows the curves for the 3 experiments compared to the experimental data.

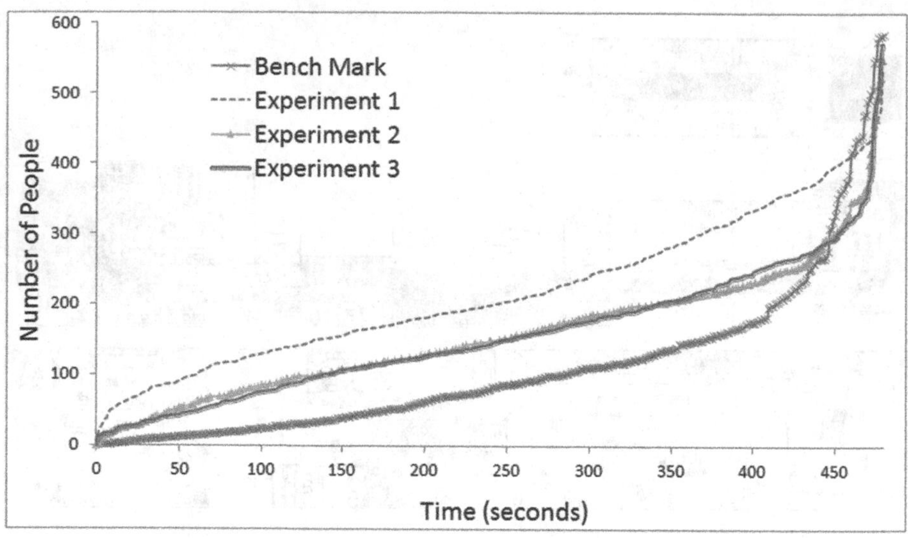

Fig. 4. Total arrival time pattern for one simulation run.

Simulation Results - Effect of Helping and Groups. In these structured simulations, the effects of helping and group formations on evacuation time, response time and number of falls are analysed. The hypotheses for the simulation experiments are: (1) evacuation time slows down when people are helping, because the helpers will need more time to evacuate; (2) groups have a higher effect in slowing down the evacuation time. Table 4 shows the experimental design.

Table 4. Factors and levels in simulation experiment helping and groups.

Level	Factor		
	Crowd density	Helping	Travelling alone
1	Low: 2 people/m^2	0%	100%
2	Medium: 4 people/m^2	50%	0% (only groups of 2 adults)
3	High: 8 people/m^2	100%	0% (only groups of 3 adults)
4			0% (only groups of 4 adults)

The factor helping has a considerable influence only for low crowd density, resulting, in average, 14% increase of total evacuation time (50 s more time to evacuate) with helping 'On'. While, for medium and high crowd densities this difference disappears. Figure 5 (left) shows the influence of helping on evacuation time. The discrepancy in results in low density environments can be explained by the few bottle-necks that are created in front of the exits. As many people stop their evacuation to help others as much is their contribution to delaying the evacuation. On the other hand, for medium and high crowd density scenarios, many people that are stuck in front of the exits are awaiting their chance of accessing the exits. In these cases, if some people stop their trajectory to help someone else, that time has few or almost no influence on the final evacuation time, since that same person will be stopped by other people close to the exit doors. The average response time (time to decide to evacuate and perform the actual evacuation for of each agent) does not present remarkable results in any of the scenarios.

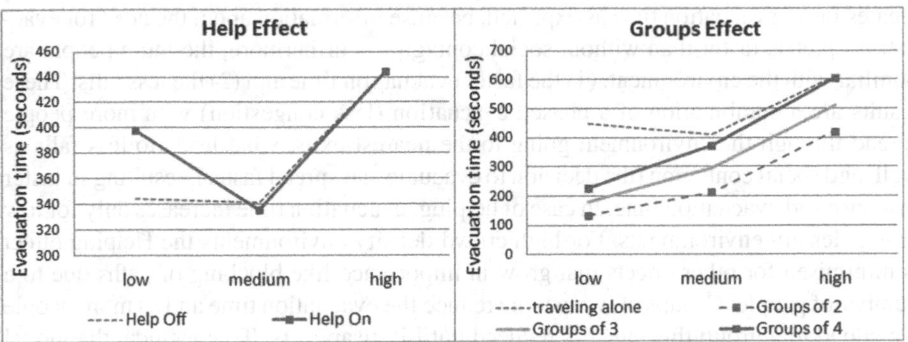

Fig. 5. Help and group factors influence over evacuation time to different crowd densities.

Contrary to expectations, the result in Fig. 5 (right) shows that groups reduced the average evacuation time. Groups of two presented the best results and the time increases when more people are added to a group. The explanation is social contagion. In groups the contagion of beliefs and emotions are spread faster than when all passengers travel alone. Therefore, groups evacuate faster than people travelling alone.

5 Discussion

This work aims to fill the gap in the research field of evacuation simulations by modelling evacuations including social and psychological factors in the model. This work is an extension of the initial proposed model in [15] and the new features were analysed and the model was validated. Existing features were refined and helping and group formation were added, approximating the model to reality in terms of evacuation time, falls and social interaction in crowded environments. Experiments were conducted to compare the outputs of both models. The current IMPACT model demonstrates clear improvements over the initial model in terms of evacuation time, falls and social influences in the agent's behaviour. Although the experiments presented in this work show the influence of social aspects individually, more experiments have to be conducted to analyse effects of combination of them. The cross relations between social effects and more complex environments might be explored, e.g. environments with pillars or multiple rooms.

A validation was conducted over a complex benchmark environment. The results are as expected and close to reality. All IMPACT model variations performed less than 7.11% to 14.77% (between 42 and 86 s) of difference from the benchmark's total evacuation time, which is a good TAT according to [13], and the curves of acceptance in Table 3 show values close to the prescribed boundaries, establishing the model's validity. For the future, it is recommended to apply new benchmarks over the model, increasing the confidence about the model's results.

To finalise, experiments that analysed the influence of helping and groups demonstrated interesting patterns useful for future security protocols. Social contagion effect creates faster evacuation time as expected, because information about the need for evacuation spreads faster than without social contagion. Furthermore, the more people are familiar with the environment: (1) the faster evacuation time and (2) the less falls. These results are a combination of a phased evacuation (less congestion) with more people spread through the environment going to the nearest exits, what leads to less falls as well, and social contagion (the decision to evacuate can spread faster), resulting in faster response and evacuation time. In case of helping, evacuation time increases only for low crowd density environments. For high crowd density environments the Helping effect is minimised for other effects that grow in importance like blocking of paths due to a number of people. Groups of two people reduce the evacuation time and as more people are added to a group this effect is reduced until it disappears. To conclude, the model has advantages over others that don't consider social effects in collective behaviour to evacuation scenarios, presenting reasonable results and can be used to predict real scenarios. As next steps, new social aspects will be incorporated and more benchmarks will be applied to it.

References

1. Challenger, R., et al.: Understanding Crowd Behaviours. Practical Guidance and Lessons Identified, vol. 1. The Stationery Office (TSO), London (2010)
2. Drury, J., Cocking, C., Reicher, S.: Everyone for themselves? A comparative study of crowd solidarity among emergency survivors. Br. J. Soc. Psychol. **48**(3), 487–506 (2009)
3. Eagly, A.H., Crowley, M.: Gender and helping behavior: a meta-analytic review of the social psychological literature. Psychol. Bull. **100**(3), 283–308 (1986)
4. Feinberg, W.E., Norris, R.J.: Firescap: a computer simulation model of reaction to a fire alarm. J. Math. Sociol. **20**(2-3), 247–269 (1995)
5. Galea, E., Deere, S., Filippidis, L.: The Safeguard Validation Data Set—sgvds1 a Guide to the Data and Validation Procedures. Fire Safety Engineering Group, University of Greenwich, London (2012)
6. Gwynne, S., Galea, E.R., Owen, M., Lawrence, P.J., Filippidis, L.: A review of the methodologies used in the computer simulation of evacuation from the built environment. Build. Environ. **34**(6), 741–749 (1999)
7. Kobes, M., et al.: Building safety and human behaviour in fire: a literature review. Fire Saf. J. **45**(1), 1–11 (2010)
8. Murakami, Y., Minami, K., Kawasoe, T., Ishida, T.: Multi-agent simulation for crisis management. In: 2002 Proceedings of Knowledge Media Networking, pp. 135–139 (2002)
9. Owen, M., Galea, E.R., Lawrence, P.J.: The exodus evacuation model applied to building evacuation scenarios. J. Fire. Prot. Eng. **8**(2), 65586 (1996)
10. Proulx, G., Fahy, R.F.: The time delay to start evacuation: review of five case studies. Fire Saf. Sci. **5**, 783–794 (1997)
11. Rao, A.S., Georgeff, M.P.: BDI agents: from theory to practice. ICMAS **95**, 312–319 (1995)

12. Reininger, B.M., et al.: Intention to comply with mandatory hurricane evacuation orders among persons living along a coastal area. Disaster Med. Public Health Prep. **7**(01), 46–54 (2013)
13. Santos, G., Aguirre, B.E.: A critical review of emergency evacuation simulation models. In: Peacock, R.D., Kuligowski, E.D. (eds.) Proceedings of the Workshop on Building Occupant Movement During Fire Emergencies, pp. 27–52, 10–11 June 2004
14. Soto, C.J., John, O.P., Gosling, S.D., Potter, J.: Age differences in personality traits from 10 to 65: big five domains and facets in a large cross-sectional sample. J. Pers. Soc. Psychol. **100**, 330 (2011)
15. Still, G.K.: Introduction to Crowd Science. CRC Press, Boca Raton (2014)
16. Templeton, A., Drury, J., Philippides, A.: From mindless masses to small groups: conceptualizing collective behavior in crowd modeling. Rev. Gen. Psychol. **19**, 215 (2015)
17. Treur, J.: Network-Oriented Modeling: Addressing Complexity of Cognitive, Affective and Social Interactions. Springer, Berlin (2016)
18. van der Wal, C.N., Formolo, D., Bosse, T.: An agent-based evacuation model with social contagion mechanisms and cultural factors. In: Proceedings of the IEA/AIE 2017 (2017) (in press)
19. van der Wal, C.N., Formolo, D., Robinson, M.A., Minkov, M., Bosse, T.: Simulating crowd evacuation with socio-cultural, cognitive, and emotional elements. Trans. Comput. Collect. Intell. (2017) (in press)
20. Wilensky, U.: NetLogo: center for connected learning and computer-based modeling. Northwestern University, Evanston (1999). http://ccl.northwestern.edu/netlogo/
21. Willis, A., Gjersoe, N., Havard, C., Kerridge, J., Kukla, R.: Human movement behaviour in urban spaces: Implications for the design and modelling of effective pedestrian environments. Environ. Plan. **31**(6), 805–828 (2004)
22. Zheng, X., Zhong, T., Liu, M.: Modeling crowd evacuation of a building based on seven methodological approaches. Build. Environ. **44**(3), 437–445 (2009)

Social Networks Based Framework for Recommending Touristic Locations

Mehdi Ellouze[1(✉)], Slim Turki[2], Younes Djaghloul[2], and Muriel Foulonneau[2]

[1] Research Group on Intelligent Machines, ENIS, University of Sfax, BP 1173, 3038 Sfax, Tunisia
mehdi.ellouze@ieee.org

[2] Luxembourg Institute of Science and Technology, 5 Avenue des Hauts-Fourneaux, 4362 Esch-sur-Alzette, Luxembourg
slim.turki@list.lu

Abstract. Tourists need tools that can help them to select locations in which they can spend their holidays. We have multiple social networks in which we find information about hotels and about users' experiences. The problem is how tourists can use this information to build their proper opinion about a particular location to decide if they should go to that place or not. We try in this paper to present a design of a solution that can be used to achieve this task. In this paper, we propose a framework for a recommender system that bases on opinions of persons on the one hand and on of users' preferences on the other hand to generate recommendations. Indeed, opinions of tourists are extracted from different sources and analyzed to finally extract how the hotels are perceived by their customers in terms of features and activities. The final step consists in matching between these opinions and the users' preferences to generate the recommendations. A prototype was developed in order to show how this framework is really working.

1 Introduction

According to United Nations World Tourism Organization, tourism contributes to about 8% of the employment in the world with nearly 260 million people work in jobs directly related to tourism. With the advent of the information technology, the management of this sector has changed enormously. The internet made the growing of this sector faster and made many countries able to promote their touristic destinations easily. Nowadays, to attract customers, destinations have to win the "Internet war". In this perspective, many organizations either public or private developed tools that can help tourists to identify quickly what they need in terms of touristic locations where they can spend their holidays. Many websites like TripAdvisor [1] or Booking.com [2] have been used by persons for a long time to plan their holidays. According to [3], in 2014, TripAdvisor attracted about 340 million of visitors per month. Currently the trend is to have smart tools that could recommend places and activities basing on users' preferences and experiences of other persons.

N.T. Nguyen et al. (Eds.): ICCCI 2017, Part I, LNAI 10448, pp. 172–181, 2017.
DOI: 10.1007/978-3-319-67074-4_17

On internet, we can find all the world destinations, we could find the potential customers and we could find the feedback of persons on these destinations. All these ingredients motivated many researchers to propose designs for tourism recommender systems. Moreover, the increasing popularity of social networks has boosted this.

As a part of this effort, we propose in this paper a new framework for a recommender system totally based on social networks for recommending touristic destinations. We worked on Tunisia destinations and especially on *Djerba* Island destination. Our motivation comes also from the fact that in 2011 the Tunisian Ministry of Tourism launched an international tender for the acquisition of smart system that can make a dashboard for the Tunisian Tourism mainly based on social networks. According to them, no effective proposal has been made. No proposal replied exactly to their requirements.

However, the main goal of our framework is not to generate dashboards but to recommend locations (Hotel, Clubs, Restaurants, etc.) to customers according to preferences that they introduce to the recommender. The only sources of information are the internet websites, portals and social networks.

To reach our goal we have to surmount many challenges: (i) How we will collect data from the sources, (ii) How we can track persons' opinions and how we can recognize their sense (positive or negative)? Finally, (iii) How we will use this data to generate recommendations?

For this reason, the framework that we proposed contains three components: the first one is dedicated to extracting from internet sources data about touristic locations and opinions related, this step is known as the crawling step, the second component is a data mining step in which we analyze, we organize and we store extracted data. The last component is dedicated to generating recommendations through finding the overlap between the user profiles and the stored destinations.

The repos of the paper will be organized as follows; in Sect. 2, we discuss works related to recommending system in tourism sector. In Sect. 3, we will state the problem, present our framework for the proposed recommender. A prototype of the proposed framework is shown in Sect. 4. In Sect. 5 we will present the results of a qualitative evaluation. We conclude with directions for future work.

2 Related Work

Recommender systems proposed in the literature may be classified into two perspectives: collaborative filtering systems, and knowledge based systems. An interesting review is made in [4].

Collaborative Filtering Systems use the experience of similar users to recommend destinations. In these systems two important elements are used to generate the recommendations: the user profile and the feedbacks to decide if we recommend a given destination or not.

A recent work called PlanTour was proposed by Cenamor et al. [5]. The proposed system aims at proposing a tourist plans using the human-generated information gathered from the Minube traveling social network. The main problem is the fact that the information gathered from Minube is really light like rating, city information, etc.

In this perspective we can cite the work of Castilloa et al. [6] in which the authors developed a system called SAMAP. In this system, we store the user profiles and preferences. These preferences are permanently updated by tracking the different visits made by the user and by analyzing past planning behavior of him and of similar users. The output is generated through a case-based reasoning and it consists of a list of touristic locations. The proposed system is even able to make a global timetable and plan for visiting the proposed destinations. All the information needed by the system is stored into an ontology defined by the authors.

In the same perspective we can cite also the work of Loh et al. [7]. The goal of the proposed system is to help travel agents in recommending touristic activities to their customers. The system analyzes textual messages exchanged between the travel agents and the customers through private Web chat, and recognizes customer's interest areas using a predefined ontology. The system will try to find in a database activities overlapping with the interest areas of the user.

Authors in [8] combine information coming from user profile, complaints, experiences of other customers and their ratings to some touristic locations in order to generate for every customer recommendations using fuzzy rules.

Currently, some collaborative works, tend to integrate the social networks in the recommending process. However, their number is not so important. For instance, in [9] the authors proposed a system that uses the actual geographic position of the user and the social networks in order to provide recommendations. Social Networks are used to gather information about the user profiles and to have feedbacks of other users on some location geographically near. Similarly, in [10] the authors propose a location based recommendation system. They handle a database containing attractive places tagged with keywords and containing also user profiles. They match between the user profiles and the attractive places via the TF-IDF techniques. The recommendations are generated according to the current location of the person. This recommendation is filtered in a second step using model-based collaborative filtering in which we use the ratings of similar profiles to eliminate some irrelevant recommendations.

In a recent work [11], recommendations are generated using also social networks. When a user asks for recommendations for a special category of location, he will be localized and the recommender looks for locations belonging to the same category and which are well rated by users similar to him. The ratings are extracted from social networks. The same thing in [12], authors proposed a framework for tracking potential customers using social networks. Trip destinations are proposed to users after analyzing their profiles in social networks and their preferences.

The second kind of works are Knowledge Based Systems. These works base on two main elements: the user profile and a set of rules in order to generate recommendations. In this perspective Garcia et al. [13] generate recommendations either for a single person or for group of persons. For a single person, the user profile is collected and a taxonomy is used to indicate what kind of places could be recommended to him. The taxonomy is a representation of typical kind of activities that a user may practice. The idea is to get exactly what kinds of activities can be recommended. In case of group of persons, the system will recommend the places that match the best with group users.

In [14] classical fuzzy sets are used to generate recommendations. Experts in tourism are consulted to define the fuzzy sets representing hotel features and the customer potential requirements. User profiles are collected and Fuzzified. After that, the fuzzy rules are evaluated in order to obtain a set of fuzzy values for the hotel features. A weight is computed to determine the matching level between the hotel features and the customer requirements. Hotels having high weights are recommended to customers.

In [15] a multi-agent system was proposed. An agent is attributed to every customer by which he can mention his preferences. The same thing for the activities an agent was attributed to each activity. A recommender agent will communicate with activity agents and user agents to generate recommendations.

In [16] authors proposed a recommender system for medical health care in Tunisia. They collect users' interests from the users directly and from social networks. The preferences are stored into an ontology. The authors prepared also a medical domain ontology containing information about medical activities, thalassotherapy cures centers, hotels SPA, travels and entertainments of Tunisia. An algorithm using a set of rules uses the two ontologies to generate recommendations.

In [17] authors realize a context analysis to generate recommendations. They handle a database in which we have potential attraction activities and the contextual features to which they are related. The construction of this database was made using the experience of some persons and this database is updated permanently by relevance feedback. The link between attraction activities and these contextual features is made through Bayesian models. When a person was introduced to a system, a recommendation is generated using the characteristics of his profile and using the appropriate Bayesian model.

Some works like [18] combines users' experiences and fuzzy rules to make recommendations. Recommendations generated from uses experiences are adjusted by fuzzy rules.

In the literature, there are also some works proposed to make forecast on tourism using information coming from social networks.

3 Proposed Approach

3.1 Framework Design

To design our framework, we started from the following assumptions. First of all, we think that to have an efficient recommender we should imperatively start from user experiences. They represent a kind of assessors and their opinions is a kind of assessments that we should collect, organize and on which we should capitalize to have a general idea on the quality of a given location. Second, we think that this information is essentially located in social Networks. Finally, we think also that Natural Language Processing are the most effective techniques that could help us to extract this information. An effective recommender system should be able through these techniques to analyze exactly what are the elements that users are evoking in their comments and what are their opinions.

On these assumptions, we built our framework. To the best to our Knowledge, this is the first work where clear Natural Language Processing Techniques are clearly presented to show how they have been used in tourism recommendation.

As shown in Fig. 1, the system that we propose is made of 3 modules: (*i*) the fetching module used to get comments from different data sources as Facebook, Twitter, Web pages. (*ii*) A data Analysis module used to analyze the content of fetched data and to extract important information. This information is stored into a database. (*iii*) Finally, a recommendation module that matches the user profile and the stored data in order to generate recommendations.

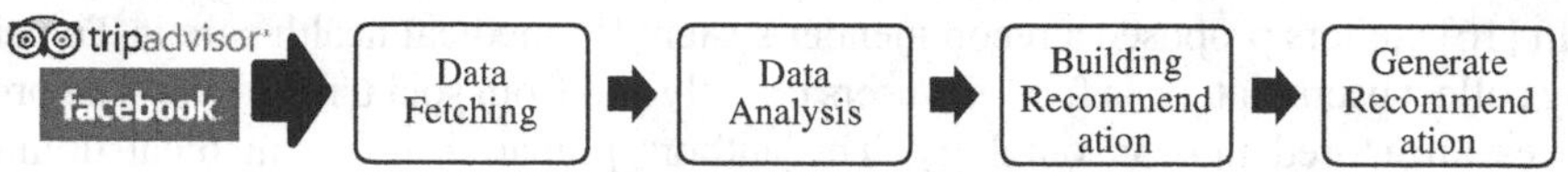

Fig. 1. Tourism recommender framework overview

3.2 Data Fetching

Data sources used to extract the comments are social networks like Facebook and Twitter and some tourism portals like Booking.com and TripAdvisor.

The data fetching consists in extracting comments, recognize the elements cited inside the comments and finally extract the sense of these comments (positive, negative, mitigated, etc.). To do that we based on Segmented Discourse Representation Theory which supposes that a given comment can be segmented into elementary units called EDU (Elementary Discourse Units) and these units are linked via rhetorical relations. Inside the EDU the smallest making of an opinion against a given concept is called and Elementary Opinion Units (EOU). We distinguish two kinds of EOU: simple and complex. For instance, in the sentence "Great Hotel", "Great" is the only word that expresses the sense. However, in the sentence "the pool was not clean", the sense is expressed by the expression "not clean".

From the portals like "TripAdvisor" or "Booking.com" we targeted people opinions dealing with the features of touristic locations (Fig. 2).

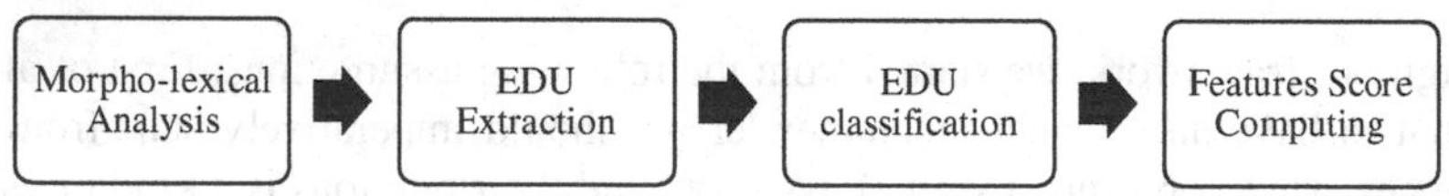

Fig. 2. Data analysis process

From the portals like "*TripAdvisor*" or "*Booking.com*" we targeted people opinions dealing with the features of touristic locations (Fig. 2).

To extract comments from "*Booking*" and "*TripAdvisor*" we used Kimono API, which is an extension of Chrome web browser and which can extract all the comments in a given site with all the necessary metadata. We have only to show what we want to extract from an excerpt of a comment and it will generalize that on all the comments.

From the social networks like Facebook or Twitter we do not targeted comments but we targeted events organized in hotels and we aim at having feedbacks about these events from persons who participated in them.

To fetch Tweets and Facebook pages we used Facepager API. This API returns the tweets and status in JSON, XML, RSS or ATOM format.

All the extract data is translated in to the English language and stored into a CSV files in order to use them to make a knowledge base for Djerba Hotels and to use it to generate recommendations.

In the CSV file we identify the source of the comment, the time, the concerned hotel, the nationality of person, and the comments in which persons evaluate the activities in the hotel, the features, the cleanness, etc.

3.3 Data Analysis

After fetching the data, the next step will be the analysis of the comments and the extraction of important parts of the discourse that can be used to generate recommendations. We are looking for Elementary Discourse Units evoking one of the elements shown, like the activities, the locations, the services, the budget, the cleanness, the rooms, the foods, etc.

The idea is to extract from the comments opinions of customers about these concepts and to store them into database in order to reuse them in the future. The process to achieve the data analysis is shown by the Fig. 2.

Morpho-Lexical Analysis. In this step, the comments are analyzed to extract sentences and identifying the categories of every word in the sentence. We use GATE software [19] which a is a text engineering machine including interesting modules that help to analyze textual information. To realize the morpho-lexical analysis we used the modules: "Document Reset", "GATE Unicode Tokenizer", "ANNIE Sentence Splitter" and the "LingPipe POS TAGGER".

Elementary Discourse Units Extraction. After categorizing words composing the comments, we should now look for EDUs that we assume that they contain Elementary Opinion Units which express opinions about some aspects related to the hotels. However, before starting the extraction of EDU we should achieve some improvements on the on the comments like the replacement of some abbreviations (LOL, ASAP), the correction of spelling mistakes, the replacement of emoticons ((:-O)), the replacement of repetitions in letters ('niiiiiiiiiiiice hotel', 'gooooooood staff', etc.), etc. After achieving all the improvement operations, we proceed now at the extraction of EDUs and we used GATE software to do that. GATE proposes JAPE language used to make rules that reflect patterns of the EDUs. We do not target all the EDUs we target only EDUs which are carrying opinions. That's why the rules that we used to extract are intended to extract only ones which are containing opinions.

Adjectives Based Rules. To express an opinion, we generally use adjectives to qualify some features in hotels. We identified 6 patterns of expressions in which we use adjectives and we made for everyone a rule to extract them:

Adverb Based Rules. In general, we use adverbs to make an exact judgement on a particular feature or we use them with past participle of verbs to make an opinion. For this reason, we decide to make 2 rules to extract EDU containing adverbs:

Predefined List Based Rules. We need to extract EDU that give information about of the hotels (location, distance from the beach, time to reach, transportation, etc.). For this reason, we defined lists (see Table 1) in which we put all distance and time units, some key locations in Djerba and transports means.

Predefined List Based Rules. We need to extract EDU that give information about of the hotels (location, distance from the beach, time to reach, transportation, etc.). For this reason, we defined lists (see Table 1) in which we put all distance and time units, some key locations in Djerba and transports means. We make 4 rules that use these predefined lists in order to extract information about the hotels.

Other Rules. Sometimes we find in the EDUs some expressions which are grammatically incorrect but they contain opinions of persons. We make 2 rules to extract them.

EDU Classification. After extracting all the EDUs We should classify them into “Positive” or “Negative”. We should also attach them to a given activity or to a given feature. This operation is known as polarity detection. As example of activities and the features that we identified we can cite “Rooms”, “Cleanness”, “Foods”, “location”, etc. So, we should detect inside the EDU the type of opinions against these features.

We compute the rate of positive or negative appreciations for every feature and we add them or we deduce them to or from the 0.5 value, taken as the reference score.

At the end of this operation we will have a score for every feature in the hotel ranging from 0 to 1. This information will be exported to CSV file.

3.4 Database Schema

To build our recommender, we have to build a database where we will store data. The database schema that we proposed. It is made of 6 entities: “User”, “Event”, “Hotel”, “Group”, “Hotel” and “Score”.

The “User” entity includes on one hand personal information about the user and on the other hand information about the user experience and about his evaluation of a specific hotel. All the hotels mentioned in the comments are referenced by the entity “Hotel”, which includes information about the hotels. “Event” is the entity referencing events that occur in the hotels. “Group” is the entity that references group of persons having similar profiles, i.e. persons who have the same nationality, the same age, the same preferences and which came to the hotel at the same period. The last entity is “Score” in which we put the scores computed for every feature.

3.5 Generate Recommendation

In this section, we will explain how user can use the proposed system to have recommendations for hotels.

Similar Group Identification. The user should type information about him and about his trip. He should also give an importance score for every feature for the targeted hotel.

Using these information, the system will try firstly to retrieve a group of persons that has been identified in the comments and which has the same profile as the user.

Of course, we can fail in retrieving a group, that matches perfectly with the user. For this reason, we can ask him to enrich the typed information. After that if the profile is retrieved we match it with the nearest one.

Compute Recommendation. At this step, the system will try to match between the scores stored in the database and which are computed per hotel for the identified group and the importance scores given by the user. A simple Euclidean distance is used to order the hotels in the recommendation List.

The next step will be the selection of the hotel and the reservation of rooms. The system should propose a web site portal proposing the lowest price.

Results Displaying. Results are displayed in a smart way. The top 5 Hotels are recommended by the system. The hotel which matches the best will be displayed in upper left corner of the page and located in Google Maps. For this hotel, a brief summary of the comments is displayed. These comments will be organized according to features. The same information will be displayed if one of the four remaining hotels will be selected.

Moreover, to have an effective idea about the hotel features, we think that we should know how they are evolving along the time. That's why we proposed another GUI in which we average the comments scores of persons for very feature and per touristic season (winter, summer, spring and autumn).

Finally, from the social networks like Facebook and Twitter we extracted events.

We display for every hotel the events that has been organized in it and we present also a general appreciation about these events computed through counting the number of positive and negative opinions and after that rating them.

4 Prototyping

A prototype of the Framework was proposed in order to evaluate it. This prototype was made in Java Language and we used in it some frameworks like JavaScript, jQuery, Html5 et CSS3. The Database that we used is MySQL.

The server on which we tested the prototype is a medium capacity server with a 2 GHZ of CPU frequency and 2 GB of RAM.

A video presentation was made online on the following URL in order to appreciate more the functionalities of the prototype:

http://www.dailymotion.com/video/k65fuq1SUS419cc9f8a.

5 Evaluation

One of the challenging problems of the Recommender Systems is their evaluation. Many works did not achieve the evaluation because of its difficulty [5, 6, 9, 10]. Indeed, to have a right evaluation for likewise systems you should present it to an important number of users with detailed evaluation criteria. Elsewhere, the evaluation will not be really effective. For this reason, we decided to evaluate our framework qualitatively by a Tourism Business Software Specialist Partner, I-WAY company.

I-Way experts recommended 4 things. First, they recommend to enrich the sources of information in addition to TripAdvisor, Booking, Facebook and twitter. They recommend also to achieve a crosscheck of comments. Indeed, some comments are not really effective and they can make a kind of noise on real comments.

The third thing that has been discussed with the experts is the use of relevance feedback. The recommend us to integrate this mechanism in the recommender system. Certainly, the return of this mechanism is not immediate but in long-term it can be interesting. The last thing is to make the framework smarter by adding the functionality of planning of the whole holidays.

6 Conclusion and Perspectives

We presented in this paper a framework for making a Tourism recommender system. The framework was mainly build on user experiences of people. Comments of these people are extracted and analyzed to be used to generate recommendation. We proposed also for this framework an implementation and a graphical user interface. The main advantage of our proposal is the fact that we present a framework with a real prototype that we hope to extend to let him mature enough to be widely used.

References

1. TripAdvisor. https://www.tripadvisor.com/ (2016). Accessed Sept 2016
2. Booking. http://www.booking.com/ (2016). Accessed Sept 2016
3. TripBarometer: Yearly TripAdvisor Report. Accessed Sept 2016
4. Borràsa, J., Morenob, A., Vallsb, A.: Intelligent tourism recommender systems: a survey. Expert Syst. Appl. **41**, 7370–7389 (2014)
5. Cenamor, I., de la Rosa, T., Núñez, S., Borrajo, D.: Planning for tourism routes using social networks. Expert Syst. Appl. **69**, 1–9 (2017)
6. Castilloa, L., Armengolb, E., Onaindíac, E., Sebastiác, L., González-Boticariod, J., Rodríguezd, A., Fernándeze, S., Ariase, J.D., Borrajo, D.: SAMAP: an user-oriented adaptive system for planning tourist visits. Expert Syst. Appl. **34**, 1318–1332 (2008)
7. Loh, S., Lorenzi, F., Saldana, R., Licthnow, D.: A tourism recommender system based on collaboration and text analysis. Inf. Technol. Tour. **6**, 157–165 (2003)
8. Schiaffino, S., Amandi, A.: Building an expert travel agent as a software agent. Expert Syst. Appl. **36**, 1291–1299 (2009)
9. Khoshnood, F., Mahdavi, M., Sarkaleh, M.K.: Designing a recommender system based on social networks and location based services. Int. J. Manag. Inf. Technol. **4**, 41–47 (2012)

10. Husain, W., Dih, L.Y.: A framework of a personalized location-based traveler recommendation system in mobile application. Int. J. Multimed. Ubiquitous Eng. **7**, 11–18 (2012)
11. Ravi, L., Vairavasundaram, S.: A collaborative location based travel recommendation system through enhanced rating prediction for the group of users. Comput. Intell. Neurosci. **2016**, 7 (2016). Article ID 1291358
12. Fernandez, Y.B., Nores, M.L., Arias, J.J.P., Duque, J.G., Vicente, M.I.M.: TripFromTV+: exploiting social networks to arrange cut-price touristic packages. In: Proceedings of IEEE International Conference on Costumer Electronics, pp. 223–224 (2011)
13. Garcia, I., Sebastia, L., Onaindia, E.: On the design of individual and group recommender systems for tourism. Expert Syst. Appl. **38**, 7683–7692 (2011)
14. García-Crespo, Á., López-Cuadrado, J.L., Colomo-Palacios, R., González-Carrasco, I., Ruiz-Mezcua, B.: Sem-fit: a semantic based expert system to provide recommendations in the tourism domain. Expert Syst. Appl. **38**, 13310–13319 (2011)
15. Batet, M., Moreno, A., Sánchez, D., Isern, D., Valls, A.: Turist@: agent-based personalized recommendation of tourist activities. Expert Syst. Appl. **39**, 7319–7329 (2012)
16. Frikha, M., Mhiri, M., Gargorui, F.: A semantic social recommender system using ontologies based approach for tunisian tourism. Adv. Distrib. Comput. Artif. Intell. J. **4**, 90–106 (2015)
17. Chang, W., Ma, L.: Personalized E-tourism attraction recommendation based on context. In: Proceedings of 10th International Conference on Service Systems and Service Management, Hong Kong, pp. 674–679 (2013)
18. Pai, P.-F., Hung, K.-C., Lin, K.-P.: Tourism demand forecasting using novel hybrid system. Expert Syst. Appl. **41**(8), 3691–3702 (2014)
19. Gate—A General Architecture for Text Engineering Documentation. http://gate.ac.uk. Accessed Sept 2016

Social Network-Based Event Recommendation

Dinh Tuyen Hoang[1], Van Cuong Tran[2], and Dosam Hwang[1(✉)]

[1] Department of Computer Engineering, Yeungnam University, Gyeongsan, South Korea
hoangdinhtuyen@gmail.com, dosamhwang@gmail.com
[2] Quang Binh University, Quang Binh, Viet Nam
vancuongqbuni@gmail.com

Abstract. The number of events generated on social networks has been growing quickly in recent years. It is difficult for users to find events that most suitably match their favorites. As a solution, the recommender system appears to solve this problem. However, event recommendation is significantly different from traditional recommendations, such as products and movies. Social events are created continuously, and only valid for a short time, so recommending a past event is meaningless. In this paper, we proposed a new even recommendation method based on social networks. First, the behavior of users be detected in order to build the user's profile. Then the users' relationship is extracted to measure the interaction strength between them. That is a fundamental factor affecting a decision of a user to attend events. In addition, the opinions about attended events are taken into account to evaluate the satisfaction of attendees by using deep learning method. Twitter is used as a case study for the method. The experiment shows that the method achieves promising results in comparison to other methods.

Keywords: Recommender system · Event recommedation · Social event

1 Introduction

In recent years, the popularity of event-based social networks, such as Meetup and Eventbrite, has significantly increased, allowing users to plan an event and share it with others. For convenience, a large number of events is generated. For example Meetup currently has 30.26 million members, 611535 monthly Meetups[1]; Eventbrite currently has two million events per year[2]. It is hard for users to find the events that suitably match their favorites. How to recommend the best-matched event to the target user is an important task.

Twitter, which is one of the most well-known online social networks, has been popular in recent years. With more than 500 million tweets per day and over 300 million users, Twitter is full of opportunities to extract information[3].

[1] https://www.meetup.com/about/.
[2] https://www.eventbrite.com/.
[3] https://about.twitter.com/company.

N.T. Nguyen et al. (Eds.): ICCCI 2017, Part I, LNAI 10448, pp. 182–191, 2017.
DOI: 10.1007/978-3-319-67074-4_18

In addition, the geo-tagged data of users and timeline information in Tweets are available through both mobile applications and social media. Therefore, based on the features that users provide, a recommender system can determine the preferences of users for certain events. It is an essential factor for generating recommendations of upcoming events that a target user might find interesting.

Recommender systems aim at producing a list of recommendations for the target user based on collaborative-based or content-based filtering. Collaborative-based filtering builds a model to predict items that the user may be interested in. Meanwhile, content-based filtering recommends items based on their characteristics and user preferences. Hybrid recommender systems combine these approaches in order to increase overall performance.

The event recommendation method is significantly different from the traditional recommendation scenarios (e.g., product recommendations), where other users have already rated the recommended items. Since events are time-varying, recommending past events is unnecessary. Moreover, users often do not clearly rate their satisfaction with the attended events. Thus, recommending useful events to a target user is a difficult task.

Some recommender systems have been developed to solve that problem. For example, Magnuson et al. [4] developed a system that connects a user's tweets with real-world events via their geo-location tags in order to create a user's interest profile about events. However, the relationships between users were not taken into account in their method. Therefore, they missed valuable information from users who can have a significant influence on a decision to attend an event. Another method [9] did not consider users' opinions about attended events. This led to some events with a low rating being recommended to the target user. In this study, a new event recommendation method is proposed by taking into account relationships between users as well as users' opinions. The relationships between users are computed based on Twitter activity, such as followers, those following, reTweets, etc. Users who have the closest relationship to the target user will be chosen for the extraction of information. This method mines the users' opinions about events via tweet content, reTweet content, and other comments by using sentiment analysis. The rest of this paper is organized as follows. In the next section, closely related work on social event recommendation systems is investigated. The proposed method is presented in Sect. 3. In Sect. 4, experiments are shown. Lastly, the conclusions and some directions for future work are presented in Sect. 5.

2 Related Works

With a large number of events published all the time in event-based social networks, such as Eventbrite, Meetup, and Plancast, it has become harder for users to find the events that best match their preferences. Therefore, event recommendation has attracted a lot of research attention in recent years. For instance, Qiao et al. [9] presented a Bayesian probability model that can wholly use the power of heterogeneous social relations, and efficiently deals with an implicit feedback feature for event recommendation. An experiment on several real-world datasets showed the utility of their method. Kang et al. [1] built a real-time event recommendation system (Eventera) from large heterogeneous online media. The system crawled large heterogeneous online media from various channels in real time and aggregated them into events. Their method also mined relationships among the events and recommended events to relevant users based on their profiles or past browsing histories. A topic modeling method [12] was applied to connect the semantic gap between events and user preferences. Latent Dirichlet allocation (LDA) was used to identify underlying latent topics to discover events that accommodate user favorites in semantics. The impact of a user's connections was also considered. The authors presented a hybrid model combining three topic modeling-based approaches. Most of the existing methods did not take into account the users' opinions about attended events from geo-tagged data of users and the timeline information available through both mobile applications and social media. These are excellent resources for finding users' preferences. On Twitter, some methods have been proposed to detect events. For example, Li et al. [3] proposed the Twitter-based Event Detection and Analysis System (TEDAS), which helps to detect new events, analyze the spatial and temporal pattern of events, and then identify their importance. Ozdikis et al. [8] presented an event detection method for Twitter based on clustering of hashtags and using semantic similarities between the hashtags. Magnuson et al. [4] built a system that connects a user's tweets with real-world events through their geo-location tags in order to generate a users' interest profile regarding events. From a Web crawl of Eventbrite, a list of events was extracted. Based on geotagged Twitter and timelines, they identified users who attended these events. From the tweets that users created, retweeted, or received, the users' opinions about an event are discovered using sentiment analysis. These are good ways to determine users' opinions. However, they did not extract relationships among users. That is a major factor that affects the decisions of users to attend events. In this paper, the relationship among users is taken into account to filter out critical information. In addition, users' opinions are analyzed in order to make better recommendations. A set of events that best match the target user will be recommended.

3 The Event Recommendation Method

This section presents how to develop a social event recommender system based on the social networks. Let $U = \{u_1, u_2, ..., u_N\}$ be a set of users, and let $E_F = \{e_1, e_2, ..., e_f\}$ be a set of future events. This method focuses on detecting a set of past social events, $Ep = \{e_1, e_2, ..., e_p\}$ of U and finds a set of future social events, $E_f (E_f \subseteq E_F)$, that target user u_m should attend. The details of the proposed method are described in the following subsections.

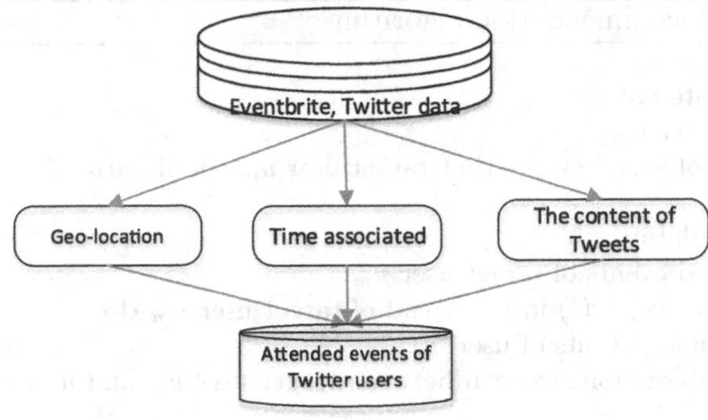

Fig. 1. The workflow of the detection stage

3.1 Detection Stage

We analyzed the content and used *time correlations* and *geo-locations* with a tweet posted to identify the past events of users, as shown in Fig. 1. In order to determine if a user has been to an event, two aspects are taken into account, as follows. First, if the user appears in an event bounding box, and second, does he/she appear in the bounding box within the time bounds of the event's lifetime? We identify that person as a participant in an event when the conditions mentioned above are satisfied. Once the user is connected to past events, we develop the user's previous event activity.

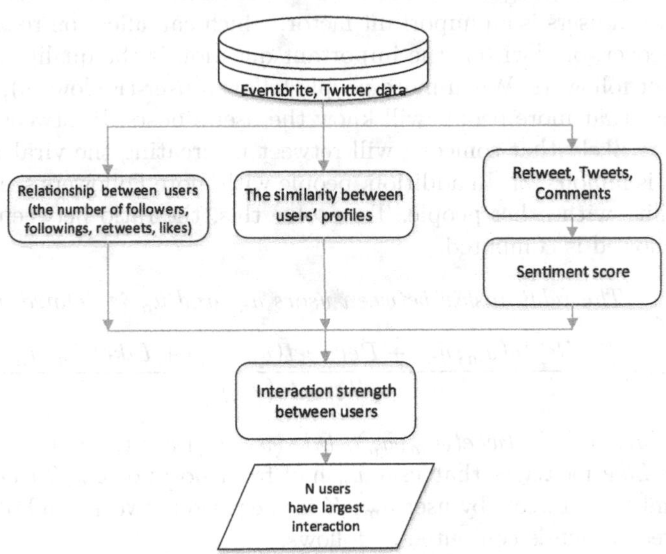

Fig. 2. The workflow of the extraction stage

Algorithm 1. Recommendation algorithm

Input: - Twitter data;
- Eventbrite data;
- Target user u_m;

Output: - A list of social events that target user u_m should attend;

Preprocessing data;
Detect attended events of target user u_m;
for each user u_i ($u_i \in U$) in list friend of target user u_m **do**
    Detect attended events of user u_i;
    Calculate interaction strength between target user u_m and user u_i;
end for;
Get N users having the largest value of interaction strength with target user u_m;
for each future event e_f in the set of future events E_F **do**
    Sentiment analysis and calculate the similarity between all attended event of N users and e_f;
    Sentiment analysis and calculate the similarity between e_f and all attended events of target user u_m;
    Calculate recommendation score;
end for;
return A list of social events have largest recommendation score;

3.2 Extraction Stage

Normally, the relationship between users can be reflected via interactivity, such as comments, likes, retweets, etc. An event favored by a user who has good relationship with the target user should be recommended. Therefore, the relationship between users is an important factor, which can affect on recommender system. Moreover, on Twitter, the important question is the quality and quantity of Twitter followers. When more people follow a user (Followers), there is a greater chance that more people will know the user's posts. If a tweet is a good one, it is more likely that someone will retweet it, creating the viral marketing that Twitter is famous for. In addition, people with more Followers seem to have more credibility with other people. To resolve this, the ratio between followers and those followed is computed.

Definition 1. *The relationship between users u_m and u_n is defined as follows:*

$$R(u_m, u_n) = \frac{1}{3} \times \left(\frac{Reply(u_m, u_n) + Retweet(u_m, u_n) + Like(u_m, u_n)}{Tweets(u_m)} \right) \times K \quad (1)$$

where $Reply(u_m, u_n)$, $Retweet(u_m, u_n)$, $Like(u_m, u_n)$ are the number of *Reply*, *Retweet* and *Like* messages that user u_n had from posts of u_m; $Tweets(u_m)$ is the total number of Tweets by user u_m; K is the ratio between the Follower and those Followed, which is computed as follows:

$$K = \frac{1}{1 + e^{-\frac{F_e}{F_i}}} \quad (2)$$

where F_e, F_i are the numbers of followers and those followed of user u_n. The value of $R(u_m, u_n)$ is ranged between 0 and 1. The greater the value of $R(u_m, u_n)$, the higher the interoperability between users.

Moreover, given that like-minded persons generally like similar items, users who regularly attend social events will likely prefer similar social events in the future. In this method, the title and textual descriptions of social events are extracted to measure the similarity between them. However, challenges are the short lengths of the titles and textual descriptions, as well as the numerous social events. In addition, the spread of social networks has brought a new way to express sentiments of individuals. A user's satisfaction with an attended event is an important factor in the decision-making process to attend similar future events. However, due to the short and informal nature of a tweet, it is hard to analyze it with sentiment analysis. To solve this problem, a word embedding model, Word2Vec [5], is used to improve the accuracy of recommendations.

Definition 2. *The similarity between the attended events of users u_m and u_n is defined as follows:*

$$S(u_m, u_n) = \sum_{e_i \in E_{u_m}} \sum_{e_j \in E_{u_n}} \frac{sim(e_i, e_j) \times Sen(e_j)}{(|E_{u_m}| \times |E_{u_n})} \tag{3}$$

where $S(u_m, u_n)$ measures the similarity between the attended events of users u_m and u_n. Function $sim(e_i, e_j)$ indicates the value of the content similarity between events e_i and e_j. The value of $sim(e_i, e_j)$ is computed by using deep learning Doc2Vec model, which obtains state-of-the-art results in document similarity [2]. $|E_{u_m}|$, $|E_{u_n}|$ are the number of all attended social events of users u_m and u_n. $Sen(e_j)$ represents the sentiment of user u_n about event e_j. Normally, the value of $S(u_m, u_n)$ ranges between 0 and 1.

The interaction strength between users is calculated by combining Eqs. (1) and (3) as follows:

$$T(u_m, u_n) = \alpha.R(u_m, u_n) + (1 - \alpha).S(u_m, u_n) \tag{4}$$

where $0 \leq T(u_m, u_n) \leq 1$, which represents the strength of the interaction between users u_m and u_n. Parameter α is a constant, which controls the rates of reflecting two importance values to the user's relationship and the similarity of attended events. By means of ranking the values of interaction strength, we obtain N users who have the greatest interaction with the target user, as shown in Fig. 2.

3.3 Recommendation Score

Normally, the best predictor of future behavior is past behavior. People tend to attend events with themes that match their personal interests. Thus, the attended events of target user u_m is analyzed, and we calculate the similarity between them and future event (e_f). In addition, sentiment analysis is mined to measure the user's satisfaction about attended events. These are two key factors that influence the decisions of target user u_m, computed as follows:

$$F_1 = \frac{\sum_{e_i \in E_{u_m}} sim(e_i, e_f) \times Sen(e_i)}{|E_{u_m}|} \tag{5}$$

where function $sim(c_i, c_f)$ is the content similarity between events e_i and e_f. The greater the value of F_1; the higher the probability that user u_m will attend future event c_f.

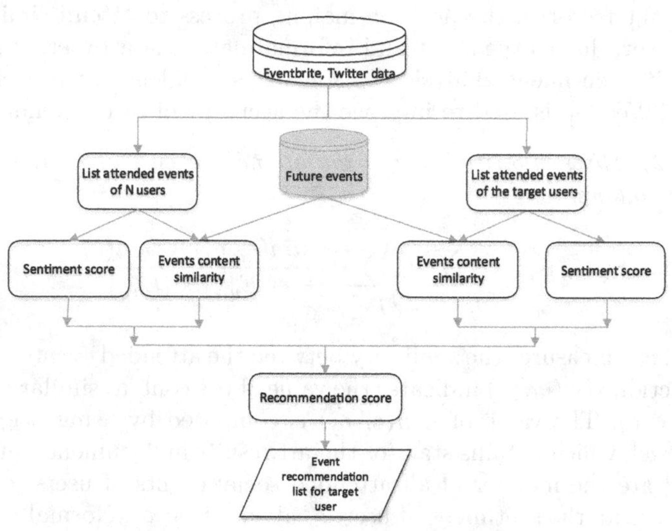

Fig. 3. The workflow of the recommendation stage

In addition, all attended events of N users who have largest interaction with the target user are computed for the similarity to future events as follows:

$$F_2 = \frac{\sum_{u_i \in U_N} \left(\frac{\sum_{e_i \in E_N} sim(e_i, e_f)}{|E_N|} \times T(u_m, u_i) \right)}{\sum_{u_i \in U_N} T(u_m, u_i)} \tag{6}$$

where U_N is the top N users who have the greatest interaction with target user u_m, and $U_N \subseteq U$; E_N shows all events that the top N users had attended.

As shown in Fig. 3, a recommendation score is created by combining Eqs. (5) and (6), as follows:

$$F(u_m, e_f) = (1 - \beta).F_1 + \beta.F_2 \tag{7}$$

where $\beta \in [0, 1]$, which controls the importance of the weights between F_1 and F_2. The value of $F(u_m, e_f)$ is within the range [0,1]. A set of recommended events for target user u_m is created based on ranking the recommendation score obtain by Eq. (7).

4 Experiments

4.1 Datasets

This study applies the proposed recommender system to Twitter and Eventbrite datasets. Twitter provides REST APIs[4], which users can use to interact with the service. For collecting the Twitter dataset, we used the Tweepy[5] library, which allowed getting to a user most recent around 3200 tweets. We collected tweets from users and remove some users who have a few tweets. We selected 2015 users to do experimentation. Tweet geolocations and timestamps were extracted for use in detection users' attended events. The Scrapy[6] library was used to crawl events from Eventbrite to extract event content. We extracted 300 events that related to the 2015 users mentioned above and identified their locations. The time span that the events were going to last, and the contents related to the events, were analyzed using topic modeling tools. The event dataset was crawled from Eventbrite[7], which supports an application programming interface (API) to access the available dataset.

4.2 Evaluation

In this work, a prediction accuracy method, which is the most popular way employed in the recommendation literature, evaluates the performance of the proposed method. A set of events was collected from Eventbrite. A target Twitter user was randomly selected, and then the recommender system predicted a set of events that this selected user should attend. We divided the dataset into two parts. The first part was used for the training dataset, and the second part was used for the testing dataset. Usually, the system recommends many events, but a target user only likes a few of them. Thus, in an offline scenario, we assumed that the recommendation results would not include events that the target users or their friends had not attended. The values of *Precision*, *Recall*, and *F_measure* are computed as follows:

$$Precision = \frac{TP}{TP + FP} \tag{8}$$

$$Recall = \frac{TP}{TP + FN} \tag{9}$$

$$F\_measure = \frac{2 \times Precision \times Recall}{Precision + Recall} \tag{10}$$

[4] https://dev.twitter.com/rest/public.
[5] http://tweepy.readthedocs.io/en/v3.5.0/.
[6] texthttps://scrapy.org/.
[7] https://www.eventbrite.com/.

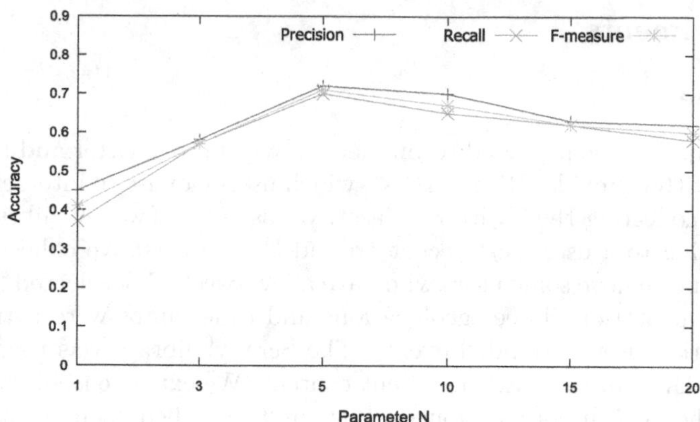

Fig. 4. The accuracy of the proposed system

4.3 Results and Discussions

We implemented the recommender system with N sets at 1, 3, 5, 10, 15, or 20, and parameters α and β were both set at 0.5. The results of the recommender system are shown in Fig. 4. As shown in Fig. 4, the accuracy of the system got the best results at $N = 5$. When N is less than 5, other users who also have good interactions might be skipped. Besides, when N is greater than 5, the interaction strength between the target users and their friends will be weaker. Thus, the recommendations generated by a small or large value of N are prone to error. In particular, the results show that the interaction strength between target users and their friends is a major factor that impacts the decisions of users to attend social events. Many events are created on social networks, but the target users might only attend a few of them. This research filtered them and focused on recommending social events that are similar to previously attended events of the target users and their friends. Therefore, this recommender system overcomes other disadvantages mentioned by Magnuson et al. [4]. In addition, the geo-tagged data of users and timeline information are available through both mobile applications and social media. Therefore, these are good resources for analyzing users' opinions about the events they attended. This is helpful in making better recommendations.

5 Conclusion and Future Work

This work proposed a new method for recommending events to target users. Firstly, the proposed method detects events attended by users to build users' profiles. Secondly, the interaction strength between users was calculated by taking into account the users' relationships and similarities in their profiles. Also, their opinions were analyzed to determine the N users who have greatest interaction strength with the target users. Lastly, recommendation scores were computed by combining the similarities between future events and attended events of the N users and the target users. In addition, the opinions of the N users and the target users about attended events were analyzed to help make better recommendations. The experiment shows that our approach achieves promising results in social network-based event recommendations. In future work, we will integrate more social network databases, such as Facebook, Instagram, and Flickr [6], in order to improve the accuracy of the recommender system. For the integration process, ontology and consensus methods should be used [7,10,11].

Acknowledgment. This research was supported by Basic Science Research Program through the National Research Foundation of Korea (NRF) funded by the Ministry of Science, ICT & Future Planning (21C000151).

References

1. Kang, D., Han, D., Park, N., Kim, S., Kang, U., Lee, S.: Eventera: real-time event recommendation system from massive heterogeneous online media. In: 2014 IEEE International Conference on Data Mining Workshop, pp. 1211–1214. IEEE (2014)
2. Le, Q.V., Mikolov, T.: Distributed representations of sentences and documents. arXiv preprint arXiv:1405.4053 (2014)
3. Li, R., Lei, K.H., Khadiwala, R., Chang, K.C.C.: TEDAS: a twitter-based event detection and analysis system. In: 2012 IEEE 28th International Conference on Data Engineering, pp. 1273–1276. IEEE (2012)
4. Magnuson, A., Dialani, V., Mallela, D.: Event recommendation using twitter activity. In: Proceedings of the 9th ACM Conference on Recommender Systems, pp. 331–332. ACM (2015)
5. Mikolov, T., Chen, K., Corrado, G., Dean, J.: Efficient estimation of word representations in vector space. arXiv preprint arXiv:1301.3781 (2013)
6. Nguyen, N.T.: Advanced Methods for Inconsistent Knowledge Management. Advanced Information and Knowledge Processing. Springer, London (2008). doi:10.1007/978-1-84628-889-0
7. Nguyen, V.D., Merayo, M.G.: Intelligent collective: some issues with collective cardinality. J. Inf. Telecommun. (2017). doi:10.1080/24751839.2017.1323702
8. Ozdikis, O., Senkul, P., Oguztuzun, H.: Semantic expansion of hashtags for enhanced event detection in twitter. In: Proceedings of the 1st International Workshop on Online Social Systems. Citeseer (2012)
9. Qiao12, Z., Zhang, P., Zhou, C., Cao, Y., Guo, L., Zhang, Y.: Event recommendation in event-based social networks. In: Proceedings of the Twenty-Eighth AAAI Conference on Artificial Intelligence. AAAI (2014)
10. Sriharee, G.: An ontology-based approach to auto-tagging articles. Vietnam J. Comput. Sci. **2**(2), 85–94 (2015)
11. Uddin, M.N., Duong, T.H., Nguyen, N.T., Qi, X.M., Jo, G.S.: Semantic similarity measures for enhancing information retrieval in folksonomies. Expert Syst. Appl. **40**(5), 1645–1653 (2013)
12. Zhang, Y., Wu, H., Sorathia, V.S., Prasanna, V.K.: Event recommendation in social networks with linked data enablement. In: ICEIS, vol. 2, pp. 371–379 (2013)

Deep Neural Networks for Matching Online Social Networking Profiles

Vicentiu-Marian Ciorbaru and Traian Rebedea(✉)

University Politehnica of Bucharest,
313 Splaiul Independentei, 060042 Bucharest, Romania
vicentiu@ciorbaru.ro, traian.rebedea@cs.pub.ro

Abstract. This paper details a novel method for grouping together online social networking profiles of the same person extracted from different sources. Name ambiguity arises naturally in any culture due to the popularity of specific names which are shared by a large number of people. This is one of the main problems in people search, which is also multiplied by the number of different data sources that contain information about the same person. Grouping pages from various social networking websites in order to disambiguate between different individuals with the same name is an important task in people search. This allows building a detailed description and a consolidated online identity for each individual. Our results show that given a large enough dataset, neural networks and word embeddings provide the best method to solve this problem.

Keywords: Profile matching · Social networking · Deduplication · Deep learning

1 Introduction

Billions of queries are performed each day using online search engines [11]. Out of all these queries, 11–17% include a person name and about 4% contain only a person name [1,2]. Extracting the correct personal information from the web pages retrieved by a search engine is a difficult process, however this task is particularly useful for people search. The most important problem in people search is name ambiguity, which makes the process of extracting personal information from a web page and assigning it to a specific individual cumbersome. There is a huge overlap in person names worldwide and the ratio of identical names to different people is around 1 to 10,000. For example, the most popular 90,000 full names (first and last names) worldwide are shared by more than 100 million individuals [3]. Thus, detecting which person an online document actually refers to poses challenges both to humans and to machines.

Lately, there has been a change in focus in people search, shifting from performing searches using generic online search engines to making use of social networking profiles [6]. Therefore, we have chosen not to focus on the general

N.T. Nguyen et al. (Eds.): ICCCI 2017, Part I, LNAI 10448, pp. 192–201, 2017.
DOI: 10.1007/978-3-319-67074-4_19

people search use case of disambiguating generic web pages containing personal information. Instead we have made use of the more reliable structure of social networking websites, from which personal data can be extracted easier. The ability to deduplicate profiles, especially within professional social networks, is important for generating relevant people search results by building a consolidated online identity for each person [12].

This paper compares several methods for identifying all the different social profiles of a given individual in a large dataset. Our proposed solution employs a binary classifier implemented with a neural network (NN) that decides whether two social profiles belong to the same person. The classifier makes use of features generated using Word2Vec [5] for the texts in each profile. We also implement several domain specific features, such as the edit distance and cosine similarity between names in each profile. The set of all profiles belonging to a single individual will be referred to as a cluster and a perfect solution should provide a perfect clustering of the social profiles. To highlight the degree of name ambiguity present in the dataset, we also provide a simple baseline by grouping all nodes that have identical or very similar names. Alternate unsupervised and supervised solutions are implemented as comparison for the proposed NN approach.

Section 2 continues with an overview of relevant related work in people search, focusing on personal web page deduplication. Our proposed approach for treating social profiles deduplication as a classification problem is presented in Sect. 3, while the next section contains the most important details about the dataset and the features employed by the classifiers. The experimental setup is detailed in Sect. 5, focusing on the NN solution which achieves a substantial improvement over all other methods. These results are highlighted in Sect. 6, followed by concluding remarks and details on future work.

2 Related Work

Deduplication is an important problem in people search and several different solutions have been proposed to solve it. Some of the most relevant ones were put forward as part of the Web People Search (WePS) competitions [1,2]. However, these contests focused on a more general problem than the one solved by us, grouping web pages that relate to the same specific person.

2.1 Personal Web Pages Deduplication

This was the main problem that the WePS 2 competition [2] aimed to solve. Given all the web pages returned by a generic search engine for an ambiguous query containing a popular name, one needed to cluster these pages such that each cluster corresponded to one specific person. There are a number of possible metrics available to assess the accuracy of this clustering. Due to their robustness to adversarial classifications, we have chosen the metrics recommended by WePS: B-Cubed precision and recall and the corresponding F1 score.

$$Correctness(e, e') = \begin{cases} 1 \iff L(e) = L(e') \text{ and } C(e) = C(e') \\ 0 \text{ otherwise} \end{cases} \quad (1)$$

$$Precision_{BCubed} = Avg_e[Avg_{e'.C(e)=C(e')}[Correctness(e,e')]] \quad (2)$$

$$Recall_{BCubed} = Avg_e[Avg_{e'.L(e)=L(e')}[Correctness(e,e')]] \quad (3)$$

In (1)–(3), $L(e)$ represents the manually assigned label (person identity) for web page e, while $C(e)$ is the cluster computed for page e.

The best solution for WePS 2 employed several techniques to produce the final clustering: feature extraction from web pages using Wikipedia concepts and Named Entity Recognition (NER), together with a bag of words (BoW) model [4]. All the features discovered in a page were then attached a weight according to a feature weighting model. Clustering is then performed based on these feature vectors using both cosine and overlap similarity to compare documents. Hierarchical Agglomerative Clustering (HAC) was selected as the best method to group pages belonging to the same person.

A pairwise approach for solving this problem was proposed by Nuray-Turan et al. [8]. The authors proposed a two step process. First, they computed the probability that two pages may refer to the same person. The second step was to perform a clustering of the pages by joining pairs that have a sufficiently high probability for representing the same person. The probability function was computed by joining multiple features including the named-entity similarity, middle name dissimilarity, and hyperlink-based similarity.

2.2 Social Networking Profiles Matching

More recent research has turned away from deduplicating general web pages containing personal information to linking social networking profiles belonging to the same individual. Zhang et al. [13] use a binary classifier to identify profiles of a given person from different online social networks. The classifier is a factor graph, called matching graph, that learns an energy function which outputs the probability that two profiles from different social networks are referring to the same real-world person. Their research highlights the difficulty in training the proposed model, given the large set of possible pairs in the factor graph. The authors make use of a vector space using BoW and TF-IDF for the text in each social profile, but also look at its social status (position in network) and connections. Since our dataset does not contain connections (e.g. friends or followers) within the same social network, our method is different as we only use the textual information in a profile. However, considering also the connections for each profile should provide further improvement to our method.

3 Proposed Approach

We have attempted to solve the social profile matching problem using two different methods: unsupervised clusterization and binary classification. These solutions have been inspired by previous research in personal web page and profile matching. However we improve these results by proposing a deep NN trained on the large dataset presented in the previous section.

For the unsupervised approach, first we generated the feature vector for each profile, followed by Hierarchical Agglomerative Clustering (HAC) using cosine distance. We varied the number of desired clusters and computed the precision and recall values for several thresholds.

For the binary classification scenario, our objective is twofold. First, we aim to detect whether two profiles refer to the same person and can be matched (pairwise matching). Second, for the graph of connected profiles discovered in the first phase we compute its connected components. Each connected component should represent the identity of a single individual (one of the clusters in the first scenario). To learn the pairwise matching function, we trained a binary classifier over a dataset constructed as shown in (4).

$$f(u,v) = \begin{cases} 1 \text{ if pages u an v refer to the same person} \\ 0 \text{ otherwise} \end{cases} \tag{4}$$

We have experimented with multiple classifiers, but the two most noteworthy results came from using a Random Forest classifier and a neural network (NN) trained using TensorFlow[1]. The best results were obtained using a NN classifier with several dense layers in a deep learning manner. Other classifiers have a rather poor performance, either due to the difficulty of training them on the full train set or by producing poor results mainly caused by underfitting.

4 Dataset and Features

4.1 Profiles Dataset

Our corpus contains a snapshot of multiple social networking profiles collected throughout the year 2015. The list of social networks used is: Academia, Code-Project, Facebook, Github, Google+, Instagram, Lanyrd, Linkedin, Mashable, Medium, Moz, Quora, Slideshare, Twitter, and Vimeo. Making use of the page structure of each website, for each profile we have extracted several/all of the following attributes: username, name (full name or distinct first and last names), gender, bio (short description), interests, publications, jobs, etc. The average number of social profiles per individual is 2.04 and the maximum is 10. Most profile pages feature a brief description (bio) of the owner. Profiles do not contain their connections, nor any posts written by the owner. The dataset is large, containing more than half a million profile pages extracted automatically from about.me[2]. These profiles are manually built by users and contain links to other social profiles of that user in different platforms, thus providing a user-annotated ground truth. The total number of about.me accounts in our dataset is 210, 967.

The entire corpus was provided by Wholi[3] and is one of the largest corpora used for social profile matching. While other datasets [9,13] have a larger number of distinct profiles, the ground truth (profiles belonging to the same

[1] https://www.tensorflow.org/.
[2] https://about.me/.
[3] https://www.wholi.com.

individual manually labeled) is one order of magnitude larger in our dataset (200,000+ compared to 10,000 items). This allows training a more complex classifier, such as the deep NN model proposed in the subsequent section. However, it is important to understand that all this data has been entered by the users of the about.me website, thus the ground truth might be incorrect in some cases (e.g. entry errors, misbehaviour of users). We can say that this dataset is a crowd-sourced corpus. This type of dataset have been widely used in natural language processing (NLP) lately.

When performing the train/test split on the dataset we took into account the following rules: (1) Train and test sets should contains different identities (persons); (2) The clusters in the training set should have no entries present in the test set in order to avoid overfitted models; (3) The test set must have the same distribution for cluster sizes as the train set to provide a relevant comparison for various sized online identities.

All the positive class pairs were generated for both datasets. Due to the size of the dataset, only a subset of negative pairs were generated for both datasets. The number of negative pairs is 10 times greater than the positive ones for the training set and 100 times greater for the test set. These ratios were chosen to be as close as possible to a real-world scenario during testing (each profile has 100 negative candidates and some positive ones when testing), but also to avoid a very imbalanced dataset when training. The final train and test sets statistics are presented in Table 1.

Table 1. Training and test datasets for profile matching as binary classification

Dataset	Total items	Positive items	Negative items	Class split
Train	4,000,000	400,000	3,600,000	1:9
Test	1,010,000	10,000	1,000,000	1:100

4.2 Profiles Features

To train the classifiers, we extracted from each profile a vector of features which were passed to the classification algorithm after applying normalization.

Domain Specific Features. Generating the feature vector was done by extracting relevant information from each profile. Several important features which are problem-specific and are discussed next.

First and last name represent the first discriminative features specific to this problem. Almost all items in the dataset feature these attributes. Although not impossible, it is unlikely that two profiles belong to the same person if the first or last names differ. Thus, we compute the edit distance between names in a candidate profile pair. Gender is also a very important signal for classification. This attribute might be flawed for some entries in the dataset, but in general gender needs to match for two profiles to belong to the same person. Thus, we treat gender as binary feature (matches or not). A matching country is another relevant feature and it is provided as a nominal attribute with 3 distinct values (countries match, no match, and information unavailable).

We also wanted to extract relevant features from all the other textual attributes in a profile (e.g. bio, publications, interests). Merely using a BoW model with TF-IDF would have provided a very sparse representation given our dataset (few words per profile, large dictionary). In turn, we generated 300-dimensional vectors (called word embeddings) through Word2Vec [5] for each word in the profiles' textual description. Then we compute the embeddings for a document as the average embedding score over all the words in a profile as shown in (5).

$$embedding(doc) = \frac{\sum_{word}^{words} Word2Vec(word)}{|words|} \qquad (5)$$

For the pairwise classifier, we needed to generate pairwise embeddings. To generate a feature vector for a profile pair we simply concatenate the embeddings of both profiles. As the pairwise features are not symmetric, the resulting vector for pair (a, b) is different from the resulting vector for pair (b, a). To counteract this problem we generated both pairs and fed them both to the classifier.

$$pair\_embedding(a, b) = concat(embedding(a), embedding(b)) \qquad (6)$$

Distance Based Features. The supervised approaches presented in the experimental section also make use of several distance-based features specific for a pair of profiles. These extra features are not employed by the unsupervised approach. The edit distance between names is a strong discriminatory feature for candidate profile pairs. For this, we convert all profile names to lowercase and then compute the Levenshtein distance between the names. Cosine similarity and euclidean distance are also computed based on the two document embeddings for each candidate profile pair.

Features Normalization. The feature values extracted from our dataset are not normalized by default. To normalize them, we employ standardization thus computing the z-scores for each feature by substracting the mean value and then dividing by the standard deviation of that feature. Finally, we perform a dimensionality reduction using Principal Components Analysis (PCA). The data's dimensionality is reduced to a 25-dimensional subspace and returned to the original feature space via the inverse transformation. This improves the performance of all classifiers; for example, for the proposed NN classifier recall boosts by 2% for the same precision.

5 Experiments

In this section we detail the methods for generating the final groups of profiles, including how each method uses the feature vectors defined in the previous section. We also introduce a simple baseline, grouping all profiles that have identical names, to highlight the level of profile ambiguity in our dataset.

5.1 Unsupervised Clustering for Profile Matching

To generate the final groups of profiles belonging to the same individual using Hierarchical Agglomerative Clustering (HAC) the first step is to compute the feature vectors as described in Sect. 4.2, less the pairwise features. Then, we compute the pairwise cosine distances between all profile pairs. The PCA transformation previously discussed is performed before generating the distances between pairs in order to remove the noise in the clustered items. HAC is performed based on these pairwise distances. The threshold for the HAC is determined using grid search to maximize the B-Cubed F1 score for the training set. Finally, the output of the HAC algorithm is compared to the gold standard.

5.2 Binary Classification for Profile Matching

The supervised approach trains a classifier which learns a separation function between pairs of profiles that need to be matched and the rest. Feature vectors are generated as for the unsupervised approach, with the addition of the pairwise features. The classifiers are trained using different weights for false positives and false negatives to counteract the imbalanced dataset. The final grouping is performed by computing the connected components for the graph having profiles as nodes and edges between pairs of profiles labeled positive by the classifier.

Random Forest Classifier. Our first supervised solution was to train a Random Forest (RF) classifier, to compare it with the unsupervised approach. Even without tuning its meta-parameters, the RF classifier surpasses both the baseline and the HAC proving that supervised profile matching achieves better results. The main advantage over the unsupervised approach is that HAC needs to have the pair-wise distances between all pairs of profiles, thus limiting its usage over subsets of the proposed test set due to its large size.

Neural Network Classifier. Given the promising results of the RF classifier, we decided to experiment with a more complex model, making incipient usage of the recent advances in deep learning for NLP tasks. Thus, we train a neural network (NN) with several dense layers that takes as input all the features, including Word2Vec embeddings, of a candidate pair and outputs the probability for the pair to refer to the same person. Using a NN is justified as we need a classifier able to model complex combinations of features and to learn nonlinearities.

The architecture of the neural net is described in Fig. 1. The first layer takes as input the features computed for the candidate profile pair. The next two layers iteratively reduce the dimensionality of the representation ending up with a denser feature space. The final layers employ RELU activation for the neurons,

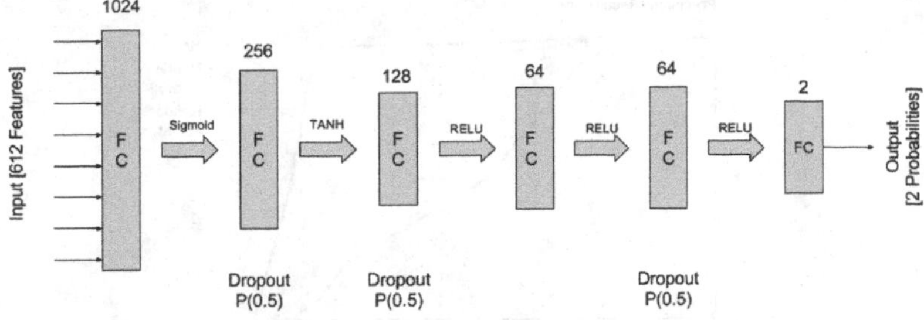

Fig. 1. Architecture of the neural network used for profile matching

as RELU units are known to provide better results for binary pairwise classification [7]. To avoid overfitting we employed dropout [10], while the output is obtained with a softmax layer. All these layers are dense (fully-connected) layers from TensorFlow. The loss function shown in (7) is defined using the cross-entropy function, with an added weight for false positives which contribute 10 times more to the loss score. This aims to penalize false connections between profiles and to counteract the imbalanced distribution.

$$L(y) = -\frac{1}{N}\sum_{n=1}^{N}[y_n * \log y'_n + (1 - y_n)\log(1 - y'_n)] * W(y_n, y'_n) \quad (7)$$

$$W(y, y') = \begin{cases} 10 & \text{if y is positive and y' is negative} \\ 1 & \text{otherwise} \end{cases} \quad (8)$$

6 Results

We have made several experiments using the training and test sets described in Table 1. All the proposed methods were assessed using an imbalanced test set with one positive profile pair for 100 negative candidates. The datasets are imbalanced to reflect a real-world scenario, where for each correct match between two profiles, one also compares tens of incorrect (but similar) candidates.

Table 2. Comparison of the proposed methods for social profiles matching

Model	Precision	Recall	F1
Baseline (same name)	0.79	0.57	0.66
Baseline (edit distance ≤ 3)	0.61	0.70	0.65
Unsupervised - HAC	0.58	0.45	0.51
Supervised - RF (only embeddings)	0.45	0.20	0.27
Supervised - RF (all features)	0.84	0.60	0.69
Supervised - Deep NN	**0.95**	**0.85**	**0.90**

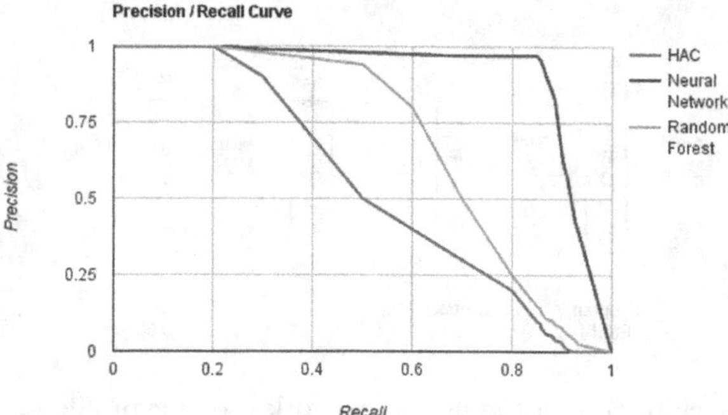

Fig. 2. Precision-Recall curves on the test dataset

The B-cubed precision and recall obtained on the test set for each method are presented in Table 2. It can be observed that the baseline achieves a good precision, with an average cluster size of 1.8 profiles. Using the same name to generate groupings, the baseline matches most of the profiles belonging to the same person correctly ($P = 0.79$). However, it generates too many groupings and misses several connections between profiles ($R = 0.57$). By allowing a larger name mismatch (edit distance ≤ 3), the baseline drops in precision significantly. Unfortunately, HAC performs worse than the baseline on our dataset, mainly because the cosine distance is not a good measure for cluster/item similarity for the proposed feature vectors.

The RF classifier performs well only when domain specific features are added to the word embeddings. The classifier identifies name distance as an important feature for classification and achieves better precision and recall compared to the baseline. Due to the large training set, we were forced to limit the complexity of the RF ($numberoftrees = 12$). The neural network overcomes this limitation thanks to mini-batch training. Thus, it is able to learn a combination between word embeddings and name distance, grouping profiles with similar embeddings and similar names. Due to the usage of embeddings, it can also correctly group profiles with different names, significantly improving the recall ($R = 0.85$).

In Fig. 2 are plotted the Area Under Curve (AUC) precision-recall graphs for the proposed methods. By varying the penalty for false negatives in the case of classifiers and the threshold for HAC, we can opt for a trade-off between precision and recall. It can be seen that while the RF can obtain good precision only for $R \leq 0.60$, the deep NN can achieve both $R \geq 0.80$ and high precision.

7 Conclusions

In this paper we have shown that deep neural networks can be successfully used for social profile matching given a sufficiently large dataset for training. The experiments showed that a deep NN can achieve a high precision ($P = 0.95$)

with a good recall rate ($R = 0.85$), surpassing other unsupervised or supervised methods for this task. As far as we know, this result outperforms existing approaches for social network profile matching. Further advancements can be made by training more complex deep learning models, using recurrent or convolutional networks, and also by adding features extracted from profile pictures.

References

1. Artiles, J., Borthwick, A., Gonzalo, J., Sekine, S., Amigó, E.: Weps-3 evaluation campaign: overview of the Web people search clustering and attribute extraction tasks. In: CLEF (Notebook Papers/LABs/Workshops) (2010)
2. Artiles, J., Gonzalo, J., Sekine, S.: Weps 2 evaluation campaign: overview of the web people search clustering task. In: 18th WWW Conference 2nd Web People Search Evaluation Workshop (WePS 2009), vol. 9. Citeseer (2009)
3. Artiles, J., Gonzalo, J., Verdejo, F.: A testbed for people searching strategies in the www. In: Proceedings of the 28th Annual International ACM SIGIR Conference on Research and Development in Information Retrieval, pp. 569–570. ACM (2005)
4. Chen, Y., Lee, S.Y.M., Huang, C.R.: Polyuhk: A robust information extraction system for web personal names. In: 18th WWW Conference 2nd Web People Search Evaluation Workshop (WePS 2009) (2009)
5. Mikolov, T., Sutskever, I., Chen, K., Corrado, G.S., Dean, J.: Distributed representations of words and phrases and their compositionality. In: Advances in Neural Information Processing Systems, pp. 3111–3119 (2013)
6. Morris, M.R., Teevan, J., Panovich, K.: What do people ask their social networks, and why? a survey study of status message q&a behavior. In: SIGCHI Conference on Human Factors in Computing Systems, pp. 1739–1748. ACM (2010)
7. Nair, V., Hinton, G.E.: Rectified linear units improve restricted boltzmann machines. In: Proceedings of the 27th International Conference on Machine Learning (ICML-10), pp. 807–814 (2010)
8. Nuray-Turan, R., Chen, Z., Kalashnikov, D.V., Mehrotra, S.: Exploiting web querying for web people search in weps2. In: 18th WWW Conference 2nd Web People Search Evaluation Workshop (WePS 2009). Citeseer (2009)
9. Perito, D., Castelluccia, C., Kaafar, M.A., Manils, P.: How unique and traceable are usernames? In: Proceedings of the 11th International Conference on Privacy Enhancing Technologies, PETS 2011, pp. 1–17 (2011). http://dl.acm.org/citation.cfm?id=2032162.2032163
10. Srivastava, N., Hinton, G.E., Krizhevsky, A., Sutskever, I., Salakhutdinov, R.: Dropout: a simple way to prevent neural networks from overfitting. J. Mach. Learn. Res. **15**(1), 1929–1958 (2014)
11. Sullivan, D.: Google now handles at least 2 trillion searches per year (2016). http://searchengineland.com/google-now-handles-2-999-trillion-searches-per-year-250247. Accessed 10 Apr 2017
12. Watts, D.J., Dodds, P.S., Newman, M.E.: Identity and search in social networks. Science **296**(5571), 1302–1305 (2002)
13. Zhang, Y., Tang, J., Yang, Z., Pei, J., Yu, P.S.: Cosnet: connecting heterogeneous social networks with local and global consistency. In: Proceedings of the 21th ACM SIGKDD International Conference on Knowledge Discovery and Data Mining, KDD 2015, NY, USA, pp. 1485–1494 (2015). http://doi.acm.org/10.1145/2783258.2783268

Effect of Network Topology on Neighbourhood-Aided Collective Learning

Lise-Marie Veillon[1(✉)], Gauvain Bourgne[2], and Henry Soldano[1,3]

[1] Université Paris 13, Sorbonne Paris Cité, L.I.P.N UMR-CNRS 7030, 93430 Villetaneuse, France
{Veillon,Henry.Soldano}@lipn.univ-paris13.fr

[2] CNRS & Sorbonne Universités, UPMC Université Paris 06, LIP6 UMR 7606, 4 Place Jussieu, 75005 Paris, France
Gauvain.Bourgne@lip6.fr

[3] Atelier de BioInformatique, ISYEB - UMR 7205 CNRS MNHN UPMC EPHE, Museum National D'Histoire Naturelle, 75005 Paris, France

Abstract. This article is about multi-agent collective learning in networks. An agent revises its current model when collecting a new observation inconsistent with it. While revising, the agent interacts with its neighbours in the community, and benefits from observations that other agents send on a utility basis. When considering the learning speed of an agent with respect to all the observations within the community, it clearly depends on the neighbourhood structure, i.e. on the network topology. A comprehensive experimental study characterizes this influence, showing the main factors that affect neighbourhood-aided collective learning. Two kinds of informations are propagated in the networks: hypotheses and counter-examples. This study also weights the impact of these propagation by considering some variants in which one kind of propagation is stopped. Our main purpose is to understand how network characteristics affect to what extent the agents learn and share models and observations, and consequently the learning speed within the community.

Keywords: Collective learning · Multi-agents learning · Agents network

1 Introduction

The role of networks, as expressing the link between intelligent entities, become prominent through the emergence of the Internet of Things, multi agent systems, online social networks, mobile assistants and so on. While artificial intelligence at the individual level still is an important area, there is a clear need to study learning phenomena at the collective level, and in particular their dependency on the network structure that connect these entities. We will use in this study a simple collective learning protocol, with learning guarantees, to observe efficiency of collective learning mechanisms with respect to both network topology and agents behaviour.

N.T. Nguyen et al. (Eds.): ICCCI 2017, Part I, LNAI 10448, pp. 202–211, 2017.
DOI: 10.1007/978-3-319-67074-4_20

SMILE (Sound Multi-agents Incremental LEarning) [2] is a multi agent learning protocol that consider learning in a community or multi-agent system (MAS) as the consequence of a set of interleaved model revisions and interactions resulting from queries from a current learning agent to other agents. It investigates interaction-based collective learning phenomena through simulations, in a way close to what is proposed in Ontañón [8]. In such a framework, agents looks primarily at revising their current model, when addressing some online learning from observations task. They also collaborate with other agents, asking queries regarding the quality of their current model with respect to observations known of other agents, and answering other agents queries by transmitting counter-examples, i.e. observations they know of that contradict the querying agent's model. From the point of view of a single agent, this is a typical query learning situation [1] where the omniscient oracle is replaced by a set of imperfect agents.

In the original SMILE protocol all agents were linked together: each agent could send a query to all other agents, in order to benefit from their counter-examples to revise its current model. Further variants of the protocol consider the case of a network of agents [4] and compensate the fact that agents only interact with their neighbours by allowing some form of propagation of models or queries within the network. However, in the present study we are interested in the effect of the network topology on the decrease in learning accuracy when there is no such compensation, i.e. when revising their current model, agents do no have access to any information outside of their neighbourhood. For that purpose we stick to the original SMILE protocol that in this case only guarantees consistency of an agent current model with observations found in its neighbourhood. Furthermore, in the original SMILE protocol, another behaviour is added to agents, that we call *adoption*. This means that whenever an agent finds its current model consistent with the observations known by the other agents, it notifies the community that a new model is available that is consistent with all observations known in the community at that moment. As the other agents don't have such a guarantee regarding their own current model, they replace it by the querying agent model. As a consequence, when no more observations are collected by agents and all interactions are finished, there is only one current model in the community.

In our experiments we consider, for various network characteristics, both the original SMILE and some variants with no adoption behaviour or with no memorization of counter-examples obtained from any other agent, and observe to what extent the restriction of interactions to the neighbourhood affect the learning accuracy. Section 2 details the system of agents and the collective learning task, describing how to use SMILE in a structured network of agents by restricting its application to the neighbourhood. Then, Sect. 3 presents the experimental setting of our study, used in Sect. 4 and 5 to investigate respectively the influence of the topology of the network on the learning speed and the influence of sharing hypotheses or examples. At last, Sect. 6 will conclude.

Note that there is rather few works on this subject. A related domain is *network epistemology* as investigated by Zollman [7] which studies hypotheses propagation and selection within a community, rather than their construction, still displaying fascinating effects of the network topology on the time needed by a community to select the correct hypothesis. Regarding the way information propagates within networks, it has been deeply investigated [6], but, again, we are more interested here in the collective learning counterpart of such study.

2 Neighbourhood-Aided Collective Learning

We consider a network of N agents represented by a graph $\mathcal{G} = \langle Ag, \mathcal{C} \rangle$ where nodes $Ag = \{a_0, \ldots, a_{N-1}\}$ represent agents, and edges $\{a_i, a_j\}$ in $\mathcal{C}$ represent bidirectional communicational links between agents a_i and a_j. Each agent a_i has an *example memory* E_i and must learn some target concept, based on the examples it possesses. Its model of this target concept will be called a *hypothesis* and denoted by H_i.

In this paper, we study the influence of the topology of the network on learning by focusing on a simple learning task: concept learning of boolean formulae. We consider a propositional language defined over a set of atoms. An *example* e in an example memory is represented by a label + or - (depending whether it belongs or not to the target concept) and a description which is a conjunctive statement, that is, a conjunction of atoms, represented here by a set of atoms. Given an example memory E_i, we denote by E_i^+ the set of examples labelled by + (positive examples) and by E_i^- those labelled by - (negative examples). A *hypothesis* will then be a disjunction of conjunctive statements, or terms, that is, $H_i = t_1 \vee \ldots \vee t_m$ where each term t_j is a conjunction of atoms. A hypothesis is thus a formula in disjunctive normal form.

A hypothesis H is said to *cover* an example e if and only if there is some term t in H such that all the atoms of t belongs to the description of e. We shall say that a hypothesis h is consistent with a set of examples E if and only if h covers all positive examples of E and does not cover any negative example of E.

SMILE [3,5] is a generic protocol ensuring that, in a fully-connected system of agents, all hypotheses H_i of the agents are consistent with the set $\bigcup_{i \in \{0,\ldots,N-1\}} E_i$ of all examples in the system. It is based on a learner critic principle, where the agent revising its hypothesis, taking the role of a learner, proposes it to all the other agents, acting as critics, which either accept the proposed hypothesis if it is consistent with their example memory or give a counter-example otherwise. The learner agent then revises its hypothesis to take into account the counter-example, and the process is iterated with this revised hypothesis until the learner produces an hypothesis that is consistent with the example memory of each of the other agents, and thus, also with the union of all theses example memories. It is then adopted by all agents, guaranteeing that each agent possesses an hypothesis consistent with all the examples.

Here, we adapt this protocol to non fully connected graphs by considering that the learner agent will only propose its hypothesis to its direct neighbours. By interacting in turn with each of its neighbours, it will be able to revise its hypothesis until it is accepted by all of its direct neighbours. In effect, it unfolds as if the learner agent applied SMILE normally in a subset of the MAS constituted only of its neighbourhood. Its hypothesis is then only guaranteed to be consistent with the set of examples in its neighbourhood, that is, after a revision by a_i, h_i is guaranteed to be consistent with $E_i \cup \bigcup_{j \in N(i)} E_j$ where $N(i) = \{j, \{a_i, a_j\} \in \mathcal{C}\}$. We shall say that H_i is *group-consistent* wrt a_i. a_i will still inform its neighbours of this group consistency, and in the default version, these neighbours will then adopt h_i as their own hypothesis. A variant without adoption will also be considered to evaluate the impact of sharing hypotheses. Without adoption, an agent never takes advantage of the hypotheses proposed by its neighbours. Note that during a revision, the learner might get some new examples from its neighbours. In a later revision where it acts as critic, it can give this example as a counter-example. As a result, given a series of revisions, a crucial counter-example might be propagated along the network. To evaluate the impact of this propagation, we shall also consider a variant called Forgetness in which, once it has finished its revision, the learner agent will forget all the examples it obtained from its neighbours. Example 1 illustrates our protocol.

Example 1. We consider four agents in a line ($Ag = \{a_0, a_1, a_2, a_3\}$ and $\mathcal{C} = \{\{a_0, a_1\}, \{a_1, a_2\}, \{a_2, a_3\}\}$). Agents are trying to learn the target concept $A \vee B$. After the reception of $e_1^+(A \wedge not\_B \wedge C)$ by a_1 (who revises its hypothesis as $H_1 = A \wedge not\_B \wedge C$ and share it with a_0 and a_2), and the reception of $e_2^+(A \wedge B \wedge not\_C)$ by a_3 (who revises its hypothesis as $H_2 = A \wedge B \wedge not\_C$ and share it with a_2), agent a_2 receives positive example $e_3^+(\neg A \wedge B \wedge C)$ (see Fig. 1). a_2 adds e_3^+ to its example memory and observe that it contradicts its current hypothesis (H_2). It revises it with its learning mechanism and gets $H_3 = B$ and proposes it to its neighbours, starting with a_1. a_1 answers with a counter-example (e_1^+), triggering a new revision by a_2 of H_3 into $H_4 = B \vee (A \wedge not\_B \wedge C)$. H_4 is in turn proposed by a_2 to a_1 who accepts it, and then to a_3, who also answers with an acceptation. Since all its neighbours have accepted its hypothesis, the local revision by a_2 is now finished: H_4 is group-consistent wrt a_2. Unless we used the variant without Adoption, a_1 and $_3$ will then adopt H_4 (but not a_0 as it is not a neighbour of a_2). If we use the Forgetness variant, a_2 will then forget the counter-examples it received during the revision, removing e_1^+ from E_2, otherwise, e_1^+ will remain in a_2's example memory.

3 Experimental Setting

We perform in the following a comprehensive experimental study to investigate the effect of neighbourhood-aided collective learning on the average accuracy of the hypotheses of the agents. This value represents the probability any agent's hypothesis will cover or not the next example accordingly to the target concept. In our setting, the accuracy of an hypothesis will be evaluated on a testing set of examples, by computing the ratio of correctly predicted test examples over the total number of test examples. As agents only share their hypotheses with their neighbours, there will be several different hypotheses in the system at a

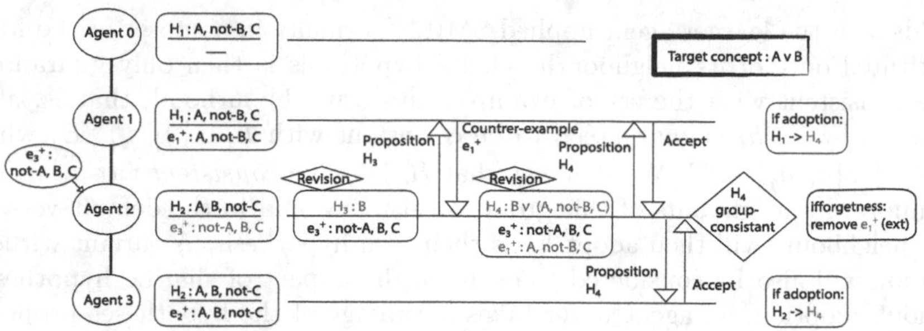

Fig. 1. Example of collective learning in a line network of 4 agents.

given time. We shall evaluate the learning speed of the system by measuring the average accuracy with respect to the total number of different examples in the system. A system of agents has a better learning speed than another if it reaches a higher mean accuracy with less examples.

An experiment is typically composed of 100 runs, each one corresponding to a sequence of examples incrementally sent to random agent in MAS. We choose to restrain this study to networks of 50 agents with a steady layout from the beginning to the end of each run. We will compare the efficiency of different networks in term of accuracy (with respect to the number of different examples in the system of agents).

Learning Problem. We choose the 11-multiplexer (M11) as the main difficult learning problem to study. M11 uses 3 address boolean attributes a_0 ,a_1 ,a_2 and 8 data boolean attributes d_0, ..., d_7. The formula is satisfied when the number coded by the 3 address attributes is the number of a data attribute whose value is 1. (ex : with $a_{1,2,3}$ value being 000 and represents 0 the formula is satisfied if $d_{0,1,2,3,4,5,6,7}$ is 1000 0110 but not 0000 0110).

Network Topologies. We are going to show the influence of the network layout on collective learning where communications are limited to the neighbourhood. We choose a large panel of communication graphs in order to find the proprieties which have a hold on the average accuracy of the system. The regular graphs are composed of k regular trees called Tree$k$, circular regular graphs (each node is linked to all $d/2$ previous ones) called Reg$d$ and a Torus whose section is a circle of 5 nodes which revolves in a circle of 10 nodes. Small-Worlds called SmWdpx, are built with Watts and Strogatz algorithm ([9]) from circular regular graphs with mean degree d and reconnection probability x of each edge. We design centralized network patterns called MultiHubs. MultiHubs main nodes, called hubs, are linked to as many secondary nodes as possible. Links are first established between hubs then from hubs to secondary nodes and at last between secondary nodes. A variant called MultiHubs separated (sep) skips the first step and do not link hubs to each other. At last `MCluster5circ(star)` is formed of ten cliques of 5 agents having one edge being rewired in order to connect them together in a circle(star) shape.

4 Characterizing the Influence of the Network Topology on the Learning Speed

We first investigate different topologies of 50 agents with the default version of the SMILE protocol (with adoption and without forgetness). We can identify several factors of decreasing importance that affect learning in such networks.

4.1 Main Factors

Density. With a fixed number of agents it is equivalent to compare density of graphs or their mean degree (with 50 agents: $density = Meandegree/49$). We compare first a set of similar graphs, Small-Worlds built from regular circular graphs with mean degree 4 to 49 (clique) and a probability of rewiring each edge of 0,5. Figure 2 (left) shows a better learning speed as density increases. Density represents how much communications between agents are limited, the smaller it gets, the less each agent can access others agents information and thus learning gets less accurate.

Changing the second parameter of Small-Worlds also have an influence on accuracy. SmW04p01 with a probability of rewiring edges of 0,1 is closer to regular circular graphs whereas SmW04p05 rewiring shortens both diameter and mean distance between pairs of agents. SmW4p05 learns faster than SmW04p01 and we'll confirm this tendency with a bigger set of graphs of this same density.

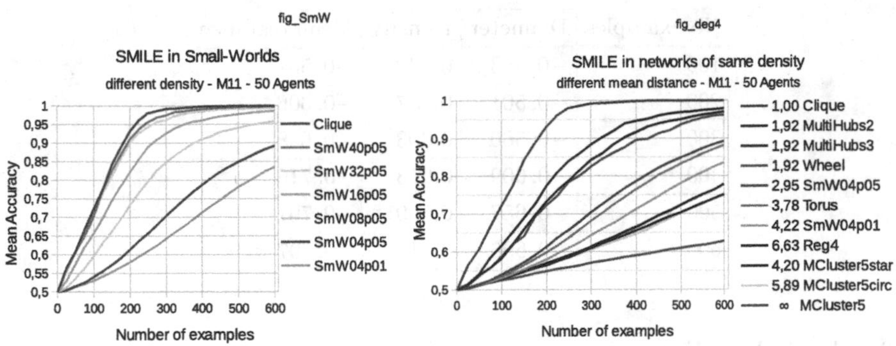

Fig. 2. Mean accuracy evolution as a function of the number of examples within the MAS. Similar structure (left): Small-Worlds with mean degree 4 to 49. Fixed density (right): influence of graph's mean distance between nodes for various graphs of mean degree 4 compared to Clique. The means distances are reported near the graph name.

Diameter and Mean Distance. Figure. 2 (right) reports the learning accuracy curves of a variety of mean degree 4 graphs. These graphs have different mean distance between pairs of agents and we may observe that the learning accuracy clearly decreases with this characteristic. All graphs are between the clique which have no limitation and MCluster5 a non-connected graph formed of ten cliques of 5 agents. Mean distance represents mean number of agents which have to perform a revision in order for information to be passed on between two random nodes. Diameter only represent the worst case of the two further nodes of one graph. Information is passed on more easily as they are both shorter, meaning better accuracy of hypotheses. Layouts with high local density but few links between groups of agents (Mcluster5*) constitute a first exception as Reg4 performs a little better with a higher mean distance. Another group composed of MultiHubs isn't fully explained with neither mean distance nor diameter as they have the same ones but shows different accuracies. We will first see how the main characteristics influence learning over time and second find another property to explain these oddities.

Evolution of This Influence During Learning. We found three main graph properties to have impact on accuracy. We made a set of graphs composed of Small-Worlds, trees and mean degree 4 graphs in order to compare the influence of these properties at different learning stages in Table 1. Mean distance and, with a lower impact, diameter, are more and more influential as learning progresses. Density unlike the other two have very high influence since the beginning but it decreases over reception of more examples.

Table 1. Correlation between hypotheses mean accuracy with respectively diameter, density and mean distance between pairs of nodes at different learning steps.

Nb examples	Diameter	Density	Mean distance
100	$-0,453$	0,848	$-0,557$
200	$-0,501$	0,787	$-0,606$
300	$-0,560$	0,693	$-0,668$
400	$-0,609$	0,613	$-0,716$
500	$-0,657$	0,559	$-0,761$
600	$-0,696$	0,517	$-0,797$

4.2 Secondary Factors

We make a focus on MultiHubs in order to find another property to explain what the three main ones couldn't. We study two groups of MultiHubs of different density with same diameter and mean distance. For density 0.08 (resp 0.2) we can have up to 3 (resp 9) hubs but they can't be full hubs, connected to every other nodes, since the presence of the 3rd (resp 6th) hub. The layout achieving

Table 2. Two different groups of MultiHubs showing the high correlation between the number of shorter paths of length 2 and accuracy (300 examples) at a fixed density.

nbr hubs	Density	Diameter	Mean dist.	nbr paths 2	Accuracy
1	0,08	2	1,92	2352	0,82
2				**4506**	**0,88**
3				3000	0,85
3	0,2	2	1,8	6198	0,92
4				7964	0,93
5				**9828**	**0,94**
6				8928	0,93
7				7732	0,92

better accuracy in the two groups is the one maximizing the number of full hubs (respectively 2 or 5 for a density of 0.08 or 0.2). Full hubs represent opportunities to resynchronise the whole MAS with their revisions as their hypotheses are verified and adopted by every agent.

A short mean distance represents the necessity of few agents' intervention in order to transmit any information to other agents in the networks. This value doesn't take into account the number of possible paths between two agents with the shortest length found. As revisions happen randomly to any agent when it gets new examples, alternative paths gives greater chance for information to be passed on. Shortest path of length 1 are represented by density as there can't be more than one edge between to agents. We then focus on the next shortest length (2) in Table 2. Graphs maximizing full hubs in each group are also maximizing the number of shortest paths. The 0,2 density, MultiHubs group (1 to 9 hubs) presents a very high correlation coefficient between number of shortest paths of length 2 and accuracy. It's value is between 0.945 and 0.97 depending on the MAS number of examples.

Table 3. Third group of MultiHubs showing the high correlation between the number of shorter paths of length 2 and accuracy (300 examples) at a fixed density.

nbr hubs	Density	Diameter	Mean dist.	nbr paths 2	accuracy
21	0,2	**3**	**2,12**	1416	0,77
21(sep)		4	2,34	**4576**	**0,87**

The number of shortest paths of length 2 is less important than the number of paths of length 1 measured with edges number through density. It can explain on the other hand a better accuracy from MultiHubs21sep even though it presents slightly worse mean distance than MultiHubs21 (Table 3).

5 Importance of Hypotheses and Examples Propagation

During a local revision with SMILE, two kinds of information are propagated:

- The *learner* agent may gather some *counter-examples* from its neighbours. In the default protocol, it will memorize them in its example memory, and will thus be able to provide them to other agent in the future when acting as a critic. If we use a Forgetness variant, the learner will delete these examples at the end of the revision, and thus a counter-example will never go beyond the neighbourhood of the agent possessing it.
- The *critic* agents will receive the *hypothesis* of the learner, and will, in the default protocol, adopt it at the end of the revision. Then, in future revision, when acting as learner, this adopted hypothesis will serve as the base of its revision (incremental learning) and some of the information it contains will thus be kept. If we use a variant without adoption, an agent will always keep its own hypothesis, and will not benefit from the revision of its neighbours.

By studying the variants with forgetness or without adoption, we can thus investigate how these two propagations impact the learning speed on different topologies. Indeed, networks with different layouts do not all benefit from hypothesis adoption and example memorization in a same way. Charts representing results for adoption: with and without forgetness are grouped on Fig. 3 and those without adoption are grouped on Fig. 4. We can extract properties from each graph's performance analysis in the four cases:

1. Hypothesis propagation through adoption is always beneficial.
2. Memorization have no effect on accuracy of diameter 2 graphs while there's already hypothesis adoption.
3. For bigger diameter graphs memorization becomes more important than adoption in the end.
4. Learning benefits more from adoption in the beginning (first 100 examples).

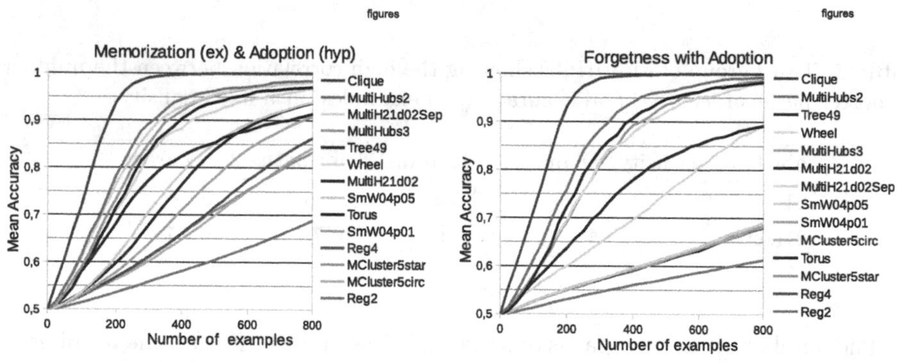

Fig. 3. Mean accuracy evolution in graphs with 50 agents on M11 problem as a function of the number of examples within the MAS; protocols with hypothesis adoption with (left) or without(right) external example memorization

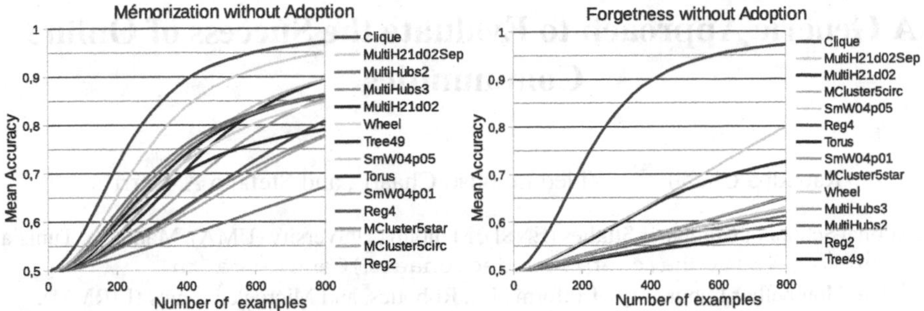

Fig. 4. Mean accuracy evolution in graphs with 50 agents on M11 problem as a function of the number of examples within the MAS; protocols without hypothesis adoption with (left) or without(right) external example memorization.

6 Conclusion

Networks of agents learn faster with high density, short mean distance between nodes and multiple paths for information to be passed on. They benefit both from example memorization and hypothesis adoption. This constitutes a preliminary study before investigating strategies to handle multiple hypotheses. Agents could either keep all the hypotheses they receive, choose between them or build their own one from them. These strategies determine, on a macroscopic level, theory formation as much as their transmission and opposition in a network.

References

1. Angluin, D.: Queries revisited. Theor. Comput. Sci. **313**(2), 175–194 (2004)
2. Bourgne, G., Bouthinon, D., El Fallah Seghrouchni, A., Soldano, H.: Collaborative concept learning: non individualistic vs individualistic agents. In: IEEE International Conference on Tools with Artificial Intelligence (ICTAI), pp. 653–657. IEEE Computer Society, Newark, USA, November 2009
3. Bourgne, G., El Fallah Seghrouchni, A., Soldano, H.: Smile: sound multi-agent incremental learning;-). In: International Conference on Autonomous Agents and Multiagent Systems (AAMAS), pp. 164–171. ACM, Honolulu, Hawaï (2007)
4. Bourgne, G., El Fallah Seghrouchni, A., Soldano, H.: Learning in a fixed or evolving network of agents. In: ACM-IAT 2009. IEEE (2009)
5. Bourgne, G., Soldano, H., El Fallah Seghrouchni, A.: Learning better together. In: Coelho, H., Studer, R., Wooldridge, M. (eds.) ECAI. Frontiers in Artificial Intelligence and Applications, vol. 215, pp. 85–90. IOS Press, The Netherlands (2010)
6. Guille, A., Hacid, H., Favre, C., Zighed, D.A.: Information diffusion in online social networks: a survey. SIGMOD Rec. **42**(2), 17–28 (2013)
7. Zollman, K.J.S.: Network Epistemology. Ph.D. thesis, University of California, Irvine (2007)
8. Ontañón, S., Plaza, E.: Multiagent inductive learning: an argumentation-based approach. In: Proceedings of ICML 2010, pp. 839–846. Omnipress (2010)
9. Watts, D.J., Strogatz, S.H.: Collective dynamics of /'small-world/' networks. Nature **393**(6684), 440–442 (1998)

A Generic Approach to Evaluate the Success of Online Communities

Raoudha Chebil[1(✉)], Wided Lejouad Chaari[1], and Stefano A. Cerri[2]

[1] National School of Computer Studies (ENSI), Manouba University (UMA), Manouba, Tunisia
{raoudha.chebil,wided.chaari}@ensi-uma.tn

[2] The Montpellier Laboratory of Informatics, Robotics, and Microelectronics (LIRMM), University of Montpellier (UM), National Center for Scientific Research (CNRS), Montpellier, France
cerri@lirmm.fr

Abstract. The success of online communities depends on different aspects and has been subject of several evaluation works. Most of the existing works in this scope are not generalizable due to their strong dependence to the considered community characteristics. Face to this finding, our aim is to propose a generic approach to evaluate and to improve the success of online communities. This paper starts by an identification of the success determinants of online communities including the participants, the technology as well as the common goal. After that, we propose a generic evaluation approach based on two levels. The first level consists in a failure detection based on a quantification of the success determinants. The second level focuses on a failure explanation and an identification of the most plausible causes of the detected failures. This approach was applied to evaluate the success of a social network-based community; it gave reliable results.

Keywords: Online community · Success evaluation · Failure detection · Failure explanation · Social networks

1 Introduction

Since several decades, technologies have increasingly invaded our lives by intervening and facilitating practically all the tasks we need to perform. Despite its obvious interest, the use of technologies in any task is not always a guarantee of its success. Consequently, the evaluation of technology-supported tasks has been a topical issue for a long period. Due to the actual widespread use of social networks producing a large number of communities, we are interested in the success evaluation of online communities. In the literature, several definitions of the term "online community" are proposed. In this work, we adopt the following definition proposed by Howard Rheingold describing an online community as "a group of people who may not meet one another face-to-face, and who exchange words and ideas through the mediation of computer bulletin boards and networks" [10].

Nowadays, online communities are emerging in several domains such as e-commerce and e-learning. Some of them progress and succeed and others fail and

N.T. Nguyen et al. (Eds.): ICCCI 2017, Part I, LNAI 10448, pp. 212–222, 2017.
DOI: 10.1007/978-3-319-67074-4_21

disappear. In several works, the proposed measures to quantify the success of online communities are not generalizable since they depend on the studied community. This finding was confirmed by the work of Jenny Preece [9] (in Sect. 5.4).

Consequently, the first aim of this work is to propose a number of measures allowing to assess the success or failure of an online community independently from its belonging domain. Our second aim consists in making the considered community more successful by explaining the previously detected failures.

This paper will be organized as follows. In Sect. 2, we summarize the existing works interested in the evaluation of online communities. In Sect. 3, we describe the suggested evaluation approach composed by two levels: a failure detection level and a failure explanation level. In Sect. 4, we describe the application of our evaluation approach on a social network based community. Finally, Sect. 5 concludes the paper.

2 Literature Review

For over forty years, the success of using a software was a direct consequence of its interface appropriateness called usability. In fact, usability means that people who use the product can do so quickly and easily to accomplish their own tasks [3]. In the literature, several works in the fields of Human-Computer Interaction (HCI) and Computer-Supported Cooperative Work (CSCW), focused on this aspect and proposed methods to evaluate and to improve usability [1, 2, 8]. More recently, the emergence of online communities imposes a more serious considering of social aspects in studying all community types. Several works, underlined this finding [6, 9]. Particularly, in the evaluation topic, Jenny Preece [9] proposed some success determinants of online communities related to both usability and sociability. The sociability determinants include the community purpose, people and policies while the usability determinants cover the dialog and social interaction support, the information design, the navigation and the access. Determinants of sociability are quantified by measures such as the number of participants in a community, the number of messages per unit of time, members' satisfaction and the amount of reciprocity. Measures of usability include the number of errors, the speed of learning and several others.

Both the sociability and the usability measures are interesting for some communities but they can be completely insignificant for others [9]. For example, in an e-commerce community, a high rate of messages can express a great interest to the product in question but in a learning community, it is not always a success indicator; it depends on the community policies.

Hsiu Fen Lin [6] developed and tested a virtual community success model by using and adapting the Information System Success Model proposed by DeLone and McLean's. The obtained model focuses on two particular success determinants consisting of: system characteristics (e.g. Information and system quality) and social factors (e.g. Trust and social usefulness). The authors confirmed [6] (in Sect. 6.4) that the obtained results have limited generalizability.

According to these findings and especially the dependence of the proposed success determinants not only to the field but also to the considered community, we propose in what follows, a generic approach to evaluate the success of online communities. With the aim of improving the evaluated online community, the proposed approach is not limited to a success evaluation: it also explains the eventually detected success problems called failures. Consequently, our approach is based on two different levels: a failure detection and a failure explanation.

3 Evaluation

In this section, we start by identifying the determinants of success in online communities, after that we describe the two evaluation levels of the proposed approach.

3.1 Determinants of Success in Online Communities

In fact, an online community is a group of people called "participants or members" who interact with each other exchanging ideas and information related to the context of the community. Thanks to technologies, these members may not meet one another face-to-face; their interactions are supported by technological tools providing various means of communication. In addition to members and technology, an online community is usually built around a common goal.

According to this vision and inspired by the work of Jenny Preece [9], we focus, in this work, on three key components contributing to the success of online communities and consisting of: participants, technology and the common goal. These features are called in the following, determinants of success in online communities.

3.2 First Evaluation Level

As previously explained, several measures have been proposed to assess the success of online communities. For example, if we count the number of participants as proposed by [9], which range of values expresses positive assessment about the success of an online community? The response to this question depends on the considered community: there is no general values referring to the community success. To provide more generalizable measures; we propose to base our success measurement on the comparison between the obtained and the expected values that quantify the previously identified determinants of success. For example, we compute the rate between the expected and the effective participants number instead of simply computing the number of participants to the community. In fact, the success is more intense when the correspondence between the two values is greater. Due to this thinking, the approach application supposes the availability of information about the expected features.

The proposed measures called success indicators are the following:

- The participation rate *PR* expressing the correspondence between the expected and the effective numbers of active members;
- The satisfaction rate *SR* showing the correspondence between the expected and the obtained satisfaction from the technological support;

- The accomplishment rate *AR* expressing the correspondence between the expected and the estimated accomplishment degree of the overall goal.

The proposed success indicators are computed according to the following equations:

$$\mathrm{PR} = \frac{\mathrm{PN}}{\mathrm{EPN}} * 100 \tag{1}$$

$$SR = \frac{AS}{ESD} * 100 \tag{2}$$

$$AR = \frac{AD}{EAC} * 100 \tag{3}$$

Where:

- *PN*: expresses the effective number of active participants;
- *EPN*: expresses the expected number of active participants;
- *AS*: is the average satisfaction degree of members from the technological support;
- *ESD*: is the expected satisfaction degree of the community members from the technological support;
- *AD*: is the average of the estimated accomplishment degrees of the common goal retrieved from the community members;
- *EAC*: expresses the expected accomplishment degree of the common goal.

To compute the success indicators, some values (*AS and AD*) must be recovered from community members using a survey. To quantify the satisfaction degree from the technological support and the accomplishment degree of the common goal, we use the "Likert Scale" [7] as shown in the first part of the survey accessible in the following URL:

https://docs.google.com/forms/d/e/1FAIpQLScsgMCLBX2Tvz7B_5rsm1oKwILtnR83YbBjeWbid9z9QSDuug/viewform?c=0&w=1

The "Likert scale" is adapted to our aim and is widely used in several domains including marketing, health and social psychology. According to this scale, we set the expected rates of the satisfaction from the technological support (*ESD*) as well as the accomplishment degree (*EAC*) to 100%: they do not need to be retrieved from the evaluator. The effective number of active participants is directly retrieved from the technological support while the expected number of participants must be retrieved from the evaluator.

In the following, the term "success failure" expresses an important mismatching between the expected and the effective states of success determinants. It is expressed by a low value of the success indicator. To identify success failures, the evaluator is also asked about the minimal rate of correspondence (*MC*) between the expected and the effective values under which a success failure is detected.

This evaluation level is a preliminary step allowing to measure the success of online communities and eventually detect success failures. It is supposed to be deepened by rigorous explanations permitting to find the causes of each detected failure. This work will be the subject of the next evaluation level.

3.3 Second Evaluation Level

Since this level focuses on the explanation of the previously detected failures, we performed a review of literature works that are interested to failure explanation.

Further to this research, we decided to inspire from two interesting works focused on failure explanation [4, 12]. These two works are based on the human reliability error analysis method entitled CREAM (Cognitive Reliability and Error Analysis Method) and proposed by Hollnagel [5].

CREAM Method Application. The choice of this method is made due to the following advantages:

- The CREAM method is compatible with our context since it is focused on the analysis and assessment of a human participation;
- The method is very concise, well-structured and independent from the application domain;
- It allows to deduce the possible causes explaining the occurrence of a detected failure as well as the causal links connecting them;
- It is based on an interesting classification of the detected causes into three different categories consisting of: persons, organization and technology. This causes classification is adapted to the studied environments.

The CREAM method is composed by three steps:

The first step consists in describing the community progress conditions using the Common Performance Conditions named CPC [5]. The CPC is a set of items fixed by the CREAM method and supposed to capture the principle aspects affecting the progress of the considered environment (the online community in this work). They consist of the following items [5]: Adequacy of organization, Working conditions, Adequacy of Man-Machine Interface and operational support, Availability of procedures and plans, Number of simultaneous goals, Available time, Time of day, Adequacy of training and expertise, Crew collaboration quality. For each CPC the CREAM method gives several descriptors (three or four) as well as their effect on performance reliability.

The second step consists in identifying the occurred failures.

The third step is an iterative step consisting in determining the possible causes of a failure occurrence as well as their causal links basing on states of the previously described CPC. In each iteration, a cause (also called an antecedent) and an effect (also called consequence) are identified; in the next iteration, the last cause becomes an effect.

Hollnagel [5] divides the different antecedents into three categories according to their relation to persons (category P), to technology (category T) or to the organization (category O).

In the following, we explain the adjustment of the CREAM method to the studied context of online communities and to the proposed evaluation.

To capture the CPC, several questions are added to the previously described survey intended to community members constituting its second part. As we can notice, the second step of the CREAM method consisting in failure identification has already been carried out in the first level of our approach. Consequently, this step will be skipped during the application CREAM method.

Thanks to the obtained responses to the diffused questionnaire allowing to retrieve the CPC, we are able to apply the third step of the CREAM method. The implementation of this step, gives several inferences for each detected problem. Each inference is constituted by a set of explanations related by a causal-effect relationship.

For better visibility of the causal link, we represent the different explanations of each failure by a causal graph. Each causal path linking a graph leaf to the considered failure represents an inference explaining it. In fact, an inference is a set of linked antecedents giving a possible explanation of the failure occurrence.

The causal graph gives different explanations of a failure; but it does not show the most plausible of them. For this reason, we were interested in the application of a probabilistic approach adapted to the graph characteristics as well as the antecedent classification into categories (P, T, O). In fact, the majority of the probabilistic approaches allow assigning a probability to each explanation regardless to its category and its antecedents. The evidence theory, introduced by Dempster and Shafer [11], has the great advantage of allowing to assign a belief degree to a whole set of related assumptions representing, in our context, an inference of a failure occurrence. It is why this theory was judged the most adapted to our context and was applied to compute the plausibility of each inference.

Application of the Evidence Theory. To compute the plausibility (or belief mass) of each inference, we need to calculate the belief masses of the different graph nodes. In this purpose, we applied the principle used in the work referenced by [4], defining the mass propagation in a causal graph. In this work, the mass of an antecedent depends on its consequences masses as shown in Eq. (4).

$$\boldsymbol{m(a) = p(c(a)) \times \sum\nolimits_{\forall b \in Cons(a)} \frac{m(b)}{\sum \forall i \in P, T, O(p(i) \times nib)}} \tag{4}$$

Where:

- *m(a)* is the mass of the a;
- *C(a)* is the category of a;
- *Cons(a)* is the set of consequences of a;
- *p(i)* is the weight of category i;
- n_{ib} is the number of antecedents of b belonging to the category i.

According to this equation, each consequence of an antecedent '*a*' gives it a part of its mass modulated by the weight category of '*a*'. The addition of these different consequences parts gives the belief mass of the antecedent a.

In the current context, there is no preliminary hypothesis expressing different impacts of the antecedent categories on a failure occurrence; so we give the same weight *(1/3)* to all the three categories (P, T and O).

The mass computing process starts by affecting a mass *m* to the failure (*m* = *1*). Then an iterative mass propagation to the graph leaves using the Eq. (4) is carried out. At the end of the calculation step, we can verify that the sum of leaves belief masses is equal to 1. This propagation is fully applicable since the causal graph is acyclic.

In this section, we presented a two-level based evaluation approach. The first level ensures a failure detection. The second focuses on explaining the detected failures by combining the CREAM method, the causal graph and the evidence theory.

To make the previous description more clear, we present in what follows, an application of the proposed approach to evaluate the success of social network-based community.

4 Approach Application on a Social Network-Based Community

In order to apply and test the proposed approach, we were interested in a political online community supported by the social network "Facebook". In the last seven years, such communities have had important impacts on political events especially in countries of the "Arabic Spring" [14]. Given this finding and the actual dynamicity of political scenes in these countries, a citizen created the considered community to encourage the diffusion of political actualities, to support discussions about them and to facilitate the organization of political events to react to them. The community creator started by inviting some of his friends as well as family members supposed to invite other participants, to share and to comment recent political news.

The community presentation on its facebook page is the following: "the objective of this community consists in following and reacting to political events. Thank you for inviting persons who may benefit from our group and have interesting and efficient feedbacks which argue in favor of the country."

After a period of two weeks, we applied the proposed approach to evaluate the success of the considered community. In this purpose, we started by diffusing the already described evaluation questionnaire.

4.1 First Evaluation Level

Given the purpose of the first evaluation level; we present, in Table 1, the obtained values allowing to compute the previously described success indicators and to detect failures. Since some values are retrieved from responses to the first part of the questionnaire, we add the number of the corresponding question as well as its topic for more clarity. In this experimentation, the evaluator who is the group creator fixed the minimal correspondence value (*MC*) to 75%. According to this value, we can easily detect a goal completion failure from Table 1. To check the possible causes of the detected failure, we perform the steps described in the second evaluation level.

Table 1. Obtained measures of success

Retrieved values		Success indicators		Question	Question topic
PN	32	PR	80%		
EPN	40				
AS	84.48%	SR	84.84%	Q1	Technological satisfaction
AD	28.44%	AC	28.44%	Q3	Goal completion

4.2 Second Evaluation Level

As previously explained, the second evaluation level consists on an explanation of the detected failures. Since this explanation is mainly based on the retrieved responses to the second part of the questionnaire, we present in Table 2, a recapitulation of the obtained responses. For more readability, we present for each question in the second part of the survey (corresponding to a CPC), the obtained predominant descriptor as well as its corresponding rate in percentage.

Table 2. Recapitulation of the obtained responses to the second part of the questionnaire

Question	Question topic	Predominant descriptor	Rate
4	Work organization	Inefficient	89.7
5	Working conditions	Incompatible	69
6	Adequacy of MMI	Tolerable	57.1
7	Work specification	Inappropriate specification	89.7
8	Number of simultaneous goals	Matching current capacity	48.3
9	Available time to the work	Continuously inadequate	55.2
10	Period of the day	During the night time	62.1
11	Work adequacy according to available skills	Inadequate training and experience	51.7
12	Quality of collaboration	Deficient	55.2

The analysis of the obtained responses allowed us to detect the following antecedents behind the detected goal completion failure:

- An ambiguous understanding of the group objectives as well as the expected results, identified from large differences between responses to the question 2.
- A low motivation of community members detected especially from responses to question 5.
- An availability problem detected from responses to question 9.
- A technological problem detected in responses to question 1.

These problems can themselves have multiple antecedents: it is what justifies the iterative progress of the evaluation work in this level.

- The first problem can be caused by an inadequate supervision of the community consisting of: an insufficient explanation of the community interest as well as its objectives (can be deduced from responses to question 4) and a poor presentation of the activities to perform in the community as well as the expected results (can be deduced from responses to question 7).
- The second problem can be due to an inadequate supervision (described below) if the community interest and the expected results were not correctly highlighted. This failure can also be caused by a disinterest of several members in political topics (can be deduced from responses to question 11).
- The third problem of availability can be due to a period inadequacy which prevents members to spend time on doing the required actions and consisting in a work

overload period or on the contrary, a holiday period (can be deduced especially from responses to questions 9 and 10).

- The fourth technological problem can be caused by: an interface problem (question 6), a limited internet connectivity (question 5) or limited technical skills (question 11).

For more clarity, the causal link between the different antecedents is illustrated by the causal graph in Fig. 1. According to the Hollnagel classification scheme, the antecedents we have just distinguished belong to the three different categories: P, O and T.

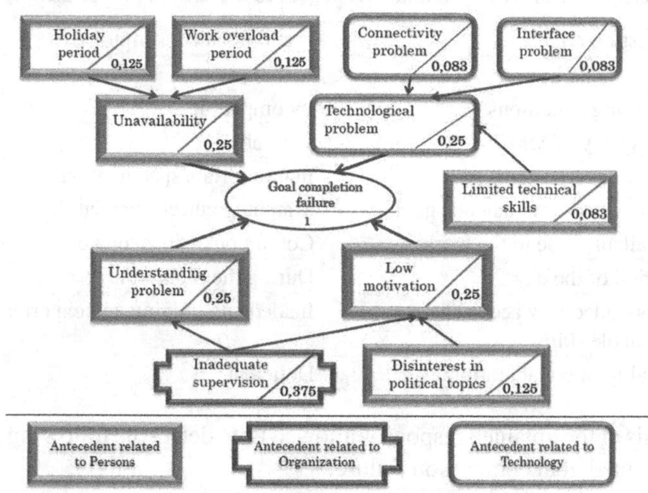

Fig. 1. Causal graph explaining the "weak goal completion"

The belief masses of the different failure explanations are calculated according to the Eq. (4).

The causal graph shown in Fig. 1 presents many available explanations of the detected failure consisting in a weak goal completion. It reveals the plausibility of the two following inferences having a belief mass equal to 0.375:

Inadequate supervision → understanding problem of the common interest and the expected results → weak goal completion: this inference supposes that the community creator realized an inappropriate supervision caused by a missing presentation of the common objective as well as the expected tasks and results. This ambiguity blocked the members dynamicity and consequently the goals completion.

Inadequate supervision → low motivation → weak goal completion: this inference supposes that the community interest presentation was not sufficiently convincing to attract members. The creator's intervention in the community and his reactions to political news was too poor to encourage members to have the expected behavior allowing the goal completion.

To check the consistency of the proposed evaluation approach, we planned to verify the correspondence between the plausible inferences and members feedbacks about the encountered problem in the considered community. For this purpose, we organized a virtual meeting in which we discussed with the considered community members about the reasons of the weak goal completion. All their responses confirm that the group creator failed in presenting the created group and consequently in motivating them to contribute and benefit from it. If we go back to the community description reported in the beginning of this section, we can easily note that it is too brief and ambiguous.

5 Conclusion

In this paper, we presented a two level-based approach to evaluate the success of online communities. The first level is based on measures that quantify the already proposed determinants of success. The obtained values from this level allow the detection of success failures. The second level focuses on the explanation of the previously detected failures. Given the impact of human participation on online communities, this level uses the CREAM method to provide explanations. It also uses the evidence theory in order to identify the most plausible explanation. The proposed approach was applied to explain the failure of a social network-based community having a political objective: it gave satisfactory results. This approach is more generalizable than the existing ones [6, 9] since it is built on the correspondence between the expected and the obtained success values.

References

1. Antunes, P., Herskovic, V., Ochoa, S.F., Pino, J.A.: Structuring dimensions for collaborative systems evaluation. ACM Comput. Surv. CSUR **44**(8), 1–28 (2012)
2. Baker, K., Greenberg, S., Gutwin, C.: Empirical development of a heuristic evaluation methodology for shared workspace groupware. In: Proceedings of the 2002 ACM Conference on Computer Supported Cooperative Work (CSCW 2002), pp. 96–105 (2002)
3. Dumas, J.S., Redish, J.: A practical guide to usability testing. Intellect books, Bristol (1999)
4. El-Kechaï, N., Després, C.: Rechercher les causes des actions erronées du formé pour les proposer au formateur. In: Actes de la conférence Environnements Informatiques pour l'Apprentissage Humain EIAH. Lausanne (2007)
5. Hollnagel, E.: Cognitive Reliability and Error Analysis Method. Elsevier Science, Oxford (1998)
6. Lin, H.F.: Determinants of successful virtual communities: contributions from system characteristics and social factors. Inf. Manag. **45**(8), 522–527 (2008)
7. Norman, G.: Likert scales, levels of measurement and the laws of statistics. Adv. Health Sci. Educ. **15**(5), 625–632 (2010)
8. Pinelle, D., Gutwin, C., Greenberg, S.: Task analysis for groupware usability evaluation: modeling shared-workspace tasks with the mechanics of collaboration. ACM Trans. Comput. Hum. Interact. TOCHI **10**, 281–311 (2003)
9. Preece, J.: Sociability and usability in online communities: determining and measuring success. Behav. Inf. Technol. **20**(5), 347–356 (2001)
10. Rheingold, H.: A slice of my life in my virtual community. High noon on the electronic frontier: conceptual issues in cyberspace, pp. 413–36 (1996)
11. Sentz, K., Ferson, S.: Combination of vidence in Dempster-Shafer theory, vol. 4015. Citeseer, Princeton (2002)

12. Thomas, P., Labat, J.-M., Muratet, M., Yessad, A.: How to evaluate competencies in game-based learning systems automatically? In: Cerri, S.A., Clancey, W.J., Papadourakis, G., Panourgia, K. (eds.) ITS 2012. LNCS, vol. 7315, pp. 168–173. Springer, Heidelberg (2012). doi:10.1007/978-3-642-30950-2_22
13. Vassileva, J.: Motivating participation in social computing applications: a user modeling perspective. User Model. User Adapt. Interact. **22**(1), 177–201 (2012)
14. Wulf, V., Misaki, K., Atam, M., Randall, D., Rohde, M.: On the ground'in sidi bouzid: investigating social media use during the tunisian revolution. In: Proceedings of the 2013 Conference on Computer Supported Cooperative Work, pp. 1409–1418. ACM (2013)

Considerations in Analyzing Ecological Dependent Populations in a Changing Environment

Kristiyan Balabanov(✉), Robinson Guerra Fietz, and Doina Logofătu

Department of Computer Science and Engineering,
Frankfurt University of Applied Sciences, 60318 Frankfurt a.M., Germany
balabano@stud.fra-uas.de

Abstract. Simulation is often used during the development of a system to study the performance of a trial design. On the basis of information gained through this process, the design may be modified and retested. Thus, software offers the convenience of introducing changes or fixes without major effort, whereas it could be quite difficult or even impossible to modify an already built hardware for instance. Moreover, scenarios exist where a simulation offers not only the cheaper, but also the safer and ethically more acceptable solution as is the case with nearly any experiment involving live organisms. This paper summarizes the work done on the subject of simulating the dynamics between populations of various organisms that share an environment. The main goal is to introduce an application that visualizes comprehensively their interaction while offering means to conduct experiments with ecological nature.

Keywords: Population dynamics simulation · Ecological simulation · Agent based modelling · Predator/Prey relation

1 Introduction

This papers presents the empirical results of a simulation modeling the population dynamics of a micro sea ecosystem. The application developed to run the tests can be seen as a mathematical game, in which the dynamic resp. spatial aspects of the species inhabiting it are the main variables. The ultimate goal is to define the boundaries for these variables for which the game can maintain self-sufficiency. Inspiration for the design of the environment has been drawn from the Wa-Tor-like world described by A.K. Dewdney's [5]. His idea encompasses a world shaped like a torus (refer to Fig. 2) and completely covered with water. Marine species are introduced to the world to simulate an actual ecosystem. Their behavior and interaction are defined by certain rules, which however reflect the real world to a certain degree. By changing the rules one can observe the respective impact on the ecosystem, if any. Emphasis is laid on the linear predator-prey relationship between the two major species – sharks (predator) and fish (prey). The sharks feed on the fish and the latter feed on imaginary

N.T. Nguyen et al. (Eds.): ICCCI 2017, Part I, LNAI 10448, pp. 223–232, 2017.
DOI: 10.1007/978-3-319-67074-4_22

plankton, i.e. the shark population depends entirely on the fish for its survival. The work described in this paper introduces a third species – imaginary whales, which feed on both, and thus, create a somewhat more complex food chain. The whales are on a theoretical level as in the real world the only species of whales that would hunt sharks are the orcas, whose actual hunting techniques are too complex for the scope of this work, hence the "imaginary" epithet. The program developed to run the simulation emphasizes more on obtaining results regarding the species' interaction rather than visual effects. Therefore, it resembles a cellular automaton similar to J.H. Conway's Game of life [9].

The design of the application follows an established model of a simulation as described by [1]. Essentially it involves a system's model, some inputs used in the model and the respective outputs produced by the model (refer to Fig. 1). In this matter of thought a model is simply an abstraction of a real world phenomenon or environment, which is too complex to theoretically describe in detail. Thus, the model is by no means perfect or accurate and its level of abstraction determines the degree to which the simulation's results would resemble the outcome of the experiment should it be conducted in the real world. Naturally, the higher accuracy of the model comes at a cost, but experience in the field has confirmed that often as not even low level abstractions can yield plausible results. The input is usually some initial state of the system and its components. The unknown variable in a simulation, i.e. the subject of interest, is the output produced by inserting the input into the given model.

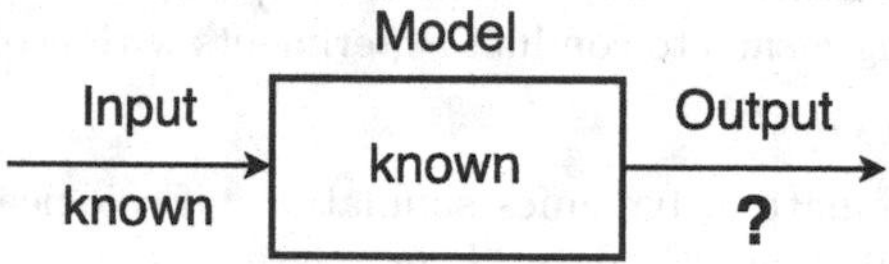

Fig. 1. The structure of a typical simulation problem as given by [1], page 9.

A variety of problems can be successfully addressed with simulation means, which are typically more economical (e.g. when testing the aerodynamics of cars [2,4]), more feasible (e.g. simulating a space rover's mission to another planet as described by [6] or when the rate of the observed phenomenon is too fast/slow in real-time as is the case with plant growth or explosions) or ethically more acceptable (e.g. performing air-bag tests or most experiments involving animals). Dewdney's proposal is a typical example of all three and more precisely the study of an ecosystem. This topic has gained popularity in the recent years as the globalization and industrialization have taken their tolls on the environment and the consequences have become evident. Species are becoming endangered or extinct and while environmentalists are coping to mitigate the damage, scientists are more concerned with determining the potential effects of the caused imbalance or lack of a species on the remaining ones. In other words, how would the predator/prey chains in an ecosystem change [3].

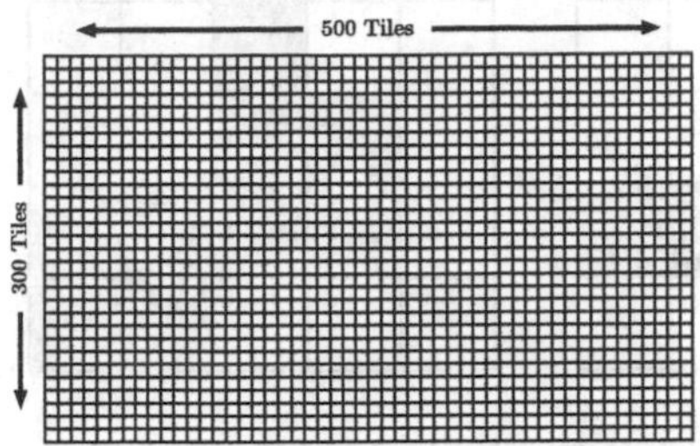

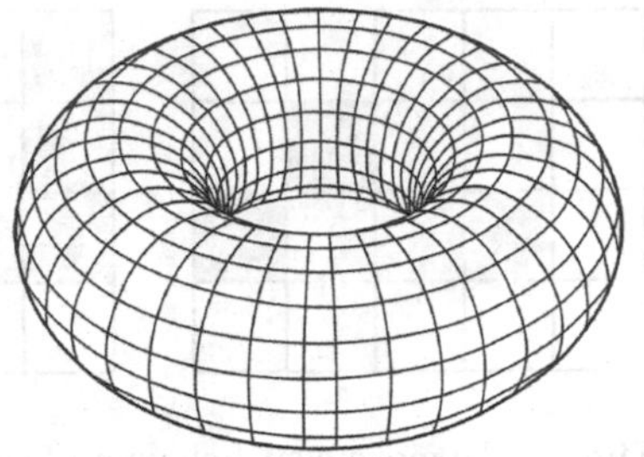

Fig. 2. The simulated world displayed in two-dimensional space as a plain (left) and in three-dimensional space as a simple torus (right). Torus image provided by Wikimedia Commons under the GFDL, CC-BY-SA license.

2 Problem Description

A major difficulty when simulating a system is to create a model of the system that is elaborate enough to yield plausible results and yet not too complicated as to require an unrealistic amount of effort. The problem of an ecosystem simulation can therefore be decomposed into several key aspects:

- Designing an adequate environment, where the system's components can interact. It should copy the essential features of the real world, while leaving out everything that has little to no effect on the simulation output.
- Create species, which would inhabit the environment and define their behavior according to their real world copies. Only actions relevant to the problem scope are to be considered. For a predator-prey populations dynamics this includes activities such as feeding, moving, reproducing, etc.
- Visualize the simulation output in a comprehensive way while also gathering ample statistical data, so that the relation between input and output can be determined.
- Ideally enable the introduction of changes, e.g. adjustable organism characteristics or environmental phenomena such as diseases, in the model as to make the analysis flexible.
- Specify the goal of the simulation, e.g. analysis (does the simulated system exhibit the expected behavior), prediction (would a species go extinct), optimization (how to change the environment so that an equilibrium is achieved) and concentrate on it.

The goals pursued in this paper can be related to the prediction whether an ecosystem is stable or not given some initial populations of inhabiting organisms, and if not, how can it be optimized for instance by controlling the populations' sizes or introducing an evolutionary component to find out the optimal characteristics of a species for it to survive under the given conditions [8].

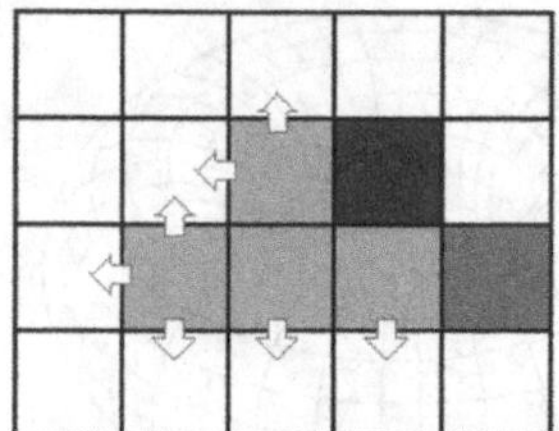
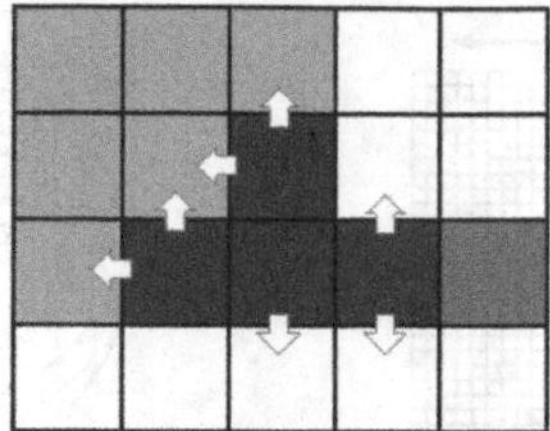
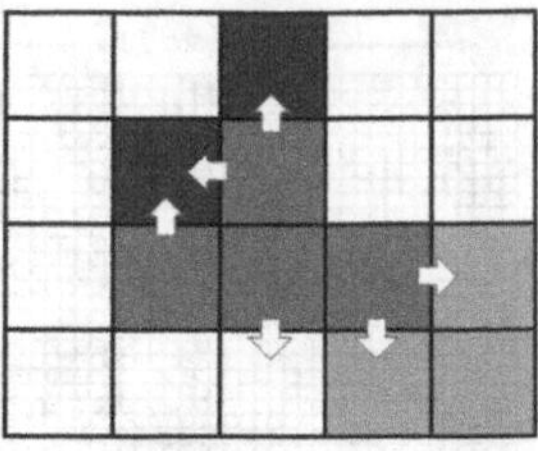

Fig. 3. The movement behavior of each species: fish (left), sharks (middle) and whales (right). Each individual chooses a random neighbor position from the available ones, if any, after considering danger from predators (for prey) resp. the opportunity to hunt prey (for predators).

3 Proposed Approaches

The actions implemented in the simulation include moving, feeding, reproducing and dying. An organism is defined by its age (fish) or remaining life energy (shark and whale), as well as reproduction maturity and energy gained after eating prey (only for predators). Each existing individual performs the aforementioned actions in a specified order and a cycle in the simulation ends when all individuals have "acted", after which the environment is updated. The prey will show awareness when moving and avoid positions already inhabited by predators. The latter will similarly display selectiveness in their movement by choosing tiles inhabited by better prey and by prey in general, avoiding empty positions, but never staying idle as this could mean starving to death.

3.1 Movement Behavior

An individual's movement consists of the horizontal or vertical, but not diagonal, displacement by one tile. All four neighbor tiles (up, down, left, right) are checked if adequate and the most optimal one is chosen. In this line of thought "adequate" suggests that moving to that tile is possible, whereas "optimal" means that moving to is not only possible, but offers some benefit as well.

The terms "optimal" and "adequate", however, differ between the various species. For a fish for instance both mean moving to a tile not inhabited by other organisms (refer to Fig. 3, left). On the other hand, a shark's optimal decision is to move to a position inhabited by a fish so that it can eat it and gain life energy, but any position not inhabited by a whale is an adequate candidate (refer to Fig. 3, middle). Similarly, a whale would always go for a neighbor containing a shark as it would give the largest energy gain, while its second best choice would be one containing a fish. The least popular decision for a whale is to simply move to a free position in search for prey (refer to Fig. 3, right). Fish do not lose energy upon moving, whereas sharks and whales do.

3.2 Reproduction Behavior

All three species have the ability to reproduce provided that certain requirements are met. These include reaching reproduction maturity and having space (adequate positions) for the offspring. Each cycle a fish's age is increased by one. Upon reaching a certain age the parent will check if viable positions for the offspring exist and produce at most two before dying. The viable positions are the same as those for movement including the position of the dead parent (refer to Fig. 4, left). The reproduction of the predator species is defined in a similar fashion, but instead of age the individuals have to possess a certain amount of energy. Only then will they create a single offspring, while losing half of their current life energy. The created child is placed in a random adequate position (refer to Fig. 4, middle resp. right).

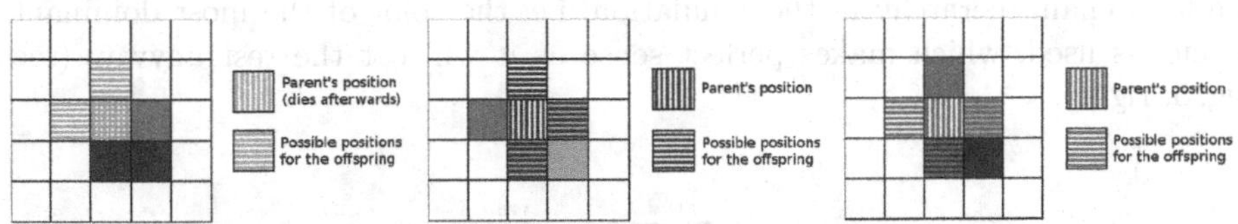

Fig. 4. The reproduction behavior of each species: fish (left), sharks (middle) and whales (right). The offspring are randomly placed at the available neighbor positions, if any, after considering danger from predators (for prey).

3.3 Feeding Behaviour

The feeding behavior of each species in the simulation follows the predefined predator-prey hierarchy, i.e. sharks eat fish, whales eat both sharks and fish. Therefore, the program checks whether a tile of the grid contains a predator, and if so whether its respective prey is also present. The latter would be eaten and so removed from the tile, and the predator would gain the specified amount of energy for consuming that type of prey. Should individuals of all three species reside on the same tile, the whale will consume both shark and fish.

3.4 An Individual's Death

An organism can die in several ways in the simulation. Fish either get eaten by a predator or die after giving birth, sharks may be killed by whales or reach a life energy of 0 (equivalent to a death due to starvation) and whales may only starve to death. Since killing prey is done during the feeding step, in the death step the simulation need only look for sharks resp. whales with life energy equal to 0.

4 Visualising the Populations

As initially specified the simulation environment is represented by a grid of square tiles, 500 tiles in length and 300 in height, effectively enabling up to 150000 organisms to inhabit it. On the screen of the device running the simulation a tile is simply a square with a side length of 2 pixels and a space of 1 pixel between neighbor tiles (refer to Fig. 5, left).

To distinguish between individuals of different species a simple approach based on colors has been used. More precisely, each species is characterized by a unique color and a tile having a certain color means that the respective species inhabits it. In the developed application for instance, cyan marks a fish, dark blue a shark and pink a whale (refer to Fig. 5). Unoccupied tiles are colored light gray. To avoid any ambiguity in the case when individuals of two or three species inhabit the same position, a color hierarchy has been defined. It corresponds to the food chain hierarchy in the simulation, i.e. the color of the most dominant species is used, which makes perfect sense as it will eat the rest anyway (see Fig. 5, right).

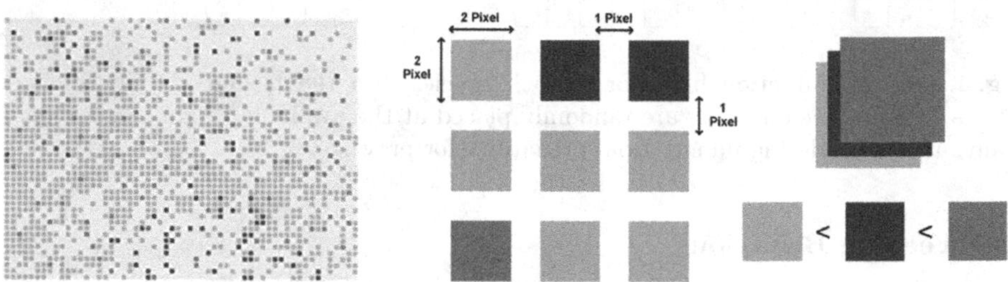

Fig. 5. The environment is visualized as a grid of squares (left) with side length 2 pixels and space of 1 pixel between neighbor tiles (middle). The color of a given tile is always that of the most dominating species, following the hierarchy: fish < shark < whale (right). Empty positions are colored gray. (Color figure online)

5 Experimental Results and Statistical Tests

Numerous tests with different combinations of initial population sizes for the three species were conducted in order to determine whether a relation between the simulated ecosystem's stability and the populations' size exists. Since the behavior of the individuals, expressed with the functions described in Sect. 3, contains a random component, it should be no surprise that a given set of start conditions can produce different results each time. For that reason, each input scenario was tested several times to assign a probability to each possible outcome. Even though the listed input combinations are just a small fraction of all possibilities, the yielded results are statistically significant enough to make certain conclusions. Table 1 summarizes the results of the simulation for some given initial populations.

1. An overpopulation of predators resp. underpopulation of prey as in the real world eventually leads to the collapse of the ecosystem as in test cases (1–5).
2. Predators are more prone to extinction than prey as they have to roam in search for it. Test cases (6–9) are good examples in which sharks and whales die out nearly always due to starvation as the world is sparsely populated.
3. Due to their less flexible diet sharks are more vulnerable than whales, which can survive eating fish as well. Moreover sharks are also prey, meaning that besides starvation, excessive hunting (should the whales be numerous) also contributes for the elimination of their species. As a proof there is only one test case (17), where the whales are more likely to die out, whereas test cases (10–16 and 18–21) all show scenarios, where sharks are.
4. The prey is crucial for the survival of the ecosystem as without it the whole food chain is disturbed. Should the fish become extinct sharks and whales die out rapidly soon after (1–4).

From the statistical data gained through the experiments and the made observations one can derive an empirical formula to approximate the system's stability factor, see (1). Obviously the probability of survival for each species should be taken into consideration, but also some species should be weighted more than others, e.g. sharks due to their vulnerable nature.

$$S = P(\text{fish}) \cdot 0.2 + P(\text{sharks}) \cdot 0.5 + P(\text{whales}) \cdot 0.3. \qquad (1)$$

The stability information gathered from all tests listed in Table 1 and from additional ones is visualized in Fig. 6 (left) as a function of the predator initial populations. It is easy to see that an increasing whale population is decreasing the stability rapidly (indicated by the sharp steps along the direction of the whale axis), whereas that of the sharks in a more gradual manner (indicted by the smaller steps along the shark axis). Beyond some point, however, the joint number of predators becomes too large for the environment to sustain.

A number of additional tests using random initial populations between 10 and 50,000 were conducted without repetition to gain insight regarding the boundaries of the set of successful combinations. Out of several hundred tests only those were chosen, where the system reached a balanced state. This subset of the tried input combinations can be seen in Fig. 6 (right). It is apparent that the fish population's size can vary, whereas that of the whales must not exceed 10,000. Similarly, the upper limit for the sharks is around 25,000. These results correspond to the information regarding the system's stability shown in Fig. 6 (left), and therefore support the approximated coefficients used in (1). To further examine the validity of the results one more test was made with initial populations taken from the subset illustrated in Fig. 6 (right). Since the subset is just an approximation of the actual one, moderate values were used taken from its center rather than its edges: 40,000 fish; 3,500 sharks: 4,500 whales; The test yielded a success rate of 200 out of 200 repetitions.

Unfortunately, the data shown so far is not very informative regarding the dynamics of the ecosystem, i.e. how do the species' populations vary over the course of the simulation. Therefore, the changing population sizes were recorded

Table 1. The effect of the three species' initial population size on the environment's stability calculated with (1).

#	Initial population			Survival probability $P(x)$			System stability
	Fish	Sharks	Whales	Fish	Sharks	Whales	
(1)	10000	50000	10000	0.0	0.0	0.0	0.0
(2)	10000	50000	50000	0.0	0.0	0.0	0.0
(3)	50000	50000	50000	0.0	0.0	0.0	0.0
(4)	50000	50000	10000	0.1	0.0	0.0	0.02
(5)	100	10000	10000	0.6	0.0	0.0	0.12
(6)	10	10	10	1.0	0.0	0.0	0.2
(7)	10	100	100	1.0	0.0	0.0	0.2
(8)	100	10	10	1.0	0.1	0.1	0.28
(9)	1000	100	10000	1.0	0.0	0.3	0.29
(10)	100	1000	10000	0.9	0.0	0.4	0.3
(11)	100	10	100	1.0	0.0	0.5	0.35
(12)	1000	10000	10000	1.0	0.0	0.6	0.38
(13)	100	100	100	1.0	0.1	0.55	0.415
(14)	1000	1000	10000	1.0	0.0	0.8	0.44
(15)	100	10000	100	1.0	0.0	0.9	0.47
(16)	100	100	1000	1.0	0.0	1.0	0.5
(17)	100	1000	100	1.0	0.6	0.1	0.53
(18)	100	10000	1000	1.0	0.1	1.0	0.55
(19)	100	1000	1000	1.0	0.2	1.0	0.6
(20)	50000	10000	10000	1.0	0.3	1.0	0.65
(21)	10000	1000	10000	1.0	0.4	1.0	0.7
(22)	10000	1000	1000	1.0	0.8	1.0	0.9
(23)	1000	100	100	1.0	1.0	1.0	1.0
(24)	1000	1000	100	1.0	1.0	1.0	1.0
(25)	1000	10000	1000	1.0	1.0	1.0	1.0

as a function of the time. A plot produced from such data can be seen in Fig. 7. Despite the simplicity of the model, one can easily distinguish the famous predator-prey relation described by the Lotka-Volterra-Model [7]. With the increase of the prey population the predators thrive and increase in numbers as well. The latter coupled with environmental factors such as overpopulation or diseases lead to a decrease in the prey population. Accordingly, soon after the predators dwindle as food becomes sparse. If the populations are big enough and no devastating disaster happens, the system will oscillate between some relatively constant boundaries, i.e. be stable. Past the 1000th cycle a similar phenomenon can be seen in Fig. 7. All three populations increase and decrease

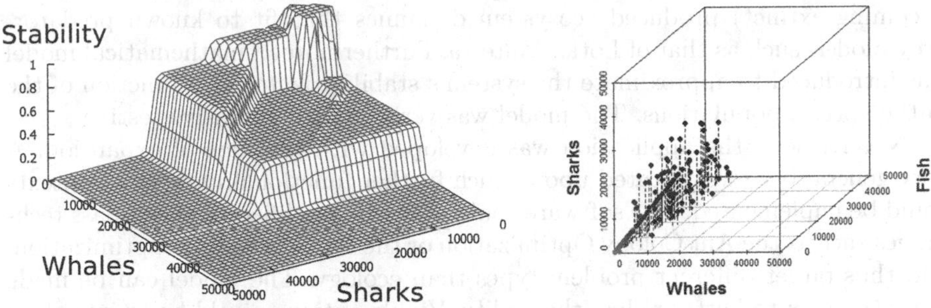

Fig. 6. The stability of the system for various initial predator populations calculated with (1) (left) and a set of initial populations for all three species, with which the system reaches a balanced state (right).

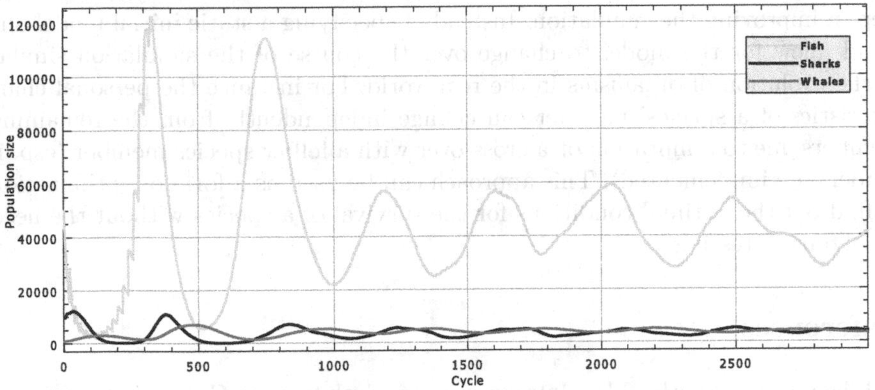

Fig. 7. A graph plotting the population sizes changing over the course of the simulation as a function of the time. The input consisted of 50000 fish (cyan), 10000 sharks (dark blue) and 1000 whales (pink). (Color figure online)

periodically with that of the sharks being slightly delayed relative to the fish population, and that of the whales relative to both fish and shark.

6 Conclusion and Future Work

The main goal of the undertaken project was to develop an application for the easy and comprehensive simulation of the interaction between objects resp. organisms. As an example a simple ecosystem was modeled and tests were performed to obtain information regarding its equilibrium state. Given the results, one can confirm that even with a simple model and basic rules the observed tendencies reflect the real world quite accurately. The input configurations that lead to the destruction of the ecosystem could easily be related with phenomena such as over- resp. underpopulation of a species. On the other hand, start configurations that allowed for the environment to reach a state of stability (no species becoming extinct) produced ecosystem dynamics that fit to known predator-prey models such as that of Lotka-Volterra. Furthermore, a mathematical model was introduced to approximate the system's stability factor as a function of the initial species populations. The model was verified with extensive testing.

Nevertheless, the application was developed with the idea of expansion. A very general base was created upon which further functionality or new contexts could be implemented. The software could easily be modified to encompass techniques such as the Ant Colony Optimization or the Particle Swarm Optimization, and thus target different problem types than ecology. The model can be made more complex to better reflect the reality. Viable options would be to introduce genders, longer movement paths or global phenomena like diseases. Implementing Collective Intelligence, so that the prey moves together and shares information regarding danger from predators and their known positions, could also lead to interesting outcomes. Evolutionary approaches would definitely be the next step in improving the simulation. Instead of specifying a static initial model one could allow for the model to change over the course of the simulation similar to the evolution of organisms in the real world. For instance the personal characteristics of a species' member can change independently from the remaining members due to a mutation or a cross-over with a fellow species member (esp. if genders are implemented). This approach can be used as a form of optimization to find out the optimal conditions for the survival of a species without the need for extensive testing.

References

1. Eiben, A.E., Smith, J.E.: Introduction to Evolutionary Computing. Springer, Heidelberg (2015)
2. Law, A.M., David, K.W.: Simulation Modeling and Analysis. McGraw-Hill Higher Education, Boston (1997)
3. Begon, M., Mortimer, M., Thompson, D.J.: Population Ecology: A Unified Study of Animals and Plants. Wiley-Blackwell, Oxford (1996)
4. Warschat, J., Wagner, F.: Einführung in die Simulationstechnik. http://www.iat.uni-stuttgart.de/lehre/lehrveranstaltungen/skripte/simulation/AltesSimulationsSkript.pdf. Accessed 22 Mar 2017
5. Dewdney, A.K.: Solutions périodiques, du computer recreations: sharks and fish wage an ecological war on the toroidal planet Wa-Tor. Scientific American (1984)
6. Norris, J.S.: Mission-critical development with open source software: lessons learned. IEEE Softw. **21**(1), 42–49 (2004)
7. Hoppensteadt, F.: Predator-prey model. Scholarpedia **1**(10), 1563 (2006)
8. Connelly, B.D., Zaman, L., McKinley, P.K.: The SEEDS platform for evolutionary and ecological simulations. In: Proceedings of the 14th Annual Conference Companion on Genetic and Evolutionary Computation, GECCO 2012, pp. 133–140. ACM (2012)
9. Gardner, M.: Mathematical Games - the fantastic combinations of John Conway's new solitaire game "life". Sci. Am. **223**, 120–123 (1970)

Automatic Deduction of Learners' Profiling Rules Based on Behavioral Analysis

Fedia Hlioui(✉), Nadia Aloui, and Faiez Gargouri

Multimedia InfoRmation System and Advanced Computing Laboratory, University of Sfax, Sfax, Tunisia
fediahlioui@gmail.com, alouinadia@gmail.com, faiez.gargouri@isims.usf.tn

Abstract. E-learning has become a more flexible learning approach thanks to the extensive evolution of the Information and Communication Technologies. A perceived focus was investigated for the exploitation of the learners' individual differences to ensure a continuous and adapted learning process. Nowadays, researchers have been oriented to use learning analytics for learner modeling in order to assist educational institutions in improving learner success and increasing learner retention. In this paper, we describe a new implicit approach using learning analytics to construct an interpretative views of the learners' interactions, even those made outside the E-learning platform. We aim to deduce automatically a learners' profiling rules independently of the learning style models proposed in the literature. In this way, we provide an innovative process that may help the tutors to profile learners and evaluate their performances, support the courses' designer in their authoring tasks and adapt the learning objects to the learners' needs.

Keywords: Behavioral indicator · Learning analytics · Learning profiling rules · Leaner model

1 Introduction

The interactivity between the E-learning actors (learners, tutors and course's designer) has undoubtedly been an important part to enhance the quality of any learning process. In a traditional learning classroom, the tutor is face to face with the learners. This allows him to accurately recognize their behaviors from their facial expressions, questions and interactions. The tutor is, therefore, able to interact at the right moment for a constructive intervention. Regrettably, this has not been the case of E-learning environment. The tutor cannot have the same feedback from the learners, owing to the absence of human contact. Additionally, learners can lose concentration and motivation easily; especially when the learning process is not tailored to their needs.

From this angle, several researchers focus on understanding the variability of the learners' individual differences to improve learning outcomes. Most of studies apply an explicit method that aim to use a direct and obvious way to identify the learners' personality. This is done through the filling of some psychological questionnaires [1]. In this context, many learning style models [2, 3], standards [4, 5] and ontologies [6, 7]

N.T. Nguyen et al. (Eds.): ICCCI 2017, Part I, LNAI 10448, pp. 233–243, 2017.
DOI: 10.1007/978-3-319-67074-4_23

were proposed. Despite the frequent adaptation of this method, it was confirmed by [8], that theses questionnaires are typically too long and contain a huge range of data elements. Consequently, the acquisition of this data consumes an overplayed time and the learner gets quickly bored, stressed and unmotivated for completing such task [9]. Furthermore, Stash [10] pointed out that most of learners are unconscious about their own preferred learning methods and they tend to provide unreliable sources of information about themselves. The social and psychological aspects such as the learners' beliefs about how people should behave can influence their answers [11]. In addition, the learning scenarios' modeling will be a big challenge and a tedious task for authors by considering all learners' personalities to achieve one single learning objective [12].

To overcome these issues, studies became conducted for modeling the personalization parameters implicitly. This approach aims to collect the information about learners in a concealed and unobtrusive way. Principally, this is done by analyzing the learners' behavior regarding to their interactions with the E-learning environment [9]. These traces reflect in-depth details of the learning activities. They can enhance the awareness of the tutor and the course's designer, about the ongoing activity. Consequently, these two actors will be able to regulate and adapt the learning scenario in order to overcome failure and errors. Additionally, the gathered traces allow the learner to visualize his/her evolution and to understand his/her progress during the learning session.

To reach this objective, we conduct a detailed comparative study of the most distinguished existing E-learning systems based on learning analytics approach, in order to point out their main advantages and shortcomings. From this study, we propose a new adaptation approach for an automatic deduction of learners' profiling rules based on behavioral analysis. Our contribution allows us to exploit all the learners' interactions, even those made outside the E-learning platform.

In this paper, we first start by presenting the learning analytics approach. Then, we present a comparative study between E-learning systems according to some common behavioral indicators. Subsequently, we specify the dependency between the learning analytics approaches and the learning styles proposed in the literature. In the third section, we detail our contributions. Finally, we discuss the benefits of our approach and we precise some perspectives.

2 Research Context and Related Works

The learner's personality does not only interest psychologists, but also pedagogues, course's designer and tutors. They search to understand the human personality and to find a systematic way to measure it. Consequently, there has been a growing interest in the analysis of learners' interaction for enhancing the learning experience. A research area referred to recently as learning analytics has emerged as a very challenging target for the field of technology enhanced learning. The learning analytics was presented in [13] as "the use of intelligent data, learner-produced data, and analysis models to discover information and social connections, and to predict and advise on learning". Based on this definition, Chatti et al. [14] defined the learning analytics process as an iterative cycle divided into three phases: (i) data collection and pre-processing,

(ii) analytics and action, and (iii) post-processing. The first phase is a fundamental part in any learning analytics approach. The relevance of the discovered patterns depend of the collected data quality. These data may be too large and/or imply many irrelevant attributes. For that reason, they must be transformed into a convenient format that can be used as input for a particular learning analytics method. The analytics and action's phase aim to explore the cleaned data in order to discover a set of high-level indicators [17]. These behavioral indicators are seen as a meta-knowledge of the traces' observations. Many researchers employed these indicators to achieve many learning objectives such as: visualization [18–20], evaluation of the learning objects' effectiveness [21], prediction of learner profile [22], generation of profiling rules [9, 23], etc. The post-processing phase is made to improve the analytics process by compiling new data from additional data sources, choosing a new analytics method, identifying new indicators, etc. [14]. Table 1 highlights a comparative study between E-learning system and behavioral indicators proposed in literature.

As shown in Table 1, most of studies explored and analyzed the learner's collaborative activities. Each of these studies defined the collaborative aspects differently. For instance, Halawa et al. [22] extracted this type of data from the learning management system Moodle and the Facebook as a social networking service. In the context of the game based learning environment, this collaborative indicator was investigated by [18] to enhance the learners' activities. Unfortunately, the learner's interaction in the client side was completely ignored, in this work. The recorded trace might not to be enough to reflect the whole learners' activities during their communication. Besides the idea of observing collaborative activity on both client and server sides, May et al. [26] compared the participation rates of the collaborative task and evaluated the productivity rates of one group in relation to another, according to the number of discussions forums, number of discussions threads, numbers of messages posted and files created and shared. In the same context, many other indicators are listed in [25] like the division of labor, Nonverbal actions indicator, active agent indicator, selected agent contribution indicator, etc. Additionally, we notice that a few studies treated the annotation activities' indicators despite the fact that an annotation is considered as a credible source of knowledge. Actually, collecting the learners' annotations is a tedious task due to the dependency to "pen-and-paper" which the learner voluntary interacts with a document in an intuitive and familiar manner. Most of studies like in [24] extracted manually the annotations traces to deduce some behavioral indicators. For that reason, Louifi et al. [28] replaced the "pen-and-paper" approach by proposing a client-side tool for tracking annotations. Nevertheless, the annotations in server-side are totally unheeded.

Moreover, most of researchers constructed their interpretative views of learners' interaction behavior with respect of a given learning style model. They convert dimensions of a given learning style model into a set of behavioral indicators to predict the learner profile. Table 2 highlights the correlation between the recent learning analytics approaches based on learning style models.

As presented in Table 2, researchers were oriented to matching the learners behavior with their respective learning style proposed in literature. For example, the indicators proposed by [23], were investigated according to some teachers' experiences and to the FSLSM model [2]. Ammor et al. [9] proposed a set of performance profiling

Table 1. Comparison between E-learning systems based on behavioral indicators

E-learning system	Behavioral indicator												
	Concentration rate	Disorientation rate	Browsing pattern	Consultation type	Contribution indicator	Division labor	Active agent indicator	Collaborative indicator	Participation group	Perseverance rate	Performance indicator	Annotation indicator	Activity indicator
Marty & Carron [18]								√					
DSLP [9]	√							√		√			
Dyckhoff et al. [20]				√				√					√
Omheni et al. [24]												√	
El haddaoui & Khaldi [16]								√					
SBT-IM [25]					√	√	√	√	√				
Aitadda & Bousbia [15]	√	√											
TRAVIS [26]								√					
IDLS [23]	√		√	√									
Campos et al. [21]				√							√		
Halawa et al. [22]								√	√		√		
Papanikolaou [17]			√	√				√			√	√	

Table 2. Comparison between learning analytics based on learning style models

E-learning system	Learning styles			
	MBTI [3]	FSLSM [2]	7 Learning profile [29]	Kolb's theory [30]
IDLS [23]		√		
DSLP [9]			√	
El heddaoui et khaldi [16]				√
Halawa et al. [22]	√			√
Graf et al. [27]	√	√		

rules based on the 7 learning profiles model [29] and only three behavior indicators (the collaborative rate, the concentration rate and the perseverance rate). In this work, the profiling rules were generated explicitly by an expert.

Despite of the complexity of psychology and cognitive science, must of the researchers didn't found correlation between the behavioral indicators and all the dimensions of the learning style model. For instance, Bousbia et al. [23] treated only the sequential/ global and active/reflective dimensions of FSLSM model and the navigation type indicator. Consequently, many behavioral indicators have not been dealt in the literature, such as the educational preferences (the autonomy, the habits, the information representations, the perception, etc.), the indicators related to information processing (the understanding process, the dependence of filed, the difficulties, etc.) and the cognitive abilities (the motivation factors, the rapidity, the emotional factors, etc.). Nevertheless, the errors committed by the learner are not handled previously. In the following section, we describe our contributions to answer the main issues, which were stated above.

3 Approach for an Automatic Deduction of Learner's Profiling Rules

In our research, we propose an approach for regulating the learning scenarios by an automatic deduction of learners' profiling rules based on the learning interactions, even those made outside the E-learning system. The aim of this study is to investigate how learners with different learning profiles use the course differently with respect to their navigational behavior. These profiling rules may help tutors to profile learners and evaluate their performance, and courses' designer to create learning objects. Our system's architecture, presented in Fig. 1, is composed of six basics components: the collection traces' component, the transformation of log file component, the fusion of log file and learning objects' component, the extraction of behavioral indicators' component, generation of profiling rules, and the regulation of learning scenarios' component.

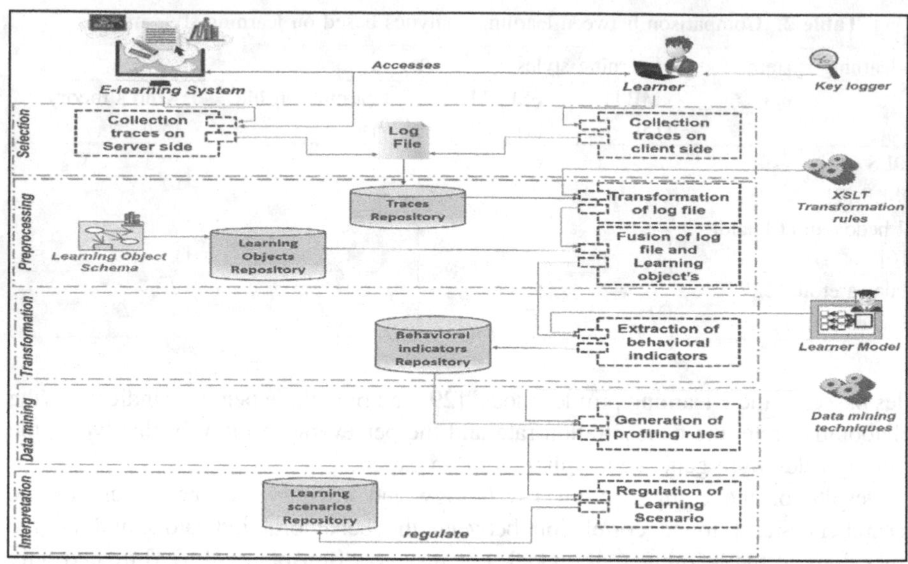

Fig. 1. Our approach for an automatic deduction of learners' profiling rules based on behavioral analysis

3.1 The Collection Traces' Component

Collecting learners' actions is the first step for discovering patterns from learners' activities. These heterogeneous data recorded from learners' interaction must be carefully chosen and handled in order to yield meaningful information and build a thorough view of learners' activity. As mention in [28], approximately 78% of E-learning systems include a trace collection tools. Table 3 highlighted a comparative study between adaptive e-learning system adopting collection tools.

Table 3. Comparison between E-learning systems according to the collection tools

E-learning system	Collection type		
	Client side	Server side	Client/Server sides
Aitadda & Bousbia [15]			√
Marty & Carron [18]		√	
DSLP [9]			√
Omheni et al. [24]		√	
El haddaoui & khaldi [16]	√		
SBT-IM [25]			√
IDLS [23]			√
Halawa et al. [22]			√

As show in Table 3, the most E-learning system adopted a collecting tool for both client and server sides. It was confirmed by [28] that approaches based on log file analysis on server side does not contain any information of any parallel activities done outside the E-learning environment. In fact, the approaches based on the client side can overcome these issues. On one hand, they provide information not only on the activity of the learner within the learning content on the E-learning environment, but also outside of it and, on the other hand; they could collect traces from heterogeneous E-learning systems. Despite the shortcoming of tools presented previously, we choose to identify the learners' navigational observations on both client and server sides. This mixed approach aims to have an overall view of learners' individual differences. For collection traces' component on client side, there are two types of traces collection tools: the key logger and the eye tracking technologies. The first technology was used frequently in many E-learning systems like [11, 15], to record each keystroke done. Recently, researchers are oriented to adopt an eye tracking technology like in [16]. They consider that the eye movements are an indication of learner interest and focus of attention. Despite all its advantages, the eye movement cannot reveal the comprehension difficulty and the thought contents. For that reason, we choose to add a Mini-Key-log software [31], a key logger that has the capability to record keystroke to a log file. The data collected from this devise indicate where the learner's attention has been caught, thus providing evidence of the learner's focus of attention over time. We choose an open-source collecting tool to prevent tracking the learner's personal data. So following the connection and the authorization of the leaner, each collection traces' component records all the observations performed on the learner's station during the learning session: visited web pages, keyboard, Mouse clicks, clipboard, data medium change, file operations and more.

3.2 The Transformation of Log Files' Component

The log files obtained as an output of the traces' collection component are not always directly exploitable, and it is sometimes necessary to go through one or more transformations. Djouad et al. [25] stated three type of traces' transformation: the selection, the re-writing and the temporal fusion. The goal of the selection process is the extraction of the relevant traces respecting some criteria. However, the re-writing aims to replace one or more traces by another trace. The temporal fusion consists to combine the traces taking into account the temporality of their observations in order to obtain a new trace.

In our case, we adopt the selection process because the log files' data are often required to remove noise (publicities, incorrect URLs, pages not found, etc.) to store them in a well-defined and expandable database format. These pre-processing operations are described in an XSLT sheets supplied as input to the traces' transformation component.

3.3 The Fusion of Log File and Learning Object's Metadata Component

The main goal of this component is to understand what a learner has done outside the learning session. It aims to combine the filtered traces and the learning object's metadata stored, in order to obtain a global vision of the learners' observations. Additionally, the computing of some behavioral indicators depends of information about learning content. For instance, the concentration rate represents the learner's level of interest in relation to the subjects studied. This behavioral indicator is calculated by the study of semantic similarity between the metadata of learning object and the web pages visited during the learning session. At the end of the fusion step, the tracking data is cleaned and transformed into an appropriate format to be exploited.

3.4 The Extraction of Behavioral Indicators Component

This component aims to calculate the learners' behavioral indicators. The input of this component are a learner model and the recorded observations. The learner model contains the list of the behavioral indicators that cover the learners' individual differences. Throughout our study, we have classified the behavioral indicators according to different aspects:

- Collaborative indicators: that reflect various aspects of social behavior in the E-learning environment such as the browsing forums structure, the replied messages, the files uploaded, the new messages, the division labor, etc.
- Cognitive indicators: that concern the interactions of learners in the learning process such as the consultation type, the browsing forms, the navigation strategy, the consultation's coherence, etc.
- Performance indicators: that reflect the product of the learner's activities such as the concentration rate, the progression level, the curiosity rate, etc.
- Temporal indicators: that concern the time spent by the learner during her connection into the E-learning platform such as the perseverance rate, the rapidity, the inactivity duration, etc.

3.5 Generation of Profiling Rules Component

Due to the vast quantities of behavioral indicators generated, it is very difficult to analyze this data manually. In recent years, there has been an increasing interest in the use of Data Mining to investigate the education sector. It was confirmed by [32] that Educational Data Mining is able to understand the learners' personalities, as well as enhancing teaching and learning processes. Data mining is the process of efficient extraction patterns from a large collection of data. Association rules are one of the distinctive rule patterns in data mining tools [33]. It generates relationships between attributes-values in databases. In our case, the attributes are the behavioral indicators stored in the repository. Therefore, we aim to discover learners' profiling rules according to these indicators using one of association rules' algorithms. An association rule $X \Rightarrow Y$ indicates that in those transactions where X occurs there is a high probability of having Y as well. X and Y represent the behavioral indicators and they are called respectively the antecedent and consequent of the rule. Each rule is measured by its confidence and support. The confidence of the rule is the percentage of transactions that contains the consequence Y in transactions that contain the antecedent X.

The support of the rule is the percentage of transactions that contains both antecedent X and consequence X in all transactions in the database [33].

3.6 Regulation of Learning Scenarios Component

For the interpretation step, the profiling rules will be used to regulate the learning scenarios. These rules will help the tutors in their monitoring and accompanying tasks, and by helping the courses' designers for modeling and adapting the learning objects according to the learner's needs.

4 Conclusions and Future Work

In this paper, we present new implicit approach using analytics instead of questionnaires based approach to identify the learners' personality. The intention of our study is to provide a new way to manage learners' traces. We describe an approach for regulating the learning scenarios by an automatic deduction of learners' profiling rules based on learning interactions. The one of particularities of our research is to analyze the learners' traces during learning session, even those made outside the E-learning system. Additionally, we present new method to discover pattern from learners' interaction independently of any learning style proposed in literature. We have already started the development of our approach. We point to validate it with a large database, in different learning sessions and with different association rules techniques. Additionally, we aim to integrate more behavioral indicators as much as possible, in order to enhance the performance of the profiling rules and to cover all the learners' individual differences. Moreover, the contention of our work is to consider the learners' emotional features using the eye tracking technology.

References

1. Tadlaoui, M.A., Aammou, S., Khaldi, M., Carvalho, R.N.: Learner modeling in adaptive educational systems: a comparative study. Int. J. Mod. Educ. Comput. Sci. **8**(13), 1 (2016)
2. Felder, R.M., Spurlin, J.: Applications, reliability and validity of the index of learning styles. Int. J. Eng. Educ. **21**(11), 103–112 (2005)
3. Catherine, B.-C., Wheeler, D.D.: The Myers-Briggs personality type and its relationship to computer programming. J. Res. Comput. Educ. **26**(13), 358–370 (1994)
4. Farance, F.: Draft standard for learning technology. Public and private information (PAPI) for learners (PAPI Learner) (2000)
5. IMS Learner Information Package Specification version 1.0 (2001). http://www.imsglobal.org/profiles/index.html
6. Messaoudi, F., Moussaoui, M., Bouchboua, A., Derouich, A.: Modeling approach to a learner based on ontology. Int. J. Soft Comput. Eng. **2**, 262–265 (2013)
7. Hlioui, F., Alioui, N., Gargouri, F.: A system for composition and adaptation of educational resources based on learner profile. In: 5th International Conference on Information & Communication Technology and Accessibility (2015)

8. Tlili, A., Essalmi, F., Jemni, M., Chen, N.-S., Kinshuk: Role of personality in computer based learning. Comput. Hum. Behav. **64**, 805–813 (2016)
9. Ammor, F.-Z., Bouzidi, D., Elomri, A.: Construction of deduction system of learning profile from performance indicators. Int. J. Inf. Educ. Technol. **3**(12), 129 (2013)
10. Stash, N.: Incorporating cognitive learning styles in a general-purpose adaptive hypermedia system, vol. 68 (2007)
11. Bousbia, N.: Analyse des traces de navigation des apprenants dans un environnement de formation dans une perspective de détection automatique des styles d'apprentissage. l'université Pierre et Marie Curie de France et l'école nationale supérieure d'informatique d'Algérie (2011)
12. Hlioui, F., Alioui, N., Gargouri, F.: A survey on learner models in adaptive E-learning systems. In: IEEE/ACS 13th International Conference of Computer Systems and Applications (2016, in press)
13. Siemens, S.: What are learning analytics, vol. 10 (2010). Accessed March
14. Chatti, M.A., Dyckhoff, A.L., Schroeder, U., Thùs, H.: A reference model for learning analytics. Int. J. Technol. Enhanc. Learn. **4**(5–6), 318–331 (2012)
15. Ait-Adda, S., Bousbia, N.: Evaluation de la désorientation de l'apprenant dans un systéme d'apprentissage. Eiah 2015, Agadir, vol. 9 (2015)
16. El Haddioui, I., Khaldi, M.: Learner behavior analysis on an online learning platform. Int. J. Emerg. Technol. Learn. IJET **7**(12), 22–25 (2012)
17. Papanikolaou, K.A.: Constructing interpretative views of learners' interaction behavior in an open learner model. IEEE Trans. Learn. Technol. **8**(12), 201–214 (2015)
18. Marty, J.-C., Carron, T.: Observation of collaborative activities in a game-based learning platform. IEEE Trans. Learn. Technol. **4**(11), 98–110 (2011)
19. May, M., George, S., Prévôt, P.: TrAVis to enhance online tutoring and learning activities: real-time visualization of students tracking data. Interact. Technol. Smart Educ. **8**(11), 52–69 (2011)
20. Dyckhoff, A.L., Zielke, D., Bùltmann, M., Chatti, M.A., Schroeder, U.: Design and implementation of a learning analytics toolkit for teachers. Educ. Technol. Soc. **15**(13), 58–76 (2012)
21. Campos, A.M.: Analyzing the effectiveness of learning objects and designs. In: 11th IEEE International Conference on Advanced Learning Technologies (ICALT) (2011)
22. Halawa, M.S., Hamed, E.M.R., Shehab, M.E.: Personalized E-learning recommendation model based on psychological type and learning style models. In: IEEE Seventh International Conference on Intelligent Computing and Information Systems (2015)
23. Bousbia, N., Gheffar, A., Balla, A.: Adaptation based on navigation type and learning style. In: International Conference on Web-Based Learning (2013)
24. Omheni, N., Mazhoud, Kalboussi, A., Kacem, A.H.: Prediction of human personality traits from annotation activities. In: WEBIST (2014)
25. Djouad, T., Mille, A., Reffay, C., Benmohammed, S.: Collaborative Activity Indicators Engineering: Using modeled traces in the context of Technology Enhanced Learning Systems. Rapport de recherche RR-LIRIS-2010-014 (2010)
26. May, M., George, S.: Using learning tracking data to support students' self-monitoring. In: CSEDU (2011)
27. Graf, S., Liu, T.-C., Kinshuk: Analysis of learners' navigational behaviour and their learning styles in an online course. J. Comput. Assist. Learn. **26**(2), 116–131 (2010)
28. Louifi, A., Bousbia, N., Azouaou, F., Merzoug, F.: ESITrace: a user side trace and annotation collection tool. In: International Conference on Web-Based Learning (2012)
29. Michel, J.-F.: Les 7 profils d'apprentissage, Editions Eyrolles (2013)

30. Petchboonmee, P., Phonak, D., Tiantong, M.: A comparative data mining technique for David Kolb's experiential learning style classification. Int. J. Inf. Educ. Technol. **5**(9), 672 (2015)
31. Mini Key Log, Version: 6.13. http://www.blue-series.com/en/products-en/mini-key-log-en.html
32. Halawa, M.S., Hamed, E.M.R., Shehab, M.E.: Predicting student personality based on a data-driven model from student behavior on LMS and social networks. In: Fifth International Conference on Digital Information Processing and Communications (2015)
33. Garcia, E., Romero, C., Ventura, S., Calders, T.: Drawbacks and solutions of applying association rule mining in learning management systems. In: Proceedings of the International Workshop on Applying Data Mining in E-learning, Crete, Greece (2007)

Predicting the Evolution of Scientific Output

Antonia Gogoglou(✉) and Yannis Manolopoulos

Department of Informatics, Aristotle University, 54124 Thessaloniki, Greece
{agogoglou,manolopo}@csd.auth.gr

Abstract. Various efforts have been made to quantify scientific impact and identify the mechanisms that influence its future evolution. The first step is the identification of what constitutes scholarly impact and how it is measured. In this direction, various approaches focus on future citation count or h-index prediction at author or publication level, on fitting the distribution of citation accumulation or accurately identifying award winners, upcoming hot research topics or academic rising stars. A plethora of features have been contemplated as possible influential factors and assorted machine-learning methodologies have been adopted to ensure timely and accurate estimations. Here, we provide an overview of the field challenges, as well as a taxonomy of the existing approaches to identify the open issues that are yet to be addressed.

Keywords: Scientometrics · Bibliographic data · Predictive modeling

1 Introduction

With the extensive recording of scientific progress on the Web and the emergence of large scale open source as well as proprietary databases of bibliographic data (Google Scholar, Web of Science, Scopus, etc.), the quantification and evaluation of scientific impact, the "science of science" [2], has attracted significant attention. In particular, research institutions, universities and even countries have been adopting research policies that emphasize "excellence" and "impact". In this direction, rigorous efforts have been made to extract meaningful and actionable information from the abundance of bibliometric data to produce rankings, aid decision making and assist peer review. This is evident by the plethora of bibliometric indices that have been proposed in the past decade, since the seminal paper by Hirsch introducing the h-index [13], and all attempts to quantify different aspects of scientific impact [39]. The high level of correlation amongst the majority of these indices has been extensively investigated [4,32]. However, the focus is now turning towards the quantification of future impact and rising influence, instead of measuring existing output in different ways.

In his preliminary work, Price deduces that current visibility, publishing venue and age highly influence a publication's future outreach [35]. In today's fast paced, ever growing and interdisciplinary research world what determines future influence? Is it possible to provide early estimation using current data? These are

N.T. Nguyen et al. (Eds.): ICCCI 2017, Part I, LNAI 10448, pp. 244–254, 2017.
DOI: 10.1007/978-3-319-67074-4_24

intriguing questions for all stakeholders of the scientific community, from individual scholars to publishers and from funding agencies to hiring committees, as current decisions on tenure, grand allocation and publishing are based on an inherent estimation of future evolution. Identification of future trends, advancing topics and trend shapers or influentials in the science world are examples of efficient utilization of predictive analytics and, therefore, they are attracting attention from both public and private sector, with the number of related publications rising every year. Thompson Reuters, a game changer corporation in publishing, has created the InCite platform (https://incites.thomsonreuters.com) for mapping and ranking scientists and their output, as well as identifying "hot" publications or up and coming research avenues. In any case, efficient and meaningful approximation of future trends can provide invaluable tools to stakeholders of the scientific world, to better coordinate research endeavors, utilize funds and create connections that will improve visibility and productivity.

Hitherto, due to difficulties in obtaining reliable and abundant data, the scientific community was largely reliant on judgement of experts to evaluate future potential of a publication or a scholar. However, with the increasing data availability and advances in big data mining, the need for computerised support in decision making has come up, given that peer review can prove to be costly and time-consuming. In addition, peers will use their own knowledge and expertise to formulate judgement leading to more conservative views not receptive to novelty. On the other hand, data intelligence, which has been utilized in various disciplines like marketing, business, security, etc. [9], can overcome personalized criteria and provide evidence based valuable insights to assist in strategy planning. Figure 1 demonstrates a workflow describing the general process for deriving actionable data intelligence from available bibliographic data.

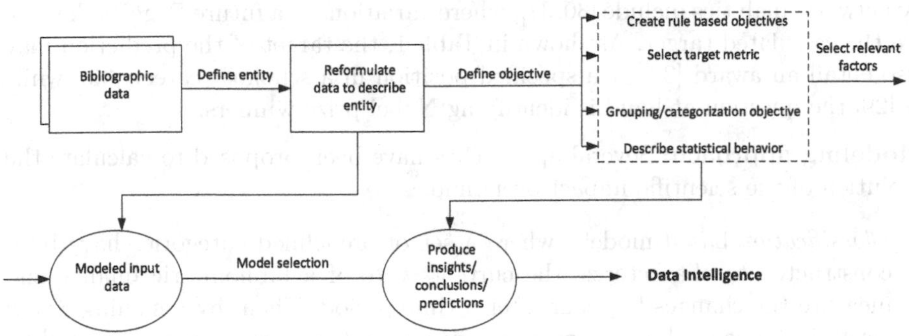

Fig. 1. Workflow from available bibliographic data to actionable data intelligence.

In the present work, we focus on the question: *Is scientific progress quantifiable and predictable?* To riddle this question, we provide an overview of existing approaches and a taxonomy of them based on their common qualities. Additionally, we identify the remaining open research issues and challenges in this area as well as the dangers that arise from the quantification of scientific evolution.

2 Taxonomy of Approaches

There exist multiple approaches to quantify the evolution of scientific impact; they can be categorized with regards to the *scientific entity concerned*, their *modeling approach* and their *target metric* (see Tables 1 and 2 in the next).

Scientific entity: An initial category stems from the entity under evaluation: *publication, author, venue* or *institution.* Most efforts focus on publications, because for the other three categories one needs to aggregate the respective entire portfolio of publications (author, venue or institution), thus increasing the calculation complexity. Also, complete information about a publication is usually available at several online databases, whereas for the other entities there is a high disambiguation amongst different online entities to ensure that the complete records are retrieved (e.g. names, abbreviations, etc.).

Target variable: Another approach is related to the target variable: e.g. the *citation count* taken as a proxy for impact or a bibliometric index, such as the *h-index.* Due to the exponential distribution of these quantities and the debate regarding their crude limiting nature, a set of works has defined the prediction problem in an alternative way to mitigate the skewness of the predicted output. For instance, in [6] the yearly rise in citation is estimated, in [11] the predictive question is whether a publication will contribute to the rise of the first author's *h*-index, whereas Garner et al. foresee how quickly the first citation of a publication will occur [12]. To avoid the heavy tailed distribution of target variables, which often inhibits the effectiveness of the model, the relative rank of a scientific entity in a network can be predicted instead. In [21] the rank of a publication is compared to all other publications in the same discipline, while in [5] it is compared against the journal publications of the same year. Approaches inspired by network analytics include [30,37], where variations of a future PageRank value are the calculated target. As shown in Table 1, the target of the prediction may also entail an award [33] or a specified position in a scholar's career [36], while in [29] the question at hand is identifying Nobel prize winners.

Modeling approach: Several approaches have been proposed to calculate the evolution of the scientific impact over time.

- *Classification* based models, where a set of predefined categories have been constructed to characterize the current state of a bibliometric quality and measure the changes to occur after a time period. Then, by assigning a new entity into one of the existing categories, its future state is approximated by that of the entire category. Even though this approach manages the diversity of scientific patterns and distinguishes amongst them effectively, placing an entity in a particular cohort only establishes how its current behavior resembles its peers; limited predictability is offered for its future state, which may significantly deviate from the group (e.g. sleeping beauties).
- *Regression* based approaches have been introduced with the seminal work by Acuna [1] and others [20,36]. This methodology has been criticized since

Table 1. Classification of approaches for estimating future impact

Categorization attribute	Examples of each category	Related work
Scientific entity	Publication	[3,5–8,10–12,16,21,25,37,40]
	Author	[1,17,20,23,28,29,31,34,36,37,41]
	Venue	[6]
	Institution	[36]
Target variable	*h*-index	[1,31]
	Citation count	[7,12,16,20,23,28,40]
	Increase in *h*-index/citations	[11]
	Shift in impact group	[6,34]
	Relative ranking	[3,5,8,17,21,31,34]
	Rank position in a network	[21,25,37,41]
	Award or promotion	[29,33,36]
Model	Classification	[5,6,8,11,23]
	Regression	[1,3,12,16,17,20,21,36]
	Statistical modeling	[29,34]
	Time series	[28]
	Citation networks	[25,37,41]
	Combination of the above	[7,10,40]

its predictability depends on the aggregation of career data across multiple age cohorts, leading to unfair models towards young researchers or "late bloomers" [24]. Therefore, recent endeavors combine entities to subsequently calculate regression coefficients individually for each group [7].

- *Statistical modeling*, inspired by social networks evolution and Web modeling, attempts to fit bibliometric quantities to existing distributions, thus approximating the mechanism they will continue to evolve over time. Logarithmic and exponential distributions have been fitted to the evolution of productivity and impact over a scholar's career, whereas parameter thresholds have been utilized to predict impact shifts [29,40]. In [34] Sinatra et al. produced the random impact model, according to which the highest impact publication may occur randomly at any point of a career and future popularity can be calculated based on a multiplicative process of the impact exponent. Although interpretable, statistical modeling approaches require an abundance of past data to calculate the model parameters, thus discouraging the quantification of future evolution for young researchers or publications. In general they oversimplify when characterizing the complex process of citation dynamics using a distribution model alone even with a plethora of parameters. Thus, the challenge becomes prominent at the author level, where the interactions amongst different models produce the final output.
- *Time series prediction* constitutes another alternative, since citation acquisition is a temporal process. By viewing scientific entities as spatio-temporal

objects [23,28] one can approximate its future trajectory based on an abundance of data from various past time slots.
- *Network approaches*, where one can consider a citation network, where each link represents a vote of confidence between researchers or publications, and therefore determining the future state of such a network constitutes a *link prediction* problem [30,41].
- *Combining* two or more of the previous methodologies has proven to yield increased performance, like [10], where a time series classification of publications occurs or [5] where classification of publications is combined with threshold based distribution modeling.

To estimate future evolution, of pivotal importance are the selected features that shape scientific impact. Redundant or irrelevant factors can cause overfitting or add unnecessary complexity to the produced model, while on the other hand failing to account for crucial factors the effectiveness, accuracy and usability of the prediction are threatened. These factors are grouped into six categories: *author centric*, *publication centric*, *content related*, *venue-centric*, *socially derived* and *temporal* ones. The outreach of any scientific contribution is determined in part by who is working to make it, his/her scientific track record, how s/he is trained as scientists and how long s/he is engaged with research. Other features that influence his/her output include the gender, the country of origin and the faculty position. Furthermore, with an increasing productivity, one increases the chances for scientific recognition. The same effect is achieved with interdisciplinary research that merges different domains together [38].

A number of features can determine the future of a publication, ranging from its topic allocation to the number of used keywords or the time of appearance. A high quality work may end up under-appreciated if it gets published in a year that ground-breaking achievements are happening in the same field or analogously if a field has started losing its overall popularity. Based on the "standing on the shoulders of giants" motto, it is expected that high quality works cite other high quality works, thus making the number of cited references a relevant predictor. Also, a high number of co-authors often results in a wider dissemination increasing its probability to be cited. A rising interest has focused recently in mining the actual content of a publication: e.g. the terms used, the position of references, the ordering of author names as well as the originality and diversity of the subject in an effort to provide more detailed and specific predictions.

In today's prestige-based interconnected world the social characteristics surrounding a publication and its authors are also highly defining factors of future impact, with the authority and networking power of the author being the most popular social features utilized in predictive modeling. It is understood that a well-connected scholar, with a large collaboration network, who also refers to other seminal works or is part of a highly respected institution will be able to better publicize his/her work. The same holds for the venue where a publication is released, with top rated venues attracting usually high quality publications and also providing a broader audience for the released work. However, a large variety of publishing patterns occur in research, as mentioned previously, raising

Table 2. Categories of features to characterize scientific impact and its evolution.

Feature origin	Features
Author-centric	Popularity and productivity
	Academic age and gender
	Affiliation and academic position
	Rank based on bibliometric indices
	Disciplines/domains
Publication-centric	Popularity and age
	Number of references and keywords
	Topic allocation and number of authors
	Relative ranking in portfolio or field
Content-related	Popularity of topic and novelty
	Positioning of references and author ordering
	Diversity of topic and MeSH terms
	Abstract content
Venue-centric	Journal Impact Factor (JIF)
	Popularity of publishing house
	Ranking in international lists (JCR, Scimago, etc.)
	Number of issues/volumes
	Open access
Social	Collaboration network
	Citation network
	Co-citation/bibliographic coupling
Temporal	Differences from past state
	Normalization over time frame
	Rate of activity within a time frame

the need for temporal evaluation of scientific output. To identify rising stars and upcoming trends, there needs to be an accurate prediction based on the timing of the discovery and not only its calculated magnitude. Also, considering the rate at which the status of a scholar or publication rises can provide more insight into the future output, than his/her static current state. The categorization of these features and examples of each category of factors are presented in Table 2.

3 Challenges

With the rising abundance and complexity in bibliographic data the science of science is focusing on measurable quantities regarding scientific output: for instance citations to past work, timing of scientific discoveries and events in career trajectories such as promotions, awards, reaching top percentile of impact amongst a group and many others. Using computational tools one can identify quantitative patterns in these events that present a straightforward metric to predict, but also raise controversy regarding fair and meaningful predictions. Many of these quantities, like the number of citations or the h-index, are heavily subject to *preferential attachment*, meaning that the majority of the scientific community achieves low scores in these metrics, with a selected few attracting significant attention. Due to the Matthew effect in citation counting [22] a number of scientists have altered the definition of the prediction problem at hand, aiming for example to predict whether a publication will contribute to a scholar's rise in h-index values [11] or his/her relative ranking amongst a group of peers [21, 37], instead of addressing the future citation count prediction per se. Scientists also argue that the inert property of citations and citation based metrics to be always accumulating creates false self-fulfilling predictive models [31]. Consequently, adjusted metrics have been utilized, like the number of citations each publication receives every year which tends to be a decreasing quality [30].

Another challenge is the prediction of the timing in which a shift in the citation pattern will occur (e.g. a scientific discovery, a seminal publication, etc.) as opposed to predicting the magnitude of one's impact. Studies have concluded that the timing of scientific output is rather random compared to the more predictable impact of this output [34]. Given that extraordinary temporal patterns appear, like "sleeping beauties" [27] indicating publications that receive recognition after a long period of time or "premature discoveries" [18], the provocative issue of the ageing of scientific output rises: Does an abrupt boost in citations mean recognition and how long before scientific work becomes obsolete? In [15, 27] threshold based and parameter free methods are proposed respectively to differentiate different ageing patterns for publications, while in [8] six categories of publication trajectories are identified with individual predictive models trained for each one of them and achieving different levels of prediction performance [7].

Studies [14, 17, 34] also suggest that the early years in a scientist's career provide the appropriate circumstances for one's seminal publication. However, this pattern could be highly related to the tendency of younger researchers to be more productive compared to more mature ones. While most of the proposed models address future impact of existing work, the problem of predicting future impact of future works that have not yet been published, is a real controversy [20]. Similarly to the link prediction problems in complex networks [19], predicting a link to an existing node is a challenging but addressable issue, while predicting the introduction of a new node and its connections is progressively harder [16, 25]. In the same direction, predicting highly cited scholars or publications is often a different problem from identifying the truly innovative ones that will break new ground in a field or will shape a new research domain. It has been pointed out that publications that conform to the mainstream within a field get cited more often than novel original works [16].

In general, many different publishing patterns are present across various disciplines, countries and academic levels, thus a predictive model that would be fair towards all groups of scientists is hard to create [24]. A set of works has undertaken the creation of different models for individual groups, e.g. publications of a specific journal [5], scholars of a given domain, academic age or position [17,36] to limit the variety of social processes that can lead to increased scientific output. These publishing patterns do not remain steady over time and a model created on a specific dataset in a limited time frame may contain significant bias.

The identification of such patterns becomes increasingly difficult given the interaction of the complex networks formulated in the scientific life, collaboration amongst scholars, affiliation of scholars, topics of a publication, citation links between publications and authors to name a few. These networks are interconnected and their evolution is co-dependent, affecting the evolution of science in non-obvious ways [26]. An additional challenge arises when considering long term vs. short term impact, with [3,23] explaining that different factors influence early predictions, whereas long term success is more complicated. However, it has been argued that short term impact is more important since it may influence the whole career trajectory of a scholar or the fate of a new publication [3].

4 Discussion and Future Research Directions

Despite the wide range of proposed approaches to quantify the future of science, a unified framework for all levels and patterns of scholarly impact is still missing. Instead of focusing on a single metric, the various aspects of scientific output need to be taken into account to produce an overall fair framework accounting for young scientists, for truly novel out of the ordinary research ideas and under-represented fields. Additionally, such a predictive framework needs to mitigate the drawbacks of the aforementioned approaches by combining them effectively and creating a resulting model that is robust to manipulation, meaning it should not encourage greed and strategic networking over the advancement of science.

There is a rising belief in the scientometric community that more timely and accurate predictions can occur from incorporating context-specific data, such as the length and content of papers, the terms used and their relationship to the topic. Also, integration with online presence and social media dissemination (posts, number of downloads, number of views, etc.) has given birth to the rising field of Altmetrics, which measures scientific outreach in today's digital era more effectively compared to accumulated citations. Most importantly, the data bias introduced by each online database, with its different properties and coverage range, significantly hinders the comparisons amongst introduced approaches. The need for a detailed diverse ground truth dataset that is widely accepted by the scientific community is imminent.

Finally, the performance of predictive frameworks relies heavily on what constitutes proven high impact, an award, a high h-index or a tenure position? A set of selected criteria accepted by the computing community and peer review that efficiently evaluate recognition needs to be utilized as a target variable for proposed frameworks. Given that a great deal of controversy has surrounded both bibliographic data and impact metrics during the past decades, the challenge to create a common basis for evaluation of predictive efforts in the science of science remains open.

References

1. Acuna, D.E., Allesina, S., Kording, K.P.: Future impact: predicting scientific success. Nature **489**(7415), 201–202 (2012)
2. Börner, K., Dall'Asta, L., Ke, W., Vespignani, A.: Studying the emerging global brain: analyzing and visualizing the impact of co-authorship teams. Complexity **10**(4), 57–67 (2005)
3. Bornmann, L., Leydesdorff, L., Wang, J.: How to improve the prediction based on citation impact percentiles for years shortly after the publication date? J. Inf. **8**(1), 175–180 (2014)
4. Bornmann, L., Mutz, R., Hug, S.E., Daniel, H.P.: A multilevel meta-analysis of studies reporting correlations between the h index and 37 different h index variants. J. Inf. **5**(3), 346–359 (2011)
5. Brizan, D.G., Gallagher, K., Jahangir, A., Brown, T.: Predicting citation patterns: defining and determining influence. Scientometrics **108**(1), 183–200 (2016)
6. Cao, X., Chen, Y., Liu, K.R.: A data analytic approach to quantifying scientific impact. J. Inf. **10**(2), 471–484 (2016)
7. Chakraborty, T., Kumar, S., Goyal, P., Ganguly, N., Mukherjee, A.: Towards a stratified learning approach to predict future citation counts. In: Proceedings 14th ACM/IEEE-CS Joint Conference on Digital Libraries (JCDL), pp. 351–360 (2014)
8. Chakraborty, T., Kumar, S., Goyal, P., Ganguly, N., Mukherjee, A.: On the categorization of scientific citation profiles in computer science. Commun. ACM **58**(9), 82–90 (2015)
9. Chaudhuri, S., Dayal, U., Narasayya, V.: An overview of business intelligence technology. Commun. ACM **54**(8), 88–98 (2011)
10. Davletov, F., Aydin, A.S., Cakmak, A.: High impact academic paper prediction using temporal and topological features. In: Proceedings 23rd ACM International Conference on Conference on Information and Knowledge Management (CIKM), pp. 491–498 (2014)
11. Dong, Y., Johnson, R.A., Chawla, N.V.: Can scientific impact be predicted? IEEE Trans. Big Data **2**(1), 18–30 (2016)
12. Garner, J., Porter, A.L., Newman, N.C.: Distance and velocity measures: using citations to determine breadth and speed of research impact. Scientometrics **100**(3), 687–703 (2014)
13. Hirsch, J.E.: An index to quantify an individual's scientific research output. Proc. Natl. Acad. Sci. **102**(46), 16569–16572 (2005)
14. Jones, B.F., Weinberg, B.A.: Age dynamics in scientific creativity. Proc. Natl. Acad. Sci. **108**(47), 18910–18914 (2011)
15. Ke, Q., Ferrara, E., Radicchi, F., Flammini, A.: Defining and identifying sleeping beauties in science. Proc. Natl. Acad. Sci. **112**(24), 7426–7431 (2015)
16. Klimek, P.S., Jovanovic, A., Egloff, R., Schneider, R.: Successful fish go with the flow: citation impact prediction based on centrality measures for term-document networks. Scientometrics **107**(3), 1265–1282 (2016)
17. Laurance, W.F., Useche, D.C., Laurance, S.G., Bradshaw, C.J.: Predicting publication success for biologists. Bioscience **63**(10), 817 (2013)

18. Li, J., Shi, D., Zhao, S.X., Ye, F.Y.: A study of the "heartbeat spectra" for "sleeping beauties". J. Inf. **8**(3), 493–502 (2014)
19. Lü, L., Zhou, T.: Link prediction in complex networks: a survey. Phys. A Stat. Mech. Appl. **390**(6), 1150–1170 (2011)
20. Mazloumian, A.: Predicting scholars' scientific impact. PLoS ONE **7**(11), 1–5 (2012)
21. McNamara, D., Wong, P., Christen, P., Ng, K.S.: Predicting high impact academic papers using citation network features. In: Li, J., Cao, L., Wang, C., Tan, K.C., Liu, B., Pei, J., Tseng, V.S. (eds.) PAKDD 2013. LNCS, vol. 7867, pp. 14–25. Springer, Heidelberg (2013). doi:10.1007/978-3-642-40319-4_2
22. Merton, R.K.: The Matthew effect in science. Science **159**(3810), 56–63 (1968)
23. Nezhadbiglari, M., Gonçalves, M.A., Almeida, J.M.: Early prediction of scholar popularity. In: Proceedings 16th ACM/IEEE-CS on Joint Conference on Digital Libraries (JCDL), pp. 181–190 (2016)
24. Penner, O., Pan, R.K., Petersen, A.M., Fortunato, S.: The case for caution in predicting scientists' future impact. Phys. Today **66**(4), 8 (2013)
25. Pobiedina, N., Ichise, R.: Citation count prediction as a link prediction problem. Appl. Intell. **44**(2), 252–268 (2016)
26. Pradhan, D., Paul, P.S., Maheswari, U., Nandi, S., Chakraborty, T.: C3-index: revisiting author's performance measure. In: Proceedings 8th ACM Conference on Web Science (WebSci), pp. 318–319 (2016)
27. van Raan, A.F.J.: Sleeping beauties in science. Scientometrics **59**(3), 467–472 (2004)
28. Revesz, P.Z.: A method for predicting citations to the scientific publications of individual researchers. In: Proceedings 18th International Database Engineering and Applications Symposium (IDEAS), pp. 9–18 (2014)
29. Revesz, P.Z.: Data mining citation databases: a new index measure that predicts Nobel prizewinners. In: Proceedings 19th International Database Engineering and Applications Symposium (IDEAS), pp. 1–9 (2015)
30. Sayyadi, H., Getoor, L.: Futurerank: ranking scientific articles by predicting their future pagerank. In: Proceedings SIAM International Conference on Data Mining (SDM), pp. 533–544 (2009)
31. Schreiber, M.: How relevant is the predictive power of the h-index? A case study of the time-dependent Hirsch index. J. Inf. **7**(2), 325–329 (2013)
32. Sidiropoulos, A., Gogoglou, A., Katsaros, D., Manolopoulos, Y.: Gazing at the skyline for star scientists. J. Inf. **10**(3), 789–813 (2016)
33. Sidiropoulos, A., Manolopoulos, Y.: A citation-based system to assist prize awarding. ACM SIGMOD Rec. **34**(4), 54–60 (2005)
34. Sinatra, R., Wang, D., Deville, P., Song, C., Barabási, A.: Quantifying the evolution of individual scientific impact. Science **354**(6312), aaf5239 (2016)
35. de Solla Price, D.J.: Networks of scientific papers. Science **149**(3683), 510–515 (1965)
36. Vieira, E.S., Cabral, J.A., Gomes, J.A.: How good is a model based on bibliometric indicators in predicting the final decisions made by peers? J. Inf. **8**(2), 390–405 (2014)
37. Wang, S., Xie, S., Zhang, X., Li, Z., Yu, P.S., He, Y.: Coranking the future influence of multiobjects in bibliographic network through mutual reinforcement. ACM Trans. Intell. Syst. Technol. **7**(4), 64:1–64:28 (2016)
38. Way, S.F., Morgan, A.C., Clauset, A., Larremore, D.B.: The misleading narrative of the canonical faculty productivity trajectory. CoRR abs/1612.08228 (2016)

39. Wildgaard, L., Schneider, J.W., Larsen, B.: A review of the characteristics of 108 author-level bibliometric indicators. Scientometrics **101**(1), 125–158 (2014)
40. Xiao, S., Yan, J., Li, C., Jin, B., Wang, X., Yang, X., Chu, S.M., Zhu, H.: On modeling and predicting individual paper citation count over time. In: Proceedings 25th International Joint Conference on Artificial Intelligence (IJCAI), pp. 2676–2682 (2016)
41. Zhang, J., Ning, Z., Bai, X., Wang, W., Yu, S., Xia, F.: Who are the rising stars in academia? In: Proceedings 16th ACM/IEEE-CS on Joint Conference on Digital Libraries (JCDL), pp. 211–212 (2016)

Data Mining Methods and Applications

Enhanced Hybrid Component-Based Face Recognition

Andile M. Gumede, Serestina Viriri(✉), and Mandlenkosi V. Gwetu

School of Mathematics, Statistics and Computer Science,
University of KwaZulu-Natal, Pietermaritzburg, South Africa
{211513796,viriris,gwetum}@ukzn.ac.za

Abstract. This paper presents a hybrid component-based face recognition. Can face recognition be enhanced by recognizing individual facial components: forehead, eyes, nose, cheeks, mouth and chin? The proposed technique implements texture descriptors Grey-Level Co-occurrence (GLCM) and Gabor Filters, shape descriptor Zernike Moments. These descriptors are effective facial components feature representations and are robust to illumination changes. Two classification techniques have been used and compared: Support Vector Machines (SVM) and Error-Correcting Output Code (ECOC). The experimental results obtained on three different facial databases, the FERET, FEI and CMU, show that component-based facial recognition is more effective than whole-face recognition.

Keywords: Face recognition · Facial components · Shape descriptors · Texture descriptors

1 Introduction

Biometric technology has greatly improved in the past few years, making facial solutions more accurate than ever. The performance of a face recognition system largely depends on a variety of factors such as illumination, facial pose, expression, age span, hair, facial wear, and motion [12]. Considering these factors, can the face recognition be performed at component level?

The main idea behind using components is to compensate for pose changes by allowing a flexible geometrical relation between the components in the classification stage, however the main challenge is the selection of components; in addition, the face is more likely to be occluded and affected by abnormal lighting conditions. There has been a significant amount of research to date in various aspects of facial recognition, holistic-based methods, feature-based methods and hybrid-based methods.

Holistic-based methods perform well on images with frontal view faces and they are characterized by using the whole face image for recognition. However, they are computationally expensive as they require third-party algorithms for dimensionality reduction such as Eigen face techniques, which represents holistic matching of faces by the applications of PCA [17].

N.T. Nguyen et al. (Eds.): ICCCI 2017, Part I, LNAI 10448, pp. 257–265, 2017.
DOI: 10.1007/978-3-319-67074-4_25

Feature-based approaches are much faster and robust against face recognition challenges. They are pure geometric methods; and they extract local features from facial landmarks [7]. Hybrid methods combines holistic and feature based methods to overcome the shortcomings of the two methods and give more robust performance.

Recently, hybrid-based methods have shown promising results in various object detection and recognition tasks such as face detection, face recognition etc. They compensate for pose changes and allow flexible geometrical relation among the face components in the classification stage [6,8,14].

Dargham et al. [9] proposed a hybrid component based face recognition system that recognises faces using three main facial components i.e. eyes, nose and mouth. The system is dysfunctional on faces rotated 45° from frontal view. The face contains rich information, focusing only on these three components might not be ideal in low lighting conditions.

In [14], a component based face recognition system is presented. It employs two-level Support Vector Machines (SVM) to detect and validate facial components. Learned face images are automatically extracted from 3-D head models that provide the expected positions of the components. These expected positions were employed to match the detected components to the geometrical configuration of the face and provide the expected positions of the components. These expected positions were employed to match the detected components to the geometrical configuration of the face.

Component-based face recognition studies are found at lower frequency in the literature. Even methods which compute similarity measures at specific facial landmarks, such as Elastic-Bunch Graph Matching (EBGM) [5] do not operate in a per-component manner. This work focuses on face recognition at component level. A robust hybrid component based strategy for facial recognition is proposed. The strategy seeks to utilize any successfully detected components to recognize and verify the identity of a person.

The rest of the paper is organized as follows: Sect. 2 describes the overall methodology of the study, Sect. 3 describe the overall face recognition system. Section 4 outlines the experiments, results and discusses the outcome of the study, and Sect. 5 concludes the paper and discusses possible extensions and the future work.

2 Methods and Techniques

2.1 Preprocessing

Image pixel values are first converted to grey scale and the contrast is enhanced using histogram equalization with adaptive parameters. This is defined as follows: if we let f be a given image represented as a m_r by m_c matrix of integer pixel intensities ranging from 0 to $L - 1$, where L is the number of possible intensity values, often 256. Let p denote the normalized histogram of f with a bin for each possible intensity, then

$$p_n = \frac{n}{n_t} \tag{1}$$

where n is number of pixels with intensity n and n_t is the total number of pixels. The image g with adjusted intensities is defined by

$$g_{i,j} = floor((L-1)\sum_{n=0}^{f_{i,j}} p_n) \tag{2}$$

where **floor()** rounds down to the nearest integer. Histogram equalization has simplified the detection and recognition of image in low lighting conditions.

2.2 Facial Components Detection

In component based facial recognition, the most challenging task is to locate the components from the face. The Viola and Jones [18] algorithm is one of the powerful algorithms to perform this, although it does not cover all the components. A cascade detector developed using this algorithm is capable of detecting, the eyes, nose and the mouth. However, we have further trained the cascade detector to detect the cheeks, chin and forehead, per-component. These three additional components have been presumably considered to be distinguishing components for our facial recognition model. Figure 1 depicts eight detected components from the face.

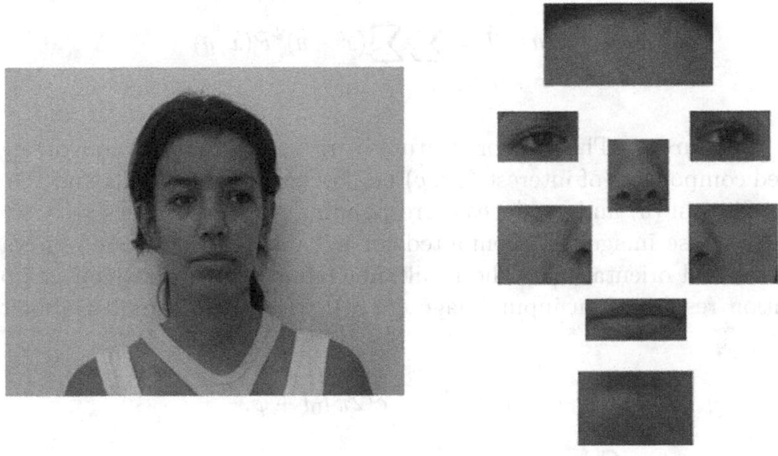

Fig. 1. Eight facial components detected individually.

3 Face Recognition

3.1 Representation of Features

Grey-Level Co-Occurrence Matrix (GLCM). was proposed by Haralick in the 1970s [4]. It is regarded as a statistical method that considers the spacial relationship of pixels. They characterize the texture of an image by calculating how often pairs of pixels with specified spacial relationship occur in an image [10,16]. We have used GLCM to extract statistical features from each face components. A single GLCM might not be enough to describe the textural features of the input image, hence, for each component, we have created multiple GLCMs for a single input image. To achieve this, we had an array of offsets. These offsets define pixel relationships of varying direction and distance.

There are directions in a matrix representing an image, (horizontal, vertical, and two diagonals) corresponding to 0°, 90°, 45°, 135° and four distances. The offsets are specified as a p-by-2 array of integers. Each row in the array is a two-element vector, [row-offset , col-offset], that specifies one offset. Row-offset is the number of rows between the pixel of interest and its neighbour, and Col-offset is the number of columns between the pixel of interest and its neighbour. [0 1], [−1 1], [−1 0], [−1 −1] defined as one neighbouring pixel in the possible four directions. We have used the following three properties of the GLCM, Energy(3), Entropy(4), and Contrast(5).

$$Energy, E = \sum_x \sum_y P(x,y)^2, \tag{3}$$

$$Entropy, S = \sum_x \sum_y P(x,y) \log P(x,y) \tag{4}$$

$$Contrast, I = \sum_x \sum_y (x-y)^2 P(x,y) \tag{5}$$

Gabor Features. The Gabor features are computed by convolving the extracted components of interest $\xi(x, y)$ i.e. forehead, eyes, cheeks and the chin, with the filter in (6) and produces corresponding response images r_ξ.

The response images are computed for a bank of filters tuned on various frequencies and orientations. The resultant Gabor feature thus consists of the convolution results of an input image $\xi(x, y)$. with all of the 40 Gabor filters (Fig. 2):

$$\psi = e^{-\alpha^2(t-t_0)^2} e^j 2\pi f_0 t + \varphi. \tag{6}$$

A feature matrix G is defined by

$$G_{m,n}(x,y) = \frac{1}{M \times N} \sum_{m=0}^{M-1} \sum_{n=0}^{N-1} I(x-m, y-n)\xi(m,n). \tag{7}$$

where ξ is the filter mask of $m \times n$ and G_{mn} the matrix of Gabor coefficients of the same size as the image I(x, y).

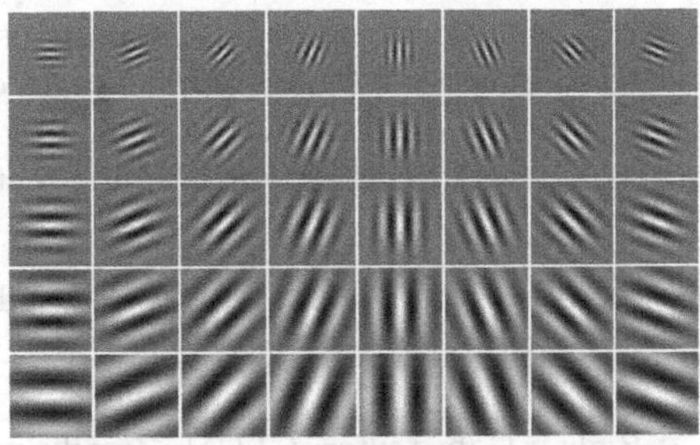

Fig. 2. Response image r_ξ, after convolving the filter in the component.

Zernike Moments. In this work, Zernike moments are used to compute the shape features of both the nose and the mouth. The face components are first scaled and normalized to maintain the same dimensions. The magnitudes of Zernike moments are rotation and reflection invariant [3] and can be easily constructed to an arbitrary order. Although higher order moments carry more fine details of an image, they are also more susceptible to noise. Therefore we have experimented with different orders of Zernike moments to determine the optimal order for our problem. The partial feature vectors of these two facial components are concatenated and normalized using the standard deviation and mean. Zernike moments of order n with l repetitions are given by:

$$A_{nl} = \frac{n+1}{\pi} \sum_{i=0}^{M-1} \sum_{j=0}^{N-1} I(i,j).R_{nl}.e^{-il\theta_{ij}} \tag{8}$$

where polar coordinates r_{ij} and θ are defined as follows:

$$r_{ij} = \sqrt{x_i^2 + x_j^2} \text{ and } \theta_{ij} = \arctan({}^{y_i}/_{x_j}) \tag{9}$$

The Cartesian coordinates x_j and y_i are given by:

$$x_j = c + \frac{j.(d-c)}{N-1} \quad 0 \leq j \leq N-1 \text{ and } y_i = d + \frac{j.(d-c)}{N-1} \quad 0 \leq i \leq M-1 \tag{10}$$

where c and d are real numbers chosen according to whether the image function is mapped outside or inside a unit circle, that is:

$$Outside\ a\ circle : c = -1 \text{ and } Inside\ a\ circle : c = {}^{-1}/_{\sqrt{2}}$$

In this paper, the mouth and the nose are mapped inside a circle. Real values of radial polynomials $R_{nl}(r)$, in a unit circle are given by:

$$R_{nl}(r) = \sum_{s=0}^{\frac{n-|l|}{2}} (-1)^s \frac{(n-s)!}{s!(\frac{n+|l|}{2}-s)! \times (\frac{n-|l|}{2}-s)!} \tag{11}$$

where $|l| \leq n$ and $n - |l|$ is always even.

Zernike moments with the highest order are mainly used to capture the detailed shape information in images. We have applied it in our case to capture the shapes properties of the facial components. Table 1, provides a summary on the set of components extracted and their corresponding features.

Table 1. The extracted face components with corresponding features.

No.	Component	Texture	Shape	Order
1	Cheeks (Left & Right)	Gabor filter, GLCM	Zernike moments	4
2	Forehead	Gabor filter, GLCM	Zernike Moments	4
3	Chin	Gabor filter, GLCM	Zernike Moments	4
4	Nose	–	Zernike moments	10
5	Mouth	–	Zernike moments	10
6	Eyes (Left & Right)	Gabor filter, GLCM	–	–

3.2 Feature Vector and Normalization

One of the most critical issues in using a vector of geometrical features is that of proper normalization. The two partial feature vectors are fused together to form a global feature vector f in (13). The main purpose of feature normalization is to modify the location and scale parameters of individual feature values to transform the value into a common domain. The feature can be normalized via various normalization schemes like min-max, z-score, tanhand median absolute [2]. We used the Min-max normalization scheme due to its robustness to outliers. The implementation of min-max normalization technique results in modified feature vectors. Let x and $\bar{x}$ denote a feature value before and after normalization. The min-max technique computes f as

$$\bar{x} = \frac{x - \min(F_x)}{\max(F_x) - \min(F_x)} \tag{12}$$

where F_x represent the function that generates x, and $\min(F_x)$ and $\max(F_x)$ represents the minimum and maximum values respectively for all possible x.

$$f = \{\bar{x}_1, \bar{x}_2, \ldots, \bar{x}_n\} \tag{13}$$

4 Experimental Results and Discussions

Three different facial databases have been used to evaluate our approach, i.e. FERET, FEI and CMU. The FERET and CMU databases are chosen for their significant facial expression variations and there are common in testing facial recognition algorithms, however, there are complex databases to work with, regarding variations in background and light, particularly, the FERET facial database consist of clustered faces with various poses.

Table 2. Recognition rates on permutation of facial components.

Components	Average accuracy (%)		
	FERET	CMU	FEI
Eyes + mouth + nose	88.4	83.4	94.8
Eyes + mouth + forehead	80	83	90
Eyes + mouth + chin	77.4	80	86
Eyes + mouth + cheeks	83	83	97
Eyes + mouth + nose + forehead	83	83.4	90
Eyes + mouth + nose + chin	80	90.3	97
Eyes + mouth + nose + chin + cheeks	95	98	90

Table 3. Comparison of facial recognition techniques

Method	Databases		
	FERET	CMU	FEI
PCA [1]	93.8	92.23	71.66
BEMD [11]	90.22	-	-
PCA-SVM [13]	93.08	-	-
SQI [15]	77.94	-	-
MSR [15]	62.07	-	-
View-based and modular eigenspaces [19]	83	-	-
Our Approach	**95**	**98**	**90**

Table 2 provides the permutation of facial components. Each of the components are paired together to evaluate their effectiveness for the recognition of the face. It can be seen that if all components are combined together, we achieved significantly positive results. Some components like the chin do not carry much information about the face, although they constitute of shape and texture information. Some images from the FERET and CMU had low light, as a result when preprocessed using the method described in Subsect. 2.1. Table 3 compares the performance our approach with the state of the art techniques PCA (Principal Component Analysis), BEMD (Bi dimensional Empirical Mode Decomposition), PCA-SVM (Principal Component Analysis and Support Vector Machines), SQI and MSR [1,11,13,15] by using the three face data-bases.

The results demonstrate that the facial components are effective for facial recognition. We have developed hybrid component based face-recognition system that uses component features. The method that is independent of pose, occlusion as well as invariant to rotation, scaling, and translation. The method is tolerant to shape distortion and adaptive. Using Zernike moments, Gabor features and GLCM as component features, two classification methods, namely SVM and ECOC have been compared. The Support Vector Machines (SVM) and

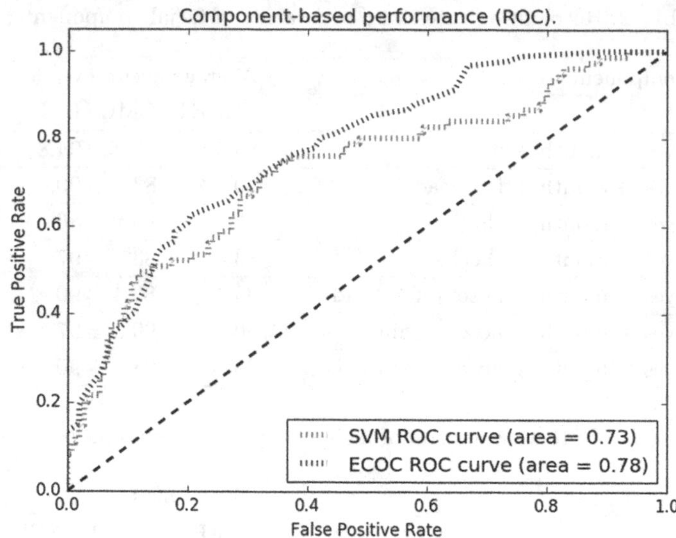

Fig. 3. The performance of the SVM and ECOC classifiers.

error-correcting output code (ECOC) classifiers based on the eight facial components, i.e., forehead, eyes, nose, mouth, cheeks and chin. Based on these face components, we have measured the performance of our algorithm. The behaviour of the two classifiers is shown in Fig. 3. The figure shows the ROC curves for both classifiers on the component based face recognizer. The performance varies in some sort, due to some images that suffered deeply on illumination and pose.

5 Conclusion

A component-based technique for facial recognition has been presented. This model implements texture descriptors Grey-Level Co-occurrence (GLCM) and Gabor Filters, shape descriptor Zernike Moments on the successfully detected facial components to recognize an individual. The model is a capable to recognize faces on various orientations under controlled illumination environment and it has achieves an overall accuracy recognition rate of 94.3%. As for future work, statistical fusion schemes for facial components feature extraction is envisaged.

References

1. Abdi, H., Williams, L.J.: Principal component analysis. Wiley Interdiscip. Rev. Comput. Stat. **2**(4), 433–459 (2010)
2. Bhardwaj, S.K.: An algorithm for feature level fusion in multi-modal biometric system. Int. J. Adv. Res. Comput. Eng. Technol. (IJARCET) **3**, 3499–3503 (2014)

3. Bailey, R.R., Srinath, M.: Orthogonal moment features for use with parametric and non-parametric classifiers. IEEE Trans. Pattern Anal. Mach. Intell. **18**(4), 389–399 (1996)
4. Beliakov, G., James, S., Troiano, L.: Texture recognition by using GLCM and various aggregation functions. In: IEEE World Congress on Computational Intelligence, IEEE International Conference on Fuzzy Systems, FUZZ-IEEE 2008, pp. 1472–1476. IEEE, June 2008
5. Bolme, D.S.: Elastic bunch graph matching (Doctoral dissertation, Colorado State University) (2003)
6. Bonnen, K., Klare, B.F., Jain, A.K.: Component-based representation in automated face recognition. IEEE Trans. Inf. Forensics Secur. **8**(1), 239–253 (2013)
7. Celik, N., Manivannan, N., Balachandran, W., Kosunalp, S.: Multimodal biometrics for robust fusion systems using logic gates. J. Biom. Biostat. **6**(1), 1 (2015)
8. Chan, C.H., Tahir, M.A., Kittler, J., Pietikäinen, M.: Multiscale local phase quantization for robust component-based face recognition using kernel fusion of multiple descriptors. IEEE Trans. Pattern Anal. Mach. Intell. **35**(5), 1164–1177 (2013)
9. Dargham, J.A., Chekima, A., Hamdan, M.: Hybrid component-based face recognition system. In: Omatu, S., De Paz Santana, J., González, S., Molina, J., Bernardos, A., RodrÍguez, J. (eds.) Distributed Computing and Artificial Intelligence. Advances in Intelligent and Soft Computing, vol. 151, pp. 573–580. Springer, Heidelberg (2012). doi:10.1007/978-3-642-28765-7_69
10. Eleyan, A., Demirel, H.: Co-occurrence based statistical approach for face recognition. In: 24th International Symposium on Computer and Information Sciences, ISCIS 2009, pp. 611–615, September 2009
11. Gumus, E., Kilic, N., Sertbas, A., Ucan, O.N.: Evaluation of face recognition techniques using PCA, wavelets and SVM. Expert Syst. Appl. **37**(9), 6404–6408 (2010)
12. Jain, A.K., Li, S.Z.: Handbook of Face Recognition. Springer, New York (2011)
13. Shakthipriya, V.S., Naidu, V.P.S.: Multi sensor image fusion using empirical mode decomposition. Int. J. Adv. Res. Electr. Electron. Instrum. Eng. 252–260 (2013)
14. Taheri, S., Patel, V.M., Chellappa, R.: Component-based recognition of faces and facial expressions. IEEE Trans. Affect. Comput. **4**(4), 360–371 (2013)
15. Tan, X., Triggs, B.: Enhanced local texture feature sets for face recognition under difficult lighting conditions. IEEE Trans. Image Process. **19**(6), 1635–1650 (2010)
16. The GLCM texture tutorial. http://www.fp.ucalgary.ca. Accessed 24 Feb 2017
17. Turk, M.A., Pentland, A.P.: Face recognition using eigenfaces. In: IEEE Computer Society Conference on Computer Vision and Pattern Recognition, Proceedings CVPR 1991, pp. 586–591, June 1991
18. Viola, P., Jones, M.: Rapid object detection using a boosted cascade of simple features. In: Proceedings of the 2001 IEEE Computer Society Conference on Computer Vision and Pattern Recognition, CVPR 2001, vol. 1, pp. I-I (2001)
19. Pentland, A., Moghaddam, B., Starner, T.: View-based and modular eigenspaces for face recognition. In: IEEE Conference on Computer Vision and Pattern Recognition (1994)

Enhancing Cholera Outbreaks Prediction Performance in Hanoi, Vietnam Using Solar Terms and Resampling Data

Nguyen Hai Chau(✉)

Faculty of Information Technology,
VNUH University of Engineering and Technology, Hanoi, Vietnam
chaunh@vnu.edu.vn

Abstract. A solar term is an ancient Chinese concept to indicate a point of season change in lunisolar calendars. Solar terms are currently in use in China and nearby countries including Vietnam. In this paper we propose a new solution to increase performance of cholera outbreaks prediction in Hanoi, Vietnam. The new solution is a combination of solar terms, training data resampling and classification methods. Experimental results show that using solar terms in combination with ROSE resampling and random forests method delivers high area under the Receiver Operating Characteristic curve (AUC), balanced sensitivity and specificity. Without interaction effects the solar terms help increasing mean of AUC by 12.66%. The most important predictor in the solution is Sun's ecliptical longitude corresponding to solar terms. Among the solar terms, `frost descent` and `start of summer` are the most important.

Keywords: Cholera outbreaks prediction · Solar terms · Resampling

1 Introduction

Cholera - an acute diarrhea disease - remains a global threat, especially in developing countries. The World Health Organization (WHO) estimates that every year, about 3–5 millions people are affected by cholera worldwide [1]. Prediction of cholera outbreaks helps mitigating their consequences. At present, cholera outbreaks prediction is still a difficult problem. Researchers have found relationships of cholera outbreaks and environmental factors [2] in Bangladesh, China and Vietnam. The factors are used as predictors for cholera outbreaks prediction models.

In Bangladesh, researchers have found that the number of cholera cases strongly associated with local temperature and sea surface temperature (SST) [3]; local weather, southern oscillation index (SOI) and flooding condition [4]; and ocean chlorophyll concentration (OCC) [5].

In China, statistical evidences showed that precipitation, temperature and location altitude, relative humidity, atmospheric pressure [6]; SST, sea surface height (SSH) and OCC [7] are linked to number of cholera cases.

N.T. Nguyen et al. (Eds.): ICCCI 2017, Part I, LNAI 10448, pp. 266–276, 2017.
DOI: 10.1007/978-3-319-67074-4_26

In Nha Trang and Hue of Vietnam, precipitation [8]; SST and river height [5] are correlated with number of cholera cases. Recently, there are attempts to predict cholera outbreaks in Hanoi, Vietnam [9,10] using machine learning models with local weather and SOI data. In these models, temperature and relative humidity are the most important factors to predict cholera outbreaks.

A common difficulty for prediction of cholera outbreaks in Hanoi is an imbalanced data set [9,10]. To deal with this difficulty, Le et al. [9] only use monthly data for cholera-present years of 2004 and 2007–2010. This approach makes good prediction models, but only for cholera-present years. For cholera-free years, these models are not suitable. Another approach is to use the number of cholera cases in the past days of a district and its geographical neighbours to make daily prediction models [10]. These models performance is good during outbreaks but still moderate when predicting the first cases of outbreaks.

In this paper, we propose a new model to enhance the performance of daily cholera outbreaks prediction in Hanoi. We resample our training data sets to overcome disadvantages of imbalanced data set. Furthermore, we use additional season information to increase prediction performance of classification models.

2 Study Area and Data Sets

2.1 Study Area

Our study area is Hanoi, the capital city of Vietnam. Hanoi is located at 21°01′N, 105°51′E. In 2016, Hanoi's population is about 7.5 millions. Hanoi's weather is warm humid subtropical, classified as *Cwa* in Köppen climate system.

2.2 Cholera Data Set

We obtained a raw cholera data set from Hanoi Grant 01C-08/-8-2014-2 project [9]. The data set consists of observed cholera cases in Hanoi from Jan 01, 2001 to Dec 31, 2012. Each observation contains date, patient's name, age, gender and home address. From the raw data set we aggregated and created a derived data set where each observation has only date and number of cholera cases in Hanoi (`epi` variable). The number of cholera cases is transformed to `yes`/`no` levels.

2.3 Local Weather Data Set

The local weather data set contains daily weather data from Jan 01, 2001 to Dec 31, 2012. Each record of the data set contains average temperature in Celsius degree, average relative humidity in percentage, daily sun hours, daily average wind speed in m/s and daily precipitation in mm. Corresponding variables names are `tavg`, `havg`, `sun`, `wind` and `precip`, respectively.

2.4 Southern Oscillation Index Data Set

We obtain SOI data from a website of Queensland government, Australia [11]. The data set contains daily SOI measurement (`soi` variable) from 1991 to date.

2.5 Solar Terms Data Set

A solar term [12,13] is an ancient Chinese concept describing a point of change in seasonal cycles. Solar terms are 15° separated along the apparent path of the Sun on the celestial sphere and are used in lunisolar calendars to synchronize with the seasons.

Although originated from China, solar terms are used in Japan, Korea and Vietnam. In Vietnam, solar terms are named ***tiết khí*** . Solar terms' Gregorian date and other corresponding climatology information from 1947 to date is available at Hong Kong Observatory [14]. We obtain solar terms information of years 2001 to 2012 from the observatory and use it as additional season data. The data includes solar terms names, corresponding Gregorian date and ecliptical longitude (EC) in degree. We describe solar terms in Table 1.

Table 1. List of solar terms.

Solar term	Vietnamese	Sun's EC	Gregorian date
Start of spring	Lập xuân	315°	Feb 3–5
Rain water	Vũ thủy	330°	Feb 18–20
Awakening of insects	Kinh trập	345°	Mar 5–7
Vernal equinox	Xuân phân	0°	Mar 20–21
Clear and bright	Thanh minh	15°	Apr 4–6
Grain rain	Cốc vũ	30°	Apr 19–21
Start of summer	Lập hạ	45°	May 5–7
Grain full	Tiểu mãn	60°	May 20–22
Grain in ear	Mang chủng	75°	Jun 5–7
Summer solstice	Hạ chí	90°	Jun 21–22
Minor heat	Tiểu thử	105°	Jul 6–8
Major heat	Đại thử	120°	Jul 22–24
Start of autumn	Lập thu	135°	Aug 7–9
Limit of heat	Xử thử	150°	Aug 22–24
White dew	Bạch lộ	165°	Sep 7–9
Autumnal equinox	Thu phân	180°	Sep 22–24
Cold dew	Hàn lộ	195°	Oct 8–9
Frost descent	Sương giáng	210°	Oct 23–24
Start of winter	Lập đông	225°	Nov 7–8
Minor snow	Tiểu tuyết	240°	Nov 22–23
Major snow	Đại tuyết	255°	Dec 6–8
Winter solstice	Đông chí	270°	Dec 21–23
Minor cold	Tiểu hàn	285°	Jan 5–7
Major cold	Đại hàn	300°	Jan 20–21

3 Data Preprocessing

We merge the cholera, local weather, and SOI into one (refers as MDS) with reference to the `date` variable as the primary key. A sample of the MDS with `date` variable is in Table 2. We use MDS for cholera outbreaks prediction. The `epi` variable, indicating cholera status of a particular date (`yes` or `no`), is the outcome. Predictors are `tavg`, `havg`, `precip`, `sun`, `wind` and `soi` variables. The `date` variable is not used directly. We derive seasonal information from `date` to make new variables. The new variables are `week` (week number in a year), `month` (month) and `solarterm`, `ec` (solar terms and corresponding Sun's ecliptical longitude, available at Hong Kong observatory [13,14]). For convenience in reference we abbreviate `solarterm`, `ec` variables as `solar`.

Table 2. A sample of MDS with `date` variable in a cholera outbreak of 2004.

Date	tavg	havg	precip	sun	wind	soi	epi
2004-05-01	28.20	82	0.00	4.40	2.30	17.48	No
2004-05-02	28.60	84	0.00	3.70	2.30	−3.74	Yes
2004-05-03	28.90	82	0.00	5.20	2.50	−15.84	Yes
2004-05-04	24.00	83	67.00	0.10	1.80	3.16	Yes
2004-05-05	21.80	71	0.00	0.00	1.80	18.25	Yes
2004-05-06	22.30	77	0.00	0.80	1.30	11.13	Yes

There are 24 solar terms. Therefore the `solarterm` and `ec` variables in MDS have a lot of null values. We process null values for `solarterm` and `ec` as follows: Each null value of `solarterm` is set to the closest solar term in the past. Not null values of `ec` (or reference values) are $0, 15, 30, \ldots, 360$ (refer to Table 1). Each null value of `ec` in a date d is set to its closest reference value in the past in a date d_0 plus $d - d_0$. A sample of MDS with seasonal variables is in Table 3.

Table 3. The same sample as in Table 2 with seasonal variables `week`, `month` and `solar`. A reference value of `ec` is 45 and corresponding `solarterm` is `start of summer`.

Week	Month	ec	solarterm	tavg	havg	precip	sun	wind	soi	epi
18	5	41	Grain rain	28.20	82	0.00	4.40	2.30	17.48	No
18	5	42	Grain rain	28.60	84	0.00	3.70	2.30	−3.74	Yes
18	5	43	Grain rain	28.90	82	0.00	5.20	2.50	−15.84	Yes
18	5	44	Grain rain	24.00	83	67.00	0.10	1.80	3.16	Yes
18	5	45	Start of summer	21.80	71	0.00	0.00	1.80	18.25	Yes
19	5	46	Start of summer	22.30	77	0.00	0.80	1.30	11.13	Yes

4 Design and Analysis of Experiments

4.1 Design of Experiments and Measure Selection

We need to perform classification tasks to predict the `epi` outcome variable of a day $d + 1$ from predictors `tavg`, `havg`, `precip`, `sun`, `wind`, `soi` of day d in the MDS data set. The MDS is imbalanced with 4.2% positive (`yes` or present of cholera cases) observations. Classification methods on imbalanced data set often give very high specificity (or true negative rate) and low sensitivity (or true positive rate) [15]. Common approaches to deal with imbalanced data sets are resampling the train data set, using class-weighted classification methods and collect additional data if possible [15,16]. In this paper, we use a new approach that combines additional solar terms data and resampling methods.

To assess performance of classifiers, researchers use common measures including precision, recall, F1, area under the Receiver Operating Characteristic curve (refers as AUC) and Cohen's Kappa. Jeni et al. [17] found effects of imbalanced data sets to the measures. They defined skewness of an imbalanced data set as

$$skew = \frac{negative\ observations}{positive\ observations} \tag{1}$$

and found that F1 and area under the Precision-Recall curve (APR) drop when $skew > 1$ and becomes larger; Kappa drops when $skew \neq 1$ and becomes more different. The AUC is virtually not attenuated while $skew$ changes. Therefore in this research, we choose the AUC, a non-sensitive measure to skewness, for assessment of classification models. The skew of MDS data set is 22.7.

Performance of cholera outbreaks prediction models is affected by three factors: resampling method, classification method and additional seasonal data. To compare the effects of factors to performance of models, we design a factorial experiment [18] with the above three factors. Levels of each factor are in Table 4. In the table, resampling factors are named after corresponding methods: `none, up, down, smote, rose` [16]. Classification methods are general linear model (`glm`), k-nearest neighbours (`knn`), C5.0 (`C5.0`) and random forests (`rf`). Seasonal information levels are `none, week, month, solar` meaning use MDS only, MDS and `week`, MDS and `month` and MDS and `solar`, respectively. We do not combine different types of seasonal information because they are all derived from `date` variable. The factors combination is 80. For each combination we build a prediction model and assess its performance using AUC measure.

Table 4. Factors and their levels.

Factor	Levels
Resampling	none, up, down, smote, rose
Method	glm, knn, C5.0, rf
Seasonal	none, week, month, solar

To build classification models, we first rearrange MDS with seasonal factors listed in Table 4 to obtain seasonMDS data sets. Each seasonMDS is randomly divided into training and testing data sets. The training set has 70% observations of the seasonMDS and the testing is the rest. The training data set is resampled following the above resampling methods. We then apply the above classification methods to build models on the training data set and test their performance on the test data set. This procedure is repeated 100 times for cross-validation. We collect the models performance measures to a PERF data set to compare models performance. The measures include AUC, sensitivity and specificity. We run all experiments in an R environment [19] and use **caret** package [20] for prediction modeling. In the next section, we analyze the result statistically.

4.2 Statistical Analysis of Models Performance

The purpose of our statistical analysis in this section is twofold. Firstly, we find which factors among `seasonal`, `sampling` and `method` influence AUC measure taking into account their possible interactions. Secondly, we compare mean of AUC of groups forming by combination of the three factors to find the best. The best groups must have high AUC, balanced sensitivity and specificity.

We use the analysis of variance (ANOVA) method [21] to analyze effects of the factors. To perform the ANOVA, we build a linear regression model named LM1 on the PERF data set. In the LM1 model, the three factors `seasonal`, `sampling` and `method` are independent variables. The AUC measure is a dependent variable. The LM1 model is

```
LM1 <- lm(auc ~ seasonal*sampling*method, data=PERF)
```

as writing in R syntax. The model takes interaction effects of the three factors into account. ANOVA test result of LM1 is in Table 5.

Table 5. ANOVA test for effects of factors and their interactions to mean of AUC.

	Df	Sum Sq	Mean Sq	F value	Pr(>F)
seasonal	3	16.6507	5.5502	4692.1734	0.0E + 00
sampling	4	4.8779	1.2195	1030.9513	0.0E + 00
method	3	19.8978	6.6326	5607.1960	0.0E + 00
seasonal:sampling	12	0.8860	0.0738	62.4214	1.5E − 145
seasonal:method	9	5.6684	0.6298	532.4556	0.0E + 00
sampling:method	12	11.1310	0.9276	784.1803	0.0E + 00
seasonal:sampling:method	36	1.6713	0.0464	39.2479	2.4E − 250
Residuals	7920	9.3684	0.0012		

In Table 5, all *p*-values corresponding to the factors (or variation sources) in the last column (**Pr(>F)**) are much smaller than 0.05 indicating that all the factors, their pairwise interactions (`seasonal:sampling`, `seasonal:method`, `sampling:method`) and interaction of all three (`seasonal:sampling:method`) have effects to AUC measure and the effects are all statistical significant. By comparing mean squares in the fourth column (**Mean Sq**), we see that `method` has highest value of 6.63, then `seasonal` (5.55), `sampling` (1.22), `sampling:method` (0.93). The interaction `seasonal:sampling:method` has lowest value of 0.05. It means that the `method` and `seasonal` variables have highest effects and the interaction `seasonal:sampling:method` has lowest effect to mean of AUC.

We compare mean of AUC taking into account the interactions by calculating adjusted mean of AUC of the groups and producing interaction plots. We use **phia** [22] and **ggplot2** [23] packages to perform this task. Figure 1 shows effects of pairwise interactions to mean of AUC. We refer plots in the figure by their coordinate (*row*, *column*) with reference to the top left plot. Each plot among (1,1), (2,2) and (3,3) has one curve. These curves describe main effects (no interaction) of `seasonal`, `sampling` and `method` to mean of AUC. As seen in the plots, `solar` of seasonal, `smote` of sampling and `rf` of method are the most influence to mean of AUC. Using another linear model with no interaction

```
LM2 <- lm(auc ~ seasonal+sampling+method, data=PERF)
```

we quantify main effects of the factors, described Table 6. In the table all p-values in the last column is much smaller than 0.05. It is statistically significant that `solar`, `week` and `month` of `seasonal` factors increase mean of AUC by 12.66%, 8.25% and 7.88%, respectively. These values are in the second column of Table 6. Main effects of other factors are interpreted in the same manner.

Table 6. A linear model describing main effects of the factors to mean of AUC.

	Estimate	Std. error	t value	Pr(>|t|)
(Intercept)	0.6310	0.0022	283.7711	0.0E + 00
seasonalmonth	0.0788	0.0019	41.5728	0.0E + 00
seasonalsolar	0.1266	0.0019	66.7829	0.0E + 00
seasonalweek	0.0825	0.0019	43.4977	0.0E + 00
samplingdown	0.0497	0.0021	23.4324	1.7E − 117
samplingrose	0.0614	0.0021	28.9797	1.1E − 175
samplingsmote	0.0715	0.0021	33.7317	2.6E − 233
samplingup	0.0518	0.0021	24.4478	2.3E − 127
methodC5.0	−0.0098	0.0019	−5.1705	2.4E − 07
methodknn	−0.0875	0.0019	−46.1373	0.0E + 00
methodrf	0.0521	0.0019	27.4557	1.2E − 158

The off-diagonal plots in Fig. 1 describe pairwise interactions of the factors. Plot (1,3) shows interactions of `method` and `seasonal`. We notice that combination of `rf` (method) and `solar` (seasonal) delivers the highest mean of AUC. If we change method to `knn`, mean of AUC will change different amounts depending on seasonal data. When seasonal data is `none` or `solar`, the mean of AUC will decrease the same amount of approximate 9%. However when it is `week` or

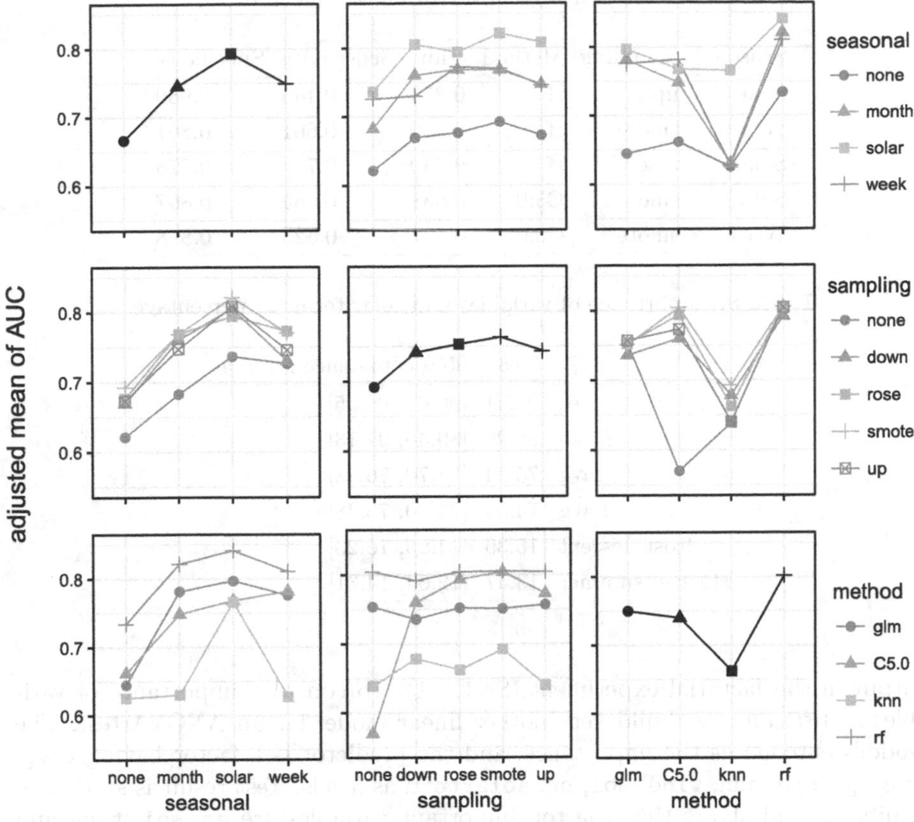

Fig. 1. Influence of pairwise interactions to mean of AUC. (Color figure online)

`month`, the mean of AUC will decrease an amount of approximate 20%. These effects are shown in the plot (1,3) by almost parallel segments of `solar` and `none` and nonparallel segments of `solar` and `month/week` with reference to `knn` and `rf` points on the x-axis. Interpretation of other interactions is similar.

To find groups those have highest means of AUC we calculate adjusted mean of AUC. The adjusted means of the best groups are in Table 7. Results in the table show that using `solar` and `rf` is recommended for high AUC. Sampling method is a choice among `up`, `smote` or `rose`. The `rose` gives balanced of sensitivity (0.746) and specificity (0.778), and therefore is chosen. The combination of (`week`, `smote`, `C5.0`) is slightly worse than (`solar`, `rose`, `rf`). However it has unbalanced sensitivity (0.525) and specificity (0.898) and is not chosen.

4.3 Importance of Variables

We have chosen (`solar`, `rose`, `rf`) as the best combination. The importance of variables in this combination is simply extracted from the random forests

Table 7. List of five groups with highest mean of AUC.

Seasonal	Sampling	Method	auc	Sensitivity	Specificity
Solar	up	rf	0.855	0.143	0.989
Solar	smote	rf	0.850	0.601	0.861
Solar	rose	rf	0.840	0.746	0.778
Solar	smote	C5.0	0.838	0.564	0.867
Week	smote	C5.0	0.837	0.525	0.898

Table 8. Importance of variables and solar terms in percentage.

	Mean	95% confidence interval
ec	97.51	[96.67, 98.35]
soi	90.29	[89.45, 91.13]
tavg	75.54	[74.70, 76.38]
havg	74.54	[73.70, 75.38]
frost descent	15.39	[14.55, 16.23]
start of summer	13.47	[12.64, 14.31]

output in the factorial experiment (Sect. 4.1). To compare importance of variables statistically, we build yet another linear model for an ANOVA test. The model's outcome is the importance; and the predictor is a factor having `tavg`, `havg`, `precip`, `sun`, `wind`, `soi`, `ec`, `solarterm` as levels. Test result is statistical significant and shows that the top important variables are `ec`, `soi`, `tavg` and `havg`. Among the solar terms, `start of summer` and `frost descent` are the most important. Table 8 lists the importance of these variables.

5 Conclusions

We have developed a new solution to enhance performance of cholera outbreaks prediction in Hanoi. Our solution is a combination of solar terms data, ROSE resampling of training data set and random forests method. The solution delivers high AUC (0.84), balanced sensitivity and specificity.

Without taking interactions of the factors into account, the solar terms data helps increasing mean of AUC by an amount of 12.66%. The most important variables in the solution are `ec`, `soi`, `tavg` and `havg`. Among the solar terms, `frost descent` and `start of summer` are the most important.

To our best knowledge, this research is the first that uses solar terms in a cholera outbreaks prediction. Our experiment results show that solar terms are worth considering as predictors for cholera modeling in Hanoi.

References

1. Jutla, A., Whitcombe, E., Hasan, N., Haley, B., Akanda, A., Huq, A., Alam, M., Sack, R., Colwell, R.: Environmental factors influencing epidemic cholera. Am. J. Trop. Med. Hyg. **89**(3), 597–607 (2013)
2. Martinez, P.P., Reiner, R.C., Cash, B.A., Rodó, X., et al.: Cholera forecast for Dhaka, Bangladesh, with the 2015–2016 El Niño: lessons learned. PLoS ONE **12**(3), e0172355 (2017)
3. Ali, M., Kim, D.R., Yunus, M., Emch, M.: Time series analysis of cholera in Matlab, Bangladesh, during 1988–2001. J. Health Popul. Nutr. **31**(1), 11–19 (2013)
4. Reiner, R.C., King, A.A., Emch, M., Yunus, M., Faruque, A.S.G., Pascual, M.: Highly localized sensitivity to climate forcing drives endemic cholera in a megacity. Proc. Natl. Acad. Sci. U.S.A. **109**, 2033–2036 (2012)
5. Emch, M., Feldacker, C., Yunus, M., et al.: Local environmental predictors of cholera in Bangladesh and Vietnam. Am. J. Trop. Med. Hyg. **78**(5), 823–832 (2008)
6. Xu, M., Cao, C.X., Wang, D.C., Kan, B., Jia, H.C., Xu, Y.F., Li, X.W.: District prediction of cholera risk in China based on environmental factors. Chin. Sci. Bull. **58**(23), 2798–2804 (2013)
7. Xu, M., Cao, C.X., Wang, D.C., Kan, B.: Identifying environmental risk factors of cholera in a coastal area with geospatial technologies. Int. J. Environ. Res. Public Health **12**, 354–370 (2015)
8. Kelly-Hope, L.A., Alonso, W.J., Thiem, V.D., et al.: Temporal trends and climatic factors associated with bacterial enteric diseases in Vietnam 1991–2001. Environ. Health Perspect. **116**(1), 7–12 (2008)
9. Le, T.N.A., Ngo, T.O., Lai, T.H.T., Le, H.Q., Nguyen, H.C., Ha, Q.T.: An experimental study on cholera modeling in Hanoi. In: Proceedings of Asian XI Conference on Intelligent Information and Database Systems 2016, pp. 230–240 (2016)
10. Chau, N.H., Ngoc Anh, L.T.: Using local weather and geographical information to predict cholera outbreaks in Hanoi, Vietnam. In: Nguyen, T.B., Do, T.V., Le Thi, H.A., Nguyen, N.T. (eds.) Advanced Computational Methods for Knowledge Engineering. AISC, vol. 453, pp. 195–212. Springer, Cham (2016). doi:10.1007/978-3-319-38884-7_15
11. Daily Southern oscillation index data set of the Queensland, Australia. https://www.longpaddock.qld.gov.au/seasonalclimateoutlook/southernoscillationindex/soidatafiles/DailySOI1887-1989Base.txt
12. Qian, C., Yan, Z., Fu, C.: Climatic changes in the twenty-four solar terms during 1960–2008. Chin. Sci. Bull. Atmos. Sci. **57**(2–3), 276–286 (2012)
13. Hong Kong Observatory's solar term introduction. http://www.weather.gov.hk/gts/time/24solarterms.htm
14. Hong Kong observatory's climatology for the 24 solar terms. http://www.weather.gov.hk/cis/statistic/ext_st_vernal_equinox_e.htm?element=0&operation=Submit
15. Kuhn, M., Johnson, K.: Applied Predictive Modeling. Springer, New York (2013)
16. He, H., Mai, Y.: Imbalanced Learning: Foundations, Algorithms and Applications. Wiley, Hoboken (2013)
17. Jeni, L.A., Cohn, J.F., Torre, F.D.L.: Facing imbalanced data recommendations for the use of performance metrics. In: ACII 2013 Proceedings of the 2013 Humaine Association Conference on Affective Computing and Intelligent Interaction (2013)
18. Montgomery, D.C.: Design and Analysis of Experiments, 8th edn. Wiley, Hoboken (2013)

19. Micheaux, P., Drouilhet, R., Liquet, B.: The R Software: Fundamentals of Programming and Statistical Analysis. Springer, New York (2013)
20. **caret** package. https://cran.r-project.org/web/packages/caret/index.html
21. Faraway, J.: Linear Models with R, 2nd edn. CRC Press, Boca Raton (2015)
22. **phia** package. https://cran.r-project.org/web/packages/phia/index.html
23. **ggplot2** package. https://cran.r-project.org/web/packages/ggplot2/index.html

Solving Dynamic Traveling Salesman Problem with Ant Colony Communities

Andrzej Siemiński(✉)

Faculty of Computer Science and Management,
Wrocław University of Science and Technology, Wrocław, Poland
Andrzej.Sieminski@pwr.edu.pl

Abstract. The paper studies Ant Colony Communities (ACC). They are used to solve the Dynamic Travelling Salesman Problem (DTSP). An ACC consists of a server and a number of client ACO colonies. The server coordinates the work of individual clients and sends them cargos with data to process and then receives and integrates partial results. Each client implements the basic version of the ACO algorithm. They communicate via sockets and therefore can run on several separate computers. In the DTSP distances between the nodes change constantly. The process is controlled by a graph generator. In order to study the performance of the ACC, we conducted a substantial number of experiments. Their results indicate that to handle highly dynamic distance matrixes we need a large number of clients.

Keywords: Dynamic Travelling Salesmen Problem · ACO parallel implementations · Scalability

1 Introduction

The aim of the paper is to learn how useful are the Ant Colony Communities (ACC) for solving the dynamic version of the Travelling Salesman Problem (DTSP). The static version of the TSP is remarkably simple: given a list of cities and the distances between each pair of cities, the task is to find the shortest possible route that visits each city exactly once. It is one of the classical problems of Artificial Intelligence. The Ant Colony Optimization (ACO) is one of the heuristics used to solve it. The paper studies the Dynamic TSP. In this version, the distances between cities change. The paper evaluates the usefulness of distributing the work to a number of cooperating Ant Colonies. Parallel implementation shortens processing so the ACC can catch up with the ever changing environment.

The paper is organized as follows. The Sect. 2 describes the related work on the DTSP and introduces the Graph Generator which is used to alter the distances. In the 3rd Section we describe the parallel implementations of the ACO concentrating on the Ant Colony Community used in experiments. During the experiments two Graph Generators were used and the performance of a considerable number of colonies was tested. The experimental setup, presentation of obtained results and their interpretation are discussed in the 4th Section. The paper concludes with an indication of future work.

N.T. Nguyen et al. (Eds.): ICCCI 2017, Part I, LNAI 10448, pp. 277–287, 2017.
DOI: 10.1007/978-3-319-67074-4_27

2 DTSP - Dynamic Traveling Salesman Problem

The simplicity of formulation of the TSP could be misleading. The running time for the brute force search lies within a polynomial factor of $O(n!)$ where n is the number of cities (nodes). Known exact algorithms for the problem could limit the time to $O(n^2 2^n)$ [1] but still, for the larger number of n suboptimal heuristic approaches are needed. In 1970's the TSP was proved to be an NP-hard problem. A detailed study of its properties can be found in [2]. A recent overview of heuristics used for solving the TSP could be found in [3, 4].

2.1 Related Work on Dynamic TSP

The dynamics of distances adds the time factor. Identifying even a near optimal path for a set of distances that had changed during the computations is certainly not much useful.

The DTSP was introduced by Psaraftis [5]. His work discussed the general properties of the dynamic TSPs, and useful performance measures but it did not propose any specific solutions. Modified distances invalidate the pheromone array. For that reason many attempts to tackle the DTSP used global and local pheromone reset strategies [6]. A global reset is activated whenever changes are detected and it resets all pheromone levels to the default value. Local reset changes only the pheromone levels for modified segments of a graph. It enables the Colony to exploit at least part of data gathered so far. Note that it is necessary to know when and where the changes have taken place. Such an information is not always available.

Other approaches apply immigrant schemes. There are three types of immigrants: randomly generated, elitism-based, and hybrid immigrants. Such approaches could use a long-term memory as in P-ACO [7]. In a more recent paper, a short-term memory is used [8]. The study showed the need for selecting a proper immigrant type for a specific changeability pattern.

The main disadvantage of the mentioned above approaches is the way the dynamics is introduced. The changes of the distance array consists just of a single node deletion/introduction or by a sequence of such operations. The changes are activated at certain intervals. Therefore the Colony could keep large part of its previous solution intact as it is not effected by limited in scope change of distances. In real world such a type of changeability is relatively rare. It could simulate properly e.g. the air traffic where airports that drop out due e.g. to weather conditions or the infrastructure of Internet where not all nodes that are always available. It is not useful e.g. for the simulation of car traffic where the cars change constantly their velocity.

Constantly changing dynamic graphs were presented in [9]. A distance array was replaced by graph generator. It used a Markov source to control distance modifications. However, the process could not guarantee that the average distance length would not change. This could certainly happen in real life but it hampered the evaluation of the tested algorithms performance.

2.2 Graph Generator

While using graph generators we cannot increase the number of the ACO iterations to improve the solution as it is possible for the static TSP. During the optimization process the ongoing changes could make the solution obsolete as it were prepared for distances that have already changed. This makes the evaluation of results even harder.

To mitigate the evaluation problem we use graph generators that preserve the average distance length. Their operation is controlled by the following three parameters:

- Total Change (*TC*): the sum of all distance changes, a real number > 0
- Change Range (*CR*): the scope limit of change, a real number in the range [0…1]
- Change Frequency (*CF*): a number of milliseconds that elapse between distance modifications.

Distance change in a sequence of modifications of a randomly selected distance. To increase or decrease a distance we use the Formulas 1 and 2 respectively:

$$Dist_{new} = Dist_{old} + \left(1 - Dist_{old}\right) * rand() * CR \tag{1}$$

$$Dist_{new} = Dist_{old} - Dist_{old} * rand() * CR \tag{2}$$

The function *rand()* generates a pseudorandom value in the range [0..1]. As you can see newly introduced distances are always in the same range. The process of updating the distances continues as long the total of all distant modifications does not surpass the *TC*.

3 ACC for TSP

The Ant Colony Optimization was inspired by the foraging behavior of real world ants and using the ACO for the TSP was a natural choice.

3.1 Basic Version of ACS for TSP

The operation of the ACO and other related metaheuristics [9] is simple: a two dimensional-array of floating point numbers represents the graph with the distances separating the cities or nodes. Another such an array stores pheromone levels. The ACO works in an iterative manner. Each iteration starts with randomly scattering the ants over the nodes. The operation of the ACO is nondeterministic but preferred are routes that have a high pheromone level and a short length. The pheromone levels are updated as the ants transverse the array and at the end of each iteration.

The selection of ACO parameters is not an easy task. An analysis of their impact on the ACO operation can be found e.g. in [10] or [11]. For the current study, it is sufficient to know that the greatest improvements are likely to occur in the early iterations. The node selection requires many floating point calculations and is therefore time-consuming. In the case of dynamic TSP the time available to deliver a solution is a crucial factor. For that reason a community of cooperating colonies was introduced.

3.2 Ant Colony Community

Researches pretty soon become aware that that is an urgent need to mitigate the long processing time of that approach. The parallelization of processing offers such a possibility. The taxonomy of parallel implementation of the ACO is presented in [12]. They could be divided into three broad groups: master-slave, cooperative and hybrid models. The ACC uses a version of the hybrid model. Two of its implementations are introduced and compared [13]. The first one uses a network of regular computers and the second is implemented in the Hadoop environment.

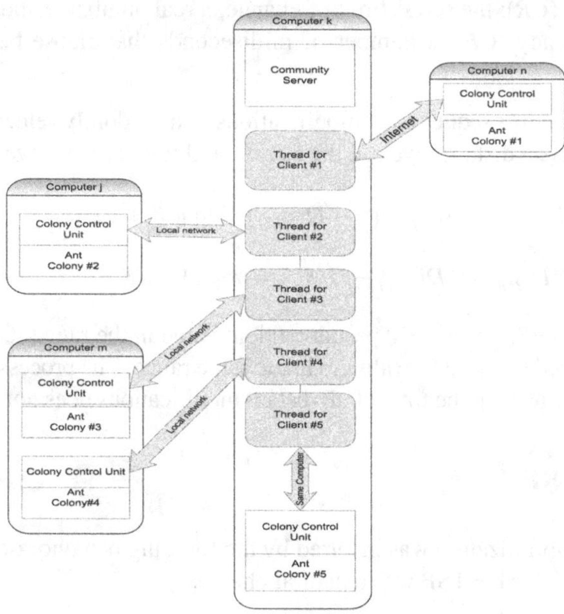

Fig. 1. A structure of an exemplary ACC.

ACC structure

The ACC consist of a single server and a set of colony clients. They communicate using the socket mechanism. The ACC components could run a single computer, computers in a local network, on internet servers or any combination of previous locations. An exemplary Community with 5 clients located on 4 computers is shown on Fig. 1.

The server creates a thread dedicated to handling communication with a single client colony. There is no practical limit on the number of clients that could be dealt with. Each client works as a separate process and contains an Ant Colony - an implementation using the ACO framework described in [14]. The parameters of work of an Ant Colony client are specified by the Community Server. Most of them have well proven default values. The only parameters that change are: number of ants and number of iterations. The communication process uses the efficient Socket mechanism and it is initialized by a client. The only information that a Client needs is the IP and port numbers of the server.

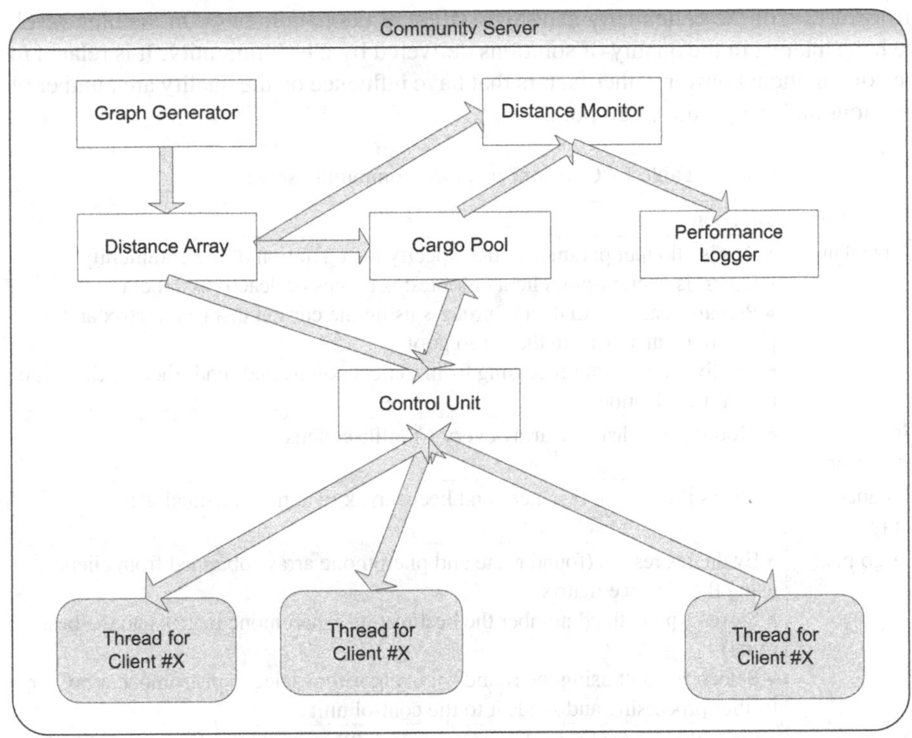

Fig. 2. Structure of community server

The structure of the server is much more complex. It is depicted in the Fig. 2. Comparing with the version used in [13] the Server had been considerably extended so it can cope with dynamic environments. Its Control Unit is responsible for sending clients packages of data to process (the so-called cargos) and for receiving results. A cargo contains among others: the distance array, the pheromone array, number of ants and iterations, the best route so far and parameters describing the operation of the ACO. The time needed to send cargo is two orders of magnitude lower than the time of the cargo processing and therefore the communication overhead had little influence upon the overall performance. Table 1 summarizes the activities performed by servers' components. All of them work concurrently as separate threads. The operation of the Community is controlled by many parameters among them by *Cf* (see Sect. 2.2) and *Uf* distance update frequency used by Distance Monitor.

3.3 Evaluation of the ACC Performance

An Ant Colony Community has a very flexible structure. It can consist of a different number of client colonies located on many computers that are controlled in a variety of ways. Therefore we need measures to evaluate the ACC performance. The low-level measures are the computer power and scalability. They depend upon the hardware infrastructure of the community and the location of client colonies. On the high level, we have interest in the quality of solutions delivered by the Community. It is related to the computational power. Other factors that have influence on the quality are number of iterations and size of the cargo pool.

Table 1. Components of the community server

Component	Function
Control unit	• Reads all input parameters that specify the operation of the community • Controls operation of client colonies: activates or deactivates them • Prepares cargo with data to process using the current distance matrix and a pheromone matrix from the cargo pool • Receives results of processing from a client colony and sends them to the cargo pool for evaluation
Graph generator	• Modifies the distance array every *Cf* milliseconds
Distance array	• Stores the current distances and keeps track of some statistical data
Cargo pool	• Evaluates results (found route and pheromone array) obtained from clients using the distance matrix • Stores a predefined number the best results (pheromone matrix and the best route) • Selects a result using the round robin algorithm select a pheromone array for further processing and sends it to the control unit
Distance monitor	• Finds the current best solution every *Uf* milliseconds. To obtain a reliable data on the performance $Cf > Uf$ • Prepares data for the performance logger
Performance logger	• Displays on-line basic data on the community operation • Performs detailed offline data analysis

Low-level measures

The most time-consuming part of an optimization process is the selection of the next node which requires comparing paths segments. The number of necessary comparisons is used as a measure of task complexity [15]. For the array of 50 nodes, 50 ants, and 50 iterations we need 3060000 comparisons. In what follows this number is denoted as *Sc* - Standard Cargo complexity. A regular computer needs around 1 s to complete such a task.

The basic low-level measure of Community performance is the *Cp* (Computational Power). For the Community *Com* and the task *T* the *Cp* is defined by the following ratio:

$$Cp(Com, T) = \frac{Sc(T) * 60}{Time(T)} \tag{3}$$

where:

Com – the tested community
Sc(*T*) – the complexity of optimization task *T* measured in *Sc* units.

Time(*T*) – the time necessary to complete the task *T* measured in seconds.

The Computational Power specifies the number of Standard Cargo tasks that could be calculated within a minute.

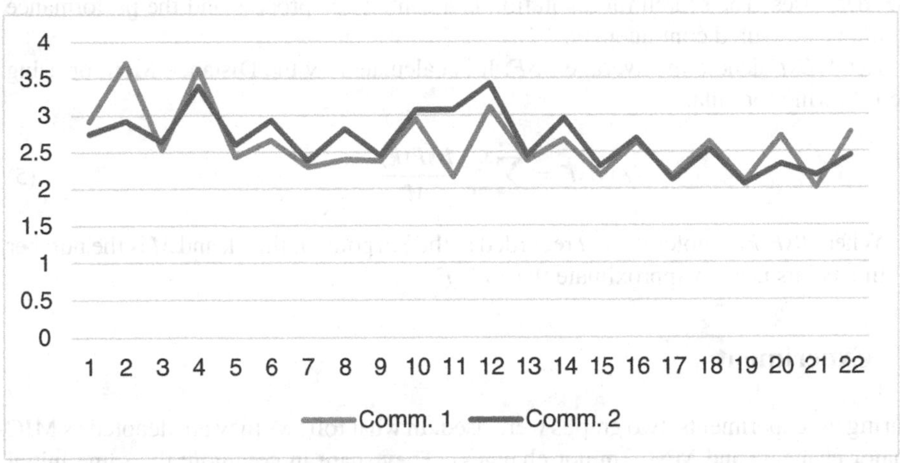

Fig. 3. Typical BSF value changes for two communities working with a graph generator.

Let *MaxPowr*(*CL*) denote the maximal power that a client *CL* can contribute to its colony. It is approximated by the Formula 4. The *MaxPowr*(*CL*) is the power of the following Community:

(a) *CL* is the only one Client in this Community.
(b) The Server is located on a LAN computer.
(c) The Server and Client computers do not run any computing intensive tasks.

The more Client Colonies are run on a single computer the longer each individual client needs compute the Standard Cargo. In order to be able to configure properly the community *Com*, we need to know its' scalability denoted by *ScalFactor*(*Com*). It is the ratio of the actual measured power while solving a task and the sum of *MaxPower* of its clients.

$$ScalFactor(Com, T) = \frac{Cp(com, T)}{\sum_{Cl \in Com} MaxPw(Cx)} \tag{4}$$

The Algorithm to optimally configure a Community for a set of computers is presented in [15]. Experiments reported there show that the *ScalFactor*(*Com*) is a good approximation of the observed performance of a Community.

High-level measure

In the case of a static graph the best measure of performance is the *BsF* - Best so Far route length valid for the last iteration. In the case of dynamic graphs, the *BsF* is not relevant: the distances change constantly and what was a decent route at a given point of time could be not a much worse solution a split of a second later. This phenomenon is well illustrated by the Fig. 3. It shows the values of the *BsF* just before and imminently after distance change for two Communities. As you can see the modification usually increases significantly the current *BsF* value. As the optimization process goes on the *BsF* decreases. The search for a solution is a continuous process and the performance should be measured continuously.

Let *AvBsF* denote the average *BSF*. It is calculated by the Distance Monitor using the following formula:

$$AvBsF = \sum_{k=1}^{M} \frac{BsF(k)}{M} \tag{5}$$

Where *BsF(k)* denotes the *Bsf* recorded at the kth point of time k and *M* is the number of time points used to approximate the *AvBsF*.

4 Experiment

During the experiments, two graphs were used. In what follows they are denoted as MJC (major changes) and MNC (minor changes). They share in common: the same initial distance matrix with 50 nodes, the total change (*TC*) equal to 20 and the change frequency (*CF*) equal to 2 s.

The only difference was their change range *CR*. It was equal to 0.1 for the MNC and 0.24 for MJC. As a result, the average number of distance modifications was equal to 390 (MJC) and 860 (MNC). This means that the MNC changed more than half of all distances but in a rather subtle way. The number of distance changes for the MJC was lower but they were 2.5 larger in scope. The difference may not look significant but it resulted in a noticeable difference in the ACC performance.

The experiment setup is summarized in Table 2.

Table 2. Experiment setup

Community property	Range
Number of physical computers	1–8
Number of client colonies on one computer	1–5
Total number of client colonies	1–30
Size of cargo pool	1–30
Iteration number	50–200
Number of ants in a colony	50
Change range	0.10–0.24
Duration of optimization run	2 min–6 min

The number of distances modifications varied from 60 to just below 200. The number of cargos with data for processing sent by the Server to Clients depended on the power of the Community but always it exceeded 50 which gives statistical credibility to the obtained results. The total number of analyzed communities exceeded well 100. The

AvBsF values for the best performing Communities working with MJC and MNC Generators are shown in Table 3.

Table 3. *AvBsF* communities working with two graph generators

MJV					MNC				
Pool No	cls	iterNo	*AvBsF*	Power	Pool No	cls	iterNo	*AvBsF*	Power
10	24	50	2.688	534.72	3	4	50	2.553	166.79
15	30	50	2.698	539.4	4	1	50	2.601	55.10
15	24	50	2.916	537.02	3	4	50	2.651	166.93
15	24	50	2.933	519.65	1	4	100	2.687	169.21
15	30	50	2.995	539.4	3	1	50	2.748	55.07
4	24	100	3.02	563.51	4	1	50	2.872	47.75
4	24	50	3.063	594.14	4	1	50	2.893	55.09
15	10	50	3.096	348.21	1	4	50	2.903	167.28
5	5	50	3.139	70.96	4	4	50	2.924	110.94
1	24	50	3.200	529.37	3	4	100	2.932	164.42
4	24	50	3.341	865.3	1	8	50	2.975	176.49
15	30	100	3.390	558.85	1	4	100	2.988	168.4
10	10	100	3.441	344.02	3	1	100	2.989	55.35
15	10	100	3.509	346.66	3	4	100	2.997	163.82
1	10	100	3.647	231.2	8	1	50	3.001	53.65

The values presented in the Table 3 are far from being coherent but still, some conclusions could be drawn. All of them perform better than the standard ACO for which the *AvBsF* was well above 7.0.

The *AvBsF* values for the MJV than for MNC generators are much alike. Note however that to achieve that level of performance the Communities working with MJC require much more client colonies then the MNC Colonies. The processing of cargos with iteration number of 200 required so much time that even the colonies with a large number of clients were unable to catch up with the changes and are not presented in the table.

The size of the pool could offset to some extend the number of Communities working with slower changing distances. The pool should contain relatively "fresh" solutions. If too many changes were introduced to the distance matrix then they contain obsolete route fragments and which is counter-productive. This is less likely when the matrix changes slowly.

5 Conclusions

The multi-population approaches gain recently much attention [16]. The paper uses it to solve the Dynamic Traveling Salesman problem. The reported experiments strongly indicate that the Community of Ant Colonies is capable of handling efficiently even fast changing graphs. They perform far better than a standard ACO.

There are however many points that need further investigation which includes among others:

- Evaluating more advanced algorithms for the selection of solutions from the pool. The currently used round-robin algorithm does not handle properly pools with several solutions for the fast changing graphs.
- The pattern of changeability that exhibit Graph Generators is rigid what does not necessary reflect real-life situations. Modifying the pattern by introducing e.g. Markov sources that control two or more independent Graph Generators is worth considering.
- The number of iterations is the parameter that is most closely related to the changeability level. The number is fixed in the current version of the ACC. Changing their number could be useful for Markov based graphs.

The research work on those topics is currently underway.

References

1. Held, M., Karp, R.M.: A dynamic programming approach to sequencing problems. J. Soc. Ind. Appl. Math. **10**(1), 196–210 (1962)
2. Applegate, D.L., Bixby, R.E., Chvatal, V., Cook, W.J.: The Traveling Salesman Problem: A Computational Study. Princeton University Press, Princeton (2011)
3. Antosiewicz, M., Koloch, G., Kamińskim, B.: Choice of best possible metaheuristic algorithm for the Travelling Salesman Problem with limited computational time: quality, uncertainty and speed. J. Theor. Appl. Comput. Sci. **7**(1), 46–55 (2013)
4. Dorigo, M.: Optimization, learning and natural algorithms. Ph.D. thesis, Politecnico di Milano, Italie (1992)
5. Psarafits, H.N.: Dynamic vehicle routing: status and prospects. Nat. Tech. Annal. Oper. Res. **61**, 143–164 (1995)
6. Guntsch, M., Middendorf, M.: Pheromone modifcation strategies for ant algorithms applied to dynamic TSP. In: EvoWorkshops 2001: Applications of Evolutionary Computation, pp. 213–222 (2001)
7. Guntsch, M., Middendorf, M.: A population based approach for ACO. In: Proceeding of 2nd European Workshop on Evolutionary Computation in Combinatorial Optimization (EvoCOP-2002), vol. 2279, pp. 72–81 (2002)
8. Mavrovouniotis, M., Yang, S.: Ant colony optimization with immigrants schemes in dynamic environments. In: Schaefer, R., Cotta, C., Kołodziej, J., Rudolph, G. (eds.) PPSN 2010. LNCS, vol. 6239, pp. 371–380. Springer, Heidelberg (2010). doi:10.1007/978-3-642-15871-1_38
9. Dorigo, M., Stuetzle, T.: Ant Colony Optimization: overview and recent advances. IRIDIA - Technical Report Series, Technical Report No. TR/IRIDIA/2009-013, May 2009
10. Siemiński, A.: TSP/ACO Partameter Optimization; Information Systems Architecture and Technology; System Analysis Approach to the Design, Control and Decision Support; pp. 151–161. Oficyna Wydawnicza Politechniki Wrocławskiej Wrocław (2011)
11. Gaertner, D., Clark, K.L.: On optimal parameters for Ant Colony Optimization algorithms. In: IC-AI, pp. 83–89 (2005)
12. Pedemonte, M., Nesmachnow, S., Cancela, H.: A survey on parallel Ant Colony Optimization. Appl. Soft Comput. **11**, 5181–5197 (2011)

13. Siemiński, A., Kopel, M.: Comparing efficiency of ACO parallel implementations. J. Intell. Fuzzy Syst. **32**(2), 1377–1388 (2017)
14. Chirico, U.: A Java framework for ant colony systems. In: Ants2004: Forth International Workshop on Ant Colony Optimization and Swarm Intelligence, Brussels (2004)
15. Siemiński, A.: Measuring efficiency of Ant Colony Communities. In: Zgrzywa, A., Choroś, K., Siemiński, Aj (eds.) Multimedia and Network Information Systems. AISC, vol. 506, pp. 203–213. Springer, Cham (2017). doi:10.1007/978-3-319-43982-2_18
16. Hong, T.-P., Peng, Y.-C., Lin, W.-Y., Wang, S.-L.: Empirical comparison of level-wise hierarchical multi-population genetic algorithm. J. Inf. Telecommun. **1**(1), 66–78 (2017)

Improved Stock Price Prediction by Integrating Data Mining Algorithms and Technical Indicators: A Case Study on Dhaka Stock Exchange

Syeda Shabnam Hasan, Rashida Rahman, Noel Mannan, Haymontee Khan, Jebun Nahar Moni, and Rashedur M. Rahman(✉)

North South University,
Plot 15, Block B, Bashundhara R/A, Dhaka 1229, Bangladesh
Shabnam.4444@yahoo.com, flgn@ymail.com,
{noel.mannan,haymontee.khan,
rashedur.rahman}@northsouth.edu, jnmoni@gmail.com

Abstract. This paper employs a number of machine learning algorithms to predict the future stock price of Dhaka Stock Exchange. The outcomes of the different machine learning algorithms are combined to form an ensemble to improve the prediction accuracy. In addition, two popular and widely used technical indicators are combined with the machine learning algorithms to further improve the prediction performance. To evaluate the proposed techniques, historical price and volume data over the past 15 months of three prominent stocks enlisted in Dhaka Stock Exchange are collected, which are used as training and test data for the algorithms to predict the 1-day, 1-week and 1-month-ahead prices of these stocks. The predictions are made both on training and test data sets and results are compared with other existing machine learning algorithms. The results indicate that the proposed ensemble approach as well as the combination of technical indicators with the machine learning algorithms can often provide better results, with reduced overall prediction error compared to many other existing prediction algorithms.

Keywords: Stock prediction · Machine learning · Regression algorithms · Time series forecast · Technical indicators

1 Introduction

The stock market is considered to be extremely important for the socio-economic stability and progress as well as the overall well-being of the people and the country. Stock investors all over the world collect historical stock price data and analyze them using automated software tools to predict the future stock price. Making accurate prediction is always extremely challenging [1], because of the high degree of non-linearity, uncertainty and volatility [2, 3] present in every stock market all over the world. The individual stocks as well as the overall stock market never follows a straight path, because they operate against a backdrop of continuous noise, news and rumor,

N.T. Nguyen et al. (Eds.): ICCCI 2017, Part I, LNAI 10448, pp. 288–297, 2017.
DOI: 10.1007/978-3-319-67074-4_28

various social, economic, political, geopolitical and miscellaneous factors that affect the stock market every day. In the backdrop of such enormous uncertainty and volatility, the prediction of a particular stock price is always challenging. The Dhaka Stock Exchange (DSE) is no exception, where situation may even become worse because many general investors invest and participate without proper plan, preparation and knowledge that often results in loss of their hard-earned valuable money.

The objective of this work is to introduce a useful guideline for the general investors of DSE on how to employ machine learning tools and algorithms along with technical indicators to predict the future stock price. We have also shown how to combine different prediction algorithms to form an effective ensemble that can predict stock price with higher accuracy. We have selected three prominent and representative stocks of DSE and demonstrated how machine learning algorithms combined with technical indicators can provide improved prediction accuracy on their future price.

The rest of this paper is organized as follows. Section 2 presents a number of recent related works on stock price prediction. Section 3 describes the proposed technique for ensemble design and two popular indicators — MACD and RSI. Section 4 describes the data source, experiment analysis. Finally Sect. 5 concludes the paper with a brief summary and suggestions for future study.

2 Existing Algorithms

To analyze and predict the future behavior of stock markets the most popular techniques employ various technical tools and parametric models in combination [6, 7]. Three most common machine learning approach are Multilayer Perceptron (MLP) or Artificial Neural Network (ANN), Support Vector Machine (SVM) and Gaussian Process based regression (GPR) [4, 5], all of which are widely used for analysis and prediction of time series data [4–8]. Although SVM based regressions (SVR) are the most widely used algorithm for time series prediction, there are some instances where MLP and Gaussian process are reported to perform better than SVR on the benchmark problems [8, 9].

The prediction made by SVR is sufficiently accurate for many prediction tasks, especially with small amount of look-ahead into the future [9–11]. An in-depth comparison between SVR and MLP is presented in [3], which is evaluated on the Hong Kong stock market to predict 1-week and 1-month price predictions. Like most other existing works, the authors in [3] made use of several technical indicators in combination with MLP and SVR.

None of the above works made use of ensemble to combine multiple machine learning algorithms. There exist very few works, e.g. [12], that emphasizes the use of ensembles to combine multiple machine learning algorithms. However, none of these papers deal with the Dhaka Stock Exchange data. We have found only a few recent works (e.g., [13, 14]) that employ machine learning algorithms on the DSE data set. In [13], the authors employed MLP and fuzzy algorithms to predict stock price in DSE, while the authors in [14] used a number of machine learning techniques for clustering, classification and regression on 30 selected stocks in DSE. However, none of them employ or emphasize an ensemble approach, as proposed in this paper. The unique contribution of this paper is to propose an ensemble architecture that combines four different machine learning algorithms and two popular technical indicators, as elaborated in the following section.

3 Proposed Methodology

The objective of our paper is to introduce a combined machine learning approach with technical indicators to predict future price of particular stocks in the Dhaka Stock Exchange. We have selected four widely used time series prediction algorithms to predict stock price and proposed four different ways to combine the outcomes of the different algorithms. We have divided our work into following three discrete parts:

3.1 Employ Existing Machine Learning Algorithms to Predict Future Stock Price

We have selected four different machine learning algorithms on time series prediction for the task of stock price prediction. The algorithms are- Support Vector Machine Based Regression (SVR), Regression by Multilayer Perceptron (MLP), Linear Regression (LNR) and Gaussian Process based Regression (GPR). These algorithms are provided by the popular machine learning software suite WEKA (Waikato Environment for Knowledge Analysis).

3.2 Ensemble Approach to Combine Multiple Predictions

In order to improve the prediction accuracy of the four machine learning algorithms, we plan to combine their predictions to form an Ensemble. The ensemble will combine the outcomes of its four component predictors using the following methods- Simple Averaging (SA), Weighted Averaging (WA), Voting by SA (VSA) and Voting by WA (VWA). Assuming the predicted stock price are O_1, O_2, O_3 and O_4, the combined output from the ensemble by the different combination methods are as follows.

Ensemble output by Simple Averaging,

$$O_{SA} = \frac{\sum_{i=1}^{n} O_i}{n}$$

Ensemble Output by Weighted Averaging,

$$O_{WA} = \frac{\sum_{i=1}^{n} w_i * O_i}{n}$$

Ensemble output from Voting by Simple Averaging,

$$O_{VSA} = O_j, \text{ where } j = \arg\min_i |O_{SA} - O_i|$$

Ensemble output by Voting by weighted averaging (O_{VWA}),

$$O_{VWA} = O_k, \text{ where } k = \arg\min_i |O_{WA} - O_i|$$

Here, n is the number of component predictors in the ensemble, which is set as $n = 4$. The w_i's in O_{WA} denote the weight values for weighted averaging, which are set as follows: the weight value w_i for a particular predictor algorithm's output O_i is set to the inverse of the Mean Absolute Error (*MAE*) value of the corresponding predictor algorithm on the training data set. Since the outcomes by SVM, MLP, LNR and GPR are denoted by O_1, O_2, O_3 and O_4 respectively, their weight values are set as:

$$w_1 = \frac{1}{MAE_{SVR}},\ w_2 = \frac{1}{MAE_{MLP}},\ w_3 = \frac{1}{MAE_{LNR}},\ w_4 = \frac{1}{MAE_{GPR}}.$$

The selection of one of the four algorithms is made for each row of the training and test sets based on which of the four algorithms produces minimum distance from the computed simple average (SA) and weighted average (WA) values of the four algorithms on the current row. This means for some rows the output of SMOReg may be picked by the ensemble, while for some other rows the output of Perceptron, Linear Regression or Gaussian Process based Regression is picked based on which of them can produce the closest output to the computed SA or WA value. This is why the result of VSA and VWA may not exactly match with one of the four component algorithms or the best performing algorithm, as seen in the Tables 1, 2 and 3. Besides, please note that average RMSE of four methods will not be equal to the RMSE of SA method. This is because for each row the average is done and the error is calculated based on this row's average. This is repeated for a set of 15 months of stock data and RMSE is calculated for SA method.

The motivation behind the ensemble approach is to improve the prediction accuracy by combining the outcomes of the different predictors. Since each predictor has its individual strengths and weaknesses, their effective combination is expected to produce improved prediction performance (i.e., reduced overall *MAE* value).

3.3 Combine Technical Indicators with Machine Learning Algorithms

There exist many technical indicators that reflect and predict different characteristics about the movements of stock price. All these indicators can be divided into two broad categories- Lagging indicators and Leading indicators. A Lagging indicator changes its value after some actual change appears it the stock price and/or volume. In contrast, the Leading indicators change their values before any actual change occurs in the stock's price and volume.

For our research, we have incorporated two technical indicators, one from the leading indicators, Relative Strength Index (RSI) and the other from lagging indicators, Moving Average Convergence Divergence (MACD), with the four machine learning algorithms. MACD is a lagging indicator that follows the ongoing price trend. The MACD value is plotted with a signal line. The crossover of MACD with the signal line gives bullish (buy) or bearish (sell) signal. Whenever MACD drops below the signal line from above, it gives a bearish signal. Similarly, when the MACD value rises above the signal line from below, it gives a bullish signal for the traders. In addition to the crossover signals, MACD may also produce trend reversal signal by divergence from the stock price. RSI is a leading indicator that measures the speed of price movements and predicts possible reversal of current trend by the change of speed of price movements. The value of RSI moves between 0 and 100. Usually a stock is considered overbought if the RSI is above 70, and oversold when RSI is below 30. If the RSI pattern diverges from the price pattern, it indicates a weakness and possible reversal of the price trend.

4 Experimental Study

4.1 Source of Data

The historical data of Dhaka Stock Exchange (DSE) is maintained by its official web site http://www.dsebd.org. However, more organized data sets are readily available at some financial websites, like the http://www.stockbangladesh.com and also, the http://www.lankabd.com. All these data sets can be readily downloaded in text (*.txt), excel (*.xls) and Comma Separated Values (*.csv) format. Currently, DSE enlists more than 400 companies from 22 different financial sectors. From them, three prominent companies have been selected to evaluate our proposed techniques, which are- ACI, Beximco, and GP. For each company, its daily data is collected for the past 15 months (since 01-Jan-2015). The daily data consists of six attributes — Date, Open, High, Low, Close, and Volume. They represent the date, daily open price, high and low price, closing price and volume of a particular stock on that day. The data set has been divided into training set (90%) and test set (remaining 10%). The proposed methods have been evaluated on both the training and test sets. A partial snapshot of the dataset of ACI is presented in Fig. 1.

	A	B	C	D	E	F
1	Date	Open	High	Low	Close	Volume
2	1/3/2016	562.1	575.4	561.9	567.5	115659
3	1/4/2016	569.6	570.9	561.1	561.3	92543
4	1/5/2016	561.7	562	549.9	555.1	192960
5	1/6/2016	557	562.4	550	550.6	127394
6	1/7/2016	550.6	561.7	550.6	557.9	92518
7	1/10/2016	558	561	548.8	551.3	79864
8	1/11/2016	552	554.6	542	544.4	67720
9	1/12/2016	550.3	550.3	540	542.1	57903
10	1/13/2016	545	548.5	538	540.3	113020
11	1/14/2016	542	557.9	540.3	555.2	95556

Fig. 1. ACI dataset over 10 working days.

4.2 Parameter Setup

The default framework of Weka does not provide any forecasting tool. We have installed an additional package (Time Series Forecasting Environment) into Weka which provides additional support for automated regression schemes by creating lagged variables and date-derived periodic variables. The package provides algorithms for Support Vector Machine based Regression, Perceptron based Regression, Linear Regression and Gaussian Process based Regression. We have used the default parameters for these algorithms for this experimental study.

4.3 Experiments and Analysis of Results

We have conducted three different set of experiments for stock price prediction using four machine learning algorithms, which are briefly described below.

Basic Machine Learning Algorithms: We have used four basic machine learning algorithms to predict the future price of three stocks – ACI, Beximco and GP. Results are presented in the Tables 1, 2 and 3, under "Basic Algorithms".

Ensemble Design: We have combined the previously used algorithms to form the ensemble in four different ways- Simple Averaging, Weighted Averaging, Voting by SA, and Voting by WA. The results are shown in Tables 1, 2 and 3 under "Ensemble".

Integrating Technical Indicators: We have included two major technical indicators – the MACD and RSI as additional columns into the data set for ACI. The resulting data set is used by the basic machine learning algorithms to predict the future price of ACI, which is shown in Tables 4 and 5. Please note that, in the 'Target Selection' list box of the 'Forecast' tab of WEKA, we have selected all of Open, High, Low and Close prices as the 'Targets' of the prediction algorithms for the results in these Tables 4 and 5, which is quite different from Tables 1, 2 and 3, where we have selected only the Close price as the 'Target' in WEKA. Since the Open, High, Low and Close prices are intricately related, predicting them together results in smaller RMSE error value, as demonstrated by the 5-day and 22-day RMSE values in Tables 4 and 5, compared to the much higher RMSE error value of the same stock over the same prediction-lengths in the previous Table 1.

Table 1. RMSE of 1-day, 5-day and 22-day-ahead predictions on ACI dataset

Data set	Forecast length	Basic algorithms				Ensemble			
		SMOReg	Perceptron	Linear regression	Gaussian process	Simple averaging	Weighted averaging	Voting by simple avg.	Voting by weighted avg.
Training set	1-day	6.5	7.4	6.5	10.8	6.6	**6.4**	6.6	6.7
	5-day	15	17.4	15.7	19.3	**12.9**	13	14	14.2
	22-day	24.9	114.3	52.5	**20.7**	36.9	24.3	36.6	22.9
Test set	1-day	13.9	27.6	18.8	40.2	**12.4**	13.2	13.7	13.7
	5-day	60.7	118.9	111.5	84.7	51.7	52.7	51	**50.9**
	22-day	120.2	142.2	2077	60.4	71.6	**35.6**	120.1	60.3
Mean RMSE		40.2	71.3	380.4	39.4	32	**24.2**	40.3	28.1
		132.8				**31.2**			

Table 2. RMSE of 1-day, 5-day and 22-day-ahead predictions on Beximco dataset

Data set	Forecast length	Basic algorithms				Ensemble			
		SMOReg	Perceptron	Linear regression	Gaussian process	Simple averaging	Weighted averaging	Voting by simple avg.	Voting by weighted avg.
Training set	1-day	0.5	0.5	**0.4**	0.6	0.5	**0.4**	0.5	**0.4**
	5-day	0.9	1.3	1.1	1.3	**0.8**	**0.8**	0.9	**0.8**
	22-day	1.3	3.6	**0.9**	1.6	1.0	**0.9**	1.0	1.0
Test set	1-day	**1.2**	5.0	1.6	3.7	1.4	1.4	1.4	1.4
	5-day	2.8	8.3	6.3	10.4	**2.0**	2.5	2.8	2.8
	22-day	8.6	8.2	27.7	5.3	4.7	**4.3**	7.6	7.6
Mean RMSE		2.6	4.5	6.3	3.8	**1.7**	**1.7**	2.4	2.3
		4.3				**2.0**			

Table 3. RMSE of 1-day, 5-day and 22-day-ahead predictions on Grameen Phone (GP) dataset

Data set	Forecast length	Basic algorithms				Ensemble			
		SMOReg	Perceptron	Linear regression	Gaussian process	Simple averaging	Weighted averaging	Voting by simple avg.	Voting by weighted avg.
Training set	1-day	2.4	3.0	2.3	8.5	2.9	2.3	2.4	**2.2**
	5-day	**4.6**	10.1	**4.6**	15.3	5.5	**4.6**	5.1	4.8
	22-day	6.9	52.5	**6.4**	10.9	14.6	7.5	7.7	7.3
Test set	1-day	7.8	14.6	4.7	4.9	5.5	6.0	**4.5**	4.6
	5-day	19.6	33.6	16.0	16.6	**15.2**	**15.2**	15.4	20.6
	22-day	23.2	122.7	33.3	**9.3**	21.9	24.1	23.2	23.2
Mean RMSE		10.8	39.4	11.2	10.9	10.9	9.9	**9.7**	10.4
		18.1				**10.3**			

Table 4. Comparison of SMOReg and Linear regression, before and after combining MACD with them.

Data set	Forecast length	Algorithm			
		SMOReg	SMOReg with MACD	Linear regression	Linear reg. with MACD
Training set	1-day	6.5	**6.4**	6.5	**6.1**
	5-day	15.0	**14.6**	15.7	**13.9**
	22-day	24.9	**22.5**	52.5	**28.3**
Test set	1-day	12.0	12.0	18.8	**13.7**
	5-day	32.3	**28.0**	111.5	**42.4**
	22-day	**79.3**	85.3	2039.0	**264.6**
MEAN RMSE		28.3	**28.1**	374.0	**61.5**

Analysis of Results: We have made a number of observations, which are summarized in the following few points.

1. Among the four basic machine learning algorithms, Support Vector Machine based Regression (SMOReg) performs overall best, especially for two stocks- Beximco and GP (Tables VI and VII, with Root Mean Squared Error, i.e., RMSE of 2.6 and 10.8 for these two stocks), followed by the Gaussian Process based Regression.

Table 5. Comparison of SMOReg and Multilayer Perceptron, before and after combining RSI with them.

Data set	Forecast length	Algorithm			
		SMOReg	SMOReg with RSI	Multilayer perceptron	Multilayer perceptron with RSI
Training set	1-day	6.5	6.5	7.4	**6.4**
	5-day	15.0	**14.7**	17.4	**12.2**
	22-day	24.9	**24.4**	114.3	**12.9**
Test set	1-day	12.0	12.0	27.6	**18.3**
	5-day	32.3	**30.5**	118.9	**91.6**
	22-day	79.3	**73.8**	101.9	**109.1**
MEAN RMSE		28.3	**27.0**	64.6	**41.8**

2. Among the four different ensemble approach, the weighted averaging performs overall best, followed by the voting by weighted averaging technique.
3. If we compare the basic algorithms with the ensemble techniques, we can easily observe much improved results by the ensemble methods. For example, ensembles produce mean RMSE = 31.2 for the ACI dataset, compared to the much larger RMSE = 132.8 by the basic algorithms. Similarly, for Beximco, and GP data sets, ensemble produce overall RMSE of only 2.0 and 10.3, compared to the much higher 4.3 and 18.1 respectively.
4. The overall better prediction performance by the ensemble approach is demonstrated. The RMSE values of the four basic algorithms are much higher compared to the smaller values of the ensemble techniques.
5. The predictions made by the four ensemble techniques closely follow the actual price pattern, as demonstrated by Fig. 2.

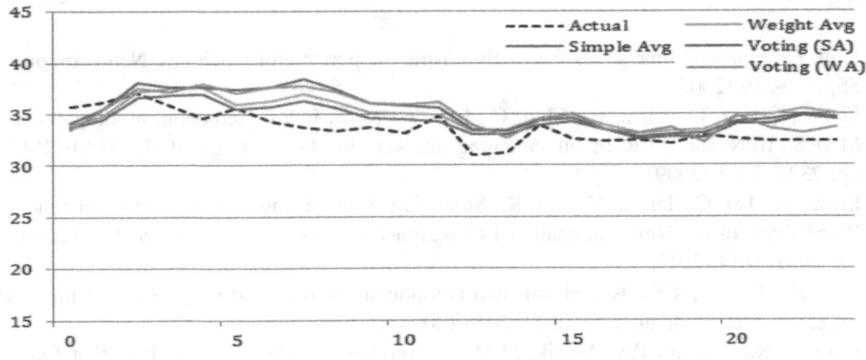

Fig. 2. Plot of actual price and predicted price of the Beximco data set by the four ensemble techniques.

6. When the technical indicators MACD and RSI are combined with the machine learning techniques, the prediction accuracy is improved, as demonstrated by lower
7. RMSE value in Tables 4 and 5. After including MACD with SMOReg and Linear Regression, the RMSE value drops to 28.1 and 61.5 from previous higher 28.3 and 374.0, respectively (Table 4). Inclusion of RSI reduces the RMSE error value into 27.0 and 41.8, compared to RMSE of 28.3 and 64.6, respectively (Table 5).
8. The improved prediction performance by including technical indicators MACD and RSI is also demonstrated. The RMSE values are much higher without MACD and RSI, but significantly smaller when MACD and RSI are included with SMOReg, Linear Regression and Multilayer Perceptron.
9. Summarizing our observations, we can say that the overall prediction performance of machine learning algorithms can be improved further by combining the outcomes of multiple algorithms into an ensemble, as well as by combining them with technical indicators for stock price.

5 Conclusion and Future Work

This study applies a number of machine learning algorithms on stock price prediction of Dhaka Stock Exchange (DSE). The experimental results with three prominent stocks of DSE show that the outcomes of different algorithms may be used to form an ensemble of predictors, which significantly improves the prediction accuracy. Further experiments reveal that inclusion of technical indicators, such as MACD and RSI, may further improve the prediction performance. Further research may include more ensemble designing techniques, use many other technical indicators and employ more machine learning algorithms, which would provide the researchers with more insights and better algorithms on stock price prediction, especially the DSE listed ones.

References

1. Kim, K.: Financial time series forecasting using support vector machines. Neurocomputing **55**, 307–319 (2003)
2. Lu, C., Chang, C., Chen, C., Chiu, C., Lee, T.: Stock index prediction: a comparison of MARS, BPN and SVR in an emerging market. In: Proceedings of the IEEE IEEM, pp. 2343–2347 (2009)
3. Lucas, K., Lai, C., James, N., Liu, K.: Stock forecasting using support vector machine. In: Proceedings of the Ninth International Conference on Machine Learning and Cybernetics, pp. 1607–1614 (2010)
4. Ince, H., Trafalis, T.B.: Kernel principal component analysis and support vector machines for stock price prediction, pp. 2053–2058 (2004)
5. Kannan, K.S., Sekar, P.S., Sathik, M.M., Arumugam, P.: Financial stock market forecast using data mining techniques. In: Proceedings of the International Multiconference of Engineers and Computer Scientists, pp. 555–559 (2010)
6. Hu, Y., Pang, J.: Financial crisis early warning based on support vector machine. In: International Joint Conference on Neural Networks, pp. 2435–2440 (2008)

7. Chen, K.-Y., Ho, C.-H.: An improved support vector regression modeling for Taiwan Stock Exchange market weighted index forecasting. In: The IEEE International Conference on Neural Networks and Brain, pp. 1633–1638 (2005)
8. Xue-shen, S., Zhong-ying, Q., Da-ren, Y., Qing-hua, H., Hui, Z.: A novel feature selection approach using classification complexity for SVM of stock market trend prediction. In: 14th International Conference on Management Science & Engineering, pp. 1654–1659 (2007)
9. Debasish, B., Srimanta, P., Dipak, C.P.: Support vector regression. Neural Inf. Process. Lett. Rev. **11**(10), 203–224 (2007)
10. Hsu, C.-W., Chang, C.-C., Lin, C.-J.: A practical guide to support vector classification. Initial version: 2003, last updated version: 2010
11. Kazema, A., Sharifia, E., Hussainb, F.K., Saberic, M., Hussaind, O.K.: Support vector regression with chaos-based firefly algorithm for stock market price forecasting. Appl. Soft Comput. **13**, 947–958 (2013)
12. Ballings, M., Van den Poel, D., Hespeels, N., Gryp, R.: Evaluating multiple classifiers for stock price direction prediction. Expert Syst. Appl. Int. J. **42**(20), 7046–7056 (2015)
13. Billah, M., Waheed, S., Hanifa, A.: Predicting closing stock price using artificial neural network and adaptive neuro-fuzzy inference system (ANFIS): the case of the Dhaka Stock Exchange. Int. J. Comput. Appl. (0975-8887) **129**(11), 1–5 (2015)
14. Shadman, A.I., Towqir, S.S., Akif, M.A., Imtiaz, M., Rahman, R.M.: Cluster analysis, classification and forecasting tool on DS30 for better investment decision. In: Akagi, M., Nguyen, T.-T., Vu, D.-T., Phung, T.-N., Huynh, V.-N. (eds.) ICTA 2016. AISC, vol. 538, pp. 197–206. Springer, Cham (2017). doi:10.1007/978-3-319-49073-1_22

A Data Mining Approach to Improve Remittance by Job Placement in Overseas

Ahsan Habib Himel, Tonmoy Sikder, Sheikh Faisal Basher, Ruhul Mashbu, Nusrat Jahan Tamanna, Mahmudul Abedin, and Rashedur M. Rahman(✉)

Department of Electrical and Computer Engineering, North South University, Plot – 15, Block – B, Bashundhara, Dhaka 1229, Bangladesh
{ahsan.himel,sikder.tonmoy,faishal.sheikh,ruhul.mashbu, nusrat.tamanna,mahmudul.abedin,rashedur.rahman}@northsouth.edu

Abstract. Remittance or foreign currency transaction plays an important role in increasing a country's financial growth. Bangladesh is a country with a reputation in manpower export and every year it receives a considerable amount of remittance. Yet the remittance can be improved further by providing the workers with the information of their future earnings. We propose a solution that will help the workers as well as the government to decide which country/countries will be best for workers in terms of earning, thus increasing the country's annual remittance. The research outcome from this paper could help the government to export the manpower to the right country and the workers who are planning to move abroad with a vision to work for the best suitable job with respect to their skill. Besides, the findings could help in reducing the unexpected returns of the workers and stop the bad experience the workers endure abroad.

Keywords: Remittance · Foreign currency · Abroad · Immigrant · Data mining · Classification

1 Introduction

The main purpose of data mining is to explore data from different fields and search for the pattern (if any) hidden among them. We have used data mining to predict a job and the job location overseas for the workers so that they can earn more money and increase the rate of remittance.

Remittance means the transfer of certain amount of funds from a country to another. To become financially stronger a country must maintain a proper balance between export and import. The goal is to maintain an export rate that is higher than the import rate. If the amount of currency going out of the country is more than earning rate, it will result in a financial collapse of a country [7].

Bangladesh is a developing country. It is important for Bangladesh to look over this sector and become financially stronger by increasing the ratio of export. Bangladesh is facing fall of remittance every year. In 2016, the rate ended up falling to 11.13% [8]. Statistics show that workers of Bangladesh are coming back to the country because of the salary they get is too low to survive in foreign country.

N.T. Nguyen et al. (Eds.): ICCCI 2017, Part I, LNAI 10448, pp. 298–306, 2017.
DOI: 10.1007/978-3-319-67074-4_29

Every worker going abroad in search of work has high hopes of providing his family with better life style [9]. Failing in their agenda, many of them return to their homeland makes them frustrated. Sometimes people end up selecting countries that are unfavorable to their skill set; but they select those countries as their relatives and ancestors went there. Eventually they end up getting depressed as they find things not going in their way because the job sector matching their skillset might not be very friendly for non-citizens [10].

Bangladesh government has a ministry called "Ministry of Expatriates Welfare and Overseas Employment". They try to approach other countries to let them know about Bangladesh's interest to send workers to their countries. Sometimes it seems that if the same man power was used in another country, there would have been an improvement in terms of profit. It is our goal to provide the government with an estimation well enough so that they can make better decisions in the future on this aspect.

Our primary objective of this paper is to help the worker and the government by giving them a recommendation about which country will be best suited for a worker and for exporting manpower with a set of skills so that the remittance record of the country can be improved.

Therefore the main contributions of our workare as follows:

1. Providing workers with information about jobs and country for which he is more likely to be suitable.
2. Giving information to government about which country is best for exporting manpower.

The organization of the paper is as follows. Section 2 briefly discusses about the related work in this area. Section 3 showcases our research methodology in detail. Section 4 presents our research findings and Sect. 5 concludes our paper.

2 Related Work

Few researches have been carried out like our area but they do not cover the area of our interest exactly. We could not find any related work directly related to our topic. However, we could find some work that uses similar approaches for solving similar type problems.

Decision tree is a widely used method in data mining for classification purpose. It uses some criteria for selecting the most important attribute in different levels of tree construction [1]. In [2], for decision making process this tree structure is used as a visual and analytical tool that helps us to predict the values and classify the data.

Divya et al. [3] showed a comparison and described the capabilities of different data mining algorithms in their paper. Decision tree is considered as one of the most powerful classifier. Because of its simplicity and high efficiency in the related field, tree based classification also has been used for predicting weather forecast [4].

Kabra and Bichkar showed that student's past academic performance can be used for predicting how a student will perform in an engineering related exam. They used decision tree for this prediction purpose [5]. It is not possible to read a student's mind to understand why he/she cut a sorry figure in the exam. So, making a classification rule and try to predict a reason behind the performance is a good decision. Though the algorithm can help us to do the main analysis work, it may take a long time to read the raw results [6].

Wei and Yuxiang [11] have worked on manpower demand forecasting using prediction technique. They have used grey system prediction in which uncertain and unknown information is black and known information are kept white where the primitive data is used to form Grey Model GM (1,1). By using the GM (1, 1) they have managed to predict the manpower demand of Chinese Strategic Emerging Industry.

3 Methodology

In this paper, we have used decision tree for government side. In decision tree, data is divided according to their attributes and at last we get the target attribute. As we want to help the government to predict a country according to manpower export and to get expected remittance. So, we think decision tree based classification would be better choice to get country name.

In Naïve Bayes classification technique, we get the highest probability result among provided dataset. Workers with different skills try to find the most suitable countries for them. So, we think independent prediction would give the best expected result. We used these two techniques in our research.

3.1 Data Collection

The datasheets we need are the ones that consist of remittance coming from foreign countries; manpower that is being exported per year, how friendly is the job sectors and income per capita for those countries.

The datasheets of remittance coming from foreign countries are obtained by The Bangladesh Bank [11]. The nominal values of countries have been categorized as Bahrain, Kuwait, and Oman etc. respectively and they were used as a target for the classification. The unique identifier is removed and the integer values are then converted into nominal values.

For our work we have organized our data in different sets and they are average data set, language_set, test_set, training_set. In average data set we keep information about country, average remittance and average immigrants from 2009–2016. In language set, country, language required for the job, job name and per capita income is recorded. In the test set year, country, existing number of immigrants, number of new, immigrants, total number of immigrants and remittance is recorded. The attributes are same for training and testing set. In average set there are 1 class, 2 attributes and 14 instances. In language set there are 1 class, 3 attributes and 101 instances. In test set 1 class, 5 attributes and 40 instances. For training set there are 66 instances (Table 1).

Datasheets containing Job demands and language used in the countries that are importing manpower from Bangladesh on a yearly basis are acquired from the Ministry of

Table 1. Collected data of remittances coming from different countries on yearly basis

Country	2009–2010	2010–2011	2011–2012	2012–2013	2013–2014
Bahrain	170.14	185.93	298.46	361.70	459.39
Kuwait	1019.18	1075.75	1190.14	1186.93	1106.88
Oman	349.08	334.31	400.93	610.11	701.08
Qatar	360.91	319.36	335.26	286.89	257.53
K.S.A	3427.05	3290.03	3684.36	3829.45	3118.88
U.A.E	1890.31	2002.63	2404.78	2829.40	2684.86
Libya	1.46	5.20	12.91	57.65	71.96
Iran	4.49	2.32	1.16	2.59	0.40
Sub total	7222.62	7215.53	8328.00	9164.72	8400.98

Expatriates and Overseas Employment. Unfortunately, these datasheets were in Portable Document Format (PDF) that is why we convert it to .xlsx format for experiments purposes.

3.2 Data Pre-processing

First, we select the data having desired attributes. The attributes are different in both datasets because it is collected from different sources. Also, both datasets do not have the same sets of data. We only take the common countries that are available in both datasets. In the first set, we have information about 19 countries. This datasheet provides us with the migration details to these countries from the year 1976 to the year 2016.

In the second data set, we have data about 18 countries. This dataset provides us with information about inward remittance from the year 2009 to the year 2016. We also took the information from 2009 to 2016 as we only had the country wise information of these years in both datasets. As very old data is not of much significance we have discarded them. In the final filtering, we remove 3 countries that take fewer than 100 immigrants each year and only take very highly skilled professionals. These ten countries are the highest importer of manpower from Bangladesh according to the government data of last eight years (2009 to 2016).

We take the average number of migrants and the average inward remittance for each of these selected 10 countries as our attributes. The total number of migrants to a country is not constant for any country. So, we take the average over last eight years. Due to the increasing number of immigrants every year, more workers are being added up in that countries work force resulting increasing inward remittance from that country. So, we take an average of the remittance over these past eight-year period.

3.3 Data Analysis

From the training data, we find out some interesting information which turns out to be beneficial for our research. From the training data set, we find out people from which profession of Bangladesh usually gets more demand in foreign countries. Other information like which country takes the most workers from Bangladesh and what percentage of the total inward remittance comes from which country.

This chart (Fig. 1) shows that in the recent years the United Arab Emirates (UAE) takes the highest number of Bangladeshi workers. Kuwait and Qatar takes the second highest number of Bangladeshi workers. Bahrain is in the 3rd position in this list.

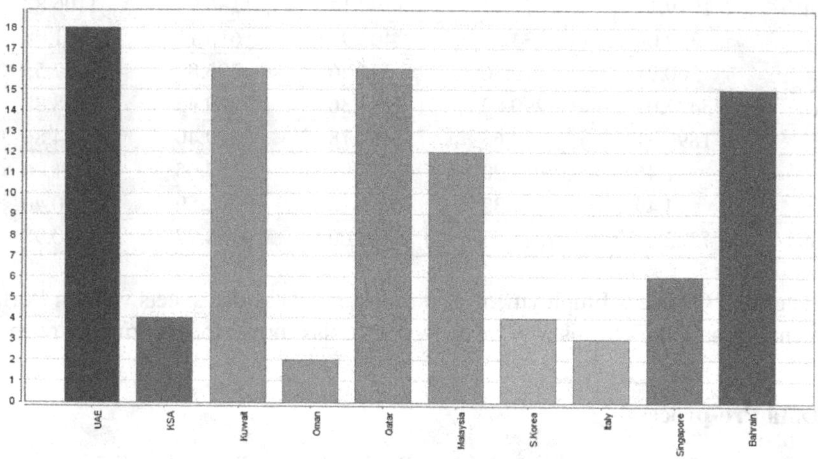

Fig. 1. Country-wise Bangladeshi manpower export

Figure 2 shows the job sectors that have comparatively more demand in these ten countries. The figure shows that the driving has the highest demand where the labor profession has the second highest demand. So, Bangladeshi workers who are skilled in these two sectors have a higher chance of getting a job in those fields and contribute to the inwards remittance. Other professions like tailor, cleaner, industrial worker have a moderate demand as well.

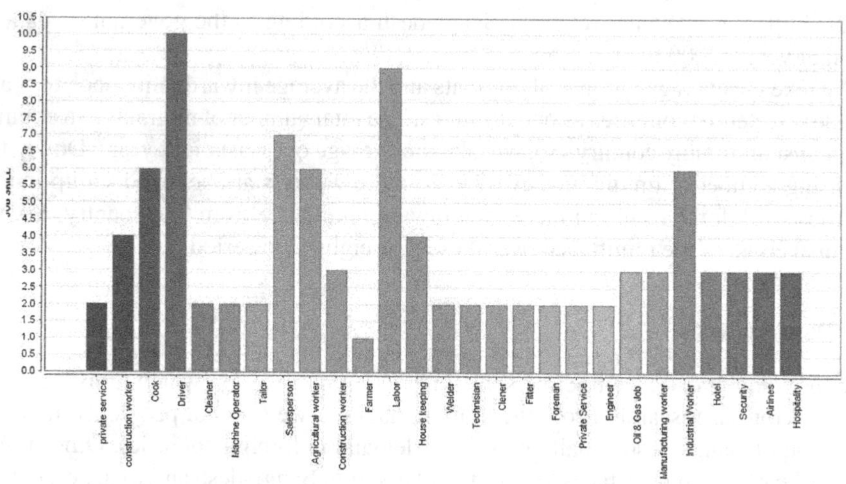

Fig. 2. Demands of jobs by country-wise

3.4 Implementation

We have to make one decision tree and one Naïve Bayes Network for predicting. Decision Tree is used for government decision making and the Naïve Bayes Network to find the probability of sending workers to appropriate country.

We have to select the columns from table and set their roles. For example which columns will be batched, which are regular attributes and which column is the class label attribute. The figure shows the attributes' roles.

For the decision tree which is for the government, we made a datasheet.

The columns of the first datasheet are: COUNTRY, Average_Imigrant, REMITTANCE(USD in Million), YEAR (Table 2).

Table 2. Roles of the attributes

COUNTRY	label
YEAR	regular
Average_Imigrant	regular
REMITTANCE(USD in Million)	regular

Now, we can start the recursive splitting process. Our program can now split the data into leaf nodes as we are using decision tree classification according to the best ratio of information gain to split information. If we want to get detailed classification rules, the support value should be smaller.

The training set will help us pre-prune the tree. When the number of samples is less than support value or class entropy is less than 0.1, the splitting process will be stopped and generates a node, which stands for a rule.

Here we have used rule induction to generate rules from the decision tree.

For validation purpose, we design our program like below which is shown in Fig. 3:

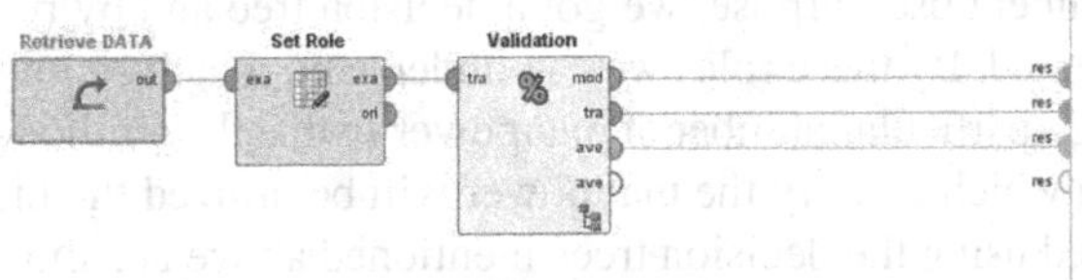

Fig. 3. Validation process part-1

The remittance data is used as a source of data for determining the country. After remittance, the algorithm looks for the average immigrants. If the remittance is related to the immigrant's number, it appears as a leaf node. The average export rate does not affect the flow of remittance. The visualization of the top portion of the Decision tree for the government obtained from the data we collected is shown in Fig. 4:

We have shown a part of our decision tree above. For the Naïve Bayes Network, useful for workers the columns of the datasheet are: Country, Language, JOB SKILL, PER CAPITA INCOME($)_2016.

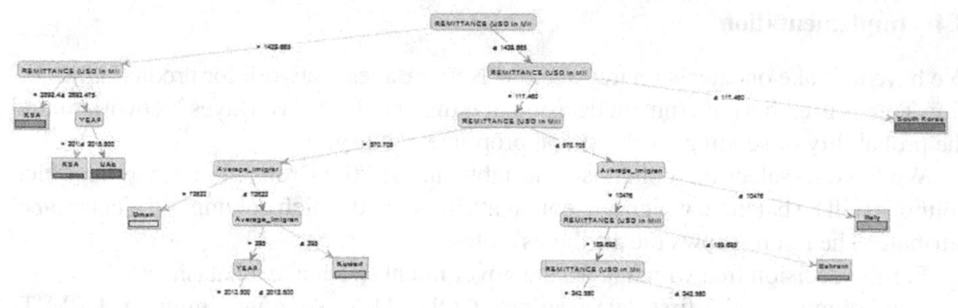

Fig. 4. Decision tree for Government

Then for our program we set rules for the COUNTRY as label class, LANGUAGE and JOB SKILL column as regular and Per CAPITA INCOME ($) _2016 as weight.

As we are predicting countries for the workers according to county's job demands and income per capita, we have used Naïve Bayes Network and Rule induction method to set rules.

After setting the pureness to 0.7 in Rule Induction we generated 28 rules. Some sample rules are given below:

if LANGUAGE = Arabic and JOB SKILL = Cook then Kuwait
if LANGUAGE = Mandarin then Singapore
if JOB SKILL = Security then Bahrain
if JOB SKILL = Driver and LANGUAGE = Arabic then Kuwait

4 Findings

After implementing the given dataset, from dataset 1 which was used to provide information for government use purpose, we got a decision tree and by rule induction a set of rules was generated. By these rules we can understand which country will be a better choice for sending a particular number of manpower so that the remittance is maximized. In other words, in which country the manpower will be utilized the most.

Rules generated using the decision trees mentioned above are shown below: (Rules are expressed in a format of If…Then…Every route from the root to leaf is a rule, former part is the attribute values and latter part is the class)

For the government, we have generated 21 rules. And all of them are correct for 79 training sets. For testing set, we have got an accuracy rate of 70.83% in Decision Tree based Classification. Criterion for both the decision tree and the rule induction gain is given below.

For Decision Tree:

- maximal depth: 20
- confidence: 0.5
- minimal gain: 0.1
- minimal leaf size: 2
- minimal size for split: 4

For Rule Induction:

- sample ratio: 0.9
- pureness: 0.9
- minimal prune benefit: 0.25

Some rules are discussed below:

Rule 1: if Average_Imigrant > 37268 and Average_Imigrant ≤ 62519.500 and REMITTANCE (USD in Million) ≤ 242.550 then Singapore

By this rule it clearly shows if the expected remittance is below 242.550 Million USD and immigrants are likely to export in the range of 37268 to 62519 then Singapore would be the best choice.

Rule 2: if REMITTANCE (USD in Million) > 474 and REMITTANCE (USD in Million) ≤ 1439.665 and Average_Imigrant ≤ 14937 then Malaysia

In rule 2 we can see if the immigrant number is below 14937 and expected remittance is greater than 474 and equal or less than 1439.665 then government should send workers to Malaysia.

By the rule set we have generated, the Government will have a knowledge about which country would be the best in terms of sending workers and bring more remittance to our country. So, when the government will make plan to send workers to a foreign country they will have a knowledge which country should be approached first. If the country refuses to accept to import the workers then the government will have a second choice.

The next experiment we have done is for the workers. We implemented our datasheet in our program and used Naïve Bayes to predict a job location and job for the workers. Then we have implemented it on the rule induction to generate a rule set.

In this experiment, we generated total 28 rules. For testing set 77 were correctly predicted among 93 records with an accuracy of 83%. For testing set, we have got accuracy rate of 65.54% Some of the rules generated using the Naïve Bayes Networks mentioned before are shown below:
Rule 1: if JOB SKILL = Agricultural worker and LANGUAGE = Arabic then UAE
It is showing that if any worker wants to work as an agriculture worker and can speak Arabic then he or she should choose UAE rather than other countries.
Rule 2: if JOB SKILL = Driver and PER CAPITA INCOME ($) _2016 ≤ 32098 then Kuwait.
If the worker wants to earn equal or less than 32098 and wants to work as a driver then he should go to Kuwait.

We have found some existing algorithms like our work; one of them was on manpower demand for some particular industries in a country by using grey prediction model based on known and unknown data [11]. Another model was about the planning of manpower control and for this the authors used square matrix and Euclidean space [12]. In our model, we have worked for both the government and the workers based on the working skill and remittance. From the decision tree we have created some rules from which we can easily predict the best suited country for workers. As we have used only known data there is less probability to fail a guess for country selection. In our model we have clearly shown the best prediction for job replacement based on remittance and this kind of work has not been done till now. By this experiment we have tried to help the worker so that they can select a country where they likely can earn enough according to their skill sets.

5 Conclusion

We have used data mining technique successfully on manpower export purposes. The work and the results above clarifies the fact that decision tree, rule induction, naïve Bayes and prediction technique in data mining can be used to determine which countries are the best choices for Bangladeshi workers for improving the rate of remittance. We hope our results and findings will help the workers to select a particular country based on their job skills. In our paper, we have tried to help both Bangladesh government and the workers. The different kinds of methods and rules in data mining play an important role in manpower export. However, we sincerely hope that our work will play a role in developing the economy and financial state of a developing country like Bangladesh.

References

1. Geetha, A., Nasira, G.M.: Data mining for meteorological applications: decision trees for modeling rainfall prediction. In: IEEE International Conference on Computational Intelligence and Computing Research. doi:10.1109/ICCIC.2014.7238481
2. Chauhan, D., Thakur, J.: Data mining techniques for weather prediction: a review. Int. J. Recent Innov. Trends. Comput. Commun. **2**(8), 2184–2189 (2014). ISSN: 2321-8169
3. Petre E.G.: A decision tree for weather prediction. In: Seria Matematică—Informatică—Fizică, pp. 77–82 (2009)
4. Kabra, R.R., Bichkar, R.S.: Performance prediction of engineering students using decision trees. Int. J. Comput. Appl. **36**(11), 8–12 (2011)
5. Wilton, W.W.T., et al.: Data mining application of decision trees for student profiling at the Open University of China. In: IEEE 13th International Conference on Trust, Security and Privacy in Computing and Communications (2014). doi:10.1109/TrustCom.2014.96
6. Rose, A.K., Ito, T.: The effects of financial crises on international trade. In: International Trade in East Asia, NBER-East Asia Seminar on Economics, vol. 14 (2003)
7. Alo, J.: Remittance sees big fall in 2016 | Dhaka Tribune. Dhaka Tribune. N.p. (2017). Web, 30 April 2017. http://www.dhakatribune.com/business/2017/01/06/remittance-sees-big-fall-2016/
8. Islam, M.: Jobs abroad for a better life. In: The Daily Star. N.p. (2017). Web, 30 April 2017. http://www.thedailystar.net/supplements/25th-anniversary-special-part-5/jobs-abroad-better-life-212662
9. Ahmed, H.S.: Neglected heroes. In: Star Weekend Magazine. Thedailystar.net. N.p. (2017). Web, 30 April 2017
10. Country-Wise Inward Remittances. Bb.org.bd. N.p. (2017). Web, 30 April 2017. https://www.bb.org.bd/econdata/wagermidtl.php
11. Yao, W., Li, Y.: Manpower demand forecasting of strategic emerging industry in China: based on grey system methodology. In: Portland International Conference on Management of Engineering and Technology (PICMET) (2015). doi:10.1109/PICMET.2015.7273042
12. Mouza, A.-M.: Application of optimal control in man power planning. Qual. Quant. **44**(2), 199–215 (2010). doi:10.1007/s11135-008-9189-4. Springer Science + Business Media B.V.

Determining Murder Prone Areas Using Modified Watershed Model

Joytu Khisha, Naushaba Zerin, Deboshree Choudhury, and Rashedur M. Rahman(✉)

Department of Electrical and Computer Engineering, North South University, Plot-15, Block-B, Bashundhara Residential Area, Dhaka, Bangladesh
{joytu.khisha,choudhury.deboshree, rashedur.rahman}@northsouth.edu, naushaba@live.com

Abstract. In this paper, we present an algorithm for cluster detection using modified Watershed model. The presented model for cluster detection works better than the *k-means* algorithm. The proposed algorithm is also computationally inexpensive compared to the *k-means*, agglomerative hierarchical clustering and DBSCAN algorithm. The clustering results can be considered as good as the results of DBSCAN and sometimes the result obtained by the proposed model is better than the DBSCAN results. The presented algorithm solves the conflicts faced by the DBSCAN in case of varying density. This paper also presents a way to reduce high dimensional data to low dimensional data with automatic association analysis. This algorithm can reduce high dimensional data to even a single dimension. Using this algorithm the challenges faced in multidimensional clustering by different algorithms such as DBSCSN is solved. This dimensionality reduction with automatic association algorithm is then applied to the Watershed model to detect cluster in Homicide Data and finding out murder prone zones and suggest a person with murder avoiding areas.

Keywords: Cluster detection · Watershed model · Association analysis · Murder prone area detection · Homicide data · Data reduction

1 Introduction

Clustering data has been a challenge in data mining field for a long time. Many problems were faced during the establishment of clustering algorithm. Popular algorithms like *k*-means, DBSCAN, and agglomerative hierarchical clustering are used mostly for cluster detection with some modifications [1]. Major researches on efficient initialization of *k*-means were also available in [2]. But algorithms which are used frequently in image processing have hardly been used for data mining. In this paper, we present a modified Watershed model for cluster detection. Watershed model is quite popular and efficiently used in image processing. Watershed model was modified for region detection in many researches [3]. Our research also presents a modified Watershed model for region detection. The clusters are determined by the set of points bounded by that region. This paper also presents a dimensionality reduction algorithm. The given algorithm tries to solve the curse of dimensionality [4]. The provided dimensional reduction algorithm can

N.T. Nguyen et al. (Eds.): ICCCI 2017, Part I, LNAI 10448, pp. 307–316, 2017.
DOI: 10.1007/978-3-319-67074-4_30

reduce high dimensional data to even a single dimension. Though the efficiency of reduction is an ongoing research, the advantage of this dimensional reduction algorithm is that it provides an automatic association analysis. Association rules are used for suggestion algorithms [5]. The dimensional reduction algorithm with automatic association algorithm is applied to Homicide data obtained from FBI available database [14] to reduce a twenty-four-dimensional data to two-dimensional data. Then the modified Watershed model for cluster detection is applied on obtained two-dimensional data. Using the available data murder prone areas are detected and murder avoiding areas are suggested using association rule analysis.

Comparing efficiency of algorithms is contradictory as it depends largely on implementations [6]. However, comparing data groupings using clustering and its efficiency of clustering can be measured in many ways [7]. Silhouette Coefficient, Cohesion and Separation based evaluation, matrix based evaluation etc. are used for data evaluation [8, 9]. In our paper, we also present a comparison between *k*-means, DBSCAN and our modified watershed algorithm. Efficiency of the algorithms is also reported.

The paper is organized as follows. In Sect. 2, we have described the related works. Section 3 describes the modified Watershed algorithm. Dimension reduction with automatic association analysis algorithm is described in Sect. 4. Section 5 compares the presented clustering algorithm with *k*-means and DBSCAN and shows the efficiency of the algorithm. Algorithms are then applied to the Homicide data and the murder prone areas and murder avoiding areas are suggested followed by the conclusion and future scopes of this research in Sect. 6.

2 Related Work

Despite more than 50 years old, the *k*-means algorithm is still a popular algorithm used for clustering [10]. The authors summarize the conventional methods of clustering, the difficulties of developing new clustering techniques, and discuss the new developments and research directions in cluster analysis. The limitation of random initial centroid selection in *k*-means algorithm is discussed in another paper [2]. It is suggested that there are better alternatives to these methods that would perform more efficiently. Watersheds in edge-weighted graphs are discussed in one of the researches, and the efficiency of the proposed methods has been proved in terms of minimum spanning forests [3]. Two new linear-time algorithms were derived, which were much more efficient and outperformed all other currently existing algorithms. The results were compared with two other methods for image segmentation, and proved to be a significant improvement over the existing methods based on watershed.

In one of the papers, an overview of the R package arules is provided and its features are discussed [5]. Before arules, there was no feature available in R for clustering and mining associations. The arules package additionally provides interfaces to C implementations of two of the popular data mining algorithms, While searching for studies involving homicides, we came across a research covering the homicides committed in the UK over a year, the types of offences, and the statistics of specific offences [11]. It also included patterns of change in the offences, possible links, the sentences condemned and the convictions, along with comparisons with international cases.

3 Modified Watershed Model Algorithm for Clustering

Watershed model is mainly used for image processing for region detection. We have modified the given Watershed model for data clustering. In Watershed model the key challenge is to find the most distinct points. This requirement is fulfilled by many techniques. Some renowned researches on this model apply distance transform to achieve the expected result. Another way to achieve the result is through special filtering. We have fulfilled this challenge by applying multivariate Gaussian filter on the obtained image plot from the cluster dataset. A brief introduction on Watershed model is given followed by our modified algorithm.

3.1 Watershed Region Detection Algorithm

The key idea of Watershed model is to find the distinct highest point. Then the regions are plotted as reverse hat. The regions are gradually flooded with points having highest values and a dam is constructed if regions of any water are in conflict with the other. The regions bounded by the dam are the selective distinct regions. There are many ways to flood the regions. One of the famous algorithms is Meyer's flooding algorithm [12]. We have used similar algorithm for flooding the region. But the key challenge is to obtain the distinct points from where the flooding will begin.

3.2 Methodology

Before approaching to watershed model we have to modify the cluster of data to be compatible with watershed model. The next part describes the algorithm step by step.

1. Take the minimum points acceptable in a cluster.
2. Plot the points in a two-dimensional image and modify the points suitable for the Watershed model.
3. Apply Watershed model to find out the regions.
4. Mark the points associated in that region and remove it.
5. Repeat Step 2–4 until every point is selected.

The novelty of this algorithm is modification of the two-dimensional data for watershed model. The vital challenge is in step 2. The first two steps of the algoritm are discussed below.

3.3 Modifying Data for Watershed Model

After plotting we can see that the obtained points are situated apart from each other. Therefore, it is not possible to apply watershed model to the given plots. For that we have to blur the image applying Gaussian filter so that there remains no space between points. After applying filter, the distance between highest points are calculated to determine the size of the highest filter. The algorithm for data modification is given below.

1. *Take the minimum points acceptable in a cluster.*
2. *Plot the two-dimensional data in an image.*
3. *Find out the maximum distance between the points.*
4. *Apply three Gaussian filters to the image. The 1st filter size equals to the minimum acceptable points, 2nd filter size equals to the average of maximum distance and minimum filter size, 3rd filter size equals to the maximum distance.*
5. *After highest points are achieved, if there is connectivity between highest distinct points, the value of any one point will be reduced to half.*
6. *Step 5 will be repeated until no connectivity is remaining between distinct points.*

After that, step 3–5 of the methodology subsection (Sect. 3.2) will be applied to the algorithm. Flow diagram of the algorithm is shown in Fig. 1.

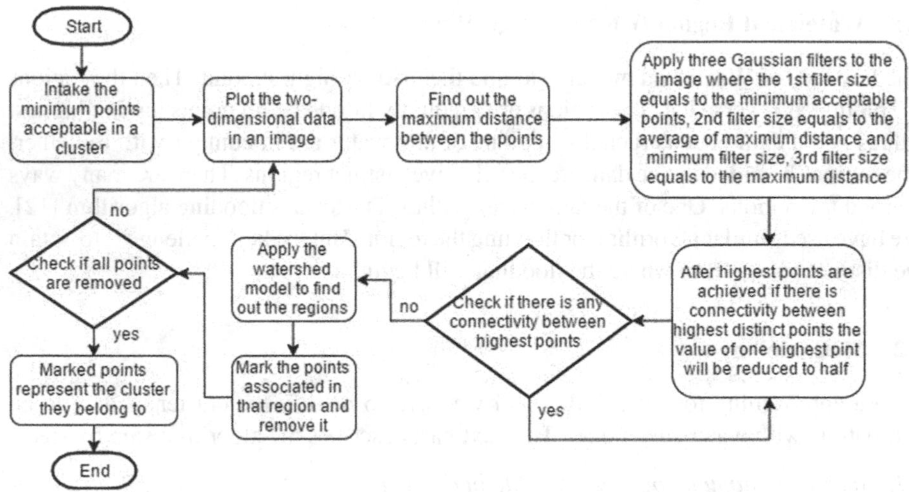

Fig. 1. Flow diagram of modified watershed algorithm for cluster analysis.

4 Dimensionality Reduction

Dimension reduction has been a great challenge in every field. Principal component analysis (PCA) is a popular technique that works like feature extraction for dimensionality reduction [13]. Rather than feature extraction or feature selection we have focused on dimension representation and plotting it in a single variable. In case of data mining most of the data we obtain can be divided to small number of classes. For example, sex could be either male or female, that is we can represent this data into binary 0 or 1. Again let us consider another data for month where there are twelve distinct data. Representing these data would require 64 bits to 128 bits. But we can simply signify the data using 4 bits. 4 bits are able to represent the distinct 16 data. In this paper, we present a homicide data. Homicide data contains 15322920 data where there are 24 distinct attributes containing 638455 rows. Computing any algorithm on such large dataset is very expensive. But using our algorithm we have reduced the data into two-dimensional data. For clarity let us take five attributes into considerations which are City, Crime type, Victim, Month, Crime solved. There are 231 distinct cities which can be represented by 8 bits. Two Crime Types namely 'Murder' and 'Manslaughter by negligence'. These attributes can be represented by single bit. Victim has two types, male and female, which can also be represented by a single bit. Month can be represented by 4 bits and crime solved 'yes' or 'no' can be represented by single bit (Table 1).

Table 1. Table for dimensionality reduction

Classes	City	Crime type	Victim	Month	Crime solved
Bit required for representation	8	1	1	4	1
Example	Laos	Murder	Male	June	Yes
Representation	00000001	0	1	0110	1

Using 15 bits we are able to represent five data. The above binary 000000010101101 represents 173. This 173 data will be associated with the id representation making the data into two dimensional data. The *k*-means, DBSCAN algorithms are not strong on solving multidimensional data. This algorithm can be used to solve this problem. But it requires Meta data for representation. To find out the metadata the algorithm has to be upgraded. The algorithm is discussed below.

1. Divide the given attributes according to two suitable groups where focused attributes for priority will obtain the most significant bits and attributes with less priority will gain least significant bits.
2. Find the unique value of attributes for each group.
3. Find the bits required representing this value using the following formula where b represent the number of bits required and x represent unique values.

$$b = \lceil log_2 x \rceil$$

4. Record the data representing the bits and store them in a metadata.

The classes are our focuses; so we keep them in the most significant bits. For example, we want to plot the cities according to murder prone areas. Then the city will have the most significant bits and we do not care very much about whether the crimes are solved or not. It will be in the least significant bits. The change in MSB will significantly change the value. 100001 and 100000 will have no difference in plotting in 2D value but 000001 and 100001 has a very significant difference.

5 Application of Algorithm in Homicide Data

The Homicide data obtained from FBI database contained 24 attributes. Those are Record Id, Agency Code, Agency Name, City, State, Year, Month, Number of incident, Crime Type, Crime Solved, Victim Sex, Victim Age, Victim Race, Victim Ethnicity, Perpetrator Sex, Perpetrator Age, Perpetrator Race, Perpetrator Ethnicity, Relationship, Victim Count, Perpetrator Count, Record Source, City Latitude and City Longitude. Out of these twenty-four attributes Record ID, Agency Code, Agency name, Record Source can be ignored. As the record ID is gradually increasing from 1 so it can be obtained any time and Record source is not required. That leaves us twenty different attributes. We can divide these twenty attributes into equal number of bits. Again, we will have same latitude and longitude for the same City so City data can also be ignored. As city is inside a state, state could also be ignored which leaves us 18 attributes. Now out of these 18 attributes we have prioritized the attributes as follows. In our 1st category we have 28 bits and 2nd category contains 39 bits. Our priority would be city latitude and city longitude if we intent to represent in geo-location. From the above data we have plotted the value in a normalized 800 by 400 pixel image in Fig. 2 (Tables 2, 3 and 4).

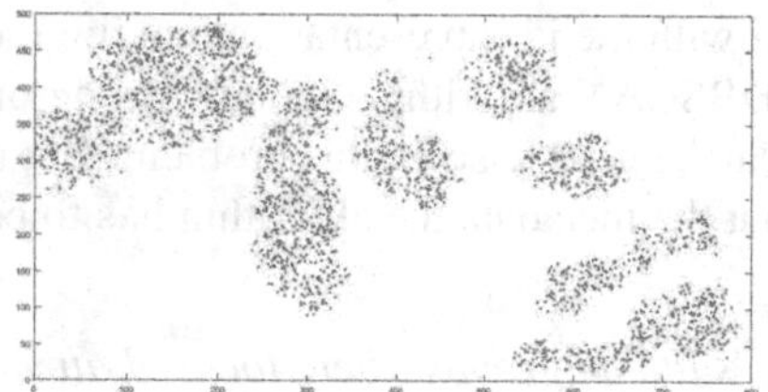

Fig. 2. Homicide detection prioritizing location vs murder rate

Fig. 3. Murder rate data image after applying Gaussian filter

Table 2. Table of category 1 atrributes

Attributes	Latitude	Longitude	Victim sex	Victim age	Victim race	Victim ethnicity
Unique values	231	231	2	99	4	3
Bit required	8	8	1	7	2	2

Table 3. Table of category 2 attributes

Attributes	Number of incident	Year	Month	Crime type	Crime solved	Perpetrator sex	Perpetrator count
Unique values	56	56	12	2	2	2	4
Bit required	6	6	4	1	1	1	2

Table 4. Table of category 2 attributes

Attributes	Perpetrator age	Perpetrator race	Perpetrator ethnicity	Relationship	Victim count
Unique values	89	4	3	28	4
Bit required	7	2	2	5	2

In order to represent the data we have to build our custom variable of 28 bits and 39 bits respectively. After that we have applied our modified watershed algorithm and got the following results. Figure 3 captures the 1st iteration where the Gaussian filter of size 8 is applied to get an image result. After the completion of Gaussian image we see individual seven regions are discovered using modified Watershed which are designated by distinct white regions in Fig. 4.

Fig. 4. Regions in image after applying modified watershed model.

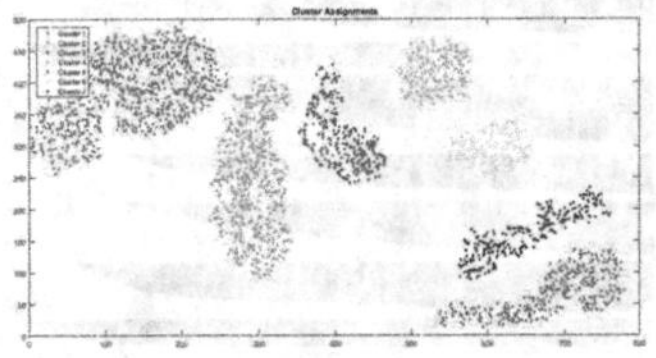

Fig. 5. Result from modified watershed model

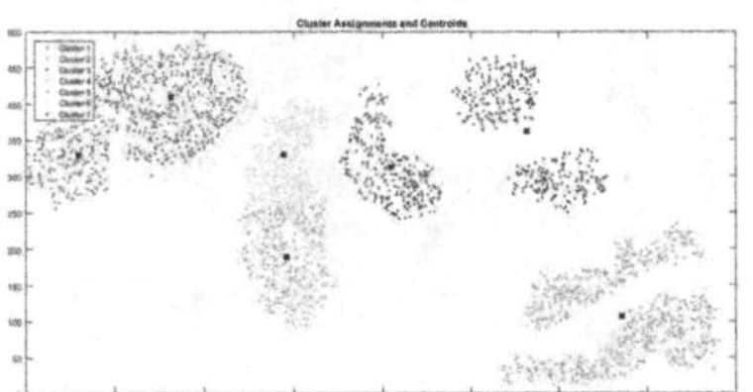

Fig. 6. Cluster analysis using K-means on homicidal data

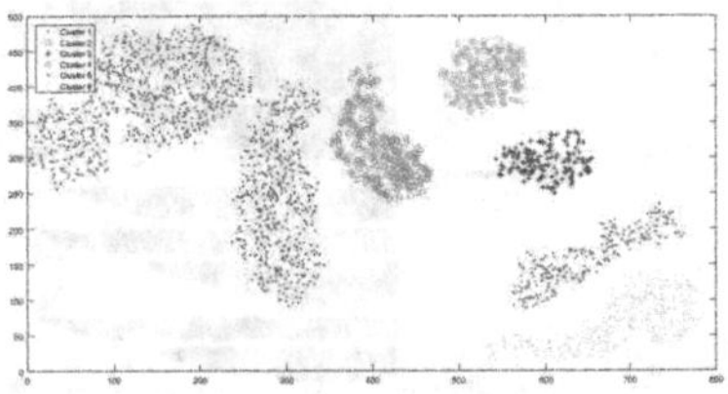

Fig. 7. Cluster analysis of homicide rate analysis using DBSCAN.

After finding out the regions we plot the points individually in the respective areas to get their original points. After applying original points we get the final result of our Watershed model to find out murder prone areas in United States. Our modified watershed algorithm produced the results that is depicted in Fig. 5. The algorithm is able to successfully detect seven clusters. Though the algorithm is efficient but representing an image containing around 700000 data is not possible. So we have to normalize the result in an image of 800 by 600 points. In x axis, the location is plotted and in y axis murder count is plotted in Fig. 2. On the same data if we apply *k*-means algorithm to find out seven clusters we get the result as described in Fig. 6.

As we can see that *k*-means is not able to successfully identify the seven individual clusters for the homicidal data. It is not possible to find pattern from the received clusters through *k*-means. Again, we have applied DBSCAN algorithm on the same amount of data set. Though DBSCAN could successfully identify clusters, it had slight problems differentiating two clusters. The epsilon value for DBSCAN for this individual experiment is 20 and minimum data inside this area is 5. Due to close proximity of points it is unable to determine the distinct seven clusters in Fig. 7. We have also measured the respective silhouette values of *k*-means, DBSCAN and our Watershed model focusing on Euclidean distance and plotted them in graphs. Figure 8 shows us the silhouette plot of *k*-means, Figs. 9 and 10 show us the silhouette plot of DBSCAN and our modified Watershed model respectively. Here the positive value signifies the measure of acceptance it belongs to its own cluster and negative value shows us the measure of rejection from its own cluster.

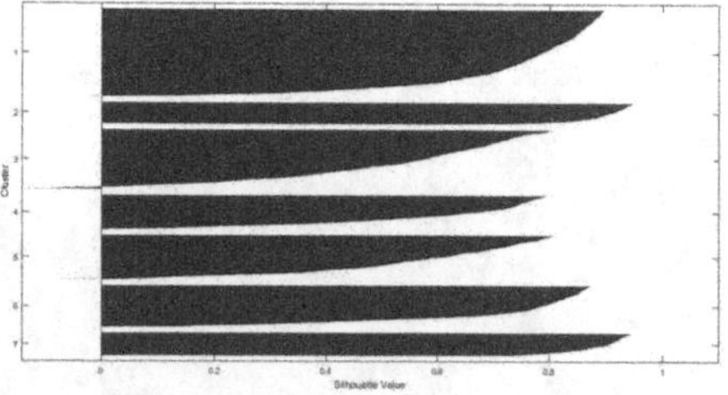

Fig. 8. Silhouette plot of *k*-means algorithm

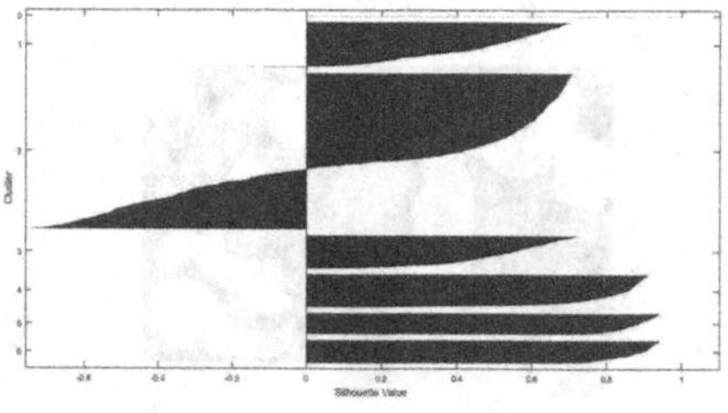

Fig. 9. Silhouette plot of DBSCAN

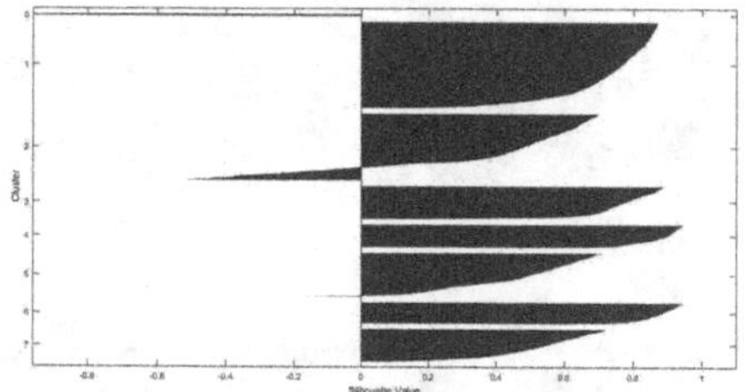

Fig. 10. Silhouette plot of modified Watershed model

Fig. 11. Plotting modified watershed model clusters against the United States map

We have also calculated the total amount of negative values where *k*-means have a total amount of negative value of only 1.943 and DBSCAN has total negative value of 193.2585 and Watershed model has a total negative value of 41.058. As we can see that Watershed model has less negative values than DBSCAN which represents that the Watershed model is better than DBSCAN in terms of cluster representation. Though *k*-means has the least Silhouette value but it is not acceptable as we can see that *k*-means failed to successfully distinguish the clusters in Fig. 6.

From the modified Watershed model, we have the position prioritizing along X axis where first 8 bits signify latitude and next 8 bits signify longitude. To better visualize the result, we have created a matrix having 3 dimensions: latitude, longitude and the cluster in which it belongs. Then we plot them against the United States map. The result is displayed in Fig. 11. We can see some derbies respect to the position of the states which is created due to lower significant bits. Here in order to find out patterns among the data in the map we can consider that states like Texas, Oklahoma, Mississippi, California, Alaska, Hawaii, Florida, Alabama belong to the same cluster which also signifies the most murder prone areas. On the other hand New York, Maine, New Hemisphere, Rhode Island, North Caroline, Kentucky belongs to the same cluster which determines the least homicidal areas. Washington individually belongs to a single cluster which is the second most murder prone zone. Montana, Minnesota, North Dakota, Wisconsin belongs to same cluster which is the second least murder prone areas.

6 Conclusion and Future Work

Though the clustering largely depends on the dataset but the efficiency of the provided model is encouraging. It performs better in cluster detection and outranks *k*-means easily but whether it is better than DBSCAN it is still in question. In case of space complexity, the Watershed model is better than DBSCAN. Though it highly depends on implementation yet through our research we believe that our model outranks in performance both *k*-means and DBSCAN in terms of time and space complexity. The result in terms of accuracy received by Watershed model outranks *k*-means in all respect. Sometimes it performs better than DBSCAN in special circumstances where density of clusters is a challenge for DBSCAN.

References

1. Drake, J., Hamerly, G.: Accelerated k-means with adaptive distance bounds. In: 5th NIPS Workshop on Optimization for Machine Learning, pp. 42–53 (2012)
2. Celebi, M.E., Kingravi, H.A., Vela, P.A.: A comparative study of efficient initialization methods for the k-means clustering algorithm. Expert Syst. Appl. **40**(1), 200–210 (2013)
3. Cousty, J., Bertrand, G., Najman, L., Couprie, M.: Watershed cuts: minimum spanning forests and the drop of water principle. IEEE Trans. Pattern Anal. Mach. Intell. **31**(8), 1362–1374 (2009)
4. Bellman, R.E.: Perturbation Techniques in Mathematics, Engineering and Physics. Courier Corporation, North Chelmsford (2003)
5. Hahsler, M., Grün, B., Hornik, K.: A computational environment for mining association rules and frequent item sets (2005)
6. Kriegel, H.P., Schubert, E., Zimek, A.: The (black) art of runtime evaluation: are we comparing algorithms or implementations? Knowl. Inf. Syst. **52**(2), 341–378 (2016)
7. MacQueen, J.: Some methods for classification and analysis of multivariate observations. In: Proceedings of the Fifth Berkeley Symposium on Mathematical Statistics and Probability, vol. 1, no. 14, pp. 281–297, June 1967
8. Rousseeuw, P.J.: Silhouettes: a graphical aid to the interpretation and validation of cluster analysis. J. Comput. Appl. Math. **20**, 53–65 (1987)
9. Stevens, W.P., Myers, G.J., Constantine, L.L.: Structured design. IBM Syst. J. **13**(2), 115–139 (1974)

10. Jain, A.K.: Data clustering: 50 years beyond K-means. Pattern Recognit. Lett. **31**(8), 651–666 (2010)
11. Richards, P.: Homicide Statistics, Research Paper 99/56, Social And General Statistics Section, House Of Commons Library, 27 May 1999
12. Barnes, R., Lehman, C., Mulla, D.: Priority-flood: an optimal depression-filling and watershed-labeling algorithm for digital elevation models. Comput. Geosci. **62**, 117–127 (2014)
13. Pearson, K.: LIII. On lines and planes of closest fit to systems of points in space. Lond. Edinb. Dublin Philos. Mag. J. Sci. **2**(11), 559–572 (1901)
14. Homicide Data. https://www.kaggle.com/murderaccountability/homicide-reports. Accessed 9 June 2017

Comparison of Ensemble Learning Models with Expert Algorithms Designed for a Property Valuation System

Bogdan Trawiński[1(✉)], Tadeusz Lasota[2], Olgierd Kempa[2], Zbigniew Telec[1], and Marcin Kutrzyński[1]

[1] Faculty of Computer Science and Management, Wrocław University of Science and Technology, Wrocław, Poland
{bogdan.trawinski,zbigniew.telec}@pwr.edu.pl

[2] Department of Spatial Management, Wrocław University of Environmental and Life Sciences, Wrocław, Poland
{tadeusz.lasota,olgierd.kempa}@up.wroc.pl

Abstract. Three expert algorithms based on the sales comparison approach worked out for an automated system to aid in real estate appraisal are presented in the paper. Ensemble machine learning models and expert algorithms for real estate appraisal were compared empirically in terms of their accuracy. The evaluation experiments were conducted using real-world data acquired from a cadastral system maintained in a big city in Poland. The characteristics of applied techniques for real estate appraisal are discussed.

Keywords: Machine learning · Ensemble models · Sales comparison approach · Expert algorithms · Property valuation · Mass appraisal

1 Introduction

Numerous methods and models are proposed for automated computer systems supporting single property valuation as well as mass appraisal of groups of properties. They range from multiple regression analysis [1] to intelligent techniques including neural networks [2], decision trees [3], fuzzy systems [4], and hybrid approaches [5, 6]. Most real estate appraisal models are based on the sales comparison approach. They are applied in computer systems called Automated Valuation Models (AVM) and Computer Assisted Mass Appraisal (CAMA) [7–9].

Various machine learning methods such as ensemble techniques [10–12], hybrid approaches including evolutionary fuzzy systems [13] and evolving fuzzy systems [14] were examined by the authors and utilized to generate data driven models. Advanced methods for feature selection are also very important [15, 16]. We devised several techniques for creating regression models for property valuation based on ensembles of genetic fuzzy systems [17, 18] and evolving fuzzy systems [19, 20]. We employed ensembles of genetic fuzzy systems and neural networks to predict from a data stream of real estate sales transactions [21, 22]. We proposed also methods for merging different urban areas into uniform zones to obtain homogenous areas embracing greater

N.T. Nguyen et al. (Eds.): ICCCI 2017, Part I, LNAI 10448, pp. 317–327, 2017.
DOI: 10.1007/978-3-319-67074-4_31

number of records which allow for building more reliable models for property valuation [23, 24].

Comparative analysis of three expert algorithms and six machine learning models developed for an automated system to aid in real estate appraisal are presented in this paper. All expert algorithms are based on the sales comparison approach and implement different mechanisms of selecting similar properties to estimate the value of a given property. The performance of the expert algorithms was experimentally examined employing real-world data derived from a cadastral system and registry of real estate transactions. Moreover, the accuracy of the algorithms was compared with six machine learning models including single models of decision trees and neural networks as well as their bagging and boosting ensembles which were created in the WEKA data mining system [25]. Statistical analysis of the experimental results was carried out using parametric and nonparametric tests. The details of this study were presented in three engineer's theses [26–28] and preliminary experimental results in [29].

2 Expert Algorithms for Real Estate Appraisal

Three expert algorithms for residential real estate appraisal implementing the sales comparison approach are presented in the paper. To estimate the price of a particular apartment, all algorithms calculate an average price of a number of similar apartments to that real estate. One of the first steps of each algorithm is determining similar properties to a given apartment. The experts selected four main attributes of apartments and partitioned their values by ranges. The ranges for each attribute defined the classes of similarity. The list of similarity classes settled for residential real estate of a given Polish city is placed in Table 1. The list was elaborated based on the dataset of sales transactions available. Four main attributes were employed: usable area of an apartment

Table 1. Similarity classes of apartments within a city determined by the experts

Feature	Class	Values
Area	1-small	Under 40 m^2
	2-medium	40–60 m^2
	3-large	Over 60 m^2
Year	1-very old	Before 1918
	2-old	1919–1945
	3-medium	1946–1975
	4-new	1976–1995
	5-newest	After 1995
Storeys	1-low-rise	1–2
	2-mid-rise	3–5
	3-high-rise	More than 5
Rooms	1-small	1–2
	2-medium	3–4
	3-large	More than 4

(*Area*), year of a building construction (*Year*), number of storeys in a building (*Storeys*), number of rooms in an apartment including a kitchen (*Rooms*). Two apartments are considered similar if the values of all their main attributes belong to the same similarity classes.

The first expert algorithm is named *NST: N-Nearest Similar (apartments) Transactions.* It computes the estimated price as an average price of N-nearest apartments in terms of the Euclidean distance belonging to the same similarity classes as the valuated one does. The pseudocode of the *NST* algorithm is shown in Table 2. The second expert algorithm is called *LTA: N-Latest Transactions in an Area.* It estimates the predicted price as an average price of the N-latest transactions of similar apartments within the same area, i.e. cadastral region or expert zone. The pseudocode of the *LTA* algorithm is presented in Table 3. The third expert algorithm was named

Table 2. Expert algorithm *NST*: *N-Nearest Similar (apartments) Transactions*

Step	Action
1	Take the next apartment X_i to price
2	Determine the number *N* of apartments needed to set the price of apartment X_i
3	Determine all apartments similar to X_i falling to the same similarity classes
4	If the number of apartments similar to X_i is less than *N*, then omit X_i and go to *Step 1*
6	Select the *N* nearest apartments from among all similar ones to X_i
7	Calculate the expected price of X_i as the average price per square metre of *N* apartments found

Table 3. Expert algorithm *LTA*: *N-Latest Transactions in an Area*

Step	Action
1	Calculate the average price per square metre for each region/zone Z_j: $AvgP_j$ *for* $j=1,2,..,NZ$
2	Take the next apartment X_i to price
3	Determine the number *N* of apartments needed to set the price of the apartment X_i
4	Determine all apartments similar to X_i falling to the same similarity classes
5	If the number of apartments similar to X_i is less than *N*, then omit *Xi* and go to *Step 2*
6	Determine the region/zone Z_i including the apartment to be priced
7	If the number of apartments similar to X_i is greater or equal to *N*, then select the *N* latest transactions and go to *Step 11*
8	Repeat Steps 9-10 until *N* transactions are found
9	Find the next region/zone Z_k where the deviation $\lvert avgP_k - avgP_i \rvert$ is the smallest
10	Select the *m* latest transactions from Z_k to complement the number *N*
11	Calculate the expected price of X_i as the average price per square metre of *N* transactions found

RST: N-Random Similar (apartments) Transactions. It appraises the cost of an apartment by averaging the prices of *N* apartments drawn randomly from among all similar ones. The pseudocode of the *RST* algorithm is presented in Table 4.

Table 4. Expert algorithm *RST*: *N-Random Similar (apartments) Transactions*

Step	Action
1	Take the next apartment X_i to price
2	Determine the number *N* of apartments needed to set the price of apartment X_i
3	Determine all apartments similar to X_i falling to the same similarity classes
4	If the number of apartments similar to X_i is less than *N*, then omit X_i and go to *Step 1*
5	Draw randomly *N* apartments from among all similar ones to X_i
6	Calculate the expected price of X_i as the average price per square metre of *N* apartments drawn

Two types of market areas were analysed in the paper, namely cadastral regions of a city and quality zones delineated by an expert. Cadastral regions constitute an administrative partition of the city and are outlined by the boundaries of housing estates, parks, and forests as well as by routes of main streets, rivers, and railway lines, etc. The regions could be considered as market areas with uniform levels of real estate prices. The second partition was devised by an professional appraiser who based on his expert's knowledge and long experience proposed 21 zones characterized by similar behaviour of real estate prices. The partition of a city area into 69 cadastral regions and 21 expert zones is depicted in Fig. 1.

Fig. 1. Partition of a city area into 69 cadastral regions (left) and 21 expert zones (right)

3 Setup of Evaluation Experiment

The main goal of an evaluation experiment was to compare the performance of expert algorithms with machine learning models built over the cadastral regions and expert zones of a Polish city. Moreover, the parameter *N*, i.e. the number of apartments taken to calculate the price of an appraised apartment, was examined.

Six machine learning models were built for each cadastral region and expert zone using data mining system WEKA [25]. They include a decision tree and neural network as a single model as well as parallel form of an ensemble such as bagging and sequential form of an ensemble such as additive regression, which is a sort of a boosting learning technique. Bagging enhances the performance by reducing model variance. It averages the results provided by a number of models built over bootstrap replicates, called bags. In turn, additive regression enhances the performance of a single regression base learner (regressor) using sequential learning of regressors. Each iteration fits a model to the residuals left by the regressor on the previous iteration. Prediction is achieved by adding the predictions of each regressor. Thus, the following models were used:

M5P-S – single model of the *Pruned Model Tree* which employs procedures for constructing M5 model trees. The algorithm applies decision trees, however, in place of values at tree nodes, it retains multivariate linear regression models there.

M5P-B – bagging ensemble with the pruned model tree (*M5P*) as the base learner.

M5P-A – additive regression with the pruned model tree (*M5P*) as the base learner.

MLP-S – single model of the *Multilayer Perceptron* which is a feed-forward neural network with the backpropagation learning algorithm. For regression problems, i.e. when classes are numeric, the output nodes turn into unthresholded linear units.

MLP-B – bagging ensemble with multilayer perceptron (MLP) as the base learner.

M5P-A – additive regression with the pruned model tree (*M5P*) as the base learner.

Each machine learning model was built using four input features pointed out by the experts as the main price drivers of an apartment. They were available in the dataset applied to evaluate the *NST, LTA*, and *RST* algorithms. There were following features: usable area of an apartment (*Area*), year of a building construction (*Year*), number of storeys in a building (*Storeys*), number of rooms in an apartment (*Rooms*). In turn, price per square metre (*Price*) was taken as the output variable.

The experiment was conducted employing real-world data about sales transactions derived from a cadastral system and a public registry of real estate transactions. The dataset contained 12,439 records of sales transactions of apartments completed within 16 years from 1998 to 2013. In order to compensate the changes of real estate prices in the course of time, all the prices were updated for the last day of 2013 using trend functions. The dataset was split randomly into 70% training set and 30% test set. The *NST, LTA,* and *RST* expert algorithms were examined in the following way. The prices of 100 apartments randomly drawn from the test set were estimated using individual algorithms. Then, the accuracy measure *MAE* was computed for each run according to

Formula 1, where P_i^a and P_i^p stand for the actual and predicted prices respectively and n denotes the number of apartments appraised in each run.

$$MAE = \frac{1}{n} * \sum_{i=1}^{n} |P_i^p - P_i^a| \quad (1)$$

This procedure was repeated 150 times providing 150 values of *MAE* for each algorithm. In this way the number of points of observation allowed for carrying out tests of statistical significance.

The following procedure for evaluating machine learning models was applied to produce comparable results with the outcome of the expert algorithms. The same split into 70% training set and 30% test set was employed. Then, the training set was partitioned into 69 subsets containing the transactions accomplished within individual cadastral regions and into 21 subsets corresponding individual expert zones. Machine learning models were generated only for those cadastral regions and expert zones which embraced at least 50 transactions. The regions and zones with smaller number of training samples were discarded because they did not allowed for generating reliable models. The main steps of the procedure to obtain expected prices of apartments using machine learning models are shown in Table 5. Again, the same 150 sets of 100 apartments, randomly drawn from the test dataset, were applied to machine learning models to predict the prices of apartments. Thus, 150 values of *MAE* were obtained for each model.

Table 5. Procedure to obtain expected prices of apartments using machine learning models

Step	Action
1	Take the next apartment X_i to price
2	Determine the region/zone Z_i including the apartment to be priced
3	If machine learning models were not generated for the region/zone Z_i, then omit X_i and go to *Step 1*
4	Take the output of individual models generated for the region/zone Z_i as the expected prices of X_i

4 Analysis of Experimental Results

The first series of experiments consisted in the examination of expert algorithms accuracy depending on the parameter *N*, i.e. the number of similar apartments taken to estimate the price of the apartment being evaluated. Each algorithm was tested for N = 11, 20, 30, 40, 50, and 100. Due to the fact that not all observations had normal distribution, based on the Shapiro-Wilk test, the nonparametric approach was applied. Average rank positions of *MAE* produced by the *NST, LTA,* and *RST* algorithms for different values of *N* determined by the Friedman test are shown in Tables 6, 7, and 8 respectively, where the lower rank value the better version of an algorithm. For comparative analysis with machine learning models *N11-NST*, *N30-LTA*, and *N40-RST*

Table 6. Rank positions of *NST* accuracy in terms of *MAE* produced by the Friedman test

	1st	2nd	3rd	4th	5th	6th
Cad. region	N11-NST	N20-NST	N30-NST	N40-NST	N50-NST	N100-NST
Avg. rank	1.71	2.14	3.49	4.12	4.41	5.13
Exp. zone	N11-NST	N20-NST	N30-NST	N40-NST	N50-NST	N100-NST
Avg. rank	1.71	2.14	3.49	4.12	4.41	5.13

Table 7. Rank positions of *LTA* accuracy in terms of *MAE* produced by the Friedman test

	1st	2nd	3rd	4th	5th	6th
Cad. region	N30-LTA	N40-LTA	N50-LTA	N20-LTA	N11-LTA	N100-LTA
Avg. rank	2.80	2.93	3.31	3.39	4.19	4.37
Exp. zone	N30-LTA	N50-LTA	N40-LTA	N20-LTA	N100-LTA	N11-LTA
Avg. rank	3.08	3.18	3.30	3.39	3.89	4.16

Table 8. Rank positions of *RST* accuracy in terms of *MAE* produced by the Friedman test

	1st	2nd	3rd	4th	5th	6th
Cad. region	N50-RST	N40-RST	N100-RST	N30-RST	N20-RST	N11-RST
Avg. rank	2.88	2.98	3.05	3.23	3.80	5.06
Exp. zone	N100-RST	N40-RST	N50-RST	N30-RST	N20-RST	N11-RST
Avg. rank	2.97	2.98	3.06	3.40	3.79	4.81

algorithms were chosen. The first two algorithms were ranked in the first position and the third one was in second place for both cadastral regions and expert zones. However, there were no statistically significant differences among the three versions *N40-RST*, *N50-RST*, and *N100-RST* according to the Wilcoxon test.

The mean values of *MAE* in terms of Polish currency *PLN* provided by six machine learning models and three selected expert algorithms are presented in Fig. 2. The values of *MAE* ranged from 367 to 834 *PLN* for cadastral regions and from 391 to 841 *PLN* for expert zones. Bagging ensembles for both *M5P* decision trees and *MLP* neural networks revealed significantly better accuracy than single models and boosting ensembles. For the best algorithm *M5P-B* the mean *MAE* values were equal to 498 and 512 *PLN* for cadastral regions and expert zones respectively. In turn, *N11-NST* revealed the best performance from among the expert algorithms.

The Friedman test conducted for 150 points of observation revealed significant differences between some algorithms and models. Average rank positions of model accuracy in terms of *MAE* produced by the Friedman test are presented in Table 9, where the lower rank value the better model. The Shapiro-Wilk tests disclosed that all results for both cadastral regions and expert zones were normally distributed. Therefore, parametric paired sample t-tests could be applied to comparative analysis of the model and algorithm performance. The parametric t-test is more powerful than the nonparametric Wilcoxon test. The zero hypothesis assumed that there were not significant differences in accuracy, in terms of *MAE*, between individual pairs of results.

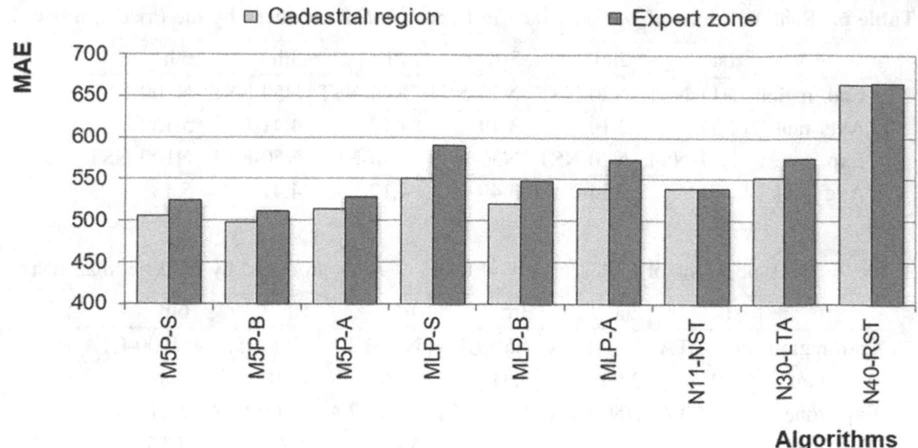

Fig. 2. Mean of MAE provided by machine learning models and expert algorithms

Table 9. Rank positions of all models and algorithms produced by the Friedman test

	1st	2nd	3rd	4th	5th	6th	7th	8th	9th
Cad. region	M5P-B	M5P-S	M5P-A	MLP-B	N11-NST	MLP-A	N30-LTA	MLP-S	N40-RST
Avg. rank	1.75	2.69	3.47	4.02	5.43	5.59	6.27	6.85	8.95
Exp. zone	M5P-B	M5P-S	M5P-A	N11-NST	MLP-B	MLP-A	N30-LTA	MLP-S	N40-RST
Avg. rank	1.71	2.85	3.16	3.89	4.49	6.17	6.27	7.53	8.93

Table 10. T-test results for pairwise comparison of models and algorithms for cadastral regions

	M5P-S	M5P-B	M5P-A	MLP-S	MLP-B	MLP-A	N11-NST	N30-LTA	N40-RST
M5P-S		−	+	+	+	+	+	+	+
M5P-B	+		+	+	+	+	+	+	+
M5P-A	−	−		+	+	+	+	+	+
MLP-S	−	−	−		−	−	−	≈	+
MLP-B	−	−	−	+		+	+	+	+
MLP-A	−	−	−	+	−		≈	+	+
N11-NST	−	−	−	+	−	≈		+	+
N30-LTA	−	−	−	≈	−	−	−		+
N40-RST	−	−	−	−	−	−	−	−	

The results of t-tests for cadastral regions and expert zones are shown in Tables 10 and 11 respectively. In Tables 10 and 11 + denotes that the model in the row outperformed significantly the one in the corresponding column, − indicates that the model in the row provided significantly worse accuracy than the one in the corresponding column. In turn, ≈ means that both results were statistically equivalent. The significance level for the null hypothesis rejection was set to 0.05.

Table 11. T-test results for pairwise comparison of models and algorithms for expert zones

	M5P-S	M5P-B	M5P-A	MLP-S	MLP-B	MLP-A	N11-NST	N30-LTA	N40-RST
M5P-S		−	+	+	+	+	+	+	+
M5P-B	+		+	+	+	+	+	+	+
M5P-A	−	−		+	+	+	+	+	+
MLP-S	−	−	−		−	−	−	−	+
MLP-B	−	−	−	+		+	−	+	+
MLP-A	−	−	−	+	−		−	≈	+
N11-NST	−	−	−	+	+	+		+	+
N30-LTA	−	−	−	+	−	≈	−		+
N40-RST	−	−	−	−	−	−	−	−	

5 Conclusions and Future Work

Three expert algorithms based on the sales comparison approach worked out for an automated system to aid in real estate appraisal were presented in the paper. The first algorithm, *NST*, processed a predefined number of the nearest similar apartments to a given apartment to estimate its price. The second one, *LTA*, took a predefined number of the latest transactions of similar apartments within the same market area in which an appraised apartment was located. The third algorithm, *RST*, accomplished estimations by averaging the prices of apartments drawn randomly from among similar ones.

The impact of the parameter *N* on the accuracy of algorithms was experimentally examined. Three versions of these algorithms, which revealed the best performance, namely *N11-NST*, *N30-LTA*, and *N40-RST* were selected for comparison with six machine learning models including single models of a decision trees *M5P* and neural networks *MLP* and their bagging and boosting ensembles.

The experiments were conducted using real-world data of sales transactions of residential premises derived from a cadastral system and a public registry of real estate transactions. Machine learning procedures were derived from the WEKA data mining system. Mean absolute error (*MAE*) was employed as the measure of accuracy.

Following main conclusions can be drawn from the results of experiments. Bagging ensembles for both *M5P* decision trees and *MLP* neural networks gave better accuracy, i.e. the lower values of *MAE*, than single models and boosting ensembles. In turn, the *N11-NST* version of the *N-Nearest Similar Transactions* revealed the best performance from among the expert algorithms. The significantly worst accuracy provided the *N40-RST* version of the *N-Random Similar Transactions*. Nevertheless, all analysed models and algorithms, but *RST*, could be employed in an automated system to support professional valuators in real estate appraisal. Moreover, the machine learning algorithms could be applied only for those market areas where sufficient number of transactions is available to generate reliable models. The expert algorithms, in turn, could be used even if only a few transaction records are available.

Further study is planned to examine the performance of expert algorithms with the smaller number of similar properties considered to estimate price of a given property. Different market areas of a city will be investigated. Moreover, a few more machine learning techniques including random forests and support vector machines will be employed for the comparative analysis.

References

1. Zurada, J., Levitan, A.S., Guan, J.: A comparison of regression and artificial intelligence methods in a mass appraisal context. J. Real Estate Res. **33**(3), 349–388 (2011)
2. Peterson, S., Flangan, A.B.: Neural network hedonic pricing models in mass real estate appraisal. J. Real Estate Res. **31**(2), 147–164 (2009)
3. Antipov, E.A., Pokryshevskaya, E.B.: Mass appraisal of residential apartments: an application of random forest for valuation and a CART-based approach for model diagnostics. Expert Syst. Appl. **39**, 1772–1778 (2012)
4. Kusan, H., Aytekin, O., Özdemir, I.: The use of fuzzy logic in predicting house selling price. Expert Syst. Appl. **37**(3), 1808–1813 (2010)
5. Kontrimas, V., Verikas, A.: The mass appraisal of the real estate by computational intelligence. Appl. Soft Comput. **11**(1), 443–448 (2011)
6. Musa, A.G., Daramola, O., Owoloko, A., Olugbara, O.: A neural-CBR system for real property valuation. J. Emerg. Trends Comput. Inf. Sci. **4**(8), 611–622 (2013)
7. Jahanshiri, E., Buyong, T., Shariff, A.R.M.: A review of property mass valuation models. Pertanika J. Sci. Technol. **19**(S), 23–30 (2011)
8. McCluskey, W.J., McCord, M., Davis, P.T., Haran, M., McIlhatton, D.: Prediction accuracy in mass appraisal: a comparison of modern approaches. J. Prop. Res. **30**(4), 239–265 (2013)
9. d'Amato, M., Kauko, T. (eds.): Advances in Automated Valuation Modeling AVM: After the Non-agency Mortgage Crisis. Studies in Systems, Decision and Control, vol. 86. Springer, Cham (2017)
10. Woźniak, M., Graña, M., Corchado, E.: A survey of multiple classifier systems as hybrid systems. Inf. Fus. **16**, 3–17 (2014)
11. Krawczyk, B., Woźniak, M., Cyganek, B.: Clustering-based ensembles for one-class classification. Inf. Sci. **264**, 182–195 (2014)
12. Burduk, R., Walkowiak, K.: Static classifier selection with interval weights of base classifiers. In: Nguyen, N.T., Trawiński, B., Kosala, R. (eds.) ACIIDS 2015, Part I. LNCS, vol. 9011, pp. 494–502. Springer, Cham (2015)
13. Fernández, A., López, V., José del Jesus, M., Herrera, F.: Revisiting evolutionary fuzzy systems: Taxonomy, applications, new trends and challenges. Knowl. Based Syst. **80**, 109–121 (2015)
14. Lughofer, E.: Evolving Fuzzy Systems—Methodologies, Advanced Concepts and Applications. STUDFUZZ, vol. 266. Springer, Heidelberg (2011)
15. Hong, T.-P., Liou, Y.-L., Wang, S.-L., Vo, B.: Feature selection and replacement by clustering attributes. Vietnam J. Comput. Sci. **1**, 47–55 (2014)
16. Pham, T.-N., Nguyen, V.-Q., Tran, V.-H., Nguyen, T.-T., Ha, Q.-T.: A semi-supervised multi-label classification framework with feature reduction and enrichment. J. Inf. Telecommun. **1**(2), 141–154 (2017)
17. Lasota, T., Telec, Z., Trawiński, B., Trawiński, K.: Exploration of bagging ensembles comprising genetic fuzzy models to assist with real estate appraisals. In: Yin, H., Corchado, E. (eds.) IDEAL 2009, LNCS, vol. 5788, pp. 554–561. Springer, Heidelberg (2009)
18. Krzystanek, M., Lasota, T., Telec, Z., Trawiński, B.: Analysis of bagging ensembles of fuzzy models for premises valuation. In: Nguyen, N.T., et al. (eds.) ACIIDS 2010, LNAI, vol. 5991, pp. 330–339. Springer, Heidelberg (2010)
19. Lasota, T., Telec, Z., Trawiński, B., Trawiński, K.: Investigation of the eTS evolving fuzzy systems applied to real estate appraisal. J. Mult. Valued Logic Soft Comput. **17**(2–3), 229–253 (2011)

20. Lughofer, E., Trawiński, B., Trawiński, K., Kempa, O., Lasota, T.: On employing fuzzy modeling algorithms for the valuation of residential premises. Inf. Sci. **181**, 5123–5142 (2011)
21. Trawiński, B.: Evolutionary fuzzy system ensemble approach to model real estate market based on data stream exploration. J. Univers. Comput. Sci. **19**(4), 539–562 (2013)
22. Telec, Z., Trawiński, B., Lasota, T., Trawiński, G.: Evaluation of neural network ensemble approach to predict from a data stream. In: Hwang, D., et al. (eds.) ICCCI 2014, LNAI, vol. 8733. Springer, Heidelberg (2014)
23. Lasota, T., Sawiłow, E., Trawiński, B., Roman, M., Marczuk, P., Popowicz, P.: A Method for merging similar zones to improve intelligent models for real estate appraisal. In: Nguyen, N.T., Trawiński, B., Kosala, R. (eds.) ACIIDS 2015. LNCS, vol. 9011, pp. 472–483. Springer, Cham (2015). doi:10.1007/978-3-319-15702-3_46
24. Lasota, T., Sawiłow, E., Telec, Z., Trawiński, B., Roman, M., Matczuk, P., Popowicz, P.: Enhancing intelligent property valuation models by merging similar cadastral regions of a municipality. In: Núñez, M., Nguyen, N.T., Camacho, D., Trawiński, B. (eds.) ICCCI 2015. LNCS, vol. 9330, pp. 566–577. Springer, Cham (2015). doi:10.1007/978-3-319-24306-1_55
25. Witten, I.H., Frank, E., Hall, M.A., Pal, C.J.: Data Mining: Practical Machine Learning Tools and Techniques, 4th edn. Morgan Kaufmann, Burlington (2016)
26. Krasnoborski, J.: Management of a cadastral map on a mobile platform (in Polish). Engineer's thesis. Wroclaw University of Science and Technology, Wrocław (2015)
27. Piwowarczyk, M.: Web application to aid in real estate appraisal (in Polish). Engineer's thesis. Wroclaw University of Science and Technology, Wrocław (2015)
28. Talaga, M.: Mobile application to aid in real estate appraisal (in Polish). Engineer's thesis. Wroclaw University of Science and Technology, Wrocław (2015)
29. Trawiński, B., Telec, Z., Krasnoborski, J., Piwowarczyk, M., Talaga, M., Lasota, T., Sawiłow, E.: Comparison of expert algorithms with machine learning models for a real estate appraisal system. In: 2017 IEEE International Conference on Innovations in Intelligent Systems and Applications (INISTA 2017). IEEE (2017)

Multi-agent Systems

Multiagent Coalition Structure Optimization by Quantum Annealing

Florin Leon[1(✉)], Andrei-Ştefan Lupu[2], and Costin Bădică[3]

[1] Faculty of Automatic Control and Computer Engineering, “Gheorghe Asachi” Technical University of Iaşi, Iaşi, Romania
florin.leon@tuiasi.ro

[2] School of Electronics and Computer Science, University of Southampton, Southampton, UK
asl1u12@soton.ac.uk

[3] Faculty of Automatics, Computers and Electronics, University of Craiova, Craiova, Romania
cbadica@software.ucv.ro

Abstract. Quantum computing is an increasingly significant area of research, given the speed up that quantum computers may provide over classic ones. In this paper, we address the problem of finding the optimal coalition structure in a small multiagent system by expressing it in a proper format that can be solved by an adiabatic quantum computer such as *D-Wave* by quantum annealing. We also study the parameter values that enforce a correct solution of the optimization problem.

Keywords: Coalition structure · Optimization · Agents · Weighted graph game · Quantum annealing · D-Wave

1 Introduction

Finding the optimal coalition structure is a problem with applications in many fields: game theory with ramifications in political science and economics, transportation science for studying traffic flow and congestion, computer networks, operation research, etc. Basically, it provides a way for heterogeneous agents to form effective teams.

Given a set of agents and connections between them, the goal is to group the agents conveniently into coalitions in order to execute the assigned tasks in optimal time or with maximum payoff [6]. The generation of such coalitions over the problem space is generally an NP-hard problem [19].

Formally, it is equivalent to a set partition problem with a set of agents A, a payoff function $v : \wp(A) \rightarrow \mathbb{R}$ and a requirement to divide the A set into disjoint, mutually exhaustive partitions or coalitions C_i, such as the total sum of payoffs, $\sum_i v(C_i)$, is maximized.

Under certain mild restrictions, the problem can be represented as a weighted graph game [7], which is a coalitional game defined by an undirected weighted graph $G = (V, W)$,

N.T. Nguyen et al. (Eds.): ICCCI 2017, Part I, LNAI 10448, pp. 331–341, 2017.
DOI: 10.1007/978-3-319-67074-4_32

where V is the set of vertices and $W : V \to V$ is the set of edge weights. For any pair $(i,j) \in V^2$, w_{ij} is the weight of the edge between i and j. Each agent corresponds to a node in the graph, and the value of a coalition is: $v(C_i) = \sum_{(i,j) \in C_i} w_{ij}$ [1].

The coalition that contains all the agents is referred to as the "grand coalition".

If all the weights are positive, it is obvious that the maximum total payoff belongs to the grand coalition. Therefore, we are more interested in the case where the weights can also be negative, such that the agents would rather prefer subcoalitions.

In general, this is still a difficult problem, and exact algorithms proposed in literature, based on dynamic programming, have a complexity of $O(3^n)$ [15]. There are also heuristic procedures that yield approximate, yet often good solutions, e.g. [17], which uses a greedy approach [2] investigates this problem in weighted graph games and provides algorithms with constant factor approximations for the optimal solutions.

In this paper, we express a simple version of the multiagent coalition structure problem as an optimization problem solvable by quantum annealing. To our knowledge, this is the first time that quantum annealing is attempted for this purpose.

The article is organized as follows. In Sect. 2, several other applications of quantum annealing are presented. Section 3 briefly outlines the *D-Wave* computer architecture, and Sect. 4 describes the proposed method of mapping a weighted graph game into a form that can be handled by quantum annealing. Section 5 presents an analysis of the feasible intervals of parameters used for the mapping, and the final section contains the conclusions of the work.

2 Related Work

Simulated annealing (SA) represents a general approach to solving optimization problems. However, with the introduction of *D-Wave*, the first adiabatic quantum computer, a significant interest arose for this classic algorithm. By using the quantum uncertainty to obtain a range of probabilities over the problem space, there could be a substantial performance improvement in solving problems, through a technique named *quantum annealing* (QA) [10].

Unlike SA, where the algorithm has a single state whose position it tries to change to a location that corresponds to a minimum of the objective function, in QA the initial state simultaneously belongs to many different locations, because of the superposition of the quantum bits (*qubits*). The probability that the state belongs to any specific location gradually evolves as the algorithm progresses, such that the areas around local minima have higher probabilities. It is however possible to exit areas of local minimum by going directly through the energy barriers that define their boundaries, by a phenomenon called *quantum tunneling*. This decreases the probability that the solution remains confined to a local minimum and moves the search towards the global minimum. The result is also improved by *quantum entanglement*, which helps the discovery of correlations between the more promising locations.

Quantum annealing is believed to be superior in performance to classic simulated annealing. Despite initial skepticism of whether *D-Wave* truly exploits quantum effects, a recent study [6] has shown that it is indeed capable of using quantum tunneling effects in order to speed up computation beyond the capabilities of classic computers. It has indicated that, for a specific set of problems, the quantum version can be eight orders of magnitude faster than simulated annealing.

In [12], the authors identify multiple speed-up classes for quantum problems. The results heavily favor quantum annealing, despite the fact that random problems, such as spin glasses [16], do not seem to benefit significantly from this approach.

NP-complete and NP-hard problems have long been classic contenders for simulated annealing. It was only natural for the research interest to shift towards them, in order to demonstrate the need and the benefits of quantum computing. However, quantum computers are indeed largely different when it comes to problem implementations, compared to regular computers. From here, one is faced with the non-trivial task of expressing problems as coherent programs that can be understood by quantum computers.

In [5], the researchers propose an implementation of the classic map coloring problem tailored for *D-Wave*. They describe it as a single quantum machine instruction using a programming model named "direct embedding". Qubits, weight and strength variables are used to describe an objective function which outputs the distribution of samples. The solutions are found by minimizing the objective function.

Another example of an algorithm implemented in a quantum computer is a SAT filter [8], which is a data structure that can efficiently query a word for membership in a set. In doing so, multiple implementations are explored and measured for performance. The results show an interesting set of trade-offs, such as the one between the block sizes used and the efficiency at low false-positive rates. Therefore, it is worth noting the performance and size limitations caused by hardware availability in quantum computers.

Other authors apply quantum annealing for Bayesian network structure learning [13] and job shop scheduling [18].

3 *D-Wave* Computer Architecture

The *D-Wave* quantum computer relies on a lattice of qubits, with couplers that connect pairs of qubits, as presented in Fig. 1 (after [14]).

The programmer cannot set the values of the qubits, instead he/she sets the *weights* associated with each qubit and the *strengths* of the couplers. The *D-Wave* computer minimizes the objective function in Eq. (1) using quantum annealing, and the resulting values of the qubits represent the solution of the optimization problem:

$$O(\mathbf{q}; \mathbf{a}, \mathbf{b}) = \sum_{i} a_i q_i + \sum_{(i,j)} b_{ij} q_i q_j. \quad (1)$$

In this equation, a_i is the weight of qubit q_i, b_{ij} is the coupler strength between qubits q_i and q_j and the (i, j) notation refers to the indices of two neighbor qubits connected by a coupler. One can recognize this expression as common to many energy-based models, starting with the Ising model, further abstracted in the Hopfield auto-associative neural network, and more recently, in restricted Boltzmann machines.

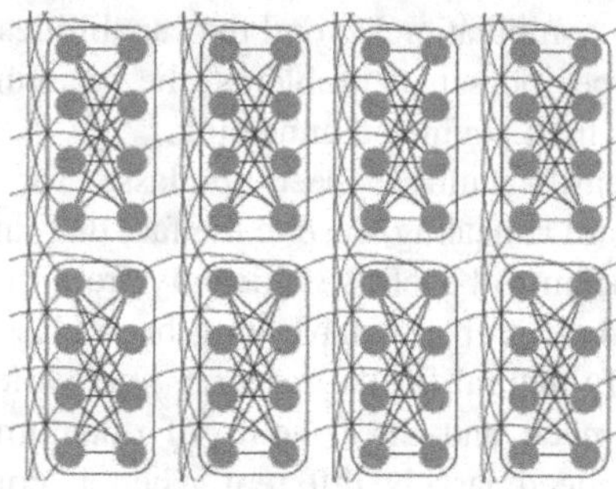

Fig. 1. The programming architecture of *D-Wave Two* (*Vesuvius*) quantum computer: a lattice of cells, each with 8 qubits, connected by couplers

The parameters a_i and b_{ij} correspond to a quantum machine instruction and *D-Wave* minimizes the objective function O such that, eventually, the qubit values q_i satisfy the given constraints.

More details about the D-Wave computer architecture can be found e.g. in [4, 9].

4 Problem Definition

In this paper, we focus on the multiagent coalition structure problem and provide a quantum machine instruction program for it, tailored for the *D-Wave* quantum computer. This study only addresses the simple scenario with four agents and two possible coalitions, because, as it will be shown, it is non-trivial to express it as an optimization problem for the quantum annealing process.

The weighted graph game that represents the problem has the structure presented in Fig. 2. However, in the *D-Wave* architecture, the qubits are placed in a lattice, therefore it is not possible to implement the overlapping links directly, such as those between A and D, and B and C, simultaneously. The approach suggested by whitepaper [5] is to use "clones". In Fig. 3, one can see the same graph with clones for the B, C and D agents. In this configuration, all the agents are directly coupled, but more constraints are introduced, i.e. all the clones of an agent should have the same value in the solution.

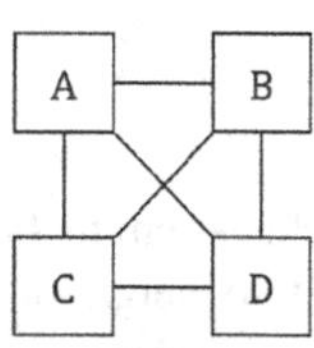

Fig. 2. The coalition structure problem considered

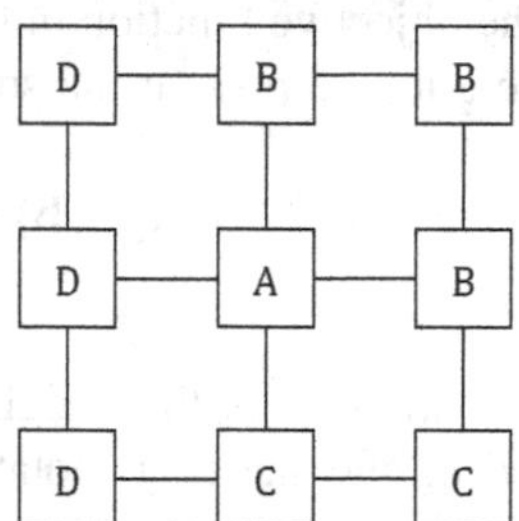

Fig. 3. The mapping of the problem on a logical *D-Wave* qubit lattice

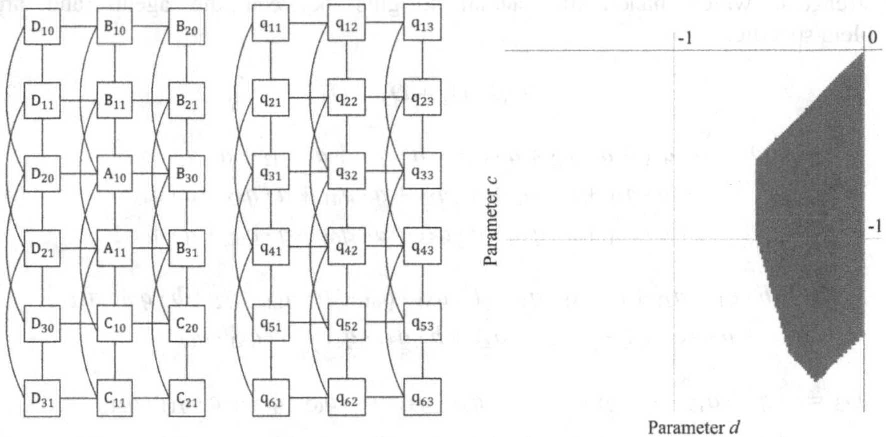

Fig. 4. The mapping on a physical qubit lattice

Fig. 5. Example of a feasible parameter region

The logical mapping in Fig. 3 must be further detailed into a physical mapping on the qubit architecture presented in Fig. 1. For each agent, two logical qubits are used. The "01" combination of values means that the agent belongs to the first coalition, and "10" means that it belongs to the second. It is not possible to use only one qubit to distinguish between the coalitions. According to Eq. (1), a qubit with a value of 0 does not count in the computation of the objective function. Therefore, at least one qubit of each agent must have a value of 1.

Using the *D-Wave* architecture, the qubits of the agents are coupled among themselves and coupled with the qubits of their neighbors. The problem is symmetric, in the sense that the identifiers of the agents can be switched without affecting the solution. Therefore, all the parameters should have a single value: $a_i = a$, $b_{ij} = b$, $c_{ij} = c$, $d_{ij} = d$, where c_{ij} are the strengths of the external couplings between the clones and d_{ij} are the strengths of the couplings between the neighbors.

The weights of the qubits a and the strengths of the couplings b are free parameters of the model. Since the problem also relies on additional parameters that specifically address the multiagent coalition structure problem, the values of these parameters can be fixed to $a = -1$ and $b = 2$, as suggested in [5].

The external couplings between the clones (parameter c) must be negative. Also, the couplings between the neighbors (parameter d) must be negative, but proportional to the values in the agent coalition graph. As it will be shown, the combination of values for these parameters that ensures a correct result is not unique.

Although lengthy, we include the complete objective function of the problem below. These expressions come from Eq. (1) applied to the structure in Fig. 4. They are important because we can distinguish between three types of couplers: the couplers with b strengths, which enforce that the complementary qubits of an agent (e.g. D_{10}/q_{11} and D_{11}/q_{21}) have opposite values, the couplers with c strengths, which enforce that the clones have the same values (e.g. B_{10}/q_{12} and B_{20}/q_{13}), and the couplers with

d strengths, which handle the actual weights between the agents and are problem-specific.

$$O = O_1 + O_2 + O_3 + O_4 \tag{2}$$

$$\begin{aligned} O_1 =& a \cdot q_{11} + a \cdot q_{12} + a \cdot q_{13} + a \cdot q_{21} + a \cdot q_{22} + a \cdot q_{23} \\ &+ a \cdot q_{31} + a \cdot q_{32} + a \cdot q_{33} + a \cdot q_{41} + a \cdot q_{42} + a \cdot q_{43} \\ &+ a \cdot q_{51} + a \cdot q_{52} + a \cdot q_{53} + a \cdot q_{61} + a \cdot q_{62} + a \cdot q_{63} \end{aligned} \tag{3}$$

$$\begin{aligned} O_2 =& b \cdot q_{11} \cdot q_{21} + b \cdot q_{31} \cdot q_{41} + b \cdot q_{51} \cdot q_{61} + b \cdot q_{12} \cdot q_{22} + b \cdot q_{32} \cdot q_{42} \\ &+ b \cdot q_{52} \cdot q_{62} + b \cdot q_{13} \cdot q_{23} + b \cdot q_{33} \cdot q_{43} + b \cdot q_{53} \cdot q_{63} \end{aligned} \tag{4}$$

$$\begin{aligned} O_3 =& c \cdot q_{12} \cdot q_{13} + c \cdot q_{22} \cdot q_{23} + c \cdot q_{52} \cdot q_{53} + c \cdot q_{62} \cdot q_{63} + c \cdot q_{11} \cdot q_{31} \\ &+ c \cdot q_{21} \cdot q_{41} + c \cdot q_{31} \cdot q_{51} + c \cdot q_{41} \cdot q_{61} + c \cdot q_{13} \cdot q_{33} + c \cdot q_{23} \cdot q_{43} \end{aligned} \tag{5}$$

$$\begin{aligned} O_4 =& d \cdot v_{BD} \cdot q_{11} \cdot q_{12} + d \cdot v_{BD} \cdot q_{21} \cdot q_{22} + d \cdot v_{AD} \cdot q_{31} \cdot q_{32} \\ &+ d \cdot v_{AD} \cdot q_{41} \cdot q_{42} + d \cdot v_{CD} \cdot q_{51} \cdot q_{52} + d \cdot v_{CD} \cdot q_{61} \cdot q_{62} + d \cdot v_{AB}/2 \cdot q_{32} \cdot q_{33} \\ &+ d \cdot v_{AB}/2 \cdot q_{42} \cdot q_{43} + d \cdot v_{AB}/2 \cdot q_{12} \cdot q_{32} + d \cdot v_{AB}/2 \cdot q_{22} \cdot q_{42} \\ &+ d \cdot v_{AC} \cdot q_{32} \cdot q_{52} + d \cdot v_{AC} \cdot q_{42} \cdot q_{62} + d \cdot v_{BC} \cdot q_{33} \cdot q_{53} + d \cdot v_{BC} \cdot q_{43} \cdot q_{63} \end{aligned} \tag{6}$$

One can notice that the weight between A and B, v_{AB}, appears in Eq. (6) as divided by 2 because in the definition of the problem from Fig. 2, there are two links between A and B.

In the following study, we ignore a further transformation of logical qubits to physical qubits, where the values of the qubit weights and the coupler strengths are divided by 2, because each logical qubit is represented by a chain of length two and the chains for each pair of logical qubits are connected by two physical couplers [5]. Since all the parameters are scaled and the objective function is linear, this division does not affect the results presented in Sect. 5.

5 Parameter Interval Analysis

In this section, we perform an analysis of the c and d parameters. We aim to detect their range of values that guarantees that quantum annealing finds a correct solution for any weighted graph game. The *D-Wave* computer allows a decent level of precision for the real value coupling strengths; for example, [11] reports a resolution of at least three decimals.

Since there are two free parameters, one should identify the feasible region of the bidimensional space that corresponds to the correct solution, i.e. inside this region, the objective function is minimum only for the correct solution.

The objective function then changes until it matches an incorrect, neighboring configuration. An example is given in Fig. 5, where the weight values are: $v_{AB} = -5$,

$v_{AC} = 0$, $v_{AD} = 7$, $v_{BC} = 6$, $v_{BD} = 1$ and $v_{CD} = -3$. In this case, since A and B, and C and D, respectively, cannot belong to the same coalition, the obvious solution is the coalition structure $C_1 = \{A, C\}$ and $C_2 = \{B, D\}$, more compactly noted as "AC–BD".

An important empirical observation is that the feasible region always contains the $(0, -1)$ point. Therefore, we can further simplify the problem by considering the c parameter to be also fixed to the value of -1.

The problem remains to find out the feasible interval for the d parameter.

The right endpoint d_r can be estimated by observing that since two agent coalitions C_1 and C_2 are preferred to the grand coalition C_G, then:

$$v(C_1) + v(C_2) > v(C_G). \tag{7}$$

This is equivalent to:

$$\sum_{\substack{(i,j)\\ i,j \in C_1}} v_{ij} + \sum_{\substack{(i,j)\\ i,j \in C_2}} v_{ij} > \sum_{\substack{(i,j)\\ i,j \in C_G}} v_{ij}. \tag{8}$$

Since d_r is the weight attributed to the payoff of the agent pairs that comprise the coalitions, then:

$$d_r \cdot \sum_{\substack{(i,j)\\ i,j \in C_1}} v_{ij} + d_r \cdot \sum_{\substack{(i,j)\\ i,j \in C_2}} v_{ij} > d_r \cdot \sum_{\substack{(i,j)\\ i,j \in C_G}} v_{ij}, \quad \forall d_r < 0. \tag{9}$$

Parameter d cannot be 0 because in this case the payoff of the agents would simply be ignored in the optimization problem. Since we are interested in the maximum value of d, it means that the solution is a negative value, as close as possible to 0:

$$d_r = -\varepsilon, \tag{10}$$

where ε is a small positive infinitesimal quantity: $\varepsilon \to 0, \varepsilon > 0$.

In order to determine the left endpoint of the interval, we rely on some experimental observations to demonstrate a procedure of computing d_l. Although a single analytical expression does not exist for all the possible coalition combinations, we show that an exact value can always be computed.

The procedure relies on the condition that, at the border between the correct solution and an incorrect one, the objective function O must be the same for the two neighboring states. However, O has different expressions for the two states, because of the different qubit values of 0 and 1. By imposing equality between these terms, one can compute the corresponding value of d_l analytically.

For example, let us consider the coalition structure AB–CD. In the qubit lattice representation from Fig. 4, this solution corresponds to the state $Q = 116487$, with the final qubit values shown in Table 1. The last six rows of the table show the terms of the objective function that are different in the two configurations under comparison. Empirically, it was observed that as d decreases below a certain threshold, the optimization problem yields an incorrect result, e.g. state $Q = 116535$. In this state, the

Table 1. Neighbor configurations for the left interval endpoint (case AB–CD) and their different terms of the objective function

Q = 116487 AB–CD	Q = 116535 (incorrect)
011	011
100	100
011	011
100	100
000	110
111	111
0	−1
0	−1
0·2	1·2
0·2	1·2
$0 \cdot d \cdot v_{CD}$	$1 \cdot d \cdot v_{CD}$
$0 \cdot d \cdot v_{AC}$	$1 \cdot d \cdot v_{AC}$

Table 2. Neighbor configurations for the left interval endpoint (case AD–BC) and their different terms of the objective function

Q = 115612 AD–BC	Q = 115677 (incorrect)
011	011
100	100
001	001
110	111
011	011
100	101
0	−1
0	−1
0·2	1·2
0·2	1·2
$0 \cdot d \cdot v_{AB}/2$	$1 \cdot d \cdot v_{AB}/2$
$0 \cdot d \cdot v_{BC}$	$1 \cdot d \cdot v_{BC}$

imposed constraints do not hold, e.g. D_{30}/q_{51} and D_{31}/q_{61} have the same value: 1. According to the O_2 partial objective in Eq. (4), because of a too large absolute value for parameter d, parameter b is no longer capable of enforcing the constraint that the two qubits that represent the coalition of an agent have different values.

At the d_l endpoint, the different terms of the objective function O are equal for the two configurations:

$$0 = 2 + d_l \cdot (v_{CD} + v_{AC}) \Rightarrow d_l = -\frac{2}{v_{CD} + v_{AC}}. \tag{11}$$

The same equality holds in case of state Q = 115612, corresponding to the AD–BC solution, which often has the incorrect state Q = 115667 as a neighbor when d decreases (Table 2). In this case, the left endpoint d_l is given by:

$$0 = 2 + d_l \cdot (v_{AB}/2 + v_{BC}) \Rightarrow d_l = -\frac{2}{v_{AB}/2 + v_{BC}}. \tag{12}$$

Similar neighboring states have been observed for the ABC–D case (state Q = 29148), with a frequent neighboring state Q = 228828:

$$0 = -4 + 8 + d_l \cdot (v_{BD} + v_{AD} + v_{AB}/2 + v_{AC}) - 1 \Rightarrow d_l = -\frac{3}{v_{BD} + v_{AD} + v_{AB}/2 + v_{AC}}. \tag{13}$$

and for the ABCD case (state $Q = 29127$), where the solution is the grand coalition, with a frequent neighboring state $Q = 228862$:

$$-2 + d_l \cdot v_{BC} = -7 + 12 + d_l \cdot (v_{BD} + v_{AD} + v_{CD} + v_{AB}/2 + v_{AC}) - 3$$
$$\Rightarrow d_l = -\frac{4}{v_{BD} + v_{AD} + v_{CD} + v_{AB}/2 + v_{AC} - v_{BC}}. \qquad (14)$$

Because of space limitations, we must omit the details regarding the qubit representation and the different terms of the objective function, which must be equal at the left endpoint of the feasible interval.

Given the number of possible neighboring states, $2^{18} - 1 = 262143$, it is possible to compute all the d_l values analytically, even if in practice only a very small subset actually matters. The left endpoint of the interval is then:

$$d_l^* = \max_i d_l^i, \qquad (15)$$

where d_l^* is of the form:

$$d_l^* = -\frac{n}{\sum_j w_j \cdot v_j}, \qquad (16)$$

with $n \in \mathbb{N}$ and $2\, w_j \in \mathbb{Z}$.

Therefore, the optimization problem is correctly solved for $a = 1$, $b = 2$, $c = -1$ and $d \in \left(d_l^*, 0\right)$.

6 Conclusions

In this paper, we showed how to express a multiagent coalition structure optimization problem as a problem solvable by quantum annealing. *D-Wave* provides a method to solve a difficult optimization problem by solving a simpler one: putting the original problem into a form that can be solved by quantum annealing. We analyzed the parameters of the model that lead to correct solutions. Since the scenario involved only four agents and two coalitions, the next natural step is to extend the model to any number of agents, which could be placed in up to four coalitions, if one takes into account the physical constraints of a *D-Wave* cell with 4 logical and 8 physical qubits from Fig. 1. But also a more general scenario, without any restrictions, can be thought of, taking as many cells into account as needed.

Another direction of research is to devise a more efficient approach to determine the neighboring states, so as to find the feasible intervals of the parameters with fewer computations. This is especially important in the general scenario, where the number of states is far larger than in the scenario presented here.

Recently, the D-Wave Systems company has released *qbsolv* [3], a tool that solves large quadratic unconstrained binary optimization (QUBO) problems by partitioning into subproblems targeted for execution on a *D-Wave* system. This tool can be used to test the results of more complex coalition structure optimization problems.

Other multiagent optimization problems, e.g. combinatorial auctions, can be addressed in the same way, i.e. by transforming them into problems that can be handled by quantum annealing.

References

1. Airiau, S.: Cooperative games: representation and complexity issues (2012). http://www.lamsade.dauphine.fr/~airiau/Teaching/CoopGames/2012/coopgames-9[8up].pdf
2. Bachrach, Y., Kohli, P., Kolmogorov, V., Zadimoghaddam, M.: Optimal coalition structures in cooperative graph games. In: Proceedings of the Twenty-Seventh AAAI Conference on Artificial Intelligence, AAAI 2013, Bellevue, Washington, pp. 81–87 (2013)
3. Booth, M., Reinhardt, S.P., Roy, A.: Partitioning optimization problems for hybrid classical/quantum execution, Technical report (2017). http://www.dwavesys.com/sites/default/files/partitioning_QUBOs_for_quantum_acceleration-2.pdf
4. Bunyk, P.I., Hoskinson, E., Johnson, M.W., Tolkacheva, E., Altomare, F., Berkley, A.J., Harris, R., Hilton, J.P., Lanting, T., Whittaker, J.: Architectural considerations in the design of a superconducting quantum annealing processor. arXiv preprint (2017). https://arxiv.org/pdf/1401.5504v1.pdf
5. Dahl, E.D.: Programming with D-Wave: map coloring problem (2013). http://www.dwavesys.com/sites/default/files/Map%20Coloring%20WP2.pdf
6. Denchev, V.S., Boixo, S., Isakov, S.V., Ding, N., Babbush, R., Smelyanskiy, V., Martinis, J., Neven, H.: What is the computational value of finite range tunneling? Phys. Rev. X **6**(3), 10–15 (2016). doi:10.1103/PhysRevX.6.031015
7. Deng, X., Papadimitriou, C.H.: On the complexity of cooperative solution concepts. Math. Oper. Res. **19**(2), 257–266 (1994). doi:10.1287/moor.19.2.257
8. Douglass, A., King, A.D., Raymond, J.: Constructing SAT filters with a quantum annealer. In: Heule, M., Weaver, S. (eds.) SAT 2015. LNCS, vol. 9340, pp. 104–120. Springer, Cham (2015). doi:10.1007/978-3-319-24318-4_9
9. D-Wave Systems: Introduction to the D-Wave quantum hardware (2017). https://www.dwavesys.com/tutorials/background-reading-series/introduction-d-wave-quantum-hardware
10. Kadowaki, T., Nishimori, H.: Quantum annealing in the transverse Ising model. Phys. Rev. E **58**(5), 53–55 (1998). doi:10.1103/PhysRevE.58.5355
11. King, A.D., Hoskinson, E., Lanting, T., Andriyash, E., Amin, M.H.: Degeneracy, degree, and heavy tails in quantum annealing. Phys. Rev. A **93**(5), 20–23 (2016). doi:10.1103/PhysRevA.93.052320
12. Mandrà, S., Zhu, Z., Wang, W., Perdomo-Ortiz, A., Katzgraber, H.G.: Strengths and weaknesses of weak-strong cluster problems: a detailed overview of state-of-the-art classical heuristics vs quantum approaches. Phys. Rev. A **94**(2), 23–37 (2016). doi:10.1103/PhysRevA.94.022337
13. O'Gorman, B., Perdomo-Ortiz, A., Babbush, R., Aspuru-Guzik, A., Smelyanskiy, V.: Bayesian network structure learning using quantum annealing. Eur. Phys. J. Spec. Top. **224** (1), 163–188 (2015). doi:10.1140/epjst/e2015-02349-9
14. Pudenz, K.L., Albash, T., Lidar, D.A.: Error-corrected quantum annealing with hundreds of qubits. Nat. Commun. **5**, Article no. 3243 (2014). doi:10.1038/ncomms4243

15. Rahwan, T., Jennings, N.R.: An improved dynamic programming algorithm for coalition structure generation. In: Proceedings of the 7th International Conference on Autonomous Agents and Multi-agent Systems, AAMAS 2008, Estoril, Portugal, pp. 1417–1420 (2008)
16. Rønnow, T.F., Wang, Z., Job, J., Boixo, S., Isakov, S.V., Wecker, D., Martinis, J.M., Lidar, D.A., Troyer, M.: Defining and detecting quantum speedup. Science **345**(6195), 420–424 (2014). doi:10.1126/science.1252319
17. Shehory, O., Kraus, S.: Methods for task allocation via agent coalition formation. Artif. Intell. **101**(1), 165–200 (1998). doi:10.1016/s0004-3702(98)00045-9
18. Venturelli, D., Marchand, D.J.J., Rojo, G.: Quantum annealing implementation of job-shop scheduling. arXiv preprint (2016). https://arxiv.org/pdf/1506.08479.pdf
19. Voice, T., Polukarov, M., Jennings, N.R.: Coalition structure generation over graphs. J. Artif. Intell. Res. **45**(1), 165–196 (2012). doi:10.1613/jair.3715

External Environment Scanning Using Cognitive Agents

Marcin Hernes(✉), Anna Chojnacka-Komorowska, and Kamal Matouk

Wrocław University of Economics, Wrocław, Poland
marcin.hernes@ue.wroc.pl

Abstract. Very significant process in business organization is an external environment scanning. It is very important for decision makers to have an understanding of the competitive position of the company. Actual and reliable information is particularly important for corporate executives and helps decision makers make quick decisions in response to competitors' actions. This knowledge will help in increasing efficiency and effectiveness of company functioning.

The aim of this paper is to develop a method for external environment scanning by using cognitive agents. The research has been performed on the example of hotel industry.

The first part of article presents the state on the art in the field. Next the problem of external environment scanning in hotel industry is presented. The method for environment scanning by using cognitive agent and the research experiment, are presented at the last part of the paper.

Keywords: Environment scanning · Cognitive agents · Decision making · Sentiment analysis

1 Introduction

Many companies are nowadays competing in a global market, not only in the domestic market. Consequently, they have to compete not only with companies from their immediate surroundings, but also from the whole region or global market. Managers of the company will look for patterns that can help them understand external environment of organization, and this may be different from what they expected. It is important for decision makers to have knowledge about the competitive position of the company. Actual and reliable information is particularly important for corporate executives and helps you make quick decisions in response to competitors' actions. This knowledge will help in improving the competitive position and improve operating efficiency of the company [1].

The aim of this paper is to develop a method for external environment scanning by using cognitive agents. The research has been performed on the example of hotel industry. The environment scanning is based on an analysis of users' opinions about hotels placed on internet web pages.

The first part of article presents the state on the art in the field. Next the problem of external environment scanning in hotel industry is presented. The method for environment scanning by using cognitive agent and the research experiment, are presented at the last part of the paper.

N.T. Nguyen et al. (Eds.): ICCCI 2017, Part I, LNAI 10448, pp. 342–350, 2017.
DOI: 10.1007/978-3-319-67074-4_33

2 Related Works

The external environment "are all events outside the company that has the potential to affect the company" [2]. Dictionary [3] defined external environment as "conditions, entities, events, and factors surrounding an organization that influence its activities and choices, and determine its opportunities and risks".

The external environment of marketing is comprises of those uncontrollable phenomena outside of the organization. These phenomena are uncontrollable because it is not possible to control them, but decision-makers can respond and adapt to them. The uncontrollable phenomena in the external environment are: competition, government policies, natural forces, social and cultural forces, demographic factors and technological changes. The external environment must effectively diagnose the following stages [4]:

- analysis of competitive environment;
- expectations for interest analysis;
- analyze the influence macro-environment;
- the development outlook of the external environment.

However, the effectiveness of these activities is determined by the use of modern IT solutions. Changes in technology and increased ability to acquire and process information lead to competing responses are more timely and effective. Information systems not only can help in environmental monitoring and control of the various activities of the company, but also can serve as a strategic weapon in the effort to gain a competitive advantage [5].

Environment scanning performed by IT systems is often based on the web pages content analysis, in other words, on text document analysis. Text document analysis methods are mainly divided to two kinds of processing [6]:

1. Shallow text analysis is defined as the analysis of the text, the effect is incomplete in relation to deep text analysis. Typically limitation is the recognition or non-recursive structures with a limited level of recursion which can be recognized with a high degree of certainty.
2. Deep text analysis is the process of computerized linguistic analysis of all possible interpretations and grammatical relationships found in natural text. Such a full analysis can be very complex.

In the shallow and deep analysis of text documents many methods are used. For example machine learning [6, 7] or the rules on the basis of which identification (annotation) pieces of text, for a specific topic, is performed [8, 9]. Such rules are based on templates, taking into account the relationship between words and semantic classes of words.

In this paper, the authors focused on the methods of competition analysis in the company environment refers to the numbers of similar competitive service/product brands' marketers in given industry, their size and market capitalizations. For environment scanning by text document analysis, it was decided to use the architecture of cognitive agent program The Learning Intelligent Distribution Agent (LIDA) developed by Cognitive Computing Research Group [10].

3 External Environment Scanning in Hotel Industry

The hotel industry on a global basis is characterized by high capital costs and a high proportion of fixed costs to total costs. The managers of hotels should concentrate on achieving the most cost-effective use of resources applied to construction, furnishing and equipment, pre-operational expenses and finance. Nowadays, the market has created a strong competition that has forced hotels to seek a competitive advantage.

The two crucial factors that enable hotels to achieve high level of competition are good location and quality of service. The second factor depends on good management and trained and motivated staff. At the same time, the right choice of location and the quality of services offered should be tailored to meet the needs of hotel guests, which are subject to the destination of travel such as work, study, sport, recreation. In this paper, the authors have focused on recreation hotels. They are located in places offering many tourist attractions or in the area providing relaxation in peace and quiet.

The hotel industry in most cities in the world provides considerable opportunity to cross-sell profitable products such as food and beverage. Tariffs are determined according to the level of differentiation achieved through location, management, staff and guest ratios and any other miscellaneous factors such as the quality of architecture or decoration [11].

The IT systems, used in hotel industry, involve the people, processes, and technologies in order to making effective decisions. For example, using data warehouses provide access to massive amounts of real time and historic data for analysis. Online analytical processing (OLAP) cubes or data mining methods allow for reporting and advanced data analysis.

There are different areas for external environment scanning by IT systems in hotel industry. One of them is supporting the booking process, which is very important especially in relation to corporate traveler. Because certain days are busier than others, IT systems can help to configure the optimal price that will drive a high occupancy rate. One of the ways that hotels help rate optimize is data mining. These systems can put historical booking trends into a formula and model to help understand what will happen if they rise or lower prices. IT systems help the hotels use their past data to make better, more efficient decisions for the future. This can be particularly helpful around the holiday season [12].

Very important is also external environment scanning in a competition area. IT systems search and automatically analyze both, web pages of competitors and users' opinions related to particular hotels and their services.

4 Method for External Environment Scanning

In order to perform external environment scanning, the Learning Intelligent Distribution Agent (LIDA) [10] architecture has been used. The architecture is developed by Cognitive Computing Research Group (CCRG). The advantage of this architecture is its emergent-symbolic character, making it possible to process information both structured (numerical and symbolic) and unstructured (stored in natural language) knowledge. The LIDA cognitive agent's architecture consists of the following modules [10]:

- sensory memory,
- perceptual memory,
- workspace,
- episodic memory,
- declarative memory,
- attentional codelets,
- global workspace,
- action selection,
- procedural memory,
- sensory-motor memory.

The basic operations, in the LIDA, are performed by the codelets (specialized, mobile programs processing information in the model of global workspace). The cognitive agent running within the frame of the cognitive cycle [13, 14].

CCRG created a Framework LIDA—software that forms the basis for the implementation of cognitive agent programs. The Framework contains object classes (implemented in Java) that perform the functions of the agent architecture (definition and methods of handling all types of memory, communication protocols, methods that enable the agent to perform operations on real-world objects, defining association between objects etc.). Our contribution is to complement the tools provided by the LIDA framework (writing the code of the program) with aspects of the specific problem domain - for example, economics, management.

Using the LIDA framework, a section of the program was written and implemented in the agent structure. The LIDA agent performs external environment scanning on the basis of users' opinions related to particular hotels and their services. The main aim of external environment scanning performed by cognitive agent is determining services ant their features which hotels shall to provide in order to satisfy customers' needs to the greatest possible extent. Detailed tasks of the analysis are as follows:

- determining overall sentiment of an opinion, i.e. determining whether an opinion is positive or negative,
- extraction of features of a service,
- determining sentiment of opinions about particular features of a service,
- selecting services and their features, which customer search for.

The agent's environment consists of a set of text documents (written in Polish language). These documents contain opinions about hotels and their services. An agent searches for opinions and then stores them in database. Opinions' analysis is performed in the following way:

1. Learning set consists of opinions on a given hotel and its services. On the basis of this set, a semantic network is created in agent's perceptual memory which contains topics (nodes), connected with a services offered by hotels, and relationships existing between them (links). The perceptual memory also stores synonyms and various forms of words.

2. Particular opinions (strings) are one by one transferred into the sensory memory.
3. Text documents' analysis is performed via codelets [15].
4. Results of analysis are stored as semantic network and transferred to the workspace. Figure 1 shows an example of results of an analysis of the following opinion "Recommended, feature1 is good, but feature2 is not good".

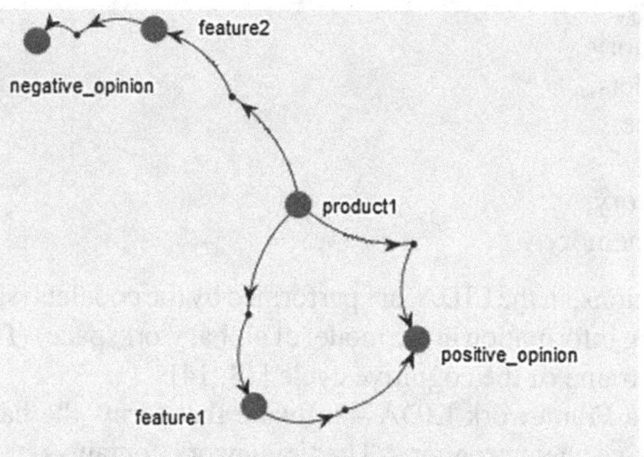

Fig. 1. An example of results of an analysis—positive opinion about the product and negative opinion about the one of the features.

5. Next the following patterns of action are automatically selected from the procedural memory: "saving results of opinion analysis into a data base" and "loading next opinion into the sensory memory".

The next part of the paper contains results of verification of developed method for environment scanning.

5 Research Experiment

In order to verify environment scanning performed by cognitive agent, a research experiment has been carried out. Results of automatic analysis of opinions about the hotels were compared with results of an analysis performed by a human (an expert), i.e. a manual analysis. The following assumptions were adopted in the experiment:

1. The opinions about hotels were received from web pages.
2. Number of analyzed opinions: 365. Because manual annotation (very time consuming process) has to be performed, the number of opinion is limited.
3. For the needs of the experiment, five features of hotels and their services were analyzed:
   - location (attractions/region),
   - rooms (apartments),
   - food (meals),
   - price,
   - personnel.

4. If analyzed opinion did not contain information about given feature's sentiment then the sentiment of that feature was the same as overall sentiment of opinion.
5. A learning set consist of 65 opinions.
6. In order to determine the accuracy of results of automatic analysis in relation to results of manual analysis, the effectiveness, the precision and the sensitivity measures was used.

Research experiment was carried out in the following way:

1. 365 randomly selected opinions about hotels found on web sites were recorded in a data base.
2. Manual annotation was performed.
3. The opinions were grouped according to the degree of difficulty (three groups—group 1—sentiment of opinions and features easy to determine, group 3—sentiment of opinions and features difficult to determine) of determining their sentiment and the sentiment of features of services offered by hotels characterized in the opinions (Table 1).

Table 1. Types of opinions according the degree of difficulty of their analysis

No.	Sample content of opinion (written in Polish language they was translated into English in this paper)
1	Very nice hotel, all right
2	Hotel very nice, clean with great animation. Food tasty and abundant. NO SERVICE at reception and canteen total floor !!!! The staff at the reception desk incompetent is the same canteen service. To this, grumpy, helpless and unhelpful! The hotel has a lot of older people, for whom bringing breakfast from the buffet to the table is a challenge, unfortunately the service does not help, on the contrary !!! Staff to exchange
3	We spent a week at the hotel named "A" last week. As nice as the service, the food is also good enough. The cleanliness of the room very good, the pool also cleans but as for the pool I am sure a reservation. Namely: Between the two jacuzzi was a railing, very dangerous for younger children and lower adults. I reported the danger to the reception, I hope it will not be downgraded! In the near future and so I plan to visit hotel named "B" so of course I will check first of all. The kid's club is quite small, its attractions are typically for children aged 0-5 years. The downside is the distance from the city to get to by car. Large parking, very spacious. I would like to commend all the bartenders from both Night Club, Lobby Bar and Eleven Club for making delicious drinks. That's it and I greet you

Source: own work

4. Next measures of performance were calculated taking into consideration particular groups of degree of difficulty.
   The opinions of the first group contain only text related to sentiment of opinions (individual features are not described), so it is easy to recognize the sentiment, however sentiment of individual features is treated as sentiment of opinions. The second group's opinions contain information about sentiment of individual features. Recognizing of a particular feature can be difficult in this case. The third group of

opinions is the most difficult to analyze as opinions belonging to the group contain descriptions of several hotels in one opinion. It is difficult to determine to which hotel sentiment of opinions should be assigned and to which hotel/service should be assigned the features characterized in the opinion.

5. Next all the 365 opinions were loaded, one by one, into the sensory memory, analyzed by a cognitive agent, and saved results of the analysis in a database.

Table 2 presents the list of obtained results.

Table 2. Results of the opinions' and features' sentiment analysis.

Group of the opinion	Measure	Sentiment of opinions	Features				
			Location	Rooms	Food	Price	Personnel
1	Effectiveness	0,8969	0,8969	0,8969	0,8969	0,8969	0,8969
	Precision	0,8969	0,8969	0,8969	0,8969	0,8969	0,8969
	Sensitivity	1,0000	1,0000	1,0000	1,0000	1,0000	1,0000
2	Effectiveness	0,8262	0,7592	0,8162	0,8021	0,7210	0,7621
	Precision	0,8731	0,6516	0,5111	0,6238	0,5322	0,6556
	Sensitivity	1,0000	0,6556	0,5934	0,6983	0,8000	0,7382
3	Effectiveness	0,6962	0,6892	0,6462	0,6912	0,7362	0,7120
	Precision	0,7231	0,5556	0,5556	0,7778	0,7300	0,5911
	Sensitivity	0,7500	0,5556	0,5556	0,6364	0,8000	0,6556
Average	**Effectiveness**	**0,8064**	**0,7818**	**0,7864**	**0,7967**	**0,7847**	**0,7903**
	Precision	**0,8310**	**0,7014**	**0,6545**	**0,7662**	**0,7197**	**0,7145**
	Sensitivity	**0,9167**	**0,7371**	**0,7163**	**0,7782**	**0,8667**	**0,7979**

On the basis on presented results we can state that effectiveness, precision and sensitivity of recognizing sentiment of opinions are high. The cause of this fact is that most often, users must obligatory determining sentiment of opinion. In case the features' sentiment analysis, the performance of this process is lower than the performance of recognizing sentiment of opinions, which means that not all words (expressions) indicating features' sentiment determined manually have been found by an agent. It caused mainly from the fact that not all of these words (expressions) were appearing in learning set. Low values of measures of group 3 opinions' sentiment analysis mean that the sentiment of the features in many cases had not been recognized correctly, i.e. many features having positive opinion had been recognized as features of a negative opinion or just the opposite.

Taking into consideration average values of particular measures one can see that the highest performance of analysis has been obtained with respect to feature "food". The lowest performance can be noted in case of the sentiment of the features "rooms" and "price".

On the basis environment scanning agent stated also that customers prefer to buy hotel services paying attention mainly to the features of:

- location,
- rooms,
- personnel,

Therefore managers of hotels must strive to improve the quality of service in these areas, as well as to take into consideration these areas during developing marketing strategies.

Environment scanning is performed by cognitive agent in near real time.

6 Conclusions

Environment scanning is a main issue in order to strengthening competitiveness. It can be performed by using IT solutions, such as cognitive agents. The method presented in this paper allows for environment scanning on the basis of users opinions about given services or products. Based on these opinions, the managers of the company can make decisions about improve the quality of the offered services or products. This can lead to acquiring new customers.

The method developed in this paper has some limitations. In many cases a sentiment of user's opinion have not been recognized property. Also quality of services/products has not been analyzed automatically. Therefore, a further work should be related to making changes in the functioning of the algorithm of codelets, as well as their configuration. Also methods for deep analysis should be implemented in LIDA cognitive agent.

Acknowledgement. This research was financially supported by the National Science Center (Decision No. DEC-2013/11/D/HS4/04096).

References

1. Hitt, M., Ireland, D., Hoskinsson, R.: Strategic Management: Competitiveness and Globalization. Thomson South-Western (2007)
2. Indris, S., Primiana, I.: Internal and external environment analysis on the performance of small and medium industries (Smes) in Indonesia. Int. J. Sci. Technol. Res. **4**(4), 188–196 (2015)
3. http://www.businessdictionary.com/definition/external-environment.html. Accessed 11 Apr 2017
4. Voiculet, A., Belu, N., Parpandel, D., Rizea, I.: The impact of external environment on organizational development strategy. MPRA Paper No. 26303, posted 3 (2010)
5. Wheelen, T., Hunger, D.: Strategic Management and Business Policy Concepts and Cases, 11th edn. Prentice Hall International USA, Upper Saddle River (2012)
6. Sebastiani, F.: Machine learning in automated text categorization. ACM Comput. Surv. (CSUR) **34**(1), 1–47 (2002)
7. Wawer, A.: Mining opinion attributes from texts using multiple kernel learning. In: IEEE 11th International Conference on Data Mining Workshops (2011)
8. Pham, L.V., Pham, S.B.: Information extraction for Vietnamese real estate advertisements. In: Fourth International Conference on Knowledge and Systems Engineering (KSE), Danang (2012)

9. Zhang, C., Zhang, X., Jiang, W., Shen, Q., Zhang, S.: Rule-based extraction of spatial relations in natural language text. In: International Conference on Computational Intelligence and Software Engineering (2009)
10. Franklin, S., Patterson, F.G.: The LIDA architecture: adding new modes of learning to an intelligent, autonomous, software agent. In: Proceedings of the International Conference on Integrated Design and Process Technology. Society for Design and Process Science, San Diego (2006)
11. Cheng, D.: Analyze the hotel industry in porter five competitive forces. J. Glob. Bus. Manag. **9**(3), 52–57 (2013)
12. Korte, D., Ariyachandra, T., Frolick, M.: Business intelligence in the hospitality industry. Int. J. Innov. Manag. Technol. **4**(4), 429–434 (2013)
13. Tran, C.: Cognitive information processing. Vietnam J. Comput. Sci. **1**(4), 207–218 (2014). Springer, Heidelberg
14. Bytniewski, A., Chojnacka-Komorowska, A., Hernes, M., Matouk, K.: The implementation of the perceptual memory of cognitive agents in integrated management information system. In: Barbucha, D., Nguyen, N.T., Batubara, J. (eds.) New Trends in Intelligent Information and Database Systems. SCI, vol. 598, pp. 281–290. Springer, Berlin (2015). doi: 10.1007/978-3-319-16211-9_29
15. Hernes, M.: Performance evaluation of the customer relationship management agent's in a cognitive integrated management support system. In: Nguyen, N.T. (ed.) Transactions on Computational Collective Intelligence XVIII. LNCS, vol. 9240, pp. 86–104. Springer, Heidelberg (2015). doi:10.1007/978-3-662-48145-5_5

OpenCL for Large-Scale Agent-Based Simulations

Jan Procházka and Kamila Štekerová(✉)

University of Hradec Králové, Rokitanského 62, 500 03 Hradec Králové, Czech Republic
{jan.prochazka,kamila.stekerova}@uhk.cz

Abstract. NetLogo is a Java-based multi-agent programmable modeling environment. Our aim is to improve the execution speed of NetLogo models with large number of agents by means of heterogeneous computing. Firstly, we describe OpenCL as a suitable computing platform. Then we propose a new NetLogo-to-OpenCL extension (NL2OCL) which encapsulates functionality of OpenCL and enables NetLogo to undertake agents' computations simultaneously on graphic processor units. The architecture of our extension is presented. An experimental flocking model with 40,000 agents is used for evaluation of NL2OCL functioning. When using NL2OCL the simulation runs more than 300-times faster than the original model which was created in NetLogo solely. It means that with NL2OLC, drawbacks in maximum size of the NetLogo model and the simulation speed are tackled. Our approach allows using standard PC configurations with suitable graphical cards for large agent-based simulations while preserving advantages of NetLogo. It is a good alternative for researchers who cannot afford high performance computational systems.

Keywords: Agent-based simulation · Heterogeneous computing · OpenCL · NetLogo

1 Introduction

Agent-Based Models (ABM) are widely applicable in exploration of complex systems which consist of large number of heterogeneous adaptive interacting individuals. Numerous software platforms for creation of ABM are available, for the recent review see [1], lists of active platforms was published in [2]. NetLogo [3] is frequently used in introductory ABM courses, but also in scientific research. It is provided under GPL, the latest version was released in March 2017. NetLogo is written in Java, using a dialect of Logo programming language. It provides an extensive library of models for experimenting with emergence phenomena, HubNet tool for participatory simulations and extensions (e.g. fuzzy, GIS, MATLAB, R, SQL, MATLAB or OWL extension) which make NetLogo even more powerful.

Unhappily, NetLogo is not suitable for large-scale simulations: with the growing number of agents and size of the environment (i.e. dimension of the grid of patches), simulations become slower, visualization/animation cannot be observed, it takes a long time to collect simulation data. Consider a simple NetLogo model with 40,000 agents operating in the grid of 200 × 200 patches, each agent calculates average values of [x, y] coordinates of neighbouring agents in 10-patches radius. Within this

N.T. Nguyen et al. (Eds.): ICCCI 2017, Part I, LNAI 10448, pp. 351–360, 2017.
DOI: 10.1007/978-3-319-67074-4_34

model, one simulation step takes approximately 8 secs on standard PC configuration (CPU Intel i5, 2.30 GHz, 8 GB RAM). Therefore, large-scale models with dozens of millions of agents such as EURACE [4] seem to be impossible in NetLogo.

A five-step strategy for building more efficient NetLogo model is suggested in [5] together with practical recommendations which are related to correct use of NetLogo commands and constructs. Our approach is different: we suggest applying parallel computations within heterogeneous computing system.

We propose an extension which is based on OpenCL platform [6]. In the rest of the paper, we present our NL2OCL extension including the demonstration of its functioning.

2 OpenCL

The advantages of using different types of computational resources within one heterogeneous system (CPU-GPU) are described in [7] where following platforms were tested: OpenCL, CUDA, Brook+, ACML, OpenMP, Pthread, Intel MKL, Intel TBB. The first two of them, OpenCL and CUDA, are the most frequently used and mutually compared. The performance of OpenCL and CUDA on a parallel implementation of the cryptographic algorithm Keccak (the 2012 winner of the new hashing algorithms competition) was equal; authors of [8] mentioned that OpenCL brings the better performance with its each new version.

OpenCL is an industry standard for cross-platform, parallel programming of diverse processors, it is open and royalty-free [9]. OpenCL encapsulates low level API and allows programmers to build programs using GPGPU (General-Purpose Computing on Graphics Processing Units) and upper programming languages. It was developed to ease programming burden when writing applications for heterogenous systems [10]. It was released in December 2008 by Khronos working group and it was soon adopted by main hardware vendors into their SDKs (while CUDA is tight to NVIDIA hardware). The latest version of OpenCL 2.2 was released in 2016 [9].

OpenCL specification defines four models of organization of parallel application: platform model, memory model, execution model and programming model, for details see [8]. OpenCL applications are portable, they should run correctly on any conformant of OpenCL implementation.

A core concept of the OpenCL is a split of the application into two parts: host and device (Fig. 1). The host is an upper level part of the application, it is written in C++ and it runs on CPU. This part is responsible for the overall control of the application, the user interaction, preparation of data for parallel processing, collecting computed results and transforming them into a form suitable for the user. The device is a lower level part of the application. It is executed on computational units (CPU or GPU or both). This part is responsible for fast parallel computations. Programs for the device are called kernels. Kernels are written in a subset of C language (OpenCL C). Interaction and communication between host and device parts are managed by OpenCL Context and Command Queue.

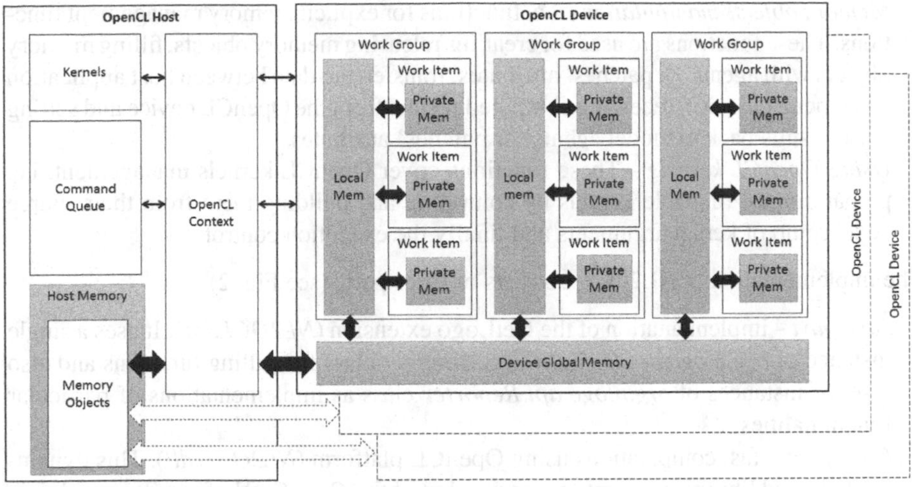

Fig. 1. OpenCL Host and Device application part with explicit memory management (Host-Global-Local-Private memory), *source*: authors.

The lifecycle of the OpenCL application involves:

- creating of context which is associated with one or more devices,
- creating command queue,
- allocating memory buffers (host and device),
- compiling and building OpenCL kernels,
- setting kernel arguments,
- running kernels,
- retrieving results from kernels (copying memory buffers from device to host),
- releasing OpenCL resources.

The OpenCL application should cover platform and device configuration, kernels preparation, execution control and explicit memory management (transferring data objects to/from devices). OpenCL requires installation of low-level software controlling graphic card (driver) and implementation of platform specific SDK which serves as API for programmer, this implementation is always vendor-dependent.

3 NetLogo-to-OpenCL Extension

Our NetLogo-to-OpenCL (NL2OCL) extension is designed to provide three sets of functionalities for:

- *Configuration of the OpenCL running environment* – these functions are responsible for OpenCL platform and devices setup and configuration, creation of OpenCL application context and preparation of OpenCL device task queue serving as an interface between host and device application parts.

- *Memory objects manipulations*, i.e. functions for explicit memory management functions. These functions are used for creating/releasing memory objects, filling memory objects with agents' or patches' attributes, transferring data between host application and OpenCL device, receiving computed results from the OpenCL device and setting these results back to model agents' or patches' attributes.
- *Using OpenCL kernels* – These functions cover OpenCL kernels management, i.e. preparation of runnable kernels by compiling and building them from their source code, setup of kernel arguments and finally the execution control.

The implementation of NL2OCL consists of three parts (see Fig. 2):

- *Java part* – implementation of the NetLogo extension (*NL2OCL.jar*). It uses a single instance of *org.nlogo.api.DefaultClassManager* class for calling functions and also a set of instances of *org.nlogo.api.Reporter* class as implementations of particular functionalities.
- *C++ part* – fast computations using OpenCL platform (*NL2OCL.dll*). This dynamically linked library represents encapsulation of the *OpenCL* platform. It provides an API for calling the functions for configuring the *OpenCL* runtime environment, for memory management and for kernels preparation, arguments setting, running and receiving results.
- *JNI part* – interface between Java and C++ parts. The interface is used to call C++ functions from Java part of the *NL2OCL* extension. It is implemented as a *singleton* which instance is used in the *Reporters*.

Interconnecting of components is managed by following interfaces:

- *I1* is the interface between OpenCL library opencl.lib and the dynamically linked library NL2OCL.
- *I2* is JNI interface between Java and C++ parts of the implementation.
- *I3* is interconnection between NetLogo models and the NL2OCL extension.
- *I4* is an optional interface to other Java agent-based simulation software, in this case there is no need to use ClassManager and Reporters.

Agents (as objects) are specified within the NetLogo model. NL2OCL enables GPU to perform fast, parallel calculations for those agents using agents' and/or patches' attributes. The explicit memory management is necessary to supply GPU device with data as well as to receive results back to the host application and finally back to the NetLogo model. NL2OCL allows to fulfil requirements of this explicit memory management in the following steps (see also Fig. 3):

- *NetLogo -> Host* – copying values from agent/patch attributes into the memory buffers of the host application.
- *Host -> Device* – transfer of memory objects from the OpenCL host into the device memory. Data must be transferred from the host application into the global memory of the device to enable GPU to work with it. This transfer is done upon the kernel arguments setup.
- *Device -> Host* – returning results back to the host application (into the prepared memory object) from the device.

– *Host -> NetLogo* – transfer of results to the original agent-based model, mapping to agents' or patches' attributes.

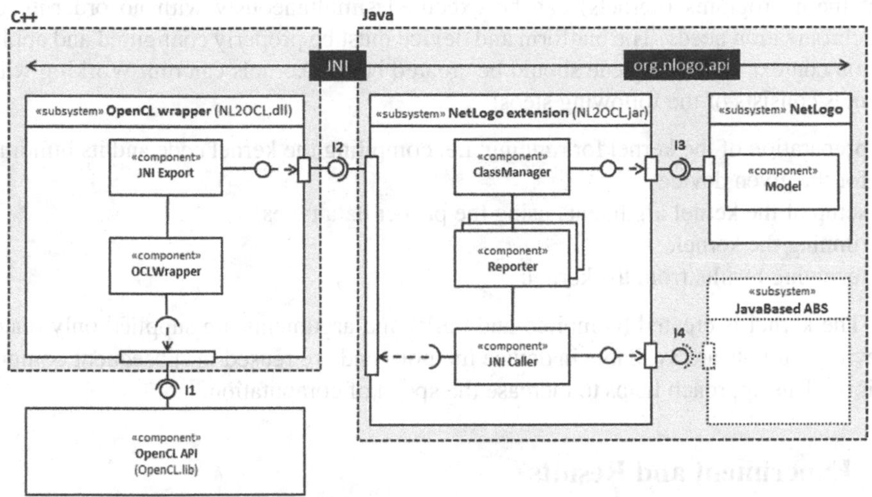

Fig. 2. NL2OCL: component diagram, *source*: authors.

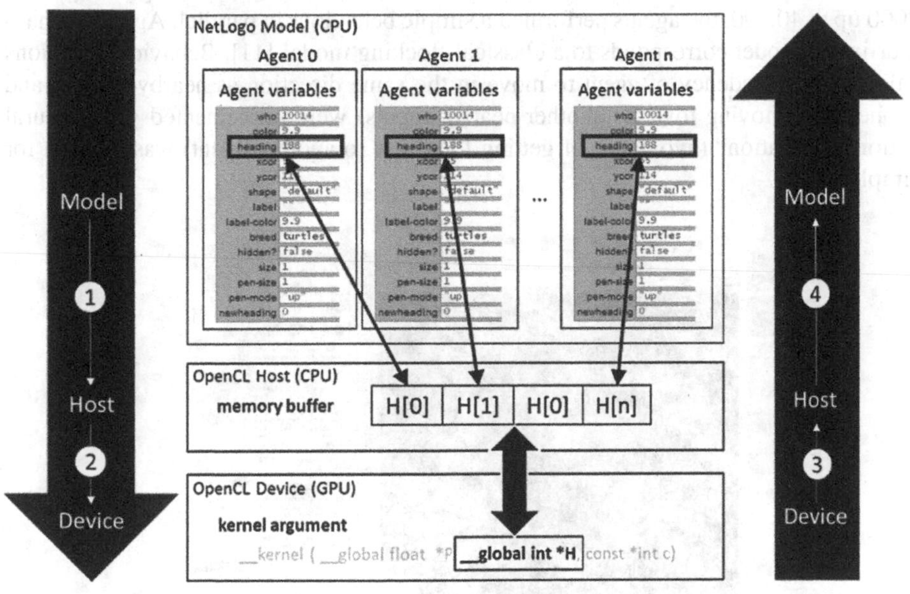

Fig. 3. Memory transfers in NL2OCL, *source*: authors.

Running kernels with arguments and receiving results are the basic operations which are executed repeatedly. NL2OCL allows running any valid (properly compiled and built) kernel with any kind of kernel arguments directly from the NetLogo environment.

The design of kernels and data structures which are used as kernel arguments is up to a model developer. However, one kernel is usually assumed to be an independent small program, it computes outputs for one agent within one simulation step. It is also assumed that these programs (kernels) can be executed simultaneously with no ordering or synchronization needs. The platform and device must be properly configured and application context and task queue should be created before kernels can run. Working with kernels consists of the following steps:

- preparation of the kernel for running, i.e. compiling the kernel code and its building for the given device,
- setup of the kernel arguments using the proper data types,
- running the kernel,
- receiving results from the kernel.

The kernel is created (compiled and built) and arguments are supplied only once, previous outputs are available in device memory and are reused in subsequent computations. This approach helps to increase the speed of computation.

4 Experiment and Results

For testing purpose, we created a micro-simulation flocking model (Fig. 4). We considered the environment of 200 × 200 patches, the number of agents was growing from 2000 up to 40,000, the agents performed a simple behaviour in parallel. Agents' behaviour in the model corresponds to a classical flocking model [11]. Behavioural actions "alignment" (tendency of agent to move in the same direction as nearby agents) and "cohesion" (moving towards another nearby agents) were implemented. Behavioural action "separation" (avoidance of getting too close to another agent) was omitted for simplification.

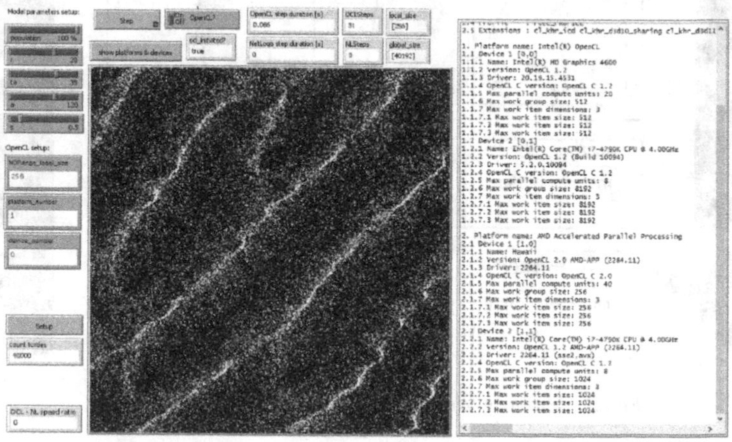

Fig. 4. Interface of the flocking model with the list of platforms and devices, *source*: authors.

The primary aim of the experimental model was to make the model computationally demanding. Each agent calculated its new direction of movement using the information about positions of agents in the agent's circle neighbourhood with 20 patches radius so the total number of elementary computations to be done for one agent and one simulation step reached 50,000,000 for the model with 40,000 agents. While NetLogo treated agent objects and their graphical representation in the model environment all calculations needed to follow the behavioural rules were undertaken by OpenCL kernel via NL2OCL extension. The kernel itself was programmed in the subset of the C language (*OpenCL C*) it served as an elemental calculation-routine for one agent. Calculations were done simultaneously for many agents at the same time in means of data parallelization approach.

The kernel works in the following way:

- Firstly, the agent's position was recognized within the global device memory, i.e. the patch was identified on which the agent was placed.
- Then the neighbouring agents in the circle radius were found.
- A new movement direction based on the information about positions and directions of the circle radius neighbour agents is calculated.
- Agent moves ahead in its new direction. A new absolute position is calculated and a target patch is identified.

The Fig. 5 shows the kernel function prototype with the following meaning of arguments: (1, 2) arrays of coordinates of agents and patches, (3) array of agents' heading, (4) array of counts of agents on patches, (5) count of agents, (6) width and height of the grid, (7) maximum x and y coordinates of the grid, (8) influence radius and its square, (9) angle of influence and turning angle, (10) agent's step length.

```
__kernel void calculateHeading(
    __global int *PX, __global float *X,          (1)
    __global int *PY, __global float *Y,          (2)
    __global int *H,                              (3)
    __global int *C,                              (4)
    const int c,                                  (5)
    const int w, const int h,                     (6)
    const float maxx, const float maxy,           (7)
    const int r, const int r2,                    (8)
    const int a, const int ta,                    (9)
    const float s)                                (10)
```

Fig. 5. Kernel function prototype and arguments, *source*: authors.

During the experiment, the number of agents grows from 2000 up to 40,000. The duration of simulation was observed (a) when computing was done solely in NetLogo, (b) when NL2OCL extension was used to parallelize computations.

When using NetLogo only, the mean value of the duration of one simulation step was 21.71 s on average and 100 steps of the simulation took more than 30 min. With NL2OCL extension, the mean value of the duration of one simulation step was 0.0066 s and the first hundred steps of the simulation took 7 s. It means that the overall speed increases when computations are parallelized. It was 100-times faster for 2000 agents and more then 300-times faster for 40,000 agents. Results are shown in Fig. 6.

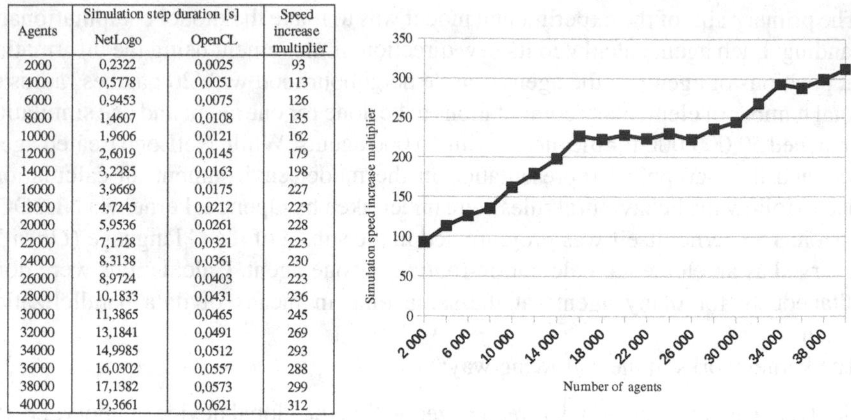

Agents	Simulation step duration [s]		Speed increase multiplier
	NetLogo	OpenCL	
2000	0,2322	0,0025	93
4000	0,5778	0,0051	113
6000	0,9453	0,0075	126
8000	1,4607	0,0108	135
10000	1,9606	0,0121	162
12000	2,6019	0,0145	179
14000	3,2285	0,0163	198
16000	3,9669	0,0175	227
18000	4,7245	0,0212	223
20000	5,9536	0,0261	228
22000	7,1728	0,0321	223
24000	8,3138	0,0361	230
26000	8,9724	0,0403	223
28000	10,1833	0,0431	236
30000	11,3865	0,0465	245
32000	13,1841	0,0491	269
34000	14,9985	0,0512	293
36000	16,0302	0,0557	288
38000	17,1382	0,0573	299
40000	19,3661	0,0621	312

Fig. 6. Experimental results, *source*: authors.

We noticed a good scalability of the parallel solution. With increasing number of agents (and increasing computational requirements) the parallelization and communication costs (especially the costs of memory transfers from NetLogo to OpenCl and back) raised significantly slower.

Visualization of emergence of flocks during the first 100 steps of the simulation run is presented in Fig. 7. The snapshots were taken in every 10th step. It shows the emergence of patterns during the simulation.

Fig. 7. Visualization of emergence of flocks during the first 100 steps of the simulation run, *source*: authors.

5 Related Work and Further Research

Our flocking model serves just as a demonstration of NL2OCL capabilities. A similar experimental flocking model (Reynolds's boids) was presented in [12] where authors explored the feasibility of the *GPU Environmental Delegation of Agent Perception* principle based on a clear separation of agents' behavior (managed by CPU) and environmental dynamics (managed by GPU). The authors used CUDA to implement a GPU module thus their implementation is tight to NVIDIA graphic cards. The GPU module is integrated to TurtleKit, a generic Java based spatial ABM. The authors reported 25% performance increase (256 × 256 patches, 1000–40,000 agents).

The approach we chose for NL2OCL implementation allows using both data and task parallelization, it means that our extension is open to a wide range of parallelization strategies. We can apply general parallelization strategies introduced by [13] where parallelization of spatial agent-based simulations is based on decomposition of models into components which can be processed independently, either as agent-parallel or as environment-parallel. The authors demonstrated the application of parallelization strategies using the predator-prey for high performance computing model as Java multithread application. We plan to adopt these parallelization strategies also for ABS parallelization based on GPU computations.

We do not want to limit NL2OCL to be only a technical solution enabling GPU computations in ABS. We want to enhance NL2OCL into a full framework for large scale ABS parallelization similar to a framework presented in the work [14]. Authors of that framework presented useful mechanisms of stochastic memory allocator for parallel agent replication, prioritization of agents' parallelly processed actions and GPU usage for statistical measures.

Next steps of transforming NL2OCL into the full ABS parallelization framework go in two directions:

- To provide NL2OCL with a set of well-defined programming patterns assisting model developers to overcome burdens of low-level aspects of GPU computations. We plan to utilize results of existing works related to a parallel application development using algorithmic skeletons, e.g. [16, 17].
- To explore possibilities of automation of agents' behavior algorithms transformation into their parallel form. We will build on results of existing work [15] in which authors presented automated searching of agents' behavior algorithms via utilizing genetic programming with GPU computations.

6 Conclusion

Parallel computations on graphic cards within standard PC configuration is a promising alternative to high performance commercial computing platforms. Our NetLogo-to-OpenCL extension is designed to enable running large agent-based simulations with limited hardware. The initial experiments proved that the extension works well. Our next research will be focused on defining general patterns and guidelines for transformation of NetLogo models, especially computational parts performed by individual agents, into OpenCL kernels. We intend to provide automatic or semi-automatic OpenCL kernels generator for preparation of OpenCL kernels directly from the NetLogo code. The NetLogo Models Library will be used for further systematic experimenting.

Acknowledgement. The financial support of the Specific Research Project *Autonomous Socio-Economics Systems* of FIM UHK is gratefully acknowledged.

References

1. Kravari, K., Bassiliades, N.: A survey of agent platforms. J. Artif. Soc. Soc. Simul. **18**(1), 11 (2015)
2. OpenABM Homepage. https://www.openabm.org/modeling-platforms. Accessed 26 Apr 2017
3. NetLogo Homepage. http://ccl.northwestern.edu/netlogo/. Accessed 26 Apr 2017
4. Deissenberg, C.H., Hoog, V.D., Dawid, H.: EURACE: massively parallel agent-based model of the European economy. http://www.eurace.org. Accessed 26 Apr 2017
5. Railsback S., Ayllón, D., et al.: Improving execution speed of models implemented in NetLogo. J. Artif. Soc. Soc. Simul. **20**(1) (2017)
6. OpenCL Homepage. https://www.khronos.org/opencl/. Accessed 26 Apr 2017
7. Mittal, S., Vetter, J.: A survey of CPU-GPU heterogenous computing techniques. ACM Comput. Surv. **47**, 69 (2015)
8. Souza, A.M., Pereira, F.D., Ordonez, E.D.M.: Exploiting heterogenous systems: Keccak on OpenCL. In: The 2013 International Conference on Parallel and Distributed, Processing Techniques and Applications (PDPTA'13) (2013)
9. Bourd, A. (ed.): The OpenCL Specification, version 2.2. https://www.khronos.org/registry/OpenCL/specs/opencl-2.2.pdf. Accessed 26 Apr 2017
10. Gaster, B.R., et al.: Heterogenous Computing with OpenCL. Elsevier, Amsterdam (2012)
11. Wilensky, U.: NetLogo flocking model. http://ccl.northwestern.edu/netlogo/models/Flocking. Accessed 26 Apr 2017
12. Hermellin, E., Michel, F.: GPU environmental delegation of agent perceptions: application to Reynolds's boids. In: Gaudou, B., Sichman, J.S. (eds.) MABS 2015. LNCS, vol. 9568, pp. 71–86. Springer, Cham (2016). doi:10.1007/978-3-319-31447-1_5
13. Fachada, N., Lopes, V.V., Martins, R.C., Rosa, A.C.: Parallelization strategies for spatial agent-based models. Int. J. Parallel Program. **45**(3), 449–481 (2017)
14. Lysenko, M., D'Souza, R.M.: A framework for megascale agent based model simulations on graphics processing units. J. Artif. Soc. Soc. Simul. **11**(4), 10 (2008)
15. Berkel, S., et al.: Automatic discovery of algorithms for multi-agent systems. In: Proceedings of the 14th Annual Conference Companion on Genetic and Evolutionary Computation (GECCO 2012), Philadelphia, USA (2012)
16. Ernsting, S., Kuchen, M.: Data parallel algorithmic skeletons with accelerator support. Int. J. Parallel Program. **45**(2), 283–299 (2016)
17. Wrede, F., Ernsting, S.: Simultaneous CPU–GPU execution of data parallel algorithmic skeletons. Int. J. Parallel Program., 1–20 (2017)

A Novel Space Filling Curves Based Approach to PSO Algorithms for Autonomous Agents

Doina Logofătu[1(✉)], Gil Sobol[2], Daniel Stamate[3], and Kristiyan Balabanov[1]

[1] Computer Science Department of Frankfurt, University of Applied Sciences, Nibelungenplatz 1, 60318 Frankfurt am Main, Germany
logofatu@fb2.fra-uas.de

[2] Industrial Engineering, Technion - Israel Institute of Technology, Haifa, Israel

[3] Department of Computing, Goldsmiths College, University of London, London SE146NW, UK

Abstract. In this work the swarm behavior principles of Craig W. Reynolds are combined with deterministic traits. This is done by using leaders with motions based on space filling curves like Peano and Hilbert. Our goal is to evaluate how the swarm of agents works with this approach, supposing the entire swarm will better explore the entire space. Therefore, we examine different combinations of Peano and Hilbert with the already known swarm algorithms and test them in a practical challenge for the harvesting of manganese nodules on the sea ground with the use of autonomous robots. We run experiments with various settings, then evaluate and describe the results. In the last section some further development ideas and thoughts for the expansion of this study are considered.

Keywords: Autonomous agents · Space filling curves · Particle swarm optimization · Deterministic leaders · Application

1 Introduction

Simultaneously with the applied research on renewable resources, it is useful to find novel ways for opening up fossil ones. As example, manganese nodules can be found on the sea bottom. A considerable application field involves rust and corrosion prevention on steel [9,10]. The degradation could be reduced substantially by collecting these manganese nodules from the sea bottom using specialized autonomous agents. Our focus in this work is to evaluate different ways in handling the movement of these agents. The experiments can be extended to cover other collecting tasks. The base for our application is a framework for simulation and improvement of swarm behavior in changing environments [1], which we redesign and extend. It simulates the swarm behavior by using the principles of Craig W. Reynolds [2]. The main purpose of the framework regarding the application is to deploy agents with a specific strategy and then to gather them. While gathering, the agents are collecting the manganese which is distributed on

N.T. Nguyen et al. (Eds.): ICCCI 2017, Part I, LNAI 10448, pp. 361–370, 2017.
DOI: 10.1007/978-3-319-67074-4_35

every position in the coordinate system. Once gathered together, there is no more movement and the simulation ends. Naturally manganese occurs in form of nodules, thus it is distributed uniformly. For the different forms of the manganese distributions, we created several benchmarks used in the results' comparison. The next step of improvement would be the collecting procedure. The greater distance the agents move, the higher is the probability to find manganese. Consequently, we intend to reach a way for passing through a larger area. The easiest solution would be to define for each agent its own path. This would probably scatter the swarm because of the bad orientation, the changing environment and the uneven surface. Most of the research works regarding swarm behavior are inspired by nature like genetic algorithms or particle swarm optimization. These outcomes focus on fish schools or bird flocks. An alternative discussion could consider, for example, a pack of wolves. A pack of wolves means actually autonomous individuals with a specific hierarchy. Not every wolf has the same power regarding decisions for the pack. Normally there is one wolf who leads the group and the others are followers [11]. This contribution aims to study this notion more closely. We intend to set one or more leaders who will move after a given route, but still be part of the swarm, and the rest calculate their new position, that means every iteration in consideration of all agents.

2 Background

This section introduces previous work the application is based on, followed by three main topics: Moving Algorithms, Particle Swarm Optimization, Hilbert and Peano Curves.

2.1 Framework for Adaptive Swarms Simulation and Optimization

The application we consider first is based on [1]. The framework is an application that runs a simulation of autonomous robots using moving algorithms Random, Square, Circle, Gauss, and Bad Centers [1]. It contains several fundamental deployment strategies used from where the moving algorithms start. The front end uses the open source framework of *processing.org* [4]. It creates the chosen deployment strategies and calculates the movement of the autonomous agents, as well as the collection of manganese nodules. In addition, the number of agents can be settled and it counts the distance in walked meters of all agents together.

2.2 Moving Algorithms

In our practical application, it is required to build a swarm of autonomous agents, where each agent individually moves forward taking into consideration the other agents of the group. There are efficient algorithms for swarm behavior and movement of agents that are implemented in the application [6]. The previous work [1] uses a simplification of the bird flock movement described by Craig W. Reynolds [2]. The contribution implemented three different algorithms that run simultaneously: cohesion, separation, alignment.

2.3 Particle Swarm Optimization

Particle Swarm Optimization (PSO) was introduced in 1995 by J. Kennedy and R. Eberhart [6]. The innovation was building swarm behavioral approaches for solving problems by iteratively improving a candidate's solution until termination criteria is satisfied [7]. It is similar to a genetic algorithm as both algorithms are initialized with a random population, in PSO called particles. The difference is that in PSO algorithms, each particle is assigned to a randomized velocity and the particles move through hyperspace. Each particle is defined by its position, velocity, current objective value and personal best value of all time. PSO also keeps track of the global best value that is the best objective value of all particles and also the corresponding position.

2.4 Space Filling Curves

A Space Filling Curve is a special function of calculus that fully covers a two or three dimensional space. Giuseppe Peano (1858–1932) discovered them first in 1890. He wanted to create a continuous mapping construction from the unit interval onto the unit square [7].

Peano Curve. Till 1890 one assumed that a constant curve with parametric function of only one variable $x = \phi(t)$ and $y = \psi(t)$, cannot reflect surjectively the unit interval onto the unit square. The reason for this was the theorem of Eugen Natto, who showed that a bijection must be unsteady to satisfy this. However, Peano found a steady function f_p, such that $f_p(I) = 2$.

Definition 1 Peano Curve [10]. *The projection* $f_p : I \to 2$ *with*

$$f_p(0_3, t_1t_2t_3t_4) = \begin{pmatrix} 0_3 \; t_1(k^{t_2}t_3)(k^{t_2+t_4}t_5) \; ... \\ 0_3 \; (k^{t_1}t_2)(k^{t_1+t_3}t_4) \quad ... \end{pmatrix}.$$

and the operator $k^{t_j} = 2 - t_j (t_j = 0, 1, 2)$, *where* k^v *is the* v*-th iteration of* k, *we call Peano Curve.*

So according to this definition we have:

$$f_p(0_3, 00t_3t_4...) = \begin{pmatrix} 0_3, \, 0\xi_2\xi_3\xi_4 \cdots \\ 0_3, \, 0\eta_2\eta_3\eta_4 \cdots \end{pmatrix}.$$

To create the Peano's curve we start at point $(0, 0)$ and finish in the diagonal corner at point $(1, 1)$. The starting point of a sub square must be the endpoint of the previous sub square. Figure 1 illustrates where the start- and endpoints are with the use of arrows.

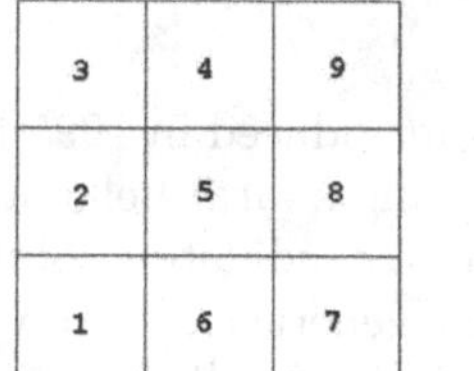

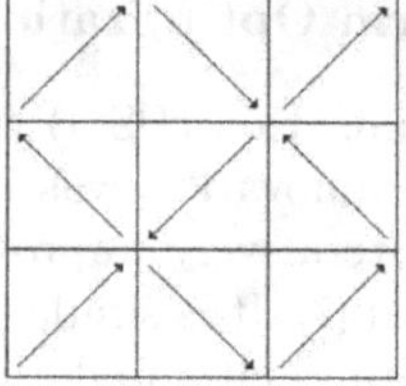

Fig. 1. Construction order and orientation for the Peano Curve.

Hilbert Curve. Peano (1890) introduced first the space-filling curves. Hilbert (1891) popularized their existence and gave an insight into their generation. His statement was that if the unit interval can be mapped steady onto the unit square, then also sub intervals can be mapped steadily onto sub squares. In the first step, Hilbert divided the unit interval into four sub intervals of the same size as well as the unit square into four equally sized sub squares, where each sub interval is mapped onto one sub square [7].

3 Implementation Details

This section describes shortly the practical changes and extensions that were necessary to implement for the experimental procedure. At first, some new classes had to be implemented to lay the basis for the new *Manganese-Nodule-Model.* These new classes help us to represent the nodules on the map and for background calculations as well as. Some of the new implemented classes and interfaces include: *DeployRing*, *DeploymentStrategy*, *ManganeseNodule*, *VisualManganeseNodule.* The used benchmarks for the manganese nodules distribution are independent files. Each line represents a y-value and each char represents an x-value in the coordinate system of the graphical user interface. The lines are filled with numbers from 0 to 7 in accordance with the size of the nodule, where zero means that no nodule can be found on this position.

3.1 Peano/Hilbert Algorithms

The Peano algorithm is implemented using a recursive function, that is called every time when the agent moves into the next unit square. The function calls change from clockwise rotation to negative rotation (counterclockwise). The implementation of the Hilbert Algorithm is analogous to the Peano Algorithm (Algorithm 1). The basic structure of how the exploring through the sub squares is done is fixed.

4 Experimental Results

This section presents relevant results we achieved with the extended implementations to the application. The distance and the collected amount of manganese

Algorithm 1. Pseudo Code Peano Algorithm

```
peanoAlgorithm(length, direction, rotation, deep){
if under lowest level then
| return
end
    peanoAlgorithm(length, direction, clockwise rotation, deep-1)
    step forward with given length and direction

    peanoAlgorithm(length, direction, counterclockwise rotation, deep-1)
    step forward with given length and direction

    peanoAlgorithm(length, direction, rotation, deep-1)
    direction turn clockwise with given rotation degree
    step forward with given length and direction

    direction turn clockwise with given rotation degree
    peanoAlgorithm(length, direction, counterclockwise rotation, deep-1)
    step forward with given length and direction

    peanoAlgorithm(length, direction, rotation, deep-1)
    step forward with given length and direction

    peanoAlgorithm(length, direction, counterclockwise rotation, deep-1)
    direction turn counterclockwise with given rotation degree
    step forward with given length and direction

    direction turn counterclockwise with given rotation degree
    peanoAlgorithm(length, direction, rotation, deep-1)
    step forward with given length and direction

    peanoAlgorithm(length, direction, counterclockwise rotation, deep-1)
    step forward with given length and direction

    peanoAlgorithm(length, direction, rotation, deep-1)
}
```

of all agents in one pass are the examined variables. The difference of agents between *Rob Total* and the sum of *Rob Hilbert* and *Rob Peano* are robots behaving according to the principles of typical Moving Algorithms.

4.1 Diamond, Square, Peano 0-50

In this experiment we increased the number of Peano Robots and ran 1000 iterations with every increase. This experiment runs with the benchmark Diamonds and the deployment strategy Square. With every increase of the number of Peano Robots, the covered distance of all robots increases by 30 000 m to 50 000 m with an average increase of 46 081.46 m. The collected manganese does not increase constantly. The global maximum of 53 080 kg is reached with a constellation of 44 Peano Robots (see Fig. 2, left). The biggest discrepancy of 15 is between the first and the second measurement, with an increase of 589 %. The fewer meters a robot has to travel for the same amount of manganese, the more efficient it is. The second diagram in Fig. 2 (right) shows this relation of average distance per kg manganese for each number of Peano Robots. The best efficiency occurs without any Peano robot in the simulation with an average distance of 3 m kg^{-1} manganese. But as we can see in the other diagram (Fig. 2, left) the total amount of manganese is very little. So we want to focus on analyzing all cases where Peano robots are involved. There are a few amounts of Peano robots with a very close distance per kg manganese. This is the case with the amount of 2, 3, 4, 5 and 6 Peano robots (average absolute deviation 3.9 m), with the amount of 16 to 21 Peano robots (average absolute deviation 3 m) or with the amount of 36 to 42 Peano robots (average absolute deviation 2.1 m). Another interesting point can be seen at 44 Peano Robots where the total manganese maximum is. The distance per kg manganese diagram shows here a local minimum of 383 m kg^{-1} manganese. This leads to the conclusion that we have a reasonably efficient constellation (Fig. 3).

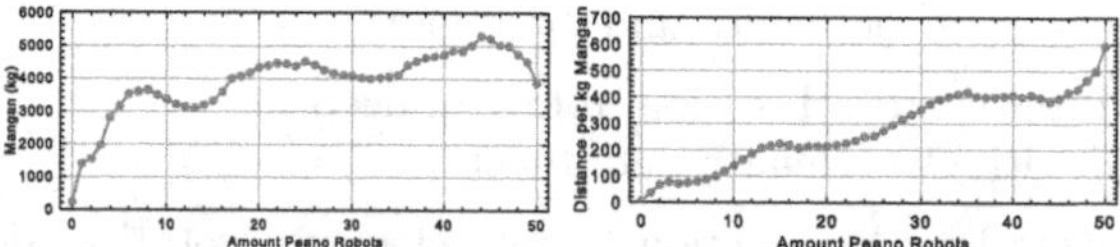

Fig. 2. Analysis of Diamonds, Square, Peano 0-50 increase: collected amount of manganese for (left); relation between the total amount of collected manganese and the distance all robots have covered.

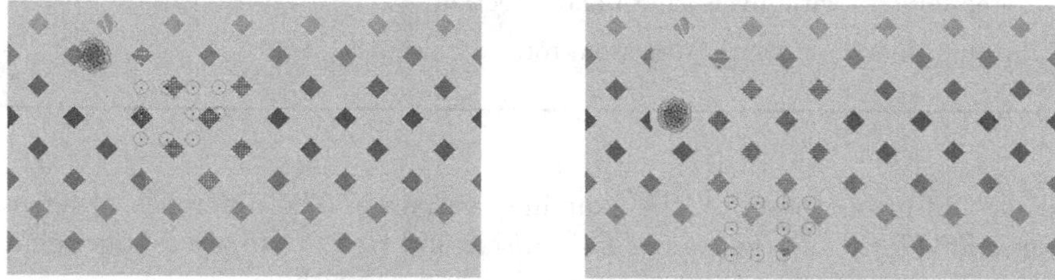

Fig. 3. Diamond, Square, Peano 0-50: Screenshots experimental procedure after 500 iterations (left) resp. after 1000 iterations (right).

To run the simulation only with Hilbert Robots brings unsatisfactory results. In this case, the total amount of manganese breaks down roughly 21 % compared to the simulation run with 49 Hilbert Robots.

4.2 Lines, Square, Peano 0-50

In this experiment we switched our benchmark to the benchmark Lines. The deployment strategy is Square and we increase the number of Peano Robots from 0-50 (Fig. 5).

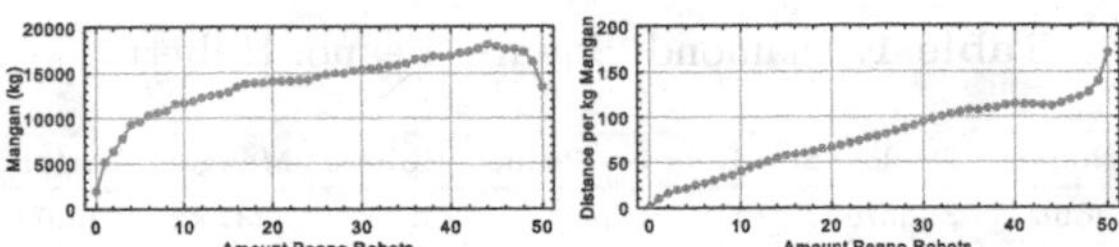

Fig. 4. Analysis of Lines, Square, Peano 0-50 increase: collected amount of manganese (left); relation between the total amount of collected manganese and the distance all robots have covered (right).

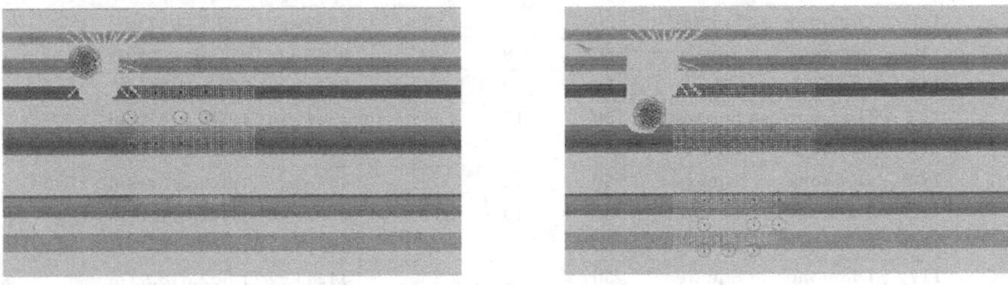

Fig. 5. Lines, Square, Peano 0-50: Screenshots experimental procedure after 500 iterations (left) resp. after 1000 iterations (right).

With every increase of the number of Peano Robots, the covered distance of all robots increases by 30 000 m to 50 000 m with an average increase of 46 081.46 m. This is identical to the results in Table 1. The collected manganese increases constantly up to an amount of 38 Peano Robots. The global maximum of 18 090 kg is reached with a constellation of 44 Peano Robots (see Fig. 4, left). With an amount of 6 Peano Robots we achieve a result of 10 341 kg total manganese. This is more than 50 % from what we achieve with our global maximum with 44 Peano Robots. That means, we have achieved half of the global maximum with an efficiency of 26.99 m kg^{-1} manganese in contrast to 112.48 m kg^{-1} manganese. In conclusion we get 100 % more manganese for 416.75 % less efficiency. That is in no reasonable relation to the benefits. Overall the distance per kg manganese increases almost linearly up to the global maximum of total manganese with 44 Peano Robots and goes steeply up afterwards. It is striking that this experiment has its maximum with the same amount of Peano Robots. The only thing that distinguishes these two experiments are the used benchmark maps.

4.3 Diamond, Square, Peano 1-25-1, Hilbert 1-49

In this experiment we mixed Hilbert and Peano Robots as well as robots using the moving Algorithms described by [2]: Cohesion, Separation and Alignment. We begin with increasing the amount of both Peano and Hilbert Robots from 0 up to 25. Then all robots are either Peano or Hilbert Robots. From this point we decrease the amount of Peano Robots and keep increasing the amount of Hilbert Robots. This experiment runs with the benchmark Diamonds and the deployment strategy Square. A total of 1000 iterations were ran with every increase. The results are presented in Table 1.

Table 1. Diamond, Square, Peano, Hilbert

Exp.	Benchmark	Deploy. St.	Robots	Peano	Hilbert	Mangan	Distance
102	Diamond	Square	50	0	0	237 kg	673.03 m
103	Diamond	Square	50	1	1	1567 kg	114,910.56 m
104	Diamond	Square	50	2	2	2133 kg	220,147.01 m
105	Diamond	Square	50	3	3	2715 kg	310,008.24 m
106	Diamond	Square	50	4	4	3222 kg	406,158.29 m
107	Diamond	Square	50	5	5	3402 kg	519,118.26 m
108	Diamond	Square	50	6	6	2976 kg	644,629.99 m
109	Diamond	Square	50	7	7	2652 kg	763,532.13 m
110	Diamond	Square	50	8	8	3068 kg	871,160.00 m
111	Diamond	Square	50	9	9	3854 kg	963,823.71 m
112	Diamond	Square	50	10	10	3944 kg	1,040,938.41 m
113	Diamond	Square	50	11	11	4597 kg	1,125,638.16 m
114	Diamond	Square	50	12	12	4628 kg	1,226,243.96 m
115	Diamond	Square	50	13	13	4587 kg	1,341,671.41 m
116	Diamond	Square	50	14	14	4480 kg	1,466,358.73 m
117	Diamond	Square	50	15	15	4424 kg	1,602,618.82 m
118	Diamond	Square	50	16	16	4631 kg	1,730,787.65 m
119	Diamond	Square	50	17	17	4792 kg	1,847,645.98 m
120	Diamond	Square	50	18	18	4971 kg	1,955,750.29 m
121	Diamond	Square	50	19	19	4978 kg	2,049,028.65 m
122	Diamond	Square	50	20	20	5155 kg	2,126,192.66 m
123	Diamond	Square	50	21	21	4866 kg	2,189,012.96 m
124	Diamond	Square	50	22	22	4972 kg	2,260,848.88 m
125	Diamond	Square	50	23	23	5190 kg	2,350,150.97 m
126	Diamond	Square	50	24	24	5102 kg	2,458,013.63 m
127	Diamond	Square	50	25	25	5386 kg	2,525,939.83 m
128	Diamond	Square	50	24	26	5494 kg	2,538,747.79 m
129	Diamond	Square	50	23	27	5439 kg	2,550,909.82 m
130	Diamond	Square	50	22	28	5583 kg	2,562,398.69 m
131	Diamond	Square	50	21	29	5640 kg	2,573,282.37 m
132	Diamond	Square	50	20	30	5646 kg	2,583,499.01 m
133	Diamond	Square	50	19	31	5709 kg	2,593,153.25 m
134	Diamond	Square	50	18	32	5709 kg	2,602,156.73 m
135	Diamond	Square	50	17	33	5646 kg	2,610,406.45 m
136	Diamond	Square	50	16	34	5709 kg	2,617,783.89 m
137	Diamond	Square	50	15	35	5625 kg	2,624,155.36 m
138	Diamond	Square	50	14	36	5691 kg	2,629,374.63 m
139	Diamond	Square	50	13	37	5679 kg	2,635,330.66 m
140	Diamond	Square	50	12	38	5631 kg	2,642,137.89 m
141	Diamond	Square	50	11	39	5640 kg	2,649,930.41 m
142	Diamond	Square	50	10	40	6078 kg	2,658,829.66 m
143	Diamond	Square	50	9	41	5826 kg	2,668,914.30 m
144	Diamond	Square	50	8	42	5590 kg	2,680,376.23 m
145	Diamond	Square	50	7	43	5157 kg	2,692,600.50 m
146	Diamond	Square	50	6	44	4784 kg	2,705,362.02 m
147	Diamond	Square	50	5	45	4674 kg	2,718,559.67 m
148	Diamond	Square	50	4	46	4771 kg	2,732,079.43 m
149	Diamond	Square	50	3	47	5165 kg	2,745,848.07 m
150	Diamond	Square	50	2	48	4994 kg	2,759,815.88 m
151	Diamond	Square	50	1	49	5084 kg	2,773,938.04 m

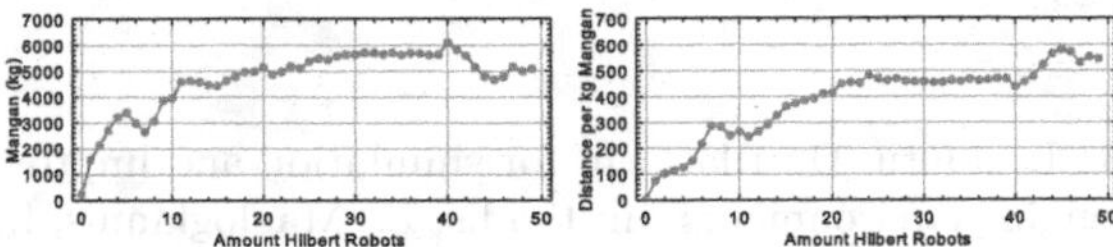

Fig. 6. Analysis of Diamonds, Square, Peano 1-25-1, Hilbert 1-49 increase: collected amount of manganese (left); relation between the total amount of collected manganese and the distance all robots have covered.

On the whole the results of this experiment are related to the two experiments we did before. The Total Mangan diagramm (Fig. 6, left) shows a rapid growth in the first 25 iterations. This is justified because every time we increase the amount of robots by two, one Hilbert Robot and one Peano Robot. For the next 15 iterations (25-10 Peano & 25-40 Hilbert) the total collected amount of mangan is more or less even. This leads to the conclusion that both algorithms have a similar efficiency, as apparently the proportion of the robots between Peano and Hilbert are distributed, as long as there is a minimum of 20% of the other programmed Robots. If we pass this 20%, the efficiency breaks down very fast and there is a result of 100 more meter per kg mangan for the proportion of 1 Peano/49 Hilbert to 10 Peano/40 Hilbert. The latter marks also the maximum of the overall amount of collected mangen with 6078 kg. These are approximately 700 kg more mangan than running only one of the algorithms.

5 Conclusions and Future Work

This work focuses only on the idea to combine swarm behavior with deterministic leaders by using space filling curves. There are still many more things to try in order to go deeper into this topic, especially using the applied research. For example, we can consider leaders with varied power (weight), i.e. the one who collected the most manganese the last time (number of iterations, seconds) will get the highest weight when calculating the next position of each autonomous agent of the group. Additionally, a distributed system could be considered and thereby the communication between the agents would be intensified. In order to get as much manganese as possible, the swarm could divide and follow different leaders or we can vary the number of agents following a specific leader. If the leader loses power, some agents can join another swarm. The leader stays on his deterministic path: thus the chance to find undetected manganese fields remains high. If a leader does not find anything for a long time, he may become a follower and join a swarm. This could also be possible the other way around. If there is a big swarm, new leaders could be chosen to search in a specific direction. Another extension would be to integrate deterministic motions with genetic algorithms, for instance build populations with different amounts of Hilbert, Peano and swarm agents, like this work already did, but develop the next generation using the principles of genetic algorithms

References

1. Canyameres, S., Logofătu, D.: Platform for simulation and improvement of swarm behavior in changing environments. In: Iliadis, L., Maglogiannis, I., Papadopoulos, H. (eds.) AIAI 2014. IAICT, vol. 436, pp. 121–129. Springer, Heidelberg (2014). doi:10.1007/978-3-662-44654-6_12
2. Reynolds, W.: Boids (simulated flocking). http://www.red3d.com/cwr/boids. Accessed 15 June 2017
3. Shyr, W.-J.: Parameters determination for optimum design by evolutionary algorithm. 10.5772/9638. Accessed 15 June 2017
4. Fry, B., Reas, C.: Processing. https://processing.org/. Accessed 15 June 2017
5. Rodriguez, F.J., García-Martínez, C., Blum, C., Lozano, M.: An artificial bee colony algorithm for the unrelated parallel machines scheduling problem. In: Coello, C.A.C., Cutello, V., Deb, K., Forrest, S., Nicosia, G., Pavone, M. (eds.) PPSN 2012. LNCS, vol. 7492, pp. 143–152. Springer, Heidelberg (2012). doi:10.1007/978-3-642-32964-7_15
6. Kennedy, J., Eberhart, R.: Particle swarm optimization. In: IEEE Conference on Neural Networks, vol. 4, pp. 1942–1948
7. Barnsley, M.F.: Fractals Everywhere. Dover Books on Mathematics, New Edition. Dover Publications Inc., Mineola (2012). ISBN 978-0486488707
8. Detailed requirements for the first prototype. http://informaticup.gi.de/fileadmin/redaktion/Informatiktage/studwett/Aufgabe_Manganernte_.pdf. Accessed 15 June 2017
9. Rossum, J.R.: Fundamentals of metallic corrosion in fresh water. http://www.roscoemoss.com/wp-content/uploads/publications/fmcf.pdf. Accessed 15 June 2017
10. Kim, M.J., Kim, J.G.: Effect of manganese on the corrosion behavior of low carbon steel in 10 wt.% sulfuric acid. Int. J. Electrochem. Sci. **10**, 6872–6885 (2015)
11. Muro, C., Escobedo, L., Spector, L., Coppinger, R.P.: Wolf-pack (Canis lupus) hunting strategies emerge from simple rules in computational simulations. Behav. Process. **88**(3), 192–197 (2011)

Multiplant Production Design in Agent-Based Artificial Economic System

Petr Tucnik, Zuzana Nemcova, and Tomas Nachazel(✉)

University of Hradec Kralove, Hradec Kralove, Czech Republic
{petr.tucnik,zuzana.nemcova,tomas.nachazel}@uhk.cz

Abstract. Management of production is a cornerstone of every economic system. This paper provides a formal description of a production unit (e.g. factory) represented by an agent in an artificial economic system. The concept of the production unit consists of several layers representing respective control processes from the operational up to the strategic level. For maintaining continuous production and optimization of the production unit performance, both the geographical (location of resources, the distance between factories) as well as economic (market structures, competition) context is important. Attention is focused primarily on the formal description of multi-plant production model with autonomous control with facilities situated in distributed geographical locations.

Keywords: Agent · Multiagent system · Supply chain management · Artificial economy · Economic models · Agent-based computational economics

1 Introduction

Effective management of production is concerned with planning and scheduling of various activities across the supply chain. As such, it is not exclusively limited to the internal functioning of a singular production "entity", but the economic and geographical context (such as proximity of other companies, local/global market properties or availability of resources) plays a crucial role here. In practice, it is common that a company has multiple facilities where stages of production take place or where goods and semi-products are stored, and effective handling of the intercompany and intracompany logistics tasks is required. With the appearance of the concept of so-called Industry 4.0 [1], the study of the procedures which would allow increased automation in these control processes gradually grows in importance. Research in this area is also necessary for economic modeling, especially in the domain of agent-based computational economics models which are the primary scope of this paper.

The problem at hand is how to effectively process and manage the inflow of workforce, materials, energy, and services required to handle production requirements on the one hand, and the outflow of the goods and services produced (and consumed) on the other. This paper describes the functioning of such autonomous production "entity" (called production unit) in a model of the artificial economics.

N.T. Nguyen et al. (Eds.): ICCCI 2017, Part I, LNAI 10448, pp. 371–380, 2017.
DOI: 10.1007/978-3-319-67074-4_36

The structure of this paper is as follows: The next section, Goals and Related Work, summarizes goals, similar projects and various research approaches relevant to the topic of this paper. Then the section Model Characteristics specifies the overall structure of the model. Finally, the section Multiplant P-agents describes the design of the proposed Production units in the model.

2 Goals and Related Work

Using agents as elementary construction blocks is reasonable for the purpose of economic modeling, especially when large-scale applications are intended for research and study [2]. Agents have a wide range of common characteristics with the economic subjects such as companies or individuals. They are capable of autonomous decision-making, are generally goal-oriented (on maximizing utility); therefore, they have a potential to make rational decisions. They are also capable of perceiving the environment and taking the rational course of action, and continuously adapt to changes in their environment. Such abilities are extremely useful in the dynamic environments, which is a common characteristic of the majority of economic systems and models.

Our model is primarily focused on the experimental study of autonomous behavior of the agent-based economy. Agents in model form up the complex adaptive systems through the bottom-up approach, establishing communities where macro-scale behavior emerges as a result of a large number of interactions between individual agents. A reasonable level of abstraction for the model design was required, due to complexity of the real-world economics, and certain economic aspects are therefore intentionally omitted (e.g. bank sector, selected financial services, foreign exchange). Incorporation of such aspects is considered for future work. Our aim is rather primarily focused on research and study of the behavior of agents in such system. Primary goals are: (1) study of **establishment and development** of the decentralized market-based economics while maintaining the **controllable experimental conditions**; (2) study of the evolution of **behavioral norms** (see also [3]); (3) the design of computational agents for the **automated markets** (see also [3]); (4) **optimizing** the overall system`s performance. This is closely connected with the adaptability of the system as a whole.

Tesfatsion in her paper [3] identifies eight research areas suitable for ACE (Agent-based Computational Economics) models. Nonetheless, some of the research topics are out of the scope of our attention at this time. The proposed agent-based economic model features primarily following properties: (a) efficient allocation of the resources; (b) efficient distribution of the goods and services; (c) adaptability to the dynamic changes in the environment; (d) bottom-up emergence of the macro-level behavior; (e) parallel goals pursue; (f) long-term sustainability.

These properties reflect the most important features of our virtual economy system. Although this virtual economy is artificial, it seems that these or very similar characteristics reflect essential pre-requisites for the real-world applications of the virtual economics, if obtained results are to be used in practice. As an example can serve Sadigh's [4] virtual enterprise framework for sustainable production. This framework is used to facilitate collaboration between SMEs (Small and Medium Enterprise), which are enabled to capture the opportunities and design the products collaboratively in a network. The virtual framework is then set in real-world, resulting in the viable real-world application.

Functioning of the economy is always related to the supply chain management, regardless if it is the real-world or virtual case. Long in his paper [5] states there is a lack of related methodological framework, allowing efficient use of agent technology and computational experiment method. Long [5] also proposes a formal framework for agent-based distributed virtual supply chain network management. To mention more resources on the subject at hand, there are also other relevant works related to supply chain management in agent-based applications, such as [6–8]. The common ground of all three papers lies in adaptability and optimization of agent-based supply chains which seems to be generally accepted common feature of such systems.

The other field, closely related to the main topic of this paper, are multi-plant production models. Usually, this domain is closely related to logistics, because it is expected that individual facilities are distributed in various geographical locations. This emphasis the need for autonomous logistics, as it is described e.g. in [9]. In accordance to modern trend of Industry 4.0, there is also possible to find combinations of virtual and physical environments, i.e. cyber-physical systems [10]. For the automation of company's internal processes, a small-scale warehouse logistics and inventory management [11] allows automated handling and micro-management on the level of individual production lines and warehouses.

3 Model Characteristics

This paper is primarily focused on the management of a **production unit** – multiagent sub-system situated in the economic environment consisting the other agents. The purpose of such production unit is the production of goods or services. The production unit in the proposed model has following characteristics, similar to manufacturing unit described by Chaib-draa & Müller [2]: (i) *Autonomy* – a company carries out the tasks by itself without external intervention and has some kind of control over its action and internal state; (ii) *Social ability* – a company in the supply chain interacts with other companies, e.g. by placing orders for products or services; (iii) *Reactivity* – a company perceives its environment, i.e. the market and the other companies, and responds in a timely fashion to changes that occur in it. In particular, each firm modifies its behavior to adapt to market and competition evolutions; (iv) *Pro-activeness* – a company does not act in a simple way in response to its environment, it can also initiate new activities, e.g. launching new products on the market.

The significant advantage of the agent-based approach is the possibility to build a complex system based upon the interactions of the individual agents, via the bottom-up approach. This allows agents to have simple, modular construction where constituent models can repeatedly be used. The complexity of such system can, in fact, rapidly increase, as more details are gradually incorporated into the model. In order to maintain the complexity in controllable boundaries, a highly modular approach is used, designing agents from similar "building blocks".

4 Multiplant P-agents

Our previous study [12] was based on the simplified variant of the F-agent (representing a factory) with a single production line, e.g. the case of a single factory F-agent. P-agents represent production agents – are responsible for the production of the goods and services. There are two subsets of P-agents: F-agents (factories, producing semi-products and goods for consumption) and S-agents (service providers). The internal construction of both types of agents is similar because the production uses the same input/output pattern. However, there are fundamental differences in the final product handling. Goods can be stored and transported; services are localized to certain areas (like cities) or bound directly to customers (companies/individual agents).

Figure 1 shows an example of two companies, each with several factories located in different geographical locations. The task of establishing the production through such production chain where the semi-products are to be transported between factories as needed is much more complex, see also [13] for detailed comparative analysis of different approaches to distributed manufacturing.

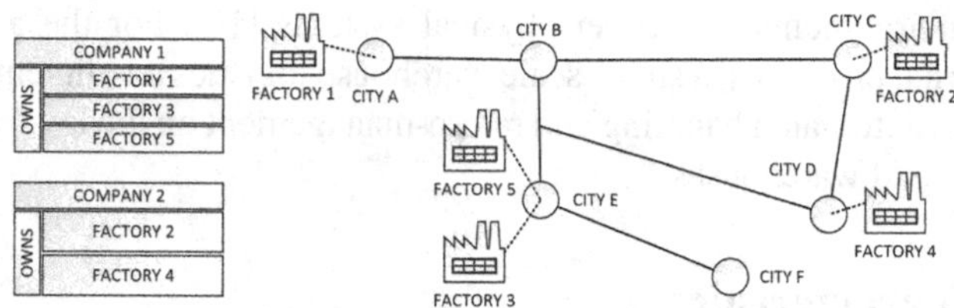

Fig. 1. Distributed F-agents with several factories.

4.1 The General Model of a Factory

The following description of the factory model is based on our previous study [12], the model is adapted to the newly arisen constrictions during the implementation to a modeling platform. The general factory model is shown in the Fig. 2. In general, each factory can produce M final products, the index m takes values $m = \{1, 2, \ldots, M\}$. The factory manufactures products by one or several - say N production lines (depicted as a number in the circle) which can be interconnected as a combination of the parallel and/or the serial manner according to the needs of individual factory type (the index n takes values $n = \{1, 2, \ldots, N\}$).

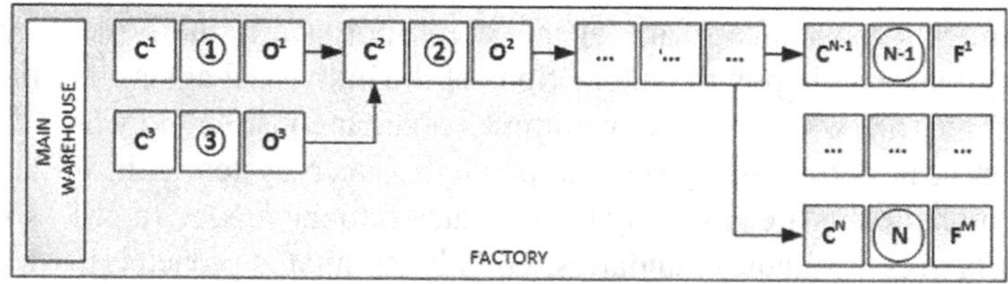

Fig. 2. The general factory model

Let us denote by C^n the set of the components (inputs) and by O^n the set of the outputs of n-th production line. The final product, marked as F^m, is the final, desired product of the factory. It is easily seen that the product F^m is a special type of output of a production line. The output O^n is the so-called semi-product. It is used for further processing in order to complete or manufacture the final product. The example of such product in the figure is output O^3, which is utilized as one of the components C^2 for a production line 2.

Together with the outputs, two more characteristics can be studied in connection with the production. The first is the waste. Waste, unlike the semi-product, does not bring additional value and the company can utilize the waste only partially and through recycling. The second characteristic is the pollution. The pollution is usually closely related to the state of the environment and is used as one of performance indicators of factories on the meta-model scale. For both characteristics it holds that more types for each of them can be defined; both are included as parameters dependent on the type of the factory.

In general, it is supposed that there are a main warehouse and hall(s) with the production lines in the factory area. Each production line has the storage space with limited capacity at its disposal. The limited capacity means particularly the space limitations. The size of storage spaces may be different for each production line. Employees, manpower for the factory, work in shifts. For simplicity, it is supposed that before the system starts working (e.g. initiation phase), the components sufficient for production of at least one work shift are prepared in the warehouse in advance. It is assumed that components for production are prepared at production lines.

Each production line of the factory has a description of its final/semi-product represented by a table in the model. The table defines the volumes of components (all volumes are in cubic meters) for the particular production of the output and also contains the list of by-products (some of them can be later classified as a waste). For example, the factory that produces furniture makes several types of products; some of them are listed in Table 1.

Table 1. Composition of the products

Product		Components		By-products	
Name	Volume	Name	Volume	Name	Volume
Particleboard	1	Timber	0.1		
		Sawdust	0.9		
Glass table	0.8	Timber	0.05	Sawdust	0.01
		Glass	0.02	Shards	0.001

It can be seen that there are three places associated with some waiting (described below) at production line i (see Fig. 3) and also three service places for processing the production can be defined. First service place, assembler, puts together the corresponding quantity of each of components required for manufacturing of the output. In general, there can be K components at each production line. The second service place is the production ("service"). Here, the assembled components are processed and after that the product falls into the third service place responsible for packing the output ("batch").

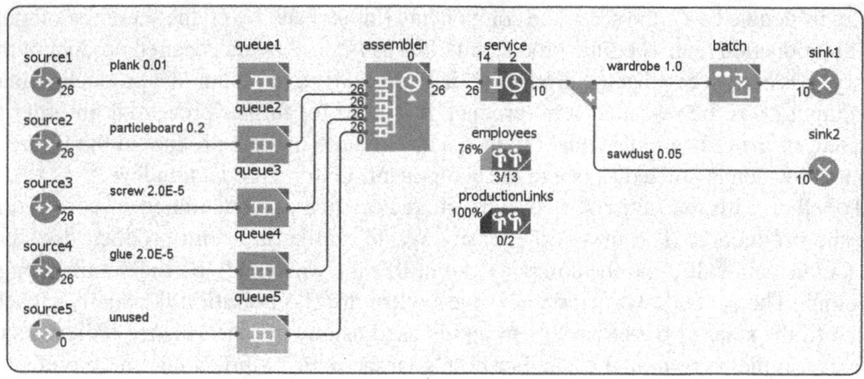

Fig. 3. Screenshot of the factory production line from the model interface (implemented in Anylogic multi-agent modeling platform)

Figure 3 shows the production line of the furniture factory that manufactures wardrobes.

The Production Time. The factory will be able to react and negotiate to supply the market demand. To decide whether to accept the contract or not, it is necessary to know if the terms of delivery are achievable. The management of the factory also needs to know the shortest time for which the supply is able to meet expected requirements. Therefore the computation of the production time of the final product is an important characteristic.

The three types of time-consuming activities at each production line i are distinguished. The total time needed for preparation of the components denoted by T_C^i, the production time denoted by T_P^i and the time for packing the product denoted by T_B^i.

The variable T_C^i stands for the total time needed for delivering the components from the main warehouse to the ith storage space. The time for preparation of particular component varies according to the type of the component, it is denoted by t_k^i. The assembler can start the work if and only if all components are available at a time, otherwise the production cannot start. The shortest time the components are prepared for assembling is then a maximum of all t_k^i. The demand to prepare next ratio of components for assembling will arise at the latest during the production of the just-assembled components. The preparation of the next batch of components can be done in parallel with the production because these two activities are separated.

The T_C^i during the production is computed by following formula:

$$T_C^i = \max\left(\max_{1 \le k \le K}\left(t_k^i\right) - T_P^i; 0\right) \tag{1}$$

Total time of production of the output O^i at ith production line, T^i, is computed as

$$T^i = T_C^i + T_P^i + T_B^i \tag{2}$$

Note that in the case of concatenated lines the possibility of parallel processing of the semi-products has to be considered. The least production time of the final product can be then computed using well-known Critical path method, see [14].

The Delivery of the Raw Materials. This subchapter deals with the delivery of the raw materials to production lines and introduces the algorithm which describes the rules for the delivery within the factory. The demand of particular production lines for raw materials in time is considered to be deterministic. It can be derived from the planned production. For simplicity, it is supposed that the main warehouse has enough raw materials for the production of at least one work shift.

Before the shift starts, necessary components are prepared in the storage space and at the lines. Every storage space (and also the main warehouse) has its space limitations. That is why every request for raw materials has to be converted into the volume units. The maximal capacity of the storage space is denoted by $CapS_{max}$ and the maximal capacity of the main warehouse by $CapMW_{max}$. The maximal capacities are given in volume units. The capacities change in time; the changes are driven by events in time. The time is denoted by t. The events are represented by movement of material.

Firstly, components at the main warehouse are denoted by C_l, $l = \{1, 2, \ldots, L\}$ and its volume by $VolC_l$. For the volume of the component C_l in time t it holds that

$$VolC_l(t) = VolC_l(t-1) + VolC_l'(t) - \sum_{\substack{i = 1 \ldots N \\ k \in W}} VolC_k^i(D) \tag{3}$$

where W is a set of demanded components D of kind C_l that have been sent to some of the production lines i and $VolC_l'(s)$ is the amount of just delivered components C_l (ordered by the main warehouse). Note that one particular component C_l can appear in the model under the one or more designations C_k^i. The orders of raw materials from the market to the main warehouse observe the minimal stock level and are made with respect to the planned production.

Due to the limited capacity of the storage space, at the moment the completed product leaves the production place, the volume of particular components at the storage is checked. When there is an amount of components for the production of the last planned product in the storage space, the production line starts to order material from the main warehouse. The orders are made at ratios sufficient for manufacturing complete products with respect to the remaining duration of the shift and limited capacity of the storage.

5 Experiments

The following section proves the proposed solution by experiments conducted with the model implementation in AnyLogic. The development of one production section was recorded while operating in the complex economic simulation. The focus of experiments is on managing resource supplies in a group of factories processing raw iron ore. Each factory has production lines to generate two products: Steel and Cast iron. Both of them need raw iron ore as a resource and employees as a workforce. Therefore, within F-agent, they compete for employees and the resource.

Two scenarios were applied on the examined factories. In the first one, the simulation includes much more sources of the resource (Iron mines) than is needed to supply the demand. This scenario shows the full potential of the factories with unlimited resources. The other scenario brings fewer sources, and the factories are not able to supply both production lines continuously.

Each experiment covers 40 days of model time. The factories are changing its settings every ten days by adding more employees or production lines. The supply of the resource for production always adjusts according to the current setting and supplies in the main warehouse. The recorded values are means of 20 factories of the same type competing in the simulation.

Figure 4 shows the distribution of the resource to production line during experiments. Figure 5 depicts units of the product manufactured per day. In both figures, left plot shows the full potential of factories, and the right one indicates the resource management during a shortage.

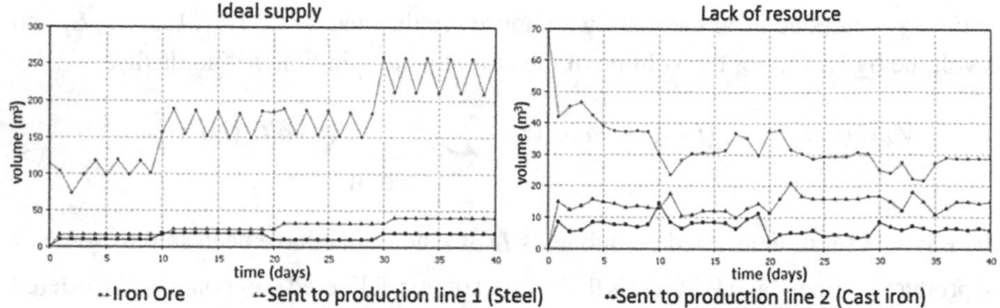

Fig. 4. Distribution of the resource to production lines

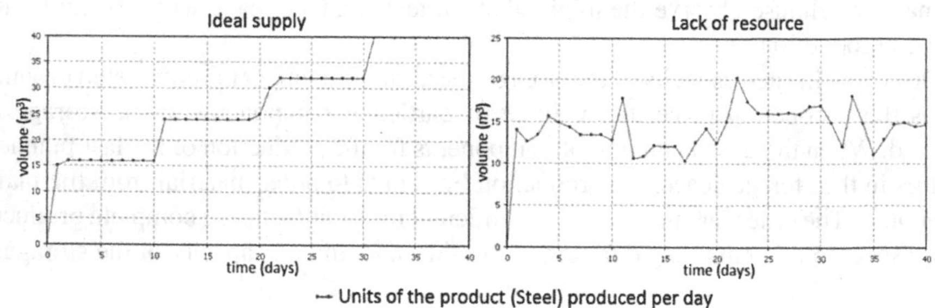

Fig. 5. Average production of factories per day

The results of the first scenario shows raise in production throughout the model run. Table 2 describes events in the scenarios and the following workforce distribution, which was autonomously managed by the F-agent during a model run. In the Fig. 4, the left plot depicts increasing amounts of the resource sent to the production lines (PLs). The only exception appeared at day 20 when PL 2 gets less resource. It is caused by the addition of one extra machine into PL 1 and the following transfer of workforce to support the extra production in prioritized PL 1.

Table 2. Internal changes in workforce distribution in the factories during scenarios

Production line	Initially 6 employees	P*	Day 10 + 4 employees	P*	Day 20 PL 1: +1 machine	P*	Day 30 + 4 employees	P*
PL 1 (Steel)	4 emp. in 2 shifts	16	6 emp. in 3 shifts	24	8 emp. in 2 shifts	32	10 emp. in 3 shifts	40
PL 2 (Cast iron)	2 emp. in 1 shift	10	4 emp. in 2 shift	20	2 emp. in 1 shift	10	4 emp. in 2 shift	20

P* - maximum production per day (measured in default model units – m^3)

Also, Fig. 4 shows the increasing amount of the resource in the main warehouse according to daily consumption. In the first scenario, there is always sufficient supply since the amount of the resource on the market is unlimited; however, in the second scenario, the lack of the resource caused very limited production through the whole model run. In this scenario, the appropriate reaction of the F-agent would be restricting the number of employees; however, managerial decision-making is not a goal of this paper. The result of the second scenario showed that the F-agent in the model can operate even under difficult conditions.

6 Conclusion and Future Work

The paper presented the concept of the P-agent production unit which is an agent sub-model of the larger ACE environment model. The main goals of this contribution were to (1) describe formal aspects of its design and (2) demonstrate its functioning on the example. Although the provided P-agent concept is only a single component in the model, due to the modularity of other agents in the system and intentional reusability of its design, it is applicable to the most of the production cycle elements.

Experiments had proved the ability of proposed model in different situations including dynamic changes during a model run. The factories and its production lines successfully transferred resource material into output material while dealing with dynamic adjustments of its internal structure and the varying availability of the input resource on the market.

The functioning of the P-agent has to be perceived in a broader context of the whole system. In order to measure its performance, the whole industrial or services segments are to be tested at the same time. This corresponds well with the progressive emergence of the system behavior through the large number of interactions between its individual components. Expansion of the experimental research towards the whole segment testing is therefore expected to be the next step in the measurement of the system performance.

Acknowledgements. The Financial support of the Specific Research Project "Autonomous Socio-Economics Systems" of FIM UHK is gratefully acknowledged.

References

1. Premm, M., Kirn, S.: A Multiagent systems perspective on industry 4.0 supply networks. In: Müller, J.P., Ketter, W., Kaminka, G., Wagner, G., Bulling, N. (eds.) MATES 2015. LNCS, vol. 9433, pp. 101–118. Springer, Cham (2015). doi:10.1007/978-3-319-27343-3_6
2. Chaib-Draa, B., Müller, J.P.: Multiagent Based Supply Chain Management. Springer, Cham (2006)
3. Tesfatsion, L.: Agent-based computational economics: modeling economies as complex adaptive systems. Inf. Sci. **149**, 262–268 (2003)
4. Sadigh, B.L., Ünver, H.Ö., Kılıç, S.E.: Design of a multi agent based virtual enterprise framework for sustainable production. In: Putnik, G.D., Cruz-Cunha, M.M. (eds.) ViNOrg 2011. CCIS, vol. 248, pp. 186–195. Springer, Heidelberg (2012). doi: 10.1007/978-3-642-31800-9_20
5. Long, Q.: An agent-based distributed computational experiment framework for virtual supply chain network development. Expert Syst. Appl. **41**, 4094–4112 (2014)
6. Aslam, T., Ng, A.: Agent-based simulation and simulation-based optimisation for supply chain management. In: Wang, L., Koh, S.C.L. (eds.) Enterprise Networks and Logistics for Agile Manufacturing, pp. 227–247. Springer, London (2010)
7. Behdani, B., van Dam, K.H., Lukszo, Z.: Agent-based models of supply chains. In: van Dam, K.H., Nikolic, I., Lukszo, Z. (eds.) Agent-Based Modelling of Socio-Technical Systems, vol. 9, pp. 151–180. Springer, Dordrecht (2013)
8. Brintrup, A.: Behaviour adaptation in the multi-agent, multi-objective and multi-role supply chain. Comput. Ind. **61**, 636–645 (2010)
9. Gehrke, J., Herzog, O., Langer, H., Malaka, R., Porzel, R., Warden, T.: An agent-based approach to autonomous logistic processes. Künstl. Intell. **24**, 137–141 (2010)
10. Greulich, C., Edelkamp, S., Eicke, N.: Cyber-physical multiagent-simulation in production logistics. In: Müller, J.P., Ketter, W., Kaminka, G., Wagner, G., Bulling, N. (eds.) Multiagent System Technologies, pp. 119–136. Springer, Cham (2015)
11. Kappauf, J., Lauterbach, B., Koch, M.: Warehouse Logistics and Inventory Management. Inventory Management, Warehousing, Transportation, and Compliance, pp. 99–213. Springer, Berlin, Heidelberg (2012)
12. Tucnik, P., Nemcova, Z.: Production unit supply management solution in agent-based computational economics model. In: Barbucha, D., Nguyen, N.T., Batubara, J. (eds.) New Trends in Intelligent Information and Database Systems. SCI, vol. 598, pp. 343–352. Springer, Cham (2015). doi:10.1007/978-3-319-16211-9_35
13. Leitão, P.: Agent-based distributed manufacturing control: a state-of-the-art survey. Eng. Appl. Artif. Intell. **22**, 979–991 (2009)
14. Lawrence, J.A., Pasternack, B.A.: Modeling, Spreadsheet Analysis, and Communication for Decision Making. Wiley, New York (2002)

Role of Non-Axiomatic Logic in a Distributed Reasoning Environment

Mirjana Ivanović[1], Jovana Ivković[1(✉)], and Costin Bădică[2]

[1] Department of Mathematics and Informatics, Faculty of Sciences, University of Novi Sad, Novi Sad, Serbia
{mira,jovana.ivkovic}@dmi.uns.ac.rs

[2] University of Craiova, Craiova, Romania
cbadica@software.ucv.ro

Abstract. The aim of this paper is to introduce the design of a novel Distributed Non-Axiomatic Reasoning System. The system is based on Non-Axiomatic Logic, a formalism in the domain of artificial general intelligence designed for realizations of systems with insufficient resources and knowledge. Proposed architecture is based on layered and distributed structure of the backend knowledge base. The design of the knowledge base makes it fault-tolerant and scalable. It promises to allow the system to reason over large knowledge bases with real-time responsiveness.

Keywords: Reasoning architecture · Non-Axiomatic Logic · Artificial General Intelligence · Big Data

1 Introduction

Big Data analytics is a hot topic that could be applied in different real life industries (e.g. [4]). Growing amounts of available information have led to the requirement that every intelligent software system must have: the ability to manipulate huge amounts of data [3]. As a consequence, several practical technologies, including, for example, graph databases and NoSQL (e.g. [11]) and the MapReduce programming model [1], have emerged.

This paper will present the idea of a new system – Distributed Non-Axiomatic Reasoning System (DNARS) [6]. This system is based on a novel distributed reasoning architecture that incorporates Big Data notions, includes fault-tolerance based on data replication, as well as a highly-scalable and distributed backend knowledge base. DNARS uses recent techniques for the purpose of large-scale distributed data processing. Our approach makes it possible for the system to operate on huge knowledge bases, and also service a large number of external clients in real-time.

The DNARS architecture consists of two central components. Its backend knowledge base is used to store the system's knowledge and experience on a large scale. To support this functionality, DNARS includes a set of algorithms packaged in the form of inference engines. In order to realize high-level reasoning

N.T. Nguyen et al. (Eds.): ICCCI 2017, Part I, LNAI 10448, pp. 381–388, 2017.
DOI: 10.1007/978-3-319-67074-4_37

capabilities, DNARS relies on the Non-Axiomatic Logic (NAL). NAL provides a formal way to support reasoning in Artificial General Intelligence (AGI) systems [14,16].

The expression "non-axiomatic" implies that there is an insufficient amount of resources and knowledge [16] available in the system, and said knowledge can also be inconsistent and uncertain. New evidence of various content can be added to the knowledge base, and thus can change the truth value of the included knowledge statements. Moreover, truth-values do not necessarily have to converge to certain limit, i.e. they can change arbitrarily. Furthermore, only a part of the available statements is usually involved in the problem-solving activities, thus providing locality to the reasoning process.

Accordingly, mechanisms in NAL support efficient handling of inconsistencies and uncertainty in available statements. They also have the ability to reduce the number of statements and thus abridge the available knowledge to enable more efficient reasoning.

The rest of the paper is organized as follows. Section 2 provides a brief introduction into NAL. Section 3 presents an overview of the proposed architecture of the DNARS system. Section 4 briefly discusses some related work, while Sect. 5 concludes the paper.

2 Non-Axiomatic Logic

Non-Axiomatic Logic (NAL) is set apart from other formalisms used in reasoning systems. It contains a set of inference rules, a symbolic grammar, and a semantic theory. NAL sentences consist of subject-copula-predicate form. Both subject and predicate can be represented by atomic (a single word) or compound (which joins atomic and/or compound terms) terms. NAL has an experience-grounded semantics [12] based on the concepts of generalization and specialization. Inheritance, formally captured as $S \rightarrow P$, is the most typical statement in NAL. Terms S and P represent subject and object, while the connector $\rightarrow$ represents the inheritance copula. The meaning of this statement is: *S is a type of P*.

NAL inference rules (syllogistic form) are used for deriving new knowledge. They support question answering or dealing with statement inconsistencies. Inference in NAL system is achieved by means of various available inference rules. Those rules are influenced by the copulas and positions of the common terms in premises.

NAL itself is hierarchically organized into 9 layers. Different inference rules, and/or new features are introduced from level to level. In the first proposal of our DNARS architecture we concentrated on the first four layers.

NAL-1 introduces inference rules on inheritance statements (deduction, induction, and abduction) [16].

NAL-2 is extended by symmetric inheritance represented in the grammar with the similarity copula: $(S \leftrightarrow P) \Leftrightarrow (S \rightarrow P) \wedge (P \rightarrow S)$. As a consequence, three new forward inference rules are introduced in this layer: comparison, analogy, and resemblance [14–16].

NAL-3 brings compound terms in form: $\{T_1\ con\ T_2\ con\ \ldots\ con\ T_n\}$ where con is the connector and $T_1 \ldots T_n$ are terms, $n \geq 1$ [14,16]. Connectors are: extensional intersection ($\backslash$), intensional intersection ($\cap$), extensional difference ($-$), and intensional difference ($\ominus$). The rule summarizes the system's experience.

NAL-4 introduces arbitrary relations among terms like product ($\times$). Inheritance between separate components of a compound term is defined in the following way [14,16]: $((S_1 \times \ldots \times S_n) \rightarrow (P_1 \times \ldots \times P_n)) \Leftrightarrow ((S_1 \rightarrow P_1) \wedge \ldots \wedge (S_n \rightarrow P_n))$. R as relational term can be defined as a term connected via inheritance to a product term, i.e. either by $(T_1 \times T_2) \rightarrow R$ or by $R \rightarrow (T_1 \times T_2)$ [14,16].

More information on NAL is available in the references mentioned in this section.

3 DNARS – Distributed Non-Axiomatic Reasoning System

Advantages and capabilities of NAL are planned to be introduced and incorporated in our novel, previously developed multi-agent middleware, named Siebog, which integrates two essential parts:

1. XJAF (Extensible Java EE-based Agent Framework) [8,13], which is a server-side multi-agent architecture that supports clustered environments.
2. A client-side multi-agent system (Radigost) [7,9], mostly based on HTML5 markup language.

Siebog is a multi-agent system that aims to support agents during their lifetime by providing several functionalities, including: maintaining the agent's life cycle, providing infrastructure for message exchange, etc. Siebog operates on top of computer clusters which yields two of its important features: load-balancing and fault-tolerance.

Additionally, it is our plan to enhance Siebog with support for DNARS, so that Siebog agents can have the capability of advanced reasoning.

DNARS (see Fig. 1) consists of the following parts:

- *Resolution engine* that provides clients with answers to their questions.
- *Forward inference engine* that is responsible for producing new knowledge from existing one.
- *Short-term memory* that only includes statements necessary for solving current problems.
- *Knowledge domain* is a sub-set of the knowledge base, consisting of closely related statements.
- *Backend knowledge base* that comprises of the system's entire knowledge base, which is an essential repository of its experience accumulated over time.
- *Event manager* that handles events triggered by changes made in the knowledge base.

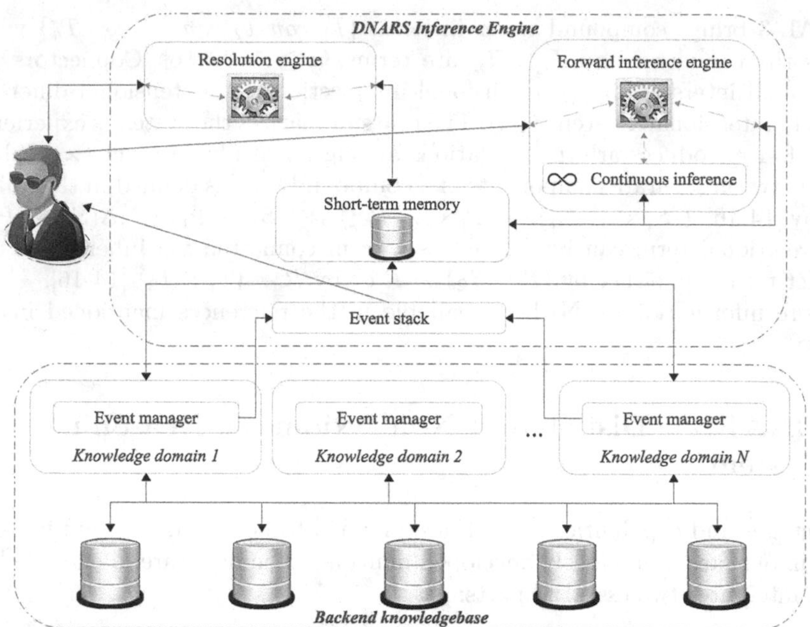

Fig. 1. The architecture of DNARS.

These aforementioned parts could be grouped into two sub-systems: *DNARS Inference engines* and *Backend knowledge base*. The *DNARS Inference engines* sub-system consists of *Resolution* and *Forward inference engines*. The *Backend knowledge base* sub-system includes *Short-term memory*, *Knowledge domain*, *Backend knowledge base* and *Event manager* components. There is a unique *Backend knowledge base* that serves all external clients, while each client is affiliated with its own set of engines. The knowledge base is designed to be scalable, and is organized in a such a way that it can support multiple clients.

3.1 DNARS Inference Engines

DNARS inference engines sub-system consists of *Resolution* (to answer questions) and *Forward inference* (to derive new knowledge) engines that support:

- Questions with "?", i.e. "*? copula P* " or "*S copula ?* ". Here, the *Resolution engine* searches through the knowledge base in order to find the most appropriate substitute for "?". The answer should be reached in real time.
- Questions like "*S copula P*". The engine will first try to find the answer in the knowledge base, but if doesn't exist, it will use backward inference rules to try and reach the required answer. The answer to this type of questions is also achievable in real time, but only if the answer is already in the knowledge base. If it doesn't exist as such then the backward inference process will be

performed asynchronoulsy [14,16] and the answer is passed on to the client when it becomes available using the *Event manager*.

Currently, the *Forward inference engine* is meant to enable DNARS with support of a subset of forward inference rules of the first four NAL layers. This approach is a starting point for implementing a first version of a system with practical reasoning abilities.

3.2 Backend Knowledge Base

DNARS is designed to manage big amounts of knowledge efficiently. The *Backend knowledge base* is made up of three layers:

- The first layer consists of the *Knowledge base* which is distributed across a computer cluster using horizontal scaling [5]. Its distributed nature allows the system to store huge amounts of data, and it also provides fault-tolerance due to the state replication of the cluster nodes.
- In the second layer the knowledge base is organized into *Knowledge domains* which allows the system to work with only a part of the knowledge base, depending on the given problem. The domains can be distributed across the cluster nodes (Fig. 1.).
- The top layer consists of the *Short-term memory* and the *Event manager*. *Short-term memory* acts as a fast memory storage. With the completion of the inference cycle, the content of the *Short-term memory* is saved to the appropriate domain. *Event manager* is based on the Observer design pattern, and its purpose is to notify clients of any relevant changes in the knowledge state.

4 Related Work

One of the main intended purposes of DNARS is to serve as an underlying reasoning engine in our multi-agent environment named Siebog [9,13]. This is a departure from the Belief-Desire-Intention (BDI) model which is common in agent technology [10].

Concepts of NAL, when compared to traditional BDI model, offer several advantages. The main characteristic of NAL statements is that they are endowed with truth-values representing confidence of belief, using the true definition of belief. On the other hand, in the BDI model, a developer has to plan for the possibility that a belief might not be true, as agents themselves cannot directly assign confidence to their beliefs.

NAL-based agents provide new features like inconsistency resolution (based on backward inference), learning (based on forward inference), and working with insufficient amount of knowledge and resources (e.g. compound terms [14,16]), while the BDI model does not offer such features.

In contrast to existing BDI systems, the main advantage that DNARS offers is the possibility of applying reasoning over huge knowledge bases.

On the other hand, BDI can be seen as another type of reasoning that can sit on top of NAL. BDI is about "practical reasoning", i.e. reasoning towards actions. Moreover, in some BDI systems, e.g. Jason, the belief base of the agent architecture is configurable. So, an agent could simply incorporate beliefs from the NAL knowledge base. Systems like that do have problems with speed and scalability, but they can describe "flexible behaviour" by mixing reactivity with proactivity, a feature that is missing from standard backward and forwards chaining reasoning.

OpenNARS [2] is an open-source implementation of NAL [14,16]. It supports the logic of all existing layers of NAL. Its main parts are the inference engine, the memory module, and lastly the control mechanism which is in charge of the reasoning cycles [14].

Both OpenNARS and DNARS are implementations of non-axiomatic reasoning and, unlike other reasoning and cognitive systems, they use NAL as a foundational formalism. According to available resources, NAL is able to handle insufficient resources and knowledge to a larger extent than other systems.

Recently, an emerging trend in data processing is characterized by different functionalities of the Big Data paradigm, especially in increasing the performance when processing large amounts of complex data. So, by combining Big Data paradigm functionalities with NAL, in DNARS we will achieve better performances when handling big and complex data with limited resources and time. A special problem that we addressed is dealing with knowledge inconsistencies only when they emerge in particular situations (e.g. when multiple answers to a question are available). Another issue that we addressed is the amount of raw information, namely we tried to decrease it by combining separate chunks of information. DNARS is designed with a NoSQL database in mind, so that it could support faster processing of large amounts of data, but at the expense of endangering consistency of the information on a temporary basis. In our proposed architecture we would like to integrate the best benefits of most important techniques and methods for big, complex and inconsistent data processing into a single architecture.

Despite their similarities, there are sharp distinctions between DNARS and OpenNARS. As it has been in development longer, OpenNARS incorporates all nine layers of NAL, and it contains more advanced control mechanisms. The main advantage of DNARS comes with its ability to reason over huge knowledge bases due to its advanced backend knowledge base organization.

5 Conclusions and Future Work

In this paper we considered different emergent technologies and their advanced functionalities, as well as the possibility to include them in our existing multi-agent system Siebog [13]. By using these technologies, we have striven towards making our system more intelligent and able to reason faster over large amounts of knowledge.

This paper has presented a new general-purpose reasoning architecture – Distributed Non-Axiomatic Reasoning System (DNARS). DNARS uses NAL as the basis for its formal reasoning. NAL is a specialized formalism that offers a well-defined syntax, experience-grounded semantics, as well as a set of inference rules. Overall, it takes into consideration the possibility of reasoning over insufficient resources and knowledge [16]. We have faced numerous challenges while designing DNARS, but we have managed to achieve an advantage when compared to other similar systems. An innovative organization of the backend knowledge base, as well as an original implementation of NAL inference rules in a highly-scalable, fault-tolerant and distributed environment makes it possible for DNARS to efficiently process large amounts of data, while being able to service a large number of external clients.

During our research, numerous challenging and open scientific questions were recognized, thus offering us several different directions of realization of high-quality reasoning services in real world environments. In the future, we plan to improve our proposed architecture by adding the remaining layers of NAL to DNARS. Although the first four layers currently considered in DNARS are sufficient for simple reasoning activity, the rest of the layers would provide DNARS with new reasoning capabilities of a higher-level.

Acknowledgments. This work was partially supported by Ministry of Education, Science and Technological Development of the Republic of Serbia, through project number OI174023: "Intelligent techniques and their integration into wide-spectrum decision support", as well as a collaboration agreement between University of Novi Sad (Serbia), University of Craiova (Romania), SRIPAS and Warsaw University of Technology (Poland).

References

1. Dean, J., Ghemawat, S.: Mapreduce: simplified data processing on large clusters. Commun. ACM **51**(1), 107–113 (2008)
2. Hammer, P., Lofthouse, T., Wang, P.: The OpenNARS implementation of the non-axiomatic reasoning system. In: Steunebrink, B., Wang, P., Goertzel, B. (eds.) AGI 2016. LNCS, vol. 9782, pp. 160–170. Springer, Cham (2016). doi:10.1007/978-3-319-41649-6_16
3. Hovy, E., Navigli, R., Ponzetto, S.P.: Collaboratively built semi-structured content and artificial intelligence: the story so far. Artif. Intell. **194**, 2–27 (2013)
4. Lee, T., Lee, H., Rhee, K.H., Shin, U.S.: The efficient implementation of distributed indexing with hadoop for digital investigations on big data. Comput. Sci. Inf. Syst. **11**(3), 1037–1054 (2014)
5. Michael, M., Moreira, J.E., Shiloach, D., Wisniewski, R.W.: Scale-up x scale-out: a case study using nutch/lucene. In: IEEE International Parallel and Distributed Processing Symposium, IPDPS 2007, pp. 1–8. IEEE (2007)
6. Mitrović, D.: Intelligent Multiagent Systems based on Distributed Non-Axiomatic Reasoning. Ph.D. thesis, University of Novi Sad, Faculty of Science (2015)
7. Mitrović, D., Ivanović, M., Bădică, C.: Delivering the multiagent technology to end-users through the web. In: Proceedings of the 4th International Conference on Web Intelligence, Mining and Semantics (WIMS 2014), p. 54. ACM (2014)
8. Mitrović, D., Ivanović, M., Budimac, Z., Vidaković, M.: Supporting heterogeneous agent mobility with alas. Comput. Sci. Inf. Syst. **9**(3), 1203–1229 (2012)

9. Mitrović, D., Ivanović, M., Budimac, Z., Vidaković, M.: Radigost: Interoperable web-based multi-agent platform. J. Syst. Softw. **90**, 167–178 (2014)
10. Rao, A.S., Georgeff, M.P., et al.: BDI agents: from theory to practice. In: ICMAS 1995, pp. 312–319 (1995)
11. Robinson, I., Webber, J., Eifrem, E.: Graph Databases: New Opportunities for Connected Data. O'Reilly Media Inc., Sebastopol (2015)
12. Rodriguez, M.A., Geldart, J.: An evidential path logic for multi-relational networks. In: AAAI Spring Symposium: Technosocial Predictive Analytics, pp. 114–119 (2009)
13. Vidaković, M., Ivanović, M., Mitrović, D., Budimac, Z.: Extensible Java EE-based agent framework – past, present, future. In: Ganzha, M., Jain, L. (eds.) Multiagent Systems and Applications. ISRL, vol. 45, pp. 55–88. Springer, Heidelberg (2013). doi:10.1007/978-3-642-33323-1_3
14. Wang, P.: Rigid Flexibility, vol. 55. Springer, Dordrecht (2006)
15. Wang, P.: Analogy in a general-purpose reasoning system. Cogn. Syst. Res. **10**(3), 286–296 (2009)
16. Wang, P.: Non-Axiomatic Logic: A Model of Intelligent Reasoning. World Scientific, Singapore (2013)

Agent Having Quantum Properties: The Superposition States and the Entanglement

Alain-Jérôme Fougères(✉)

ESTA Lab', ESTA, School of Business and Engineering, Belfort, France
ajfougeres@esta-groupe.fr

Abstract. In agent-based simulation and modelling of intelligent complex systems, the problem of decision making by agents having incomplete, uncertain, local or global, exchanged or observed information is very common. Recent studies on quantum cognition introduce in the decision process modelling and analysis, quantum properties such as superposition state, non-locality, oscillation, interference or entanglement. This paper proposes a model of quantum-like agents able to implement quantum properties of superposition state and local or non-local entanglement. A case study based on an adaptation of the Takuzu game illustrates our proposed approach of quantum agents modelling. A discussion on the interest of decomposing or not components of a system in the intelligent complex systems modelling is also proposed.

Keywords: Quantum agent · Quantum-like model · Superposition states · Entanglement · Artificial intelligence · Agent-based modelling

1 Introduction

In the field of computational intelligence, there are many scientific works and approaches for the modelling and simulation of agent-based distributed systems, mainly when agents have to make decisions based on uncertain knowledge [1–3]. Recent studies on quantum cognition also seem very interesting and promising to deal with this cognitive problem. They introduce the quantum properties of superposition, interference, locality or non-locality, ubiquity, oscillation or entanglement in cognition, and thus in the decision-making process [4, 5].

A large variety of scientific fields have been inspired by quantum theories, giving rise to many types of quantum models (often called quantum-like models). Thus, many models of cognition have been proposed on the basis of characteristics or phenomena defined in quantum theory [4]: models of decision processes, uncertain memory or cognitive measures, models for ambiguous perception or probabilistic judgments, Etc. In particular, decision-making modelling and processes, which are open, parallel, cooperative and/or competitive, benefit fully from this research on quantum probability decision models [5, 6].

Thus, after having worked for many years on the simulation and modelling of complex systems based on more or less cognitive agents [18], we began to extend our work towards the modelling of quantum agents [7] capable of implementing the quantum properties of superposition states and local or non-local entanglement.

N.T. Nguyen et al. (Eds.): ICCCI 2017, Part I, LNAI 10448, pp. 389–398, 2017.
DOI: 10.1007/978-3-319-67074-4_38

Human thinking and reasoning do not use the rules of classical logic, but those of quantum theory: things are not always well defined or determined, but intricate, ubiquitous, oscillating and superposed [8–10]. So, each time that reasoning is applied to a decision process, human decisions are typically quantum [12, 13].

Quantum properties are numerous: indeterminacy, interference, ubiquity, oscillation, entanglement between the states, superposition principle. From these properties, we have mainly retained two: (1) the superposition state for providing a very good representation of conflict, ambiguity or uncertainty that we feel when we doubt; and (2) the quantum entanglement that can be local (elements in a same locality) or non-local (elements in different localities).

Quantum entanglement is a phenomenon observed in quantum mechanics in which the quantum state of two objects must be described globally, without being able to separate an object from the other, although they can be spatially separated. When quantum objects are placed in an entangled state, there are also correlations between the observed physical properties of these objects that would not be present if these properties were local. Consequently, even if they are separated by large spatial distances, two entangled objects O_1 and O_2 are not independent and they must be considered as a single object $O' = O_1 \oplus O_2$.

In physics, the principle of locality, also known as the principle of separability, is a principle according to which distant objects cannot have a direct influence on each other. An object can be influenced only by its immediate environment. This principle is questioned by quantum physics, mainly by the phenomena of quantum entanglement. Bohm and Hiley [15] consider that there is no justification for objections to the concept of non-locality.It is then possible to establish relations between quantum entanglement and non-locality. In 1935, Einstein et al. demonstrated a phenomenon of distance action [16]: 2 entangled particles behave like a single system, whatever the distance separates them. If one of the particles is measured, the state of its twin particle will be instantaneously modified. This is called the "principle of non-locality" or "principle of non-separability". Vertesi and Brunner showed in [17] that the weakest form of quantum entanglement, called non-distillable quantum entanglement, can lead to non-local quantum correlations, the strongest forms of inseparability. According to Horodecki et al. [18] entanglement is almost "invisible" in such systems, which makes it very surprising that they can present this phenomenon of non-locality.

Multiagent is well adapted for modelling, designing and developing complex systems composed of numerous intelligent entities [20–25]. Indeed, agents are autonomous entities that can adapt to, react to, or interact with their environment [21, 22]. However, the problem of decision making by agents having local, incomplete, uncertain, exchanged or observed information in synchronous or asynchronous manner is very common. Studies on quantum cognition have proposed quantum properties such as superposition state, ubiquity, non-locality, oscillation or entanglement in the decision process for providing a solution to this cognitive problem. So, in this paper we presented a quantum-like model of agents that are able to simulate both quantum properties of state superposition and local or non-local entanglement.

This paper is structured as follows: Sect. 2 presents a quantum-like agent model; Sect. 3 develops a case study from the Takuzu game to illustrate our quantum-like agent model; finally, discussion and conclusion present and analyze our findings.

2 Quantum-like Agent Modelling

For modelling and designing complex systems with a quantum-like approach we proposed to define the following model of a quantum agent-based system M (1):

$$\mathrm{M} = \langle \mathrm{A, I, P, O}, \Psi \rangle \tag{1}$$

where A is a set of quantum agents α_i, whose state vectors are written $|\Psi_i\rangle$, and superpose several states $\{\psi_1, \psi_2, \ldots, \psi_n\} \in \Psi$; I is the set of interactions of quantum agents α_i; P is the set of roles played by quantum agents α_i; and O is the set of organizations into communities of quantum agents α_i. Moreover, knowing *card* (A) = n, the global state of M is noted $|\Psi_\mathrm{M}\rangle = |\Psi_1\rangle \otimes |\Psi_2\rangle \otimes \ldots \otimes |\Psi_n\rangle$ if the states of the n quantum agents are separable, or $|\Psi_\mathrm{M}\rangle = |\Psi_1\Psi_2\ldots\Psi_n\rangle$ if they are entangled.

Agent behaviors are often inspired by the cycle [perceive/observe/decide/act]. The behavior of a quantum agent $\alpha_i \in \mathrm{A}$ (Fig. 1) is similar to that of a common agent. So, a quantum agent can be described by the Eq. (2):

$$\alpha_i = \; < \Pi(\varepsilon_j, \pi_k), \Omega^*(\pi_k, \Sigma_{\alpha_i}, \Omega_{\alpha_i}), \Delta(\Omega_{\alpha_i}, \delta_m), \Gamma(\delta_m, \gamma_n), \mathrm{K}_{\alpha_i} > \tag{2}$$

where Π, Ω *, Δ, Γ are respectively functions of observations, interpretations*, decisions, and actions. K_{α_i} is the finite set of knowledge of α_i.

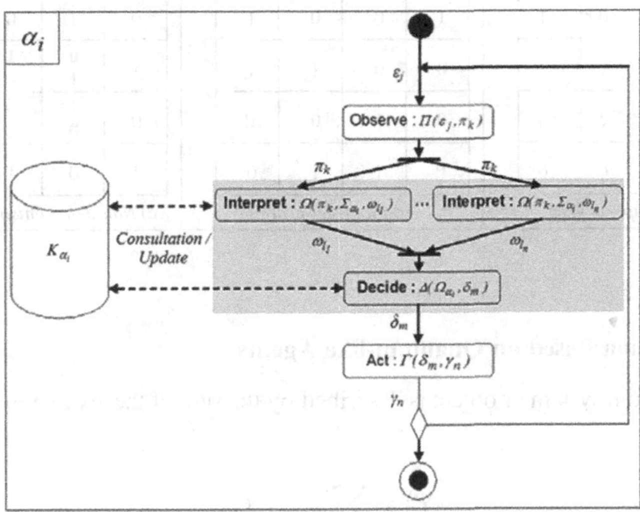

Fig. 1. Formal behaviour of a quantum-like agent α_i

3 Case Study: The Takuzu Game Based on Quantum Agents

A cellular automaton (CA) is a system consisting of a set of identical finite state automata (cells), having a limited number of states and placed on a grid. Each cell presents itself as an elementary automaton whose future state is defined by a transition function which depends on its present state and those of the cells situated in the immediate neighborhood. All cells change their states simultaneously.

Popular games such as the Sudoku or the Takuzu are kinds of CA. The grids of these games are composed of cells having values (i.e. states) dependent on values of the neighboring cells located on the same column and row. The Takuzu game is a logic-based number placement puzzle. The player's goal is to fill a Takuzu grid with two binary values (0 | 1), respecting only two rules: rule number one, an equal number of values '0' and '1' in each row and column, and rule number two, no more than two same adjacent values '0' or '1' in each row and column.

Table 1a give an illustration of a Takuzu's grid composed of 16 cells (4 × 4). According to the Takuzu's rules, cell $c_{3,3}$ must be set to 1. Compliance with rules alone is not enough to build a grid, even for simple grids with few elements (Table 1b). Takuzu's game is indeed a complex system. Moreover, the game is non-deterministic. In Table 1c we can get two valid solutions from the same initial configuration.

Table 1. Some characteristic configurations of Takuzu game grids

1	0	0	1	1	0	0	1	0	1	0	1
0	0	1	1	0	0	1	1	1	**0** **1**	**1** **0**	0
0	1	?	0	0	1	0	!!	0	**1** **0**	**0** **1**	1
1	1	0	0	!!	1	1	0	1	0	1	0
a) Example of grid				*b) Incorrect grid*				*c) Non-deterministic grid*			

3.1 Simulation Based on Quantum-like Agents

A given quantum system or object is described by the sum of these different superposed states (3):

$$|\Psi\rangle = \sum\nolimits_{i=1}^{i=n} \alpha_i |\Psi_i\rangle \tag{3}$$

where the coefficients α_i are called "probability amplitude".

For a given cell of a CA which may have only two states '0' and '1', and would have no interference with its environment, the sum of these different superposed states is given by (4) whose the different probabilities are defined in (5):

$$|cell\rangle = \frac{1}{\sqrt{2}}|0\rangle + \frac{1}{\sqrt{2}}|1\rangle \tag{4}$$

$$p(|0\rangle) = p(|1\rangle) = \left|\frac{1}{\sqrt{2}}\right|^2 = \frac{1}{2} \tag{5}$$

3.2 Quantum-like Agents for the Takuzu Game

In this section we propose a quantum-like agents model for a simulation of an adaptation of the Takuzu game. In this kind of CA, the tick of the clock is when the player fills a cell implemented by a quantum agent. Each quantum agent manages its evolution cycle time, observes the state of its neighbors and applies the two rules presented in previous section.

Quantum computation during a phase of this adapted Takuzu game is developed in three steps: (1) initial states (6), (2) transformation resulting from a player's choice (7), and (3) new probability measures (8). Each quantum agent evaluates its future state by observing its neighboring quantum agents and calculating the probabilities of its binary value '0' or '1' (9).

$$\begin{pmatrix} 1 & - & - & - \\ - & 1 & - & - \\ - & - & 1 & - \\ - & - & - & 1 \end{pmatrix} \rightarrow \begin{pmatrix} 1 & ? & ? & ? \\ ? & 1 & ? & ? \\ ? & ? & 1 & ? \\ ? & ? & ? & 1 \end{pmatrix} \rightarrow \begin{pmatrix} 1 & {}^2/_3|0\rangle + {}^1/_3|1\rangle & \alpha_{1,3}|0\rangle + \beta_{1,3}|1\rangle & \alpha_{1,4}|0\rangle + \beta_{1,4}|1\rangle \\ \alpha_{2,1}|0\rangle + \beta_{2,1}|1\rangle & 1 & \alpha_{2,3}|0\rangle + \beta_{2,3}|1\rangle & \alpha_{2,4}|0\rangle + \beta_{2,4}|1\rangle \\ \alpha_{3,1}|0\rangle + \beta_{3,1}|1\rangle & \alpha_{3,2}|0\rangle + \beta_{3,2}|1\rangle & 1 & \alpha_{3,4}|0\rangle + \beta_{3,4}|1\rangle \\ \alpha_{4,1}|0\rangle + \beta_{4,1}|1\rangle & \alpha_{4,2}|0\rangle + \beta_{4,2}|1\rangle & \alpha_{4,3}|0\rangle + \beta_{4,3}|1\rangle & 1 \end{pmatrix} \tag{6}$$

$$\begin{pmatrix} 1 & - & - & - \\ - & 1 & - & - \\ - & - & 1 & - \\ - & - & - & 1 \end{pmatrix} \rightarrow \begin{pmatrix} 1 & 0 & ? & ? \\ ? & 1 & ? & ? \\ ? & ? & 1 & ? \\ ? & ? & ? & 1 \end{pmatrix} \rightarrow \begin{pmatrix} 1 & 0 & \alpha_{1,3}|0\rangle + \beta_{1,3}|1\rangle & \alpha_{1,4}|0\rangle + \beta_{1,4}|1\rangle \\ \alpha_{2,1}|0\rangle + \beta_{2,1}|1\rangle & 1 & \alpha_{2,3}|0\rangle + \beta_{2,3}|1\rangle & \alpha_{2,4}|0\rangle + \beta_{2,4}|1\rangle \\ \alpha_{3,1}|0\rangle + \beta_{3,1}|1\rangle & \alpha_{3,2}|0\rangle + \beta_{3,2}|1\rangle & 1 & \alpha_{3,4}|0\rangle + \beta_{3,4}|1\rangle \\ \alpha_{4,1}|0\rangle + \beta_{4,1}|1\rangle & \alpha_{4,2}|0\rangle + \beta_{4,2}|1\rangle & \alpha_{4,3}|0\rangle + \beta_{4,3}|1\rangle & 1 \end{pmatrix} \tag{7}$$

$$\begin{pmatrix} 1 & 0 & - & - \\ - & 1 & - & - \\ - & - & 1 & - \\ - & - & - & 1 \end{pmatrix} \rightarrow \begin{pmatrix} 1 & 0 & ? & ? \\ ? & 1 & ? & ? \\ ? & ? & 1 & ? \\ ? & ? & ? & 1 \end{pmatrix} \rightarrow \begin{pmatrix} 1 & 0 & {}^3/_5|0\rangle + {}^2/_5|1\rangle & {}^3/_5|0\rangle + {}^2/_5|1\rangle \\ {}^2/_3|0\rangle + {}^1/_3|1\rangle & 1 & {}^2/_3|0\rangle + {}^1/_3|1\rangle & {}^2/_3|0\rangle + {}^1/_3|1\rangle \\ {}^2/_3|0\rangle + {}^1/_3|1\rangle & {}^3/_5|0\rangle + {}^2/_5|1\rangle & 1 & {}^2/_3|0\rangle + {}^1/_3|1\rangle \\ {}^2/_3|0\rangle + {}^1/_3|1\rangle & {}^3/_5|0\rangle + {}^2/_5|1\rangle & {}^2/_3|0\rangle + {}^1/_3|1\rangle & 1 \end{pmatrix} \tag{8}$$

$$\alpha_{2,1}|0\rangle + \beta_{2,1}|1\rangle \Rightarrow \frac{4}{6}|0\rangle + \frac{2}{6}|1\rangle = \frac{2}{3}|0\rangle + \frac{1}{3}|1\rangle \tag{9}$$

Table 2 illustrates this quantum computation on a Takuzu grid composed of 16 cells. The evolution of the game is not automatic, but supervised by a player who can fill the grid by viewing the different states of the cells implemented by quantum agents. The detailed sequence of the nine steps (3 cycles of decision according to the quantum computation presented below) shown in this illustration is as follows:

Table 2. Example of 3 cycles of quantum computation that a player can apply.

1)

1			
	1		
		0	
			1

2)

1	$^{2}/_{3}\|0>$ $^{1}/_{3}\|1>$		
	1		
		0	
			1

3)

1	0		
	1		
		0	
			1

4)

1	0		
	1		
		0	
			1

5)

1	0	$^{2}/_{5}\|0>$ $^{3}/_{5}\|1>$	
	1		
		0	
			1

6)

1	0	1	
	1		
		0	
			1

7)

1	0	1	
	1		
		0	
			1

8)

1	0	1	$1\|0>$ $0\|1>$
	1		
		0	
			1

9)

1	0	1	0
	1		
		0	
			1

Table 2.1 - Initial setting of the Takuzu grid;
Table 2.2 - Probabilities for the state values ($^{2}/_{3}|0> + {}^{1}/_{3}|1>$) of the cell $c_{1,2}$;
Table 2.3 - Following this observation, the player can take a decision: $c_{1,2} = 0$;
Table 2.4 - New setting of the grid according to the previous decision;
Table 2.5 - Probabilities for the state values ($^{2}/_{5}|0> + {}^{3}/_{5}|1>$) of the cell $c_{1,3}$;
Table 2.6 - Following this observation, the player can take a decision: $c_{1,3} = 1$;
Table 2.7 - New setting of the grid according to the previous decision;
Table 2.8 - Probabilities for the state values ($1|0> + 0|1>$) of the cell $c_{1,4}$;
Table 2.9 - Following this observation, the player can take a decision: $c_{1,4} = 0$.

Table 3 illustrates a local entanglement between two cells *a* and *b*. If a measure of *a* or *b* is done (cell valuation) then the other cell is also determined (valued). If no measurement is made then both cells have indeterminate states which can be '0' or '1', with the same probability. The two cells a and b can be considered as a single cell A whose probability of the state values is as follows (10):

$$p(|10\rangle) = p(|10\rangle) = \frac{1}{2} \tag{10}$$

It is then possible to consider/study the superposition of the states in A (|A >) (Table 3c) rather than the superposition of the states in a and b separately (|a > ⊗ |b >).

Table 3. Illustration of a local entanglement.

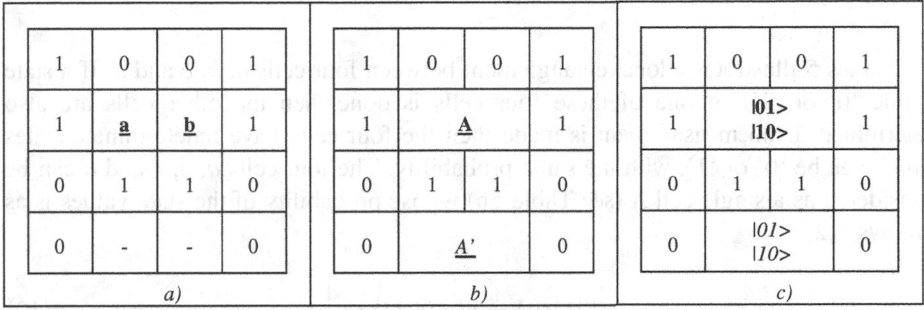

a)

1	0	0	1
1	a	b	1
0	1	1	0
0	-	-	0

b)

1	0	0	1
1	A		1
0	1	1	0
0	A'		0

c)

1	0	0	1
1	$\lvert 01\rangle$ $\lvert 10\rangle$		1
0	1	1	0
0	$\lvert 01\rangle$ $\lvert 10\rangle$		0

Table 4 illustrates a non-local entanglement between two cells a and b. If a state (value '0' or '1') of a or b is done then the other cell is also determined (even if it is distant). If no measurement is made then both cells have indeterminate states which can be '0' or '1', with the same probability. The two cells a and b can be considered as a single cell A whose probability of the state values is as follows (11):

$$p(|10\rangle) = p(|10\rangle) = \frac{1}{2} \tag{11}$$

It is then possible to consider/study the superposition of the states in A (|A >) (Table 4c) rather than the superposition of the states in a and b separately (|a > ⊗ |b >).

Table 4. Illustration of a non-local entanglement.

a)

1	0	0	1
a	0	1	b
0	1	1	0
-	-	-	-

b)

1	0	0	1
A	0	1	A
0	1	1	0
0	1	0	1

c)

1	0	0	1
$\lvert 01\rangle$ $\lvert 10\rangle$	0	1	$\lvert 01\rangle$ $\lvert 10\rangle$
0	1	1	0
1	1	0	0

Table 5. Illustration of a multiple entanglement of cells.

a)

0	1	0	1
1	a	b	0
0	c	d	1
1	0	1	0

b)

0	1	0	1
1	A		0
0			1
1	0	1	0

c)

0	1	0	1
1	|0110>		0
0	|1001>		1
1	0	1	0

Tables 5 illustrates a local entanglement between four cells *a*, *b*, *c* and *d*. If a state (value '0' or '1') of one of these four cells is done then the other cells are also determined. If no measurement is made then the four cells have indeterminate states which can be '0' or '1', with the same probability. The four cells *a*, *b*, *c* and *d* can be considered as a single cell *A* (see Table 5b) whose probability of the state values is as follows (12):

$$p(|0110\rangle) = p(|1001\rangle) = \frac{1}{2} \tag{12}$$

It is then possible to consider/study the superposition of the states in *A* (in fact two possible configurations with: $|A\rangle = \frac{1}{\sqrt{2}}|0110\rangle + \frac{1}{\sqrt{2}}|1001\rangle$, see Table 5c) rather than the superposition of the states in *a*, *b*, *c* and *d* separately $(|a> \otimes |b> \otimes |c> \otimes |d>)$.

4 Conclusion

In agent-based simulation and modelling of intelligent complex systems, the problem of decision making by agents having incomplete, uncertain, local or global, exchanged or observed information is very common. Recent studies on quantum cognition introduce in the decision process modelling and analysis, quantum properties such as superposition state, non-locality, oscillation, interference or entanglement. So, this paper proposed a model of quantum-like agents able to implement quantum properties of superposition state and local or non-local entanglement.

A quantum-like model for agents offers perspectives for the simulation and modelling of complex systems. In particular, it allows: (1) to determine the states of the components of the complex system only when a measure is to be realized (property of superposition of states) in order to represent or apprehend the indeterminism, the uncertainty and the different probabilities of states that a component might have; (2) to model and simulate components of close proximity, constantly interacting and proposing a very strong correlation of states (entanglement property); (3) to design state interactions and correlations in a context not necessarily local (principle of non-locality); and especially (4) taking into account the organization of a complex system and its necessary decomposition to model, simulate or analyze it, the entanglement of some of its components also offers the possibility of designing globalization of states.

Concerning this latter perspective, it becomes possible to consider certain groups of components like black boxes because they are entangled. The study of a complex system would then consist in determining the different levels of components, in order to identify the interactions of the lowest level, and the inseparable (even indecomposable) components whose study of interactions is irrelevant or even impossible.

A simulation of the Takuzu game with a small grid of 16 cells illustrated our approach to modelling quantum-like agents. From this case study, we have been able to show the interest of certain quantum properties such as the superposition states and the local or non-local entanglement of quantum elements for the decision-making process of an agent implementing the different cells of the game.

The integration of other quantum properties in our quantum-like agent model is our current work perspective. Thus, we are currently investigating two new quantum properties: (1) the interference between states that appears for a quantum-like agent when one of its choices made during a decision-making process interferes with the confidence it may have in the final choice; and (2) the oscillation which may reflect the hesitation of a quantum-like agent taking an individual or collective decision.

Beyond this didactic but very simplistic example, we are convinced that this modelling approach based on quantum agents can be of interest for other complex uses, developments, or applications (derived from CA or not) as varied as: neighboring algorithms in networks, intelligent transportation, studies of cell behavior in tumors, crosses and mutations in genetic algorithms, etc.

References

1. Liu, J., Zhang, S.W.: Characterizing web usage regularities with information foraging agents. IEEE Trans. Knowl. Data Eng. **40**, 7478–7491 (2004)
2. Kazemifard, M., Ghasem-Aghaee, N., Koenig, B.L., Ören, T.I.: An emotion understanding framework for intelligent agents based on episodic and semantic memories. Auton. Agent. Multi-Agent Syst. **28**(1), 126–153 (2014)
3. Scheepers, C., Engelbrecht, A.P.: Training multi-agent teams from zero knowledge with the competitive coevolutionary team-based particle swarm optimizer. Soft. Comput. **20**(2), 607–620 (2016)
4. Khrennikov, A.Y.: Ubiquitous Quantum Structure: From Psychology to Finance. Springer, Berlin (2010)
5. Wang, Z., Busemeyer, J.R., Atmanspacher, H., Pothos, E.M.: The potential of using quantum theory to build models of cognition. Top. Cogn. Sci. **5**(4), 672–688 (2013)
6. Fuss, L., Navarro, D.: Open, parallel, cooperative and competitive decision processes: a potential provenance for quantum probability decision models. Top. Cogn. Sci. **5**(4), 818–843 (2013)
7. Fougères, A.-J.: Towards quantum agents: the superposition state property. Int. J. Comput. Sci. Issues **13**(5), 20–27 (2016)
8. Aerts, D.: Quantum structure in cognition. J. Math. Psychol. **53**, 314–348 (2009)

9. Bruza, P.D., Busemeyer, J., Gabora, L.: Introduction to the special issue on quantum cognition. J. Math. Psychol. **53**, 303–305 (2009)
10. Busemeyer, J.R., Bruza, P.D.: Quantum Models of Cognition and Decision. Cambridge University Press, Cambridge (2012)
11. Aerts, D., Sozzo, S., Gabora, L., Veloz, T.: Quantum structure in cognition: fundamentals and applications. In: Proceedings of the Fifth International Conference on Quantum, Nano and Micro Technologies (ICQNM 2011), Nice, France, 21–27 August 2011
12. Wang, Z., Solloway, T., Shiffrin, R.M., Busemeyer, J.R.: Context effects produced by question orders reveal quantum nature of human judgments. PNAS **111**(26), 9431–9436 (2014)
13. Bohm, D., Hiley, B.J.: Non-locality and locality in the stochastic interpretation of quantum mechanics. Phys. Rep. **172**(3), 93–122 (1989)
14. Einstein, A., Podolsky, B., Rosen, N.: Can quantum mechanical description of reality be considered complete? Phys. Rev. **47**, 777–780 (1935)
15. Vértesi, T., Brunner, N.: Quantum nonlocality does not imply entanglement distillability. Phys. Rev. Lett. **108**(3), 030403 (2012)
16. Horodecki, R., Horodecki, P., Horodecki, M., Horodecki, K.: Quantum entanglement. Rev. Mod. Phys. **81**(2), 865 (2009)
17. Weiss, G.: Multiagent Systems: A Modern Approach to Distributed Artificial Intelligence. MIT Press, Cambridge (1999)
18. Fougères, A.-J.: Modelling and simulation of complex systems: an approach based on multi-level agents. Int. J. Comput. Sci. Issues **8**(6), 8–17 (2011)
19. Odell, J.: Agent technology what is it and why do we care? Enterp. Archit. **10**(3), 1–25 (2007). Executive report, Cutter Consortium, Arlington, MA
20. Wooldridge, M.: Agent-based software engineering. IEE Proc. Softw. Eng. **144**(1), 26–37 (1997)
21. Jennings, N.R.: On agent-based software engineering. Artif. Intell. **117**, 277–296 (2000)
22. Biswas, P.K.: Towards an agent-oriented approach to conceptualization. Appl. Soft Comput. **8**(1), 127–139 (2008)
23. Fougères, A.-J.: A modelling approach based on fuzzy agents. Int. J. Comput. Sci. Issues **9**(6), 19–28 (2013)
24. Jennings, N.R., Sycara, K., Wooldridge, M.: A roadmap of agent research and development. Auton. Agents Multi-Agent Syst. **1**(1), 7–38 (1998)
25. Macal, C.M., North, M.J.: Tutorial on agent-based modelling and simulation. J. Simul. **4**, 151–162 (2010)
26. Nielsen, M.A., Chuang, I.: Quantum Computation and Quantum Information. Cambridge University Press, Cambridge (2000)
27. Zhang, W.R.: G-CPT symmetry of quantum emergence and submergence—an information conservational multiagent cellular automata unification of CPT symmetry and CP violation for equilibrium-based many-world causal analysis of quantum coherence and decoherence. J. Quantum Inf. Sci. **6**, 62–97 (2016)

Sensor Networks and Internet of Things

A Profile-Based Fast Port Scan Detection Method

Katalin Hajdú-Szücs(✉), Sándor Laki, and Attila Kiss

Eötvös Loránd University, Budapest, Hungary
szucsk@caesar.elte.hu, lakis@elte.hu, kissae@ujs.sk

Abstract. Before intruding into a system attackers need to collect information about the target machine. Port scanning is one of the most popular techniques for that purpose, it enables to discover services that may be exploited. In this paper we propose an accurate port scan detection method that can detect port scanning attacks earlier with higher reliability than the widely used Snort-based approaches. Our method is profile-based, meaning that it does not only set a threshold on the connection attempts in a given time interval, like most of the current methods, but builds an IP profile of four features that enables a more fine-grained detection. We use the Budapest node of the FIWARE Lab community cloud as a natural honeypot to identify malicious activities in it.

Keywords: Port scan detection · FIWARE Lab · IP profile-based detection

1 Introduction

Network ports are entry points to a machine that is connected to the Internet. To exploit the vulnerabilities of standard services running on a server, hackers design port scanning methods to find those ports that are open to the world. A typical successful network penetration can be divided into five phases [12]: reconnaissance, scanning, gaining access, maintaining access and covering track. The more the attacker gets close to all phases, the stealthier is the attack. That is the reason why it is important to detect all intrusions as early as possible. A port scan activity is commonly a precursor of an intrusion attempt, generally it takes place in the first phase of the attack in order to discover all network entry points into the target system. During this method the attacker tries to connect with several ports on the destination computer and learn what services are running, what users own those services, whether anonymous logins are supported, and whether certain network services require authentication [6]. If we were able to

During this work, Dr. Laki was also with Wigner Research Centre for Physics of the Hungarian Academy of Sciences.
Dr. Kiss was also with J. Selye University, Komárno, Slovakia.

N.T. Nguyen et al. (Eds.): ICCCI 2017, Part I, LNAI 10448, pp. 401–410, 2017.
DOI: 10.1007/978-3-319-67074-4_39

reveal the malicious intention in that phase, we could prevent the hacker from causing actual damage.

Several types of port scan activities can be distinguished [10]. Based on the pattern of the explored ports we can talk about vertical scans that targets several destination ports on a single host, horizontal scans that targets the same port on several hosts and block scans that is a combination of the two former methods. Furthermore, we can make the differentiation based on the protocols and packet types they use [5]. Since substantial proportion of scanning activities are sent over TCP and the most common packet type in this traffic is TCP SYN, we focus our investigation on the detection of TCP SYN scanning [3,10,14] (also known as stealth scan or half-open scan).

To understand how this attack works, at first we need to investigate the initialisation method of TCP connections, a process known as a "three way handshake". As its first step, the initiating system sends a SYN packet to the destination, which will respond with a SYN/ACK packet, acknowledging that it has received the connection request. Finally, the first system send back an ACK packet, indicating the receipt of the answer. If all the three steps take place within a time window, the data transfer can begin. As a first step of SYN scanning, the attacker sends a TCP packet with the SYN flag on to the target system and waits for the response. If the port is closed or filtered (meaning that a filter prevents the probes from reaching the port), the host responds with an RST packet or gives no answer at all. Otherwise, when the port is open, the host will continue the handshake by sending a SYN/ACK packet. At that point the attacker can clearly know whether the given port is open, and can launch an attack immediately. Now that during this type of scanning a TCP connection is never fully established and the half-open connections are not logged by the target machine, it is easier for the attacker to stay undetected.

However there are some already existing methods for SYN scanning detection [5,9,11,16], most commercial solutions are built on the naive approach and simply set a threshold on the number of connection attempts in a given time interval. Another drawback of current solutions is that they operate with a relatively high false positive rate. One differentiating factor that distinguishes our work from previous solutions is the idea of monitoring multiple features of the communicating parties and building up IP profiles based on which scanner and benign activity can be better characterised and separated. Snort [7], one of the most well-known intrusion detection systems, also provides the possibility to detect these type of attacks. After describing our methodology in details in Sect. 3, we present our results in comparison with the performance of Snort.

2 Related Work

The first network intrusion detection system that detected scanning was the Network Security Monitor [19]. It worked with rules to alert on any source IP address that connected to more than 15 other source IP addresses in a given time window. This naive approach is still used by many commercial systems. Another early solution was the Graph Based Intrusion Detection System (GrIDS) [18]. It detected port scanning by building graphs of activity in which the nodes represented hosts, and the edges represented some network traffic between hosts. Thus a scan probe could be represented as an edge between the scanning host and the server being scanned. There is a variety of late papers concerning port scan detection as well. One of the most well known signature-based intrusion detection systems today is Snort [7], an open source lightweight network intrusion detection system based on libpcap [2]. Besides other intrusion detection tasks, Snort also supports port scan detection. This functionality is made possible by a preprocessor plug-in that keeps track of the number of destination IP addresses accessed by each source IP address in a given time window (in 3 s by default). Whenever this value is above a specified threshold (4 by default), the system raises an alarm indicating a port scan attack by the source IP address. Additionally, the port scan detector looks for single TCP packets that are not used in normal TCP operations (packets with odd combination of TCP flags). The scan will also be reported, regardless of the threshold being broken, if the packet contained an abnormal TCP flag combination. In the literature several papers can be found that utilises Snort. In [9,15] researchers implement a rule based IDS and apply some own rules to detect attacks on TCP and UDP specific protocols using Snort. The research in [21] examines the evasion technique provided by Nmap, a port scanner and Metasploit Framework, an exploit launcher against Snort. There are solutions that apply data mining techniques in order to filter out scanner activity. Authors of [16] used the number of the different TCP control packets and SYN as input for Back Propagation algorithm in order to detect port scans. Another paper that also uses the concept of neural networks is [13] in which the authors apply a Time Delay Neural Network in order to maximize the recognition rate of network attacks. A design that maintains records of event likelihood, from which the anomalousness of a given packet is approximated is described in [17]. In [11], an accurate sampling scheme is provided for the detection of SYN flooding and TCP port scan activities. Jaekwang Kim and Jee-Hyong Lee in [8] propose an abnormal traffic control framework to detect slow port scan attacks using fuzzy rules and stepwise policy, that acts as an intrusion prevention system to suspicious network traffic.

3 A Profile-Based Method

Our detection methodology is based on maintaining profiles for all the IP addresses that participate in building TCP connections. The concept is built on the assumption that a scanner leaves more initialised TCP handshakes half-open than what he finishes successfully. We classify a TCP SYN request for a connection as successful, if the initiating party sends a TCP ACK packet to the destination within 30 s (that is the default TCP handshake timeout), and we label it as a failed request otherwise. An IP profile for a specific IP address, $\hat{IP}$, is defined as $(f \rightarrow, f \leftarrow, f \downarrow, f \uparrow, Priority) \in [0,1]^5$, where

- $f \rightarrow$ is the ratio of failed connection requests sent by $\hat{IP}$ out of all connection initialisations of $\hat{IP}$,

- $f \leftarrow$ is the ratio of failed connection requests received by $\hat{IP}$ out of all received connection requests,
- $f \downarrow$ is the number of different IP addresses from which $\hat{IP}$ received failed connection requests divided by the number of all different IP addresses that sent connection requests to $\hat{IP}$,
- $f \uparrow$ is the number of different IP addresses to which $\hat{IP}$ sent failed connection requests divided by the number of all different IP addresses that received connection requests from $\hat{IP}$,
- and *Priority* is defined as $max(0, 1 - 4/f)$, where f if the actual number of failed requests sent by $\hat{IP}$.

Whenever a TCP SYN or a corresponding TCP ACK packet is sent, we update the profiles of the source and destination IPs. Table 1 summarises those extreme points of the IP profiles that indicate an attack. Any time the *Priority* value of an IP profile exceeds a predefined threshold, we calculate which is the closest extreme point to this IP that indicate an attack and we log the IP, the type of the attack, the priority value and the time-stamp corresponding to the investigated packet. By having a closer look at the IP profiles, we can also recognise which IP addresses are under attack. The last two lines of Table 1 presents the extreme points of the ip profile space that indicate victims.

Table 1. Extreme points of IP profiles that imply an attack

f→	f←	f↓	f↑	IP type
1	0	0	1	Vertical scan
1	0	0	0	Horizontal scan or SYN flood
1	0	0	0.5	Block scan
0	1	0	0	Port scan or SYN flood victim
0	1	1	0	Victim of distributed attack

4 Test Dataset

We tested the performance of the proposed model on traffic measurements carried out in the Budapest node of FIWARE Lab [1] community cloud. The FIWARE Lab, launched on 6 September 2013, is a non-commercial OpenStack-based sandbox environment where various prepared images are available for endusers to experiment with FIWARE technologies by hosting their virtual machines. It currently consists of 12 nodes in Europe and a few external nodes located in Brazil and Mexico. The system has more than 6600 users, mostly start-ups and SMEs taking part in the business acceleration programme of FIWARE. However, FIWARE Lab, similarly to Planet Lab, can only be used for non-commercial experimentation, testing, prototyping and validation, and thus the traffic mix that can be observed in such an environment is less heterogeneous than the one in real commercial clouds like Amazon or Azure. The site is connected to two external networks – the public Internet and a private management network with Multi-Domain VPN – through a physical link of 1 Gbps. The traffic was captured at the border of our network before the firewall protecting the cloud site. To analyse cloud traffic two independent traffic traces were collected. All packet headers as well as a part of raw data have been recorded. The first data set called TR-2015 was captured shortly after the launch of the cloud site from 12 May 2015 to 1st June 2015. During that time, the site only had a limited number of external users from Europe, hosting approximately 20 virtual machines. The second trace called TR-2016 was created one year later between 1–7 June 2016 when the site hosted more than 120 virtual machines with various sizes. Most of them were used by European start-ups for running their experiments and prototype services. The basic statistics of the two traffic traces are depicted in Table 2. One can observe that the nature of the traffic has slightly changed after one year. The daily average flow numbers are 0.175 and 0.34 millions for TR-2015 and TR-2016, respectively. However, the approximately two times more flows carry almost the same amount of data in each day. This is in accordance with the literature that the majority of flows are very short. We have to note that in comparison to community cloud sites the observed traffic volume is moderate, but as it will be shown in the following section, it contains enough malicious traces with attack probes for the evaluation of port scan detection performance in a live environment. Considering the main applications used during the two examination periods in the cloud, the most frequently used application is SSH (TCP port 22), a cryptographic network protocol allowing remote access to network hosts, through which 10% of the data is transacted. SSH is followed by HTTP (TCP port 80), HTTPS (TCP port 443) and HTTP-ALT (TCP port 8080). Since the cloud is supposed to process a heavy load of tunneled data and to offer services through various HTTP-based APIs, the fact that these ports carried the majority of the bytes is not surprising. This usage pattern (especially the high amount of SSH traffic) also reflects that this non-commercial cloud site acts more like an experimental platform than a commercial environment. However, since all the virtual machines are directly connected to the wild Internet, the captured traffic is expected to contain a large number of malicious activities, serving enough data for the thorough validation of our port scan detection method.

Table 2. Basic statistics of the analyzed traffic traces

Trace	Traffic volume	Flows [Million]	Packets [Million]	Time Span [Days]
TR-2015	61.7 GB	3.5	239	19
TR-2016	17.6 GB	2.4	71	7

5 Results

We have validated our scheme on the traces that were described in Sect. 4. To evaluate the performance of our method, we compared our results to the outcome of Snort port scan detector on the same dataset (the operational details of Snort can be found in Sect. 2). The first problem we had to face was the lack of ground truth in the dataset. To simplify the evaluation process, we needed to set up some assumptions about the nature of port scans in the captured traffic. Particularly, considering that all legitimate clients of the cloud were certainly located in Europe during the examination period, we treated all detected traffic that originated from outside of these areas as a result of malicious activity. To confirm the rationale behind these inferences some manual assessments have been conducted on a fraction of the dataset that approved our expectations. The daily distribution of the traffic coming from non European countries have been analysed country by country as well. The investigation showed that no daily pattern can be observed in these communications. Since the data exchange is continuous and independent from the time of the day, we suspect that the traffic may be machine generated.

After running Snort's port scan module with high sense level on TR-2015, it yielded 1006 different IP addresses as scanners. Our method on the same dataset, with Priority threshold set to 0.5, detected 1470 malicious IPs, from which 801 could be also found among the results of Snort. In TR-2016 Snort alerted on 1194 scans, while the Profile-based solution detected 1215 and there were 810 events that have been caught by both detectors. Those 801 different addresses in TR-2015 and 810 addresses in TR-2016 that were involved in such suspicious connection attempts that generated alarms by both detectors were geolocalized by WEBNet 77 [20]. From this part of the dataset we tried to weed out as many false positives as possible. Due to the large amount of the suspicious connections we narrowed the manual search for the false positive alarms to Europe, where most of the legitimate traffic could originate from. In this portion of the dataset, we found in total 15 false positive alarms, most of them caused by transmission errors in the communication of active hosts, like NATs. As it can be seen, about one-third of the addresses are located in China, whereas the second most populated area is the United States, followed by Russia as a distant third. These results are consistent with the Digital Attack Map [4] website where statistics are available on the most common locations of attack sources.

Figures 2a and b shows the location of those respectively 669 and 405 ip addresses that were detected as scanners by our algorithm but left out by Snort in the two traces. After manual investigation of the sources originating from Europe, 14 of them turned out to be false positives.

The location of those 205 IP addresses that were found by Snort but not detected by our method can be seen on Fig. 1a. Similarly, most of the traffic originated from China and the United States. In TR-2016 Snort has alerted on 384 additional addresses that were not caught by the profile-based solution. After some deeper investigation of these alerts we found that 47% of the detections

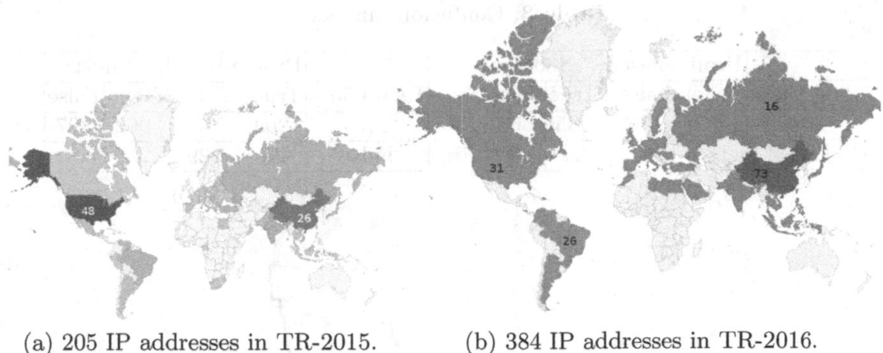

(a) 205 IP addresses in TR-2015. (b) 384 IP addresses in TR-2016.

Fig. 1. Geolocation of the IP addresses detected only by Snort.

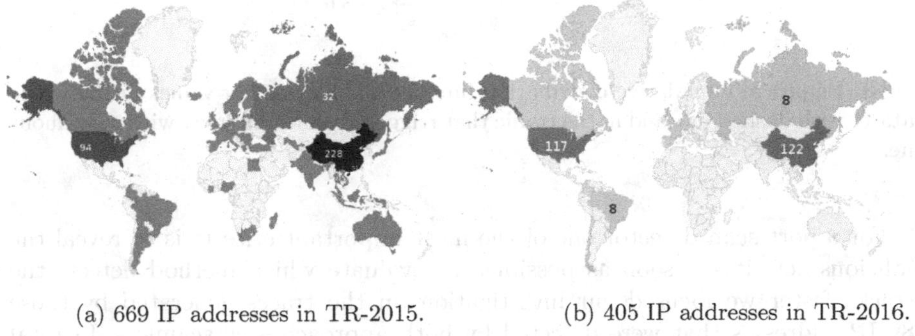

(a) 669 IP addresses in TR-2015. (b) 405 IP addresses in TR-2016.

Fig. 2. Geolocation of the IP addresses detected only by the profile-based method.

were raised on IP addresses that communicated less that 3 packets and were false positives, caused by the high sensitivity setting of Snort's port scan detection module.

Table 3a summarise the performance of the two detectors regarding the true positive, false positive, true negative and false negative rates in TR-2015. Analogously, Table 3b concludes the same measures for TR-2016. Figure 3 shows the distribution of the Priority values of the different IP profiles generated by our algorithm. As it can be seen, only around 3% of the profiles had non-zero Priority in the whole population of IPs, meaning that only a small portion of the traffic raised an alarm in our detector. However, if we narrow our investigation down to only those IP addresses that were detected by Snort, we find that ~20% of the profiles are left with zero Priority and the remaining ~80% has Priorities above 0.67, meaning that the IP profiles of these addresses have been investigated also by our method and classified either as vertical, horizontal, block scanner, or as benign if despite the increased amount of half-open connections the given IP also participated in a sufficient amount of successful connections.

Table 3. Confusion matrices

	Profile-based		Snort	
Real Class	True	False	True	False
True	1452	196	985	663
False	18	9	21	6

(a) TR-2015

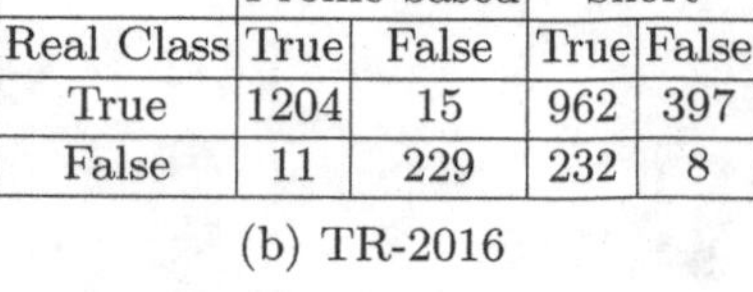

	Profile-based		Snort	
Real Class	True	False	True	False
True	1204	15	962	397
False	11	229	232	8

(b) TR-2016

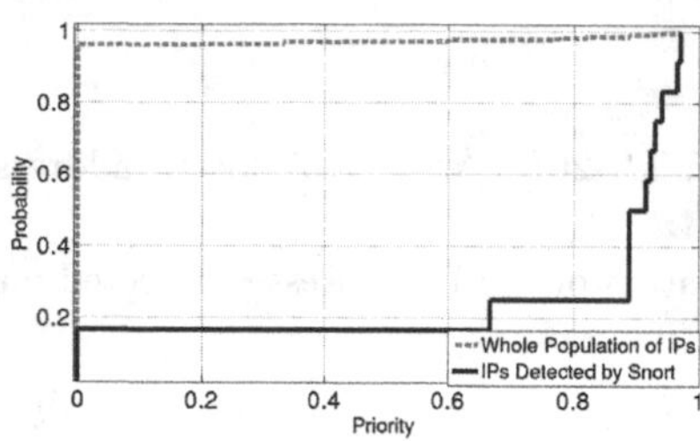

Fig. 3. Empirical cumulative distribution function of the priority values in the whole dataset with dashed line and in the traffic that triggered alarm in Snort with continuous line.

For a port scan detector one of the most important criteria is to reveal the malicious activity as soon as possible. To evaluate which method detects the attacks faster we focused our investigation on the traces generated by those 789 IP addresses that were detected by both approaches as scanners. In total there were 2752 events in TR-2015 and 1245 events in TR-2016 generated by these malicious sources. To visualise the error in the reaction time of the two designs, we calculated t_i^{dist}, the time difference between t_i^S, Snort's detection timestamp and t_i^{PB}, the profile-based detector's timestamp for the ith event, $t_i^{dist} = t_i^{PB} - t_i^S$. The resulting empirical cumulative distribution functions that can be seen on Fig. 4 show that the profile-based method is faster in average than Snort, because the area under the curve on the negative side of the x axis is larger than the area above the curve on the positive side. Note, that the x axes of these graphs are on logarithmic scale. A closer investigation on the behaviour of these curves around zero reveals that in TR-2015 more than 80% of the time differences are negative, meaning, that the profile-based solution detected faster. A small portion, less than 10% of the events have been detected at the same time and around 10% of the scans were discovered earlier by Snort. In TR-2016 the profile-based solution reacted faster in 60% of the cases. Around 30% of the activities have been detected at the same time, whereas in 10% of the cases Snort reacted faster. Although the vast majority of the differences are below 20 s, the extremums at both directions are around 3 h.

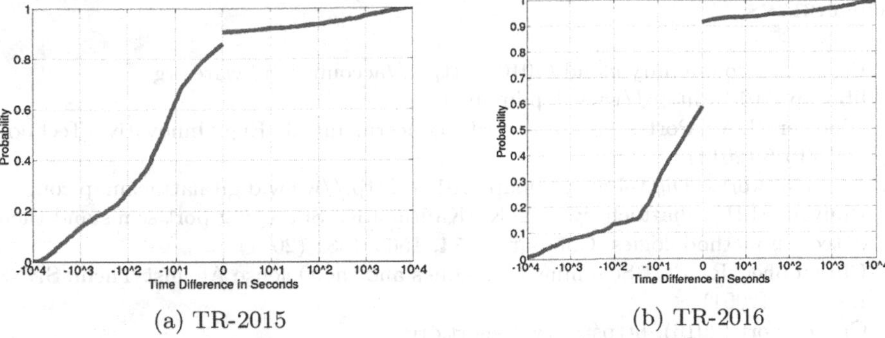

(a) TR-2015

(b) TR-2016

Fig. 4. Empirical cumulative distribution functions of time differences between the detection of our solution and Snort for the port scan activities that were detected by both methods.

6 Conclusions

We presented a new way of port scan detection that is based on building a profile for every IP participating in the communication. We used the FIWARE Lab community cloud as a natural honeypot to capture traffic that contains a significant amount of malicious traces. Due to the lack of ground truth in the dataset we geolocated the suspicious IP addresses relying on the assumption that most of the legitimated traffic must have originated from Europe. The performance of the method was evaluated in contrast to Snort's port scan detection module. We found that our solution worked reliably, with a higher true positive and a lower false negative rate. Considering the promptness of the alarms we have shown that in average our solution reacted faster. In the future we are planning to further test the performance of our solution on other labeled datasets. A larger dataset with scan labels would not only facilitate easier performance evaluation, but could open up the possibility of applying machine learning algorithms in this context.

Acknowledgment. Authors thank Ericsson Ltd. for support via the ELTE CNL collaboration. Sándor Laki also thanks the partial support of EU FP7 FI-CORE project. This publication is the partial result of the Research & Development Operational Programme for the project "Modernisation and Improvement of Technical Infrastructure for Research and Development of J. Selye University in the Fields of Nanotechnology and Intelligent Space", ITMS 26210120042, co-funded by the European Regional Development Fund.

References

1. Fiware lab community cloud (2016). https://account.lab.fiware.org
2. libpcap (2016). http://www.tcpdump.org/
3. Ahanger, T.A.: Port scan - a security concern. Int. J. Eng. Innovative Technol. (IJEIT) **3** (2014)
4. ArborNetworks: Digital attack map (2013). http://www.digitalattackmap.com
5. Bhuyan, M.H., Bhattacharyya, D.K., Kalita, J.K.: Surveying port scans and their detection methodologies. Comput. J. **54**, 1565–1581 (2011)
6. Christopher, R.: Port Scanning Techniques and the Defense Against Them. SANS Institute (2001)
7. Cisco: Snort (2016). https://www.snort.org
8. Jaekwang, K., Lee, J.-H.: A slow port scan attack detection mechanism based on fuzzy logic and a stepwise policy. In: 4th International Conference on Intelligent Environments, IET (2008)
9. Kumar, V., Sangwan, O.P.: Signature based intrusion detection system using snort. Int. J. Comput. Appl. Inf. Technol. **1**(3), 35–41 (2012). (ISSN: 2278-7720)
10. Lee, C.B., Roedel, C., Silenok, E.: Detection and characterization of port scan attacks. Univeristy of California, Department of Computer Science and Engineering (2003)
11. Maciej, K., Janowski, L., Duda, A.: An accurate sampling scheme for detecting SYN flooding attacks and portscans. In: International Conference on Communications (ICC). IEEE (2011)
12. Offensivehacking: Five phases of hacking, October 2012. https://offensivehacking.wordpress.com
13. Omar, A.-J., Arafat, A.: Network intrusion detection system using neural network classification of attack behavior. J. Adv. Inf. Technol. **6**(1) (2015)
14. Panjwani, S., et al.: An experimental evaluation to determine if port scans are precursors to an attack. In: Proceedings of the International Conference on Dependable Systems and Networks, pp. 602–611. IEEE (2005)
15. Patel, S.K., Sonker, A.: Rule-based network intrusion detection system for port scanning with efficient port scan detection rules using snort. Int. J. Future Gener. Commun. Netw. **9**(6), 339–350 (2016)
16. Soniya, B., Wiscy, M.: Detection of TCP SYN scanning using packet counts and neural network. IEEE International Conference on Signal Image Technology and Internet Based Systems SITIS 2008. IEEE (2008)
17. Stuart, S., Hoagland, J.A., McAlerney, J.M.: Practical automated detection of stealthy portscans. J. Comput. Secur. **10**(1–2), 105–136 (2002)
18. Stuart, S.-C., et al.: Grids-a graph based intrusion detection system for large networks. In: Proceedings of the 19th National Information Systems Security Conference, vol. 1 (1996)
19. Todd, H.L., et al.: A network security monitor. In: Computer Society Symposium, Proceedings. IEEE (1990)
20. WEBNet77: Multiple ip address lookup (2016). http://software77.net/geo-ip/multi-lookup/
21. Jammes, Z., Papadaki, M.: Snort IDS ability to detect Nmap and metasploit framework evasion techniques. Adv. Commun. Comput. Netw. Secur. **10**, 104 (2013)

Sensor Network Coverage Problem: A Hypergraph Model Approach

Krzysztof Trojanowski, Artur Mikitiuk(✉), and Mateusz Kowalczyk

Cardinal Stefan Wyszyński University in Warsaw, Warszawa, Poland
{k.trojanowski,a.mikitiuk}@uksw.edu.pl

Abstract. A sensor power schedule for a homogenous network of sensors with a limited battery capacity monitoring a set of points of interest (POI) depends on locations of POIs and sensors, monitoring range and battery lifetimes. A good schedule keeps the network operational as long as possible while maintaining the required level of coverage (not all POIs have to be monitored all the time). Searching for such a schedule is known as the Maximum Lifetime Coverage Problem. A new approach solving MLCP is proposed in this paper. First, in every time step, we try to achieve the required coverage level using sensors with the longest remaining working time monitoring the largest number of POIs that have not been covered yet. The resulting schedule is next used for generating a neighbour schedule by a perturbation algorithm. For experimental evaluation of our approach a new set of test cases is proposed. Experiments with these data show interesting properties of the algorithm.

1 Introduction

Multiple applications of wireless sensor networks make them a subject of great interest and wide range of research. In our paper we focus on the task of monitoring an area by a set of distributed sensors with a limited battery capacity. Precisely, at least a large, user defined percentage of a number of locations in the area has to be monitored in every time step. The aim is to find a schedule of sensor activity maximizing lifetime of a network and guaranteeing a satisfying coverage of the locations by sensing ranges of sensors. Simply, due to a large number of sensors it is not necessary to set them all in the monitoring state for obtaining a satisfying coverage of the area. Thus, effective energy management in the sensor batteries is the key issue in this case. Typically, the model of the problem assumes discrete time and the solution is represented as a schedule where rows represent control sequences for sensors over time and columns — subsequent time periods. The goal is to maximize the length of an uninterrupted sequence of columns when the coverage represents a satisfying level.

In this paper a new approach called HMA or Hypergraph Model Approach is proposed to find an effective schedule of sensor control for given locations of sensors and points to be monitored in the area. First, in every time step we try to achieve the required coverage level using sensors with the longest remaining

N.T. Nguyen et al. (Eds.): ICCCI 2017, Part I, LNAI 10448, pp. 411–421, 2017.
DOI: 10.1007/978-3-319-67074-4_40

working time monitoring the largest number of POIs that have not been covered yet. The resulting schedule is next used for generating a neighbour schedule by a perturbation algorithm. For experimental evaluation of our approach a new set of test cases is proposed.

The paper consists of six sections. Section 2 gives definition of the solved problem and all the necessary constraints. In Sect. 3 the proposed two-phase algorithm is presented. A benchmark is defined in Sect. 4 and results of experiments are discussed in Sect. 5. Section 6 concludes the paper.

2 Maximum Lifetime Coverage Problem (MLCP)

In [1] authors define a sensor coverage problem. $N_{\mathbf{S}}$ immobile sensors are randomly deployed over an area to monitor $N_{\mathbf{P}}$ points of interest, a.k.a. POIs. Each sensor has the same sensing range r_{sens} and the same battery lifetime. A sensor node can be either in a *working* mode when it consumes one unit of its battery capacity, or in a *sleeping* state when the energy consumption is negligible. It is also assumed that time is discrete, thus, the lifetime of a sensor T_{batt} is the maximal number of time steps when the sensor can be on. T_{batt} is assumed to be known in advance and always the same for all sensors. We ignore the fact that in real life effective battery capacity depends on various factors and is hard to predict. Frequent turning on and off the battery shortens its lifetime. Thus, it does matter whether the battery is on in consecutive time steps or is in every time step turned on/off but our algorithms omit this problem. Our goal is to find a schedule for the sensors activity which gives the longest time period of uninterrupted control of the given set of POIs $\mathbf{P}$. This is called the Maximum Lifetime Coverage Problem.

In the real world sensors manifest two types of activity: monitoring and communication. This may imply two classes of optimization problems: (1) scheduling for the satisfying POI coverage by working sensors over the longest time period and (2) building ad-hoc communication networks guaranteeing immediate transfer of sensed data. In the presented research we focus on the problem just from the first class and assume that the effective data transfer is always given. For this reason, the energy consumption necessary for communication is not present in the model. Moreover, it is assumed that communication graph always provides connectivity independently from the landform and localization of sensors even if some sensors are off. Literature concerning sensor networks is multiple and rich, however, publications about heuristic approaches to MLCP as defined above are not so numerous. One can find, e.g., sensor control methods [3] or schedule optimization approaches based on evolutionary algorithms [4], simulated annealing [5], or graph cellular automata [6].

An active sensor monitors all POIs located within its sensing range. One sensor can monitor a number of POIs, and on the other hand a single POI can be monitored by a number of sensors. A full coverage of a set of POIs cannot be guaranteed since it can happen that some POIs lie outside of the range of any sensor. However, full coverage is not obligatory, it is sufficient to have at least a defined percentage *cov* of the number of POIs being monitored. When the fraction of monitored POIs fits in the range $[cov, 1]$, the satisfying coverage level is met.

The aim of optimization is to maximize the lifetime of the set of sensors $\mathbf{S}$ under a satisfying coverage constraint for a given feasible instance of the problem. Therefore, the solution is represented as a schedule H where for each time step t, a slot H^t, that is, a sensor control vector, defines activity for each of the sensors. The slot is a binary vector where index i identifies the sensor ID, and the value in the i-th cell H_i^t defines the i-th sensor's activity in time t: a working mode ($H_i^t = 1$) or a sleep state ($H_i^t = 0$). The value of a schedule equals the length of the longest sequence of slots during which the satisfying coverage requirement is met.

3 Optimization Algorithm

For optimization of the sensor activity schedule a local search approach is applied. The main novelty lies in two methods: one for generation of an initial schedule and the other, for building a neighbour schedule. Both methods never generate unfeasible solutions, that is, for example, schedules including time slots with insufficient coverage of POIs, or impossible to execute due to battery load limitations. Therefore, the search process can be interrupted in any moment and the current schedule is ready to implement instantly.

3.1 Generation of the Initial Schedule

Algorithm 1 represents a greedy approach. In every time slot, it tries to achieve the required coverage level using sensors with the longest remaining working time monitoring the largest number of POIs that have not been covered yet. We remove from the model all sensors which do not monitor any POI. We call this algorithm HMA or Hypergraph Model Approach because the problem space is modeled as a hypergraph. The sensors are its nodes. The battery levels in sensors decrease as the algorithm goes and sets the sensors in the working mode in selected time slots. The hyperedges correspond to POIs. Precisely, the set of sensors able to monitor given POI forms the corresponding hyperedge.

For every time slot the algorithm

1. Turns all the sensors off.
2. For every sensor creates the list of POIs within its range (the procedure *inRange*).
3. Performs the following operations in a loop until the required level of coverage has been reached or no sensor is available to increase the level of coverage
   (a) Filters out the list of sensors leaving only those with a remaining battery lifetime $batt(S_i, t) > 0$, turned off, and able to monitor at least one POI not covered in the current time slot (the procedure $filter$). If the remaining list is empty, the algorithm ends and the current slot is not included in the schedule.

Algorithm 1. HMA: A Hypergraph Model Approach for Schedule Generation

Require: $cov, T_{\text{batt}}, \mathbf{S}, \mathbf{P}$;
Ensure: schedule H
$H \leftarrow \emptyset$; $t \leftarrow 0$ ▷ *initialization: make an empty schedule and set time to zero*
for all $S \in \mathbf{S}$ **do** $batt(S, 0) \leftarrow T_{\text{batt}}$ ▷ *initialize the sensor battery levels*
$\mathbf{M} \leftarrow inRange(\mathbf{S}, \mathbf{P})$ ▷ $\mathbf{M}$*: sum of* $\mathbf{M}_i s$ *– lists of POIs observed by sensors*
repeat ▷ *main loop of the algorithm*
    $\mathbf{M}_{\text{work}} \leftarrow \mathbf{M}$
    $t \leftarrow t + 1$
    $H^t \leftarrow 0$
    for all $S \in \mathbf{S}$ **do** $batt(S_i, t) \leftarrow batt(S_i, t-1)$ ▷ *update the sensor battery levels*
    $c \leftarrow 0$ ▷ *initial coverage level for the slot* H^t *equals zero*
    while $c < cov$ **do**
        $\mathbf{S}_{\text{work}} \leftarrow filter(\mathbf{S}, \mathbf{M}_{\text{work}})$ ▷ *filter out useless sensors*
        if $\mathbf{S}_{\text{work}} = \emptyset$ **then break;** ▷ *break the main loop with* $c < cov$
        $n \leftarrow selectBest(\mathbf{S}_{\text{work}}, \mathbf{M}_{\text{work}})$
        $H_n^t \leftarrow 1$ ▷ *turn the selected sensor* S_n *on in the current slot*
        $batt(S_n, t) \leftarrow batt(S_n, t) - 1$ ▷ *update the selected sensor battery level*
        $\mathbf{M}_{\text{work}} \leftarrow removePoi(\mathbf{M}_{\text{work}}, S_n)$ ▷ *remove POIs located within the* S_n*'s sensing range from every list of POIs* $\mathbf{M}_i \in \mathbf{M}$
        $c \leftarrow covPoi(S(H^t))$ ▷ *update current coverage level for* H^t
    if $c \geq cov$ **then** ▷ *if the coverage level for the current time slot is reached, ...*
        $H \leftarrow H + H^t$ ▷ *... attach the current time slot to the schedule*
until $c < cov$
return H

(b) Among sensors with the longest remaining battery lifetime selects the one able to monitor the largest number of POIs not covered yet (the procedure *selectBest*); when such a sensor does not exist, selects the one with remaining lifetime as long as possible.
(c) Turns the selected sensor on and decrements its battery lifetime.
(d) For every inactive sensor, removes from the list of observed POIs those which are monitored by the sensor just selected (the procedure *removePoi*).
(e) Updates the current coverage level (the procedure $covPoi(S(H^t))$, which returns the percentage of POIs covered by sensors being active in H^t).

3.2 Building a Neighbour Schedule

A method of a neighbour schedule generation is presented in Algorithm 2. First, a randomly selected slot is removed from the schedule and battery levels for sensors being on in this slot are restored. Then for each of the remaining time slots redundant sensors, that is, sensors which can be turned off without getting the coverage level below the threshold *cov* are searched for. Redundant sensors found in these slots are set off and their battery levels are restored as well. At the end the obtained slimmed schedule is the input for Algorithm 1. Due to usual

Algorithm 2. A Neighbour Schedule Generation

```
Require: H, ϵ, S;
Ensure: a neighbour schedule
for n ← 1, N_S do
    while batt(S_n, length(H)) > 0 do
        r ← randomInt(length(H))                ▷ select randomly a time slot
        if H_n^r = 0 and U(0,1) < ϵ then        ▷ U stands for a uniform distribution random number
            H_n^r ← 1                           ▷ switch the sensor on
            for i ← r, length(H) do
                batt(S_n, i) ← batt(S_n, i) − 1 ▷ update the sensor battery level
for m ← 1, length(H) do
    k ← getRedundantSensor(H^m)                 ▷ find a sensor which can be off safely
    while k ≠ −1 do
        H_k^m ← 0                               ▷ switch the sensor off
        for i ← m, length(H) do
            batt(S_k, i) ← batt(S_k, i) + 1     ▷ update the sensor battery level
        k ← getRedundantSensor(H^m)
return HMA(H)          ▷ return the outcome of the Hypergraph Model Approach
```

presence of a set of sensors which still can be on, but all together cannot give the requested coverage, one can hope that these remnant active sensors together with sensors having just restored battery levels can generate additional slots to be attached to the schedule.

4 Benchmark SCP1

A set of test cases has been prepared for experimental evaluation of the proposed HMA algorithm. It is called Sensor Coverage Problem, Set No. 1 or SCP1. Test cases have similar structure, however, generate different instances of the problem due to different values of their configuration parameters and heuristic rules of instance building. The following parameters are the same in all cases: (1) number of sensors: 2000, (2) sensing range r_{sens}: 1 distance unit (which is not one of standard units of measurements but a conventional one introduced just to show proportion between this range and other distance parameters), (3) the satisfying coverage level *cov*: 80%. The remaining parameters may vary beetween test cases. These are:

- distribution of POIs — POIs form nodes of a triangular or a rectangular grid,
- method of sensor distribution — coordinates of sensor localizations originate from (1) random, (2) Halton generator [2].
- area size — the area is a square of the side size: 13, 16, 19, 22, 25, 28 distance units.
- battery max capacity, that is, max time of sensor activity T_{batt} — it varies from 10 to 30 time steps with step 5 (we assume that for a given experiment all batteries have the same capacity).

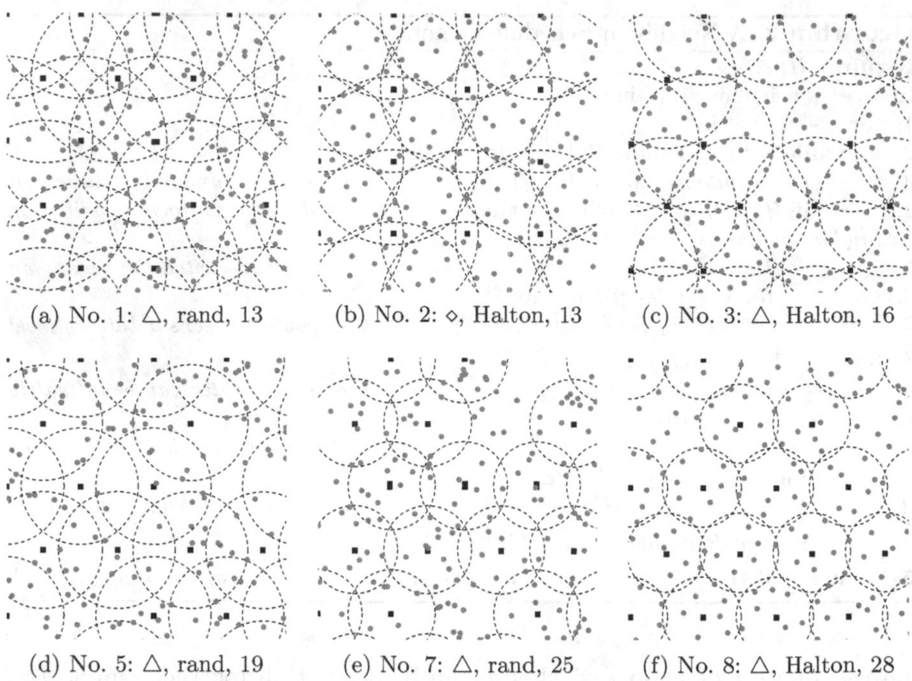

(a) No. 1: △, rand, 13 (b) No. 2: ◇, Halton, 13 (c) No. 3: △, Halton, 16

(d) No. 5: △, rand, 19 (e) No. 7: △, rand, 25 (f) No. 8: △, Halton, 28

Fig. 1. Visualizations of a monitored area for selected instances of a six among the eight test cases in SCP1: squares represent POIs, dots — sensors, circles around POIs — which sensors have in its range the POI located in the circle center.

In all test cases the number of POIs is the same while the area sizes are different. It means that the distance between POIs grows as the square side grows. To avoid full regularity in the POIs distribution, 20% of nodes in the grid is not filled with POIs. Due to nondeterministic rules of instance generation, even if the test case parameters are constant, subsequent instances differ in the number of POIs (from 199 to 240 for the triangular grid and from 166 to 221 for the rectangular grid). Simply, for each grid node a random value is generated and the POI is assigned to the node only if the value is less than threshold 0.8.

Eventually, 8 configurations of the test case have been proposed. They differ in the area side size, the type of a grid for POIs, and generator of sensor locations. A set of 40 instances has been generated for every test case. Each instance consists of two files: one with a set of POIs and one with a set of sensors.

The test cases have different diversity of the numbers of POIs covered by sensors: in the test case No. 1 there exist fractions of sensors covering 1, 2, 3, 4, and 5 POIs and some sensors cover even more that 5 POIs. For the next test cases this diversity decreases. In the test case No. 8 almost three fourths of sensors cover only a single POI. Figure 1 shows, that in the test case No. 1 the intersections between neighbor sensor monitoring areas are greater and consist of larger numbers of POIs. For the next test cases these areas decrease and the numbers of common POIs decrease as well.

5 Results of Experiments

Results of experiments with the proposed HMA algorithm and the SCP1 benchmark are presented in Table 1. For each series of experiments two series of schedule lengths are obtained: one returned by Algorithm 1 and the other – by the Local Search (LS) algorithm using proposed perturbation operator (Algorithm 2). The values in the series are paired because output schedules of Algorithm 1 were an input for LS. For every test case from SCP1, the algorithm was executed once for each of the 40 instances. Rows in Table 1 present mean lengths of schedules obtained for the 40 instances, as well as min and max lengths among them. The experiments were performed for five different values of T_{batt} varying from 10 to 30 with step 5.

To determine whether application of LS improves quality of the schedules, statistic t-tests for paired data were performed where the null hypothesis is that any differences in schedule lengths before and after LS are due to chance. All obtained p-values are less than 0.001 (Table 2), so, the null hypothesis can be rejected, that is, one can conclude with 99.9% confidence that the differences in schedule lengths before and after LS are not due solely to chance.

Obtained lengths of coverage time are higher for the cases when the intersections between the sets of POIs controlled by neighboring sensors are greater. When the sensor lifetime T_{batt} grows linearly from 10 to 30, mean lengths of schedules grow linearly as well.

One could also look at the results in the upper part of Table 1 from another perspective. We could divide the battery lifetime into shorter or longer time periods, for example 10 longer or 30 shorter time periods. Since the actual length of a time slot is irrelevant for our earlier considerations, having more shorter time slots, one could repeat several times a basic schedule generated for a smaller number of (longer) time slots. This would give us a schedule with more slots but with the same length measured in actual time units. But using shorter time slots could be beneficial because one could do better than to repeat several times some basic schedule as a building block. The results from the upper part of Table 1 show that as the number of time slots corresponding to battery lifetime (T_{batt}) grows linearly from 10 to 30, mean lengths of schedules grow a bit more than linearly. For example, for every test case the mean value for $T_{\text{batt}} = 20$ is a bit more than twice the value for $T_{\text{batt}} = 10$. Thus, using shorter time slots can slightly increase time the system is operational. The explanation of this phenomenon is obvious. When the Algorithm 1 terminates, there are usually sensors with batteries not completely discharged. When we repeat the same schedule several times, we always get the same sensors with remaining battery lifetime. But when we do not repeat the same schedule, we can get sufficient coverage for few more time slots. This observation is not true for the lower part of Table 1 because Algorithm 2 makes better use of the remaining battery lifetime.

Table 1. Mean, min and max lengths of initial schedules returned by Algorithm 1 – top part – and by the Local Search algorithm using proposed perturbation operator (Algorithm 2) – bottom part – for each of the eight test cases in SCP1 and for five values of T_{batt} from 10 to 30

No.	T_{batt}	mean	min	max	No.	T_{batt}	mean	min	max
1	10	380.3	367	392	5	10	180.0	173	185
	15	573.5	554	589		15	272.5	263	280
	20	767.1	742	788		20	366.8	354	377
	25	960.5	928	987		25	457.4	441	469
	30	1153.8	1115	1185		30	550.0	530	564
2	10	396.3	387	406	6	10	144.2	141	147
	15	598.1	585	611		15	218.0	213	222
	20	799.8	782	816		20	291.7	285	297
	25	1001.5	980	1023		25	365.5	357	372
	30	1203.4	1177	1229		30	439.2	429	447
3	10	263.8	257	271	7	10	103.3	98	107
	15	397.9	388	409		15	157.1	148	163
	20	532.1	519	547		20	210.8	199	219
	25	666.2	650	685		25	264.6	249	275
	30	800.2	780	823		30	318.3	300	330
4	10	167.0	158	172	8	10	89.5	88	91
	15	253.0	239	261		15	136.2	134	139
	20	339.1	321	350		20	182.7	180	186
	25	425.2	402	439		25	229.3	226	234
	30	511.5	484	528		30	276.0	272	281
1	10	386.7	375	396	5	10	181.8	175	187
	15	581.6	564	596		15	274.2	264	282
	20	776.7	753	796		20	366.8	354	377
	25	971.6	942	995		25	459.2	444	471
	30	1166.5	1131	1196		30	551.8	533	566
2	10	403.3	394	412	6	10	145.3	142	148
	15	606.3	593	618		15	219.0	214	223
	20	809.5	792	826		20	292.8	286	298
	25	1012.7	991	1032		25	366.5	358	373
	30	1216.0	1190	1240		30	440.3	430	448
3	10	266.4	259	274	7	10	104.4	99	108
	15	400.9	391	412		15	158.0	149	164
	20	535.5	522	550		20	211.9	200	220
	25	670.0	654	689		25	265.6	250	275
	30	804.5	785	827		30	319.3	302	331
4	10	169.0	160	175	8	10	90.6	89	92
	15	255.0	241	263		15	137.1	135	140
	20	341.3	323	353		20	183.8	181	187
	25	427.3	405	441		25	230.3	227	235
	30	513.6	486	530		30	277.0	273	282

Table 2. Results of statistical test for paired samples obtained from Algorithm 1 and the Local Search algorithm using proposed perturbation operator (Algorithm 2): p-values, std.dev.#1 for the results from Algorithm 1 and std.dev.#2 for Local Search obtained for each of the eight test cases in SCP1 and for five values of T_{batt} from 10 to 30

No.	T_{batt}	p-value	s.d.#1	s.d.#2	No.	T_{batt}	p-value	s.d.#1	s.d.#2
1	10	2.28E-030	5.93	5.35	5	10	1.29E-025	2.94	2.92
	15	3.08E-032	8.60	8.15		15	1.39E-023	4.35	4.52
	20	5.89E-033	11.30	10.67		20	1.38E-024	5.78	5.82
	25	4.58E-033	14.14	13.31		25	3.34E-023	7.22	7.36
	30	5.78E-033	16.73	15.89		30	1.66E-019	8.75	8.63
2	10	1.58E-036	4.25	3.84	6	10	4.30E-021	1.27	1.38
	15	1.67E-034	6.25	5.76		15	3.27E-022	1.93	1.98
	20	1.85E-036	8.23	7.74		20	3.64E-023	2.61	2.57
	25	7.04E-039	10.10	9.64		25	4.64E-017	3.19	3.12
	30	3.53E-038	12.14	11.47		30	4.30E-021	3.82	3.80
3	10	1.17E-028	2.91	2.84	7	10	6.90E-017	2.32	2.42
	15	6.12E-027	4.16	4.04		15	1.99E-016	3.58	3.57
	20	5.95E-030	5.52	5.36		20	8.29E-019	4.76	4.82
	25	6.91E-031	6.88	6.73		25	3.65E-020	5.96	5.97
	30	1.94E-030	8.31	8.04		30	7.55E-018	7.20	7.13
4	10	1.21E-022	3.58	3.55	8	10	3.16E-026	0.81	0.84
	15	2.63E-021	5.27	5.30		15	1.63E-018	1.22	1.21
	20	8.13E-023	6.95	7.02		20	1.60E-024	1.55	1.49
	25	5.69E-023	8.68	8.64		25	3.65E-020	1.89	1.89
	30	8.13E-023	10.35	10.48		30	2.11E-018	2.33	2.40

The coverage level generated by the algorithm in all our experiments exceeded *cov* by not more than 3%. Turning on a single sensor can hardly ever increase *cov* by more than 3%. To achieve this, one would need sensors monitoring more than 3% of all POIs. It does not seem likely in real life situations and it never happens in the selected benchmark.

Another set of experiments was performed for comparisons with results of algorithms presented in [5,6]. In this set of experiments we compared means and standard dev. of lengths of schedules obtained from LS and obtained by algorithms presented in these two papers. We selected a set of four test cases. Three of them originate from [6]. In these cases 100 POIs are located in the form of a rectangular grid on an area of size 100 × 100, and the number of randomly deployed sensors is: case#1 – 100, case#2 – 200, case#3 – 300. From [5] we selected a problem where 400 POIs are located in the form of a rectangular grid on an area of size 100 × 100, and 100 sensors are randomly deployed.

Table 3. Mean and standard dev. of lengths of schedules returned by the Local Search algorithm using proposed perturbation operator (Algorithm 2) and presented in publications (here as the reference values)

No.	Local search	Ref. value	No.	Local search	Ref. value
case#1	177.2 ± 2.24	83.0 ± 2.23	case#3	552.5 ± 2.63	248.0 ± 2.82
case#2	364.4 ± 3.18	165.0 ± 2.44	case#4	177.86 ± 2.44	49

In every case $cov = 0.9$, a sensing range $r_{\text{sens}} = 20$ and $T_{\text{batt}} = 20$. In Table 3 one can notice that using Algorithms 1 and 2 more than doubles the length of the resulting schedule. Data from Table 1 show that Algorithm 1 presents itself a good approach because using LS improves its results only by 1–2%. On the other hand results produced by algorithms described in [5,6] leave a lot of room for improvement.

6 Conclusions

In this paper we presented a new approach to solving the Maximum Lifetime Coverage Problem. Our algorithm in the first phase models problem space as a hypergraph. To obtain the required coverage for next time slot, the algorithm tries to use sensors with the longest remaining battery lifetime monitoring the largest number of POIs not covered yet in this time slot. The schedule obtained from this phase is used as a starting point for a local search algorithm using a perturbation operator.

We also proposed a set of test cases for experimental evaluation of our algorithm. Our experiments with these test cases show that when many sensors can cover multiple POIs, the system is operational longer than in cases when most of sensors is able to monitor only a single POI. When the sensor battery capacity grows linearly, mean lengths of schedules grow also linearly. Our local search algorithm when applied to schedules produced by the greedy approach gives slightly longer schedules.

References

1. Cardei, I., Cardei, M.: Energy-efficient connected-coverage in wireless sensor networks. IJSNet **3**(3), 201–210 (2008)
2. Halton, J.H.: Algorithm 247: radical-inverse quasi-random point sequence. Commun. ACM **7**(12), 701–702 (1964)
3. Tian, D., Georganas, N.D.: A coverage-preserving node scheduling scheme for large wireless sensor networks. In: Proceeding of the First ACM Int. Workshop on Wireless Sensor Networks and Applications (WSNA-2002), pp. 32–41. ACM Press (2002)
4. Tretyakova, A., Seredynski, F.: Application of evolutionary algorithms to maximum lifetime coverage problem in wireless sensor networks. In: IPDPS Workshops, pp. 445–453. IEEE (2013)

5. Tretyakova, A., Seredynski, F.: Simulated annealing application to maximum lifetime coverage problem in wireless sensor networks. In: Global Conference on Artificial Intelligence, GCAI, vol. 36, pp. 296–311. EasyChair (2015)
6. Tretyakova, A., Seredynski, F., Bouvry, P.: Graph cellular automata approach to the maximum lifetime coverage problem in wireless sensor networks. Simulation **92**(2), 153–164 (2016)

Heuristic Optimization of a Sensor Network Lifetime Under Coverage Constraint

Krzysztof Trojanowski[1], Artur Mikitiuk[1(✉)], Frédéric Guinand[1,2], and Michał Wypych[1]

[1] Cardinal Stefan Wyszyński University in Warsaw, Warsaw, Poland
{k.trojanowski,a.mikitiuk}@uksw.edu.pl
[2] Normandy University of Le Havre, Le Havre, France
frederic.guinand@univ-lehavre.fr

Abstract. Control of a set of sensors disseminated in the environment to monitor activity is a subject of the presented research. Due to redundancy in the areas covered by sensor monitoring ranges a satisfying level of coverage can be obtained even if not all the sensors are on. Sleeping sensors save their energy, thus, one can propose schedules defining activity for each of sensors over time which offer a satisfying level of coverage for a period of time longer than a lifetime of a single sensor. A new heuristic algorithm is proposed which searches for such schedules maximizing the lifetime of the sensor network under a coverage constraint. The algorithm is experimentally tested on a set of test cases and effectiveness of its components is presented and statistically verified.

1 Introduction

Different applications of sensor networks are a subject of growing interest. In our research we focus on one of them, that is, the problem of effective monitoring of activity in an environment by a set of sensors disseminated randomly over it. We define this activity with use of a set of points of interest (POIs) located in a selected area. The sensor network is regarded as operational when it is able to monitor sufficient number of POIs at a time, that is, satisfies a coverage constraint. In a considered model of the problem, we assume that a monitoring range of sensors as well as their battery capacity are limited. Fortunately, due to the high number of sensors, an operational network does not necessary have them all in active state all the time. Thus, the solution of the problem we want to address is maximizing the lifetime of an operational sensor network. Precisely, we search for the longest schedule defining sensor states over time under the coverage constraint.

For generation of sensor activity schedules we propose two new algorithms. The first one builds a schedule starting from the empty one and then constructs its subsequent components in an iterative manner. A method of a single component construction is not deterministic, however, it guarantees that the proposed settings make the sensor network operational. When another component of a schedule cannot be created, the first algorithm finishes. The outcome of this

N.T. Nguyen et al. (Eds.): ICCCI 2017, Part I, LNAI 10448, pp. 422–432, 2017.
DOI: 10.1007/978-3-319-67074-4_41

algorithm is an input for a local search algorithm which goal is to improve the obtained initial schedule. The main novelty is represented by a perturbation operator which generates neighbour schedules. Effectiveness of the proposed approach is experimentally verified on a set of newly proposed test cases.

The paper consists of six sections. Section 2 gives definition of the solved problem and all the necessary constraints. The proposed algorithms are presented in Sect. 3. A benchmark is defined in Sect. 4 and the results of experiments are discussed in Sect. 5. Section 6 concludes the paper.

2 Maximum Lifetime Coverage Problem (MLCP)

In [1] authors define a sensor coverage problem, where a homogeneous sensor network **S** comprised of $N_{\mathbf{S}}$ immobile sensors is randomly deployed over an area to cover (monitor) $N_{\mathbf{P}}$ points of interest. Each sensor has a sensing range r_{sens} and starts off its service with a fully loaded battery, that is, an initial energy E_{sens}. We assume a sensor node consumes a unit of energy per time unit for its activity. It is assumed that time is discrete, thus, the lifetime of a sensor T_{batt} is the maximal number of consecutive, or not, time steps when the sensor can be on. Once the sensors are deployed, they schedule their activity.

An active sensor monitors all POIs located within its sensing range. On the other hand, according to its position, a POI can be monitored by several active sensors during the same time period. A full coverage of a set of POIs can be obtained only if all of them lie in a range of a sensor. However, for effective monitoring full coverage of POIs is not necessary. A satisfying coverage level *cov* represents the percentage level of the number of POIs being monitored. Precisely, the satisfying coverage level is obtained if the number of monitored POIs fits in the range $[cov, cov + \delta]$, where δ represents a tolerance factor.

A schedule consisting of sensor controls giving the satisfying coverage level is a solution to this problem. During one time step a sensor can be either in a *working* mode when it uses one unit of its battery capacity, or in a *sleeping* state when the energy consumption is negligible. A schedule H is a matrix of 0 s and 1 s representing states of sensors *off* and *on* in the time slots. In the presented research communication between sensors is not a part of a solved problem, thus the energy consumption necessary for communication is not present in the model. Moreover, it is assumed that communication graph always provides connectivity independently from the landform and localization of sensors even if some sensors are off.

The optimization aim is to find a schedule for the sensor network activity which gives the longest time period of uninterrupted monitoring of the given set of POIs. This is called the Maximum Lifetime Coverage Problem. The value of a schedule equals the length of the longest sequence of slots during which the satisfying coverage requirement is met.

In spite of large number of publications devoted to theory and applications of sensor networks, publications concerning MLCP are not so numerous. One can mention, e.g., sensor control methods [3] or schedule optimization approaches based on evolutionary algorithms [4], simulated annealing [5], or graph cellular automata [6].

3 Search Algorithm

A method of searching for an effective schedule of sensor activity is presented in Algorithm 1. After the first schedule is created by *generateSchedule* (see Algorithm 2), the algorithm iteratively tries to improve it by the problem specific neighbour operator *generateNeighbour* (see Algorithm 4). When a newly found schedule is longer than the current one, the new one takes place of the current and the process continues. Both procedures never generate unfeasible solutions like, for example, solutions including time slots with insufficient coverage of POIs, or impossible to execute due to battery load limitations. Therefore, the search process can be interrupted in any moment and the current schedule is ready to implement.

Algorithm 1. Main Algorithm

Require: $\mathbf{S}, T_{\text{batt}}, cov, \delta$;
Ensure: schedule H

$H \leftarrow generateSchedule(\mathbf{S}, T_{\text{batt}}, cov, \delta)$ ▷ *initial schedule*
repeat
  $H' \leftarrow generateNeighbour(H)$
  if $F(H') > F(H)$ **then** ▷ *if a new schedule has more slots ...*
    $H \leftarrow H'$ ▷ *... the current schedule is replaced by the new one*
until termination condition met
return H

3.1 Generation of the Initial Schedule

We create an initial schedule by generation of time slots one by one as presented in Algorithm 2. First, all battery levels in the set of all sensors $\mathbf{S}$ are set to max value and a max effective coverage $cov_{S(max)}$ of the set of active sensors obtained from the *filter* procedure is calculated (lines 2–5). An outcome of $filter(\mathbf{S}, t)$ (used also in Algorithm 4) is a subset of sensors from a set $\mathbf{S}$ which still have a remaining battery lifetime in the time step t and are able to monitor at least one POI. In line 5 a procedure $covPoi(\cdot)$ is called, which returns the percentage of POIs covered by a given set of sensors. Then, the main loop starts. First, a single time slot for sensor activities is generated by *generateSingleTimeSlot* taking into account requested minimum coverage level cov, the tolerance δ and a current set of active sensors. The slot is added to the schedule and the sensor battery levels $batt(S)$ are updated respectively to the sensor activity setting proposed in the slot (lines 7–9). Next, the set of active sensors is updated and the max effective coverage $cov_{S(max)}$ is reevaluated. The loop stops when the set of remaining active sensors is not able to cover a sufficient number of POIs even if all of them are on, that is, $cov > cov_{S(max)}$.

The *generateSingleTimeSlot* procedure is presented in Algorithm 3. The procedure consists of two main phases: first, splits the set of sensors roughly

Algorithm 2. *generateSchedule*

procedure $generateSchedule(\mathbf{S}, T_{\text{batt}}, cov, \delta)$
  $H \leftarrow \emptyset$; $t \leftarrow 0$ ▷ *initialize a schedule to empty and a time counter to 0*
  for all $S \in \mathbf{S}$ **do** $batt(S) \leftarrow T_{\text{batt}}$ ▷ *sensor batteries level initialization*
  $\mathbf{S}_{\text{work}} \leftarrow filter(\mathbf{S}, t)$ ▷ *select subset of active sensors having POIs in range*
  $cov_{S(max)} \leftarrow covPoi(\mathbf{S}_{\text{work}})$ ▷ *evaluate coverage when all active sensors are on*
  repeat
    $H^t \leftarrow generateSingleTimeSlot(\mathbf{S}_{\text{work}}, cov, \delta)$
    for all $S \in \mathbf{S} \mid H^t_S = 1$ **do**
      $batt(S) \leftarrow batt(S) - 1$ ▷ *sensor batteries level update*
    $t \leftarrow t + 1$ ▷ *time counter update*
    $\mathbf{S}_{\text{work}} \leftarrow filter(\mathbf{S}, t)$ ▷ $\mathbf{S}_{\text{work}}$ *update*
    $cov_{S(max)} \leftarrow covPoi(\mathbf{S}_{\text{work}})$ ▷ $cov_{S(max)}$ *update*
  until $cov > cov_{S(max)}$
  return H

into two subsets (lines 3–7), and next, fine tunes collections of sensors in these subsets (lines 9–17).

In the first phase an important role plays *selectRand*$(\mathbf{S}, n)$ which selects a subset of sensors from a set $\mathbf{S}$. For each of the sensors in $\mathbf{S}$ a newly generated independent random value r from $U[0, 1]$ is compared with a threshold value n, and in the case $r < n$ the sensor becomes selected. The selected sensors are removed from the set $\mathbf{S}$ and returned as an outcome of the procedure. The number of $n \times 100\%$ sensors are returned on average. In the beginning, all sensors are off, and the *selectRand* procedure is called to set statistically 50% of them to on (line 5). Then the coverage level of activated sensors is compared with the requested level *cov*. When the coverage level of activated sensors does not fit the range $[cov, cov + \delta]$, the set is extended or decreased with use of *selectRand* until the coverage fits the range.

If we had a single sensor covering more than $\delta\%$ POIs, the loop in lines 3–7 could run forever turning on this sensor in one iteration (when the coverage level is below *cov*) and turning it off in next iteration (when the coverage level is above $cov + \delta$). However, this does not seem likely in real life situations and it never happens in the selected benchmark described in the next Section.

In the second phase we try to fine-tune the collection of active sensors keeping the coverage in the range $[cov, cov + \delta]$ and decreasing it as close to the lower bound of this range as possible. To obtain this, small numbers of sensors from the set of active ones are randomly selected with procedure *select* and set to off as long as the coverage of the other active sensors satisfies requirements.

3.2 Iterative Improvement of the Schedule

A schedule obtained from the *generateSchedule* procedure undergoes iterative improvement in the main loop of the algorithm (Algorithm 1, lines 2–5) where the *generateNeighbour* procedure presented in Algorithm 4 is called. In this

Algorithm 3. *generateSingleTimeSlot*

```
procedure generateSingleTimeSlot(S_work, cov, δ)
    S_on ← ∅; S_off ← S_work                          ▷ initialization of S_on and S_off
    while ¬(cov ≤ covPoi(S_on) ≤ cov + δ) do
        if covPoi(S_on) < cov then
            S_on ← S_on ∪ selectRand(S_off, 0.5)      ▷ selection and transfer of sensors
        else if covPoi(S_on) > cov + δ then
            S_off ← S_off ∪ selectRand(S_on, 0.5)     ▷ selection and transfer of sensors
    n ← 1; S_tOn ← S_on                               ▷ initialization of n and S_tOn
    while (cov ≤ covPoi(S_on) ≤ cov + δ) ∧ (n ≥ 1) do
        S_tOff ← S_off ∪ select(S_tOn, n)             ▷ selection and transfer of n sensors
        if cov ≤ covPoi(S_tOn) ≤ covPoi(S_on) then
            S_on ← S_tOn                              ▷ update S_on
            S_off ← S_tOff                            ▷ update S_off
            n ← n + 1
        else
            S_tOn ← S_on                              ▷ restore S_tOn
            n ← n/2
    for all S ∈ S_on do H^t_S ← 1
    for all S ∈ S_off do H^t_S ← 0
    return H^t                                        ▷ return single time slot for current time t
```

Algorithm 4. *generateNeighbour*

```
procedure generateNeighbour(S, H, T_max, cov, n, δ)
    for j ← 1, n do                                   ▷ remove n time slots from a schedule
        k ← rand(0, T_max)                            ▷ select a slot to remove
        for all S ∈ S | H^k_S = 1 do
            batt(S) ← batt(S) + 1                     ▷ sensor batteries level update
        H ← H − H^k; T_max ← T_max − 1                ▷ update of a schedule and its length
    S_work ← filter(S, T_max)                         ▷ select a subset of active sensors having POIs in range
    cov_S(max) ← covPoi(S_work)                       ▷ evaluate coverage when all active sensors are on
    while cov < cov_S(max) do
        H^(T_max+1) ← generateSingleTimeSlot(S_work, cov, δ)
        for all S ∈ S | H^(T_max+1)_S = 1 do
            batt(S) ← batt(S) − 1                     ▷ sensor batteries level update
        T_max ← T_max + 1                             ▷ schedule length update
        S_work ← filter(S, T_max)                     ▷ S_work update
        cov_S(max) ← covPoi(S_work)                   ▷ cov_S(max) update
    return H                                          ▷ return the neighbour schedule
```

procedure a randomly selected set of slots is removed from the schedule and for the sensors being active in these slots their battery levels are restored. Then the procedure *generateSingleTimeSlot* is called as many times as possible. The perturbation is based on observation, that the set of sensors activated again can produce new time slots which have not been present in the modified schedule and due to different distribution of sensor activities their number may be higher.

4 Benchmark SCP1

For experimental evaluation of the proposed algorithm a set of test cases has been prepared. For brevity the set is called SCP1 (Sensor Coverage Problem, Set No. 1). Every test case is controlled by a set of parameters. The parameters fixed for all test cases are: (1) number of sensors: 2000, (2) sensing range r_{sens}: 1 unit, (3) the satisfying coverage level *cov*: 80%, (4) the coverage level tolerance δ: 5%. The remaining parameters may vary between test cases. These are:

- distribution of POIs — two types of distribution: nodes of a triangular grid and nodes of a rectangular grid,
- method of sensor distribution — coordinates of sensor localization originate from (1) random, (2) Halton generator [2].
- area size — the area is a square of the side size: 13, 16, 19, 22, 25, 28 units.
- battery max capacity, that is, max time of sensor activity T_{batt}: varies from 10 to 30 time steps with step 5 (value is the same for all sensors).

It has to be stressed, that due to non-empirical nature of proposed test cases the distance unit does not represent particular number of meters or inches but its role is just to show proportion between the area size, sensing range of sensors, and distances between POIs.

Additionally, it was assumed, that the number of POIs is the same for different area sizes, which means, that the grid of POIs stretches as the square side grows. To avoid full regularity in the POIs distribution, a small fraction of nodes in the grid, precisely, 20%, is not filled with POIs. The number of POIs in subsequent test cases is similar to each other but varies slightly (from 199 to 240 for the triangular grid and from 166 to 221 for the rectangular grid). The grid nodes which do not represent POIs are randomly selected in the grid for every instance of the test case.

Eventually, 8 configurations of the test case have been proposed, which differ in the area side size, the type of a grid for POIs, and generator of sensor locations. For every test case a set of 40 instances was generated. In practice, every instance consists of two files: one with a set of POIs and one with a set of sensors. Mean numbers of sensors covering different numbers of POIs for each of the test cases are presented in Table 1.

Table 1 presents numbers of sensors which cover 1, 2, 3, 4, 5 and more POIs. One can see, that in the test case No. 1 all fractions of sensors are present, whereas, in the test case No. 8 majority of sensors (almost three fourths of them) cover just one POI and the mean number of sensors covering two POIs is really small and equals 74.7.

Additionally, Fig. 1 shows, that in the test case No. 1 the intersections between neighbor sensor monitoring areas are greater and consist of a larger number of POIs. For the next test cases these areas decrease and the numbers of common POIs decrease as well.

Table 1. Mean numbers of sensors covering 0, 1, 2, 3, 4, 5 and more that 5 POIs for the eight test cases of SCP1. △ means a triangular grid of POI locations, and ◇ - a rectangular grid.

No.	Configuration	0	1	2	3	4	5	> 5
1.	13 × 13, △, rand	5.0	58.2	234.9	559.7	691.6	327.9	122.7
2.	13 × 13, ◇, Halton	18.3	126.6	369.4	691.2	693.5	95.4	5.7
3.	16 × 16, △, Halton	24.6	211.1	679.2	902.7	182.4	0.0	0.0
4.	19 × 19, ◇, rand	135.1	763.0	951.2	128.8	21.8	0.0	0.0
5.	19 × 19, △, rand	112.7	631.7	819.7	435.8	0.0	0.0	0.0
6.	22 × 22, △, Halton	209.9	1012.6	665.9	111.6	0.0	0.0	0.0
7.	25 × 25, △, rand	340.1	1303.4	350.9	5.6	0.0	0.0	0.0
8.	28 × 28, △, Halton	450.1	1475.2	74.7	0.0	0.0	0.0	0.0

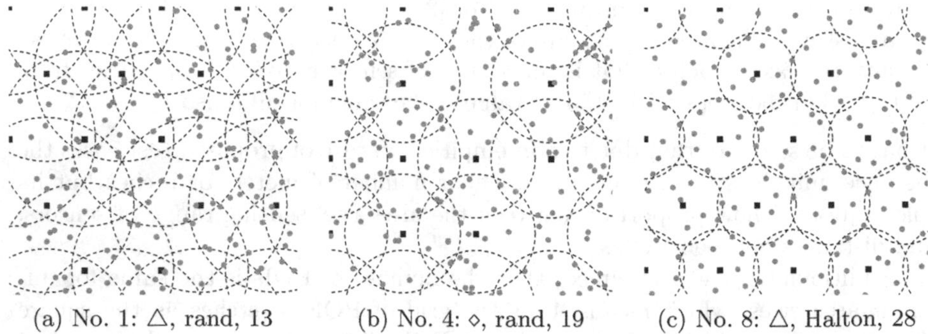

(a) No. 1: △, rand, 13 (b) No. 4: ◇, rand, 19 (c) No. 8: △, Halton, 28

Fig. 1. Visualizations of a monitored area for selected instances of a three among the eight test cases in SCP1: squares represent POIs, dots — sensors, circles around POIs — which sensors have in its range the POI located in the circle center.

5 Results of Experiments

For experimental evaluation of the proposed algorithm the SCP1 benchmark was used. For every test case from SCP1 40 instances were generated and the algorithm was executed once for each of the instances. Table rows present mean lengths of schedules obtained for respective 40 instances, as well as min and max lengths among them. The experiments were performed for five different values of T_{batt} varying from 10 to 30 with step 5.

Results of experiments with the proposed algorithm and the SCP1 benchmark are presented in Table 2. They are divided into two groups: the first one (the top half of Table 2) presents quality of initial schedules returned by *generateSchedule* (Algorithm 2), and the second one (the bottom half of Table 2) presents quality of schedules returned by the main algorithm (Algorithm 1). The values in the series are paired because output schedules of Algorithm 2 were an input for the main loop in Algorithm 1.

Table 2. Mean, min and max lengths of schedules returned by *generateSchedule* (Algorithm 2) – top part – and by the main algorithm (Algorithm 1) – bottom part – for each of the eight test cases in SCP1 and for five values of T_{batt} from 10 to 30

No.	T_{batt}	mean	min	max	No.	T_{batt}	mean	min	max
1	10	210.1	202	215	5	10	95.8	93	99
	15	316.2	309	326		15	144.5	140	149
	20	421.4	410	434		20	192.8	187	200
	25	527.5	513	544		25	241.4	234	249
	30	633.1	617	650		30	290.1	281	299
2	10	216.8	211	222	6	10	75.5	73	77
	15	325.9	319	332		15	113.5	111	116
	20	435.3	426	444		20	151.5	149	155
	25	543.6	534	559		25	189.8	184	194
	30	652.7	643	672		30	228.1	223	233
3	10	139.4	137	143	7	10	56.1	54	59
	15	209.9	204	217		15	85.0	81	88
	20	280.4	274	287		20	113.3	107	118
	25	350.5	342	360		25	141.9	135	148
	30	420.9	410	431		30	170.8	163	177
4	10	88.5	85	92	8	10	48.6	47	50
	15	133.2	126	138		15	73.6	72	76
	20	178.2	170	186		20	98.3	97	101
	25	222.7	210	230		25	123.1	120	125
	30	267.7	257	277		30	148.2	145	151
1	10	214.8	209	222	5	10	97.8	95	101
	15	321.5	312	331		15	146.9	142	152
	20	427.8	415	440		20	195.7	190	203
	25	534.2	519	548		25	244.5	237	252
	30	640.1	623	656		30	293.4	284	302
2	10	221.9	218	227	6	10	77.2	76	79
	15	332.2	327	342		15	115.8	113	118
	20	441.9	432	453		20	154.2	151	158
	25	551.6	543	567		25	192.5	187	196
	30	661.2	650	680		30	231.2	226	236
3	10	142.2	140	145	7	10	57.9	55	60
	15	213.2	207	220		15	87.0	83	90
	20	284.1	276	291		20	115.8	110	121
	25	354.7	346	365		25	144.7	137	150
	30	425.3	414	436		30	173.8	166	180
4	10	90.6	86	94	8	10	50.1	49	51
	15	135.5	129	140		15	75.4	74	77
	20	181.2	174	188		20	100.6	99	103
	25	226.0	214	234		25	125.8	123	128
	30	271.4	261	280		30	151.2	148	154

To determine whether application of LS improves quality of the schedules, statistic t-tests for paired data were performed where the null hypothesis is that any differences in schedule lengths before and after LS are due to chance. Obtained p-values are presented in Table 3. In every case the value is less than 0.001, which means that the null hypothesis can be rejected, that is, one can conclude with 99.9% confidence that the differences in schedule length before and after LS are not due solely to chance.

Additionally, for comparisons with results of algorithms presented in [5,6] another set of experiments was performed. In these experiments means and standard dev. of lengths of schedules obtained from LS are compared with means and standard dev. obtained by algorithms presented in these two papers. A set of four test cases has been selected. From [6] we selected three cases where 100 POIs are located in the form of a rectangular grid on an area of size 100×100, and the number of randomly deployed sensors is: case#1 – 100, case#2 – 200, case#3 – 300. From [5] one problem was selected where 400 POIs are located

Table 3. Results of statistical tests for paired samples obtained from Algorithm 2 and the main algorithm (Algorithm 1): p-values, std.dev.#1 for the results from Algorithm 2 and std.dev.#2 for the main algorithm obtained for each of the eight test cases in SCP1 and for five values of T_{batt} from 10 to 30

No.	T_{batt}	p-value	s.d.#1	s.d.#2	No.	T_{batt}	p-value	s.d.#1	s.d.#2
1	10	$2.47E^{-22}$	2.97	3.14	5	10	$3.04E^{-24}$	1.62	1.58
	15	$7.2E^{-22}$	4.55	4.52		15	$1.58E^{-20}$	2.46	2.49
	20	$8.22E^{-24}$	6.01	6.15		20	$4.93E^{-27}$	3.16	3.34
	25	$1.05E^{-21}$	8.02	7.83		25	$1.96E^{-23}$	3.92	4.09
	30	$4.29E^{-23}$	9.22	9.05		30	$2.89E^{-25}$	4.75	4.82
2	10	$1.24E^{-21}$	2.50	2.60	6	10	$1.11E^{-20}$	0.90	0.90
	15	$1.86E^{-21}$	3.50	3.52		15	$5.38E^{-23}$	1.38	1.17
	20	$1.88E^{-23}$	4.64	4.62		20	$6.38E^{-25}$	1.43	1.53
	25	$1.79E^{-22}$	5.36	5.82		25	$1.9E^{-26}$	2.09	2.03
	30	$1.02E^{-22}$	6.28	6.84		30	$8.93E^{-27}$	2.46	2.32
3	10	$3.01E^{-22}$	1.41	1.46	7	10	$7.29E^{-22}$	1.22	1.21
	15	$7.78E^{-23}$	2.69	2.54		15	$5.25E^{-21}$	1.92	1.83
	20	$1.78E^{-19}$	3.42	3.34		20	$3.17E^{-22}$	2.47	2.43
	25	$3.06E^{-29}$	4.06	4.10		25	$1.41E^{-23}$	2.74	2.86
	30	$4.88E^{-25}$	4.61	4.76		30	$9.55E^{-26}$	3.48	3.43
4	10	$3.06E^{-19}$	1.84	1.75	8	10	$2.82E^{-19}$	0.62	0.55
	15	$3.77E^{-20}$	2.57	2.49		15	$2.21E^{-19}$	1.03	0.84
	20	$2.37E^{-22}$	3.51	3.28		20	$5.38E^{-24}$	1.17	1.17
	25	$8.77E^{-24}$	4.45	4.57		25	$1.9E^{-26}$	1.19	1.18
	30	$6.35E^{-25}$	4.73	4.60		30	$1.2E^{-26}$	1.24	1.38

Table 4. Mean and standard dev. of lengths of schedules returned by the main algorithm (Algorithm 1) and presented in publications (here as the reference values)

No.	LS	ref. value	No.	LS	ref.value
case#1	123.5 ± 1.38	83.0 ± 2.23	case#3	358.4 ± 2.75	248.0 ± 2.82
case#2	241.3 ± 2.29	165.0 ± 2.44	case#4	118.6 ± 1.69	49

in the form of a rectangular grid on an area of size 100×100, and 100 sensors are randomly deployed. In every case $cov = 0.9$, a sensing range $r_{\text{sens}} = 20$ and $T_{\text{batt}} = 20$. 30 instances were generated for each of the four cases. Results from Table 4 show that Algorithm 1 gives schedules almost 50% longer than schedules obtained using methods from [6]. In case #4 our schedule is more than 100% longer than the reference value.

6 Conclusions

The paper presents a new heuristic approach to the Maximum Lifetime Coverage Problem. The proposed algorithm is build on the frame of the Local Search approach, and novelty is represented by two problem specific procedures. The first one generates an initial solution for the algorithm, whereas the second one is a perturbation procedure. Schedules generated by these procedures always satisfy coverage constraint.

For experimental evaluation of our algorithm a set of eight test cases is also proposed. Results of our experiments, that is, lengths of obtained schedules are divided into two groups: in the first one we evaluate effectiveness of the procedure generating initial solution, whereas in the second one effectiveness of the entire algorithm. Differences between schedule lengths obtained in the first and the second group demonstrate influence of the perturbation procedure on the quality of the results. Statistical tests confirmed our hypothesis that differences between means obtained for schedules before and after application of the iterative improvement are not due to chance, that is, the proposed perturbation can significantly improve obtained schedules. Moreover, our approach gives results much better than methods described in [5,6]

Further research will concern development of the process of a schedule iterative improvement on one hand and on the other – experimental comparisons of the proposed algorithm based on benchmarks presented in other publications.

References

1. Cardei, I., Cardei, M.: Energy-efficient connected-coverage in wireless sensor networks. IJSNet **3**(3), 201–210 (2008)
2. Halton, J.H.: Algorithm 247: radical-inverse quasi-random point sequence. Commun. ACM **7**(12), 701–702 (1964)

3. Tian, D., Georganas, N.D.: A coverage-preserving node scheduling scheme for large wireless sensor networks. In: Proceedings of the First ACM International Workshop on Wireless Sensor Networks and Applications (WSNA-02), pp. 32–41. ACM Press (2002)
4. Tretyakova, A., Seredynski, F.: Application of evolutionary algorithms to maximum lifetime coverage problem in wireless sensor networks. In: IPDPS Workshops, pp. 445–453. IEEE (2013)
5. Tretyakova, A., Seredynski, F.: Simulated annealing application to maximum lifetime coverage problem in wireless sensor networks. In: Global Conference on Artificial Intelligence, GCAI, vol. 36, pp. 296–311. EasyChair (2015)
6. Tretyakova, A., Seredynski, F., Bouvry, P.: Graph cellular automata approach to the maximum lifetime coverage problem in wireless sensor networks. Simulation **92**(2), 153–164 (2016)

Methods of Training of Neural Networks for Short Term Load Forecasting in Smart Grids

Robert Lis[1(✉)], Artem Vanin[2], and Anastasiia Kotelnikova[2]

[1] Wroclaw University of Science and Technology, 50370 Wroclaw, Poland
robert.lis@pwr.edu.pl

[2] Moscow Power Engineering Institute, Moscow 111115, Russia
Kotelnikova.anastasiia@gmail.com

Abstract. Modern systems of voltage control in distribution grids need load forecast. The paper describes forecasting methods and concludes that using of artificial neural networks for this problem is preferable. It shows that for the complex real networks particle swarm method is faster and more accurate than traditional back propagation method.

Keywords: Load forecasting · Neural network · Power quality

1 Introduction

Today in Russian distribution grids there are a variety of problems connected with power quality. The most common problem is an unacceptable level of voltage among consumers. Invalid voltage deviation leads to increased deterioration and failures of equipment, breakdowns in technological processes, incorrect operation of the control and automation systems. The common reasons of deviations are high network load, inconsistency of load curves, incorrect or insufficient regulation of voltage.

Now voltage control in distribution networks is carried out mainly using the tap changing transformers at power supply centers of 110–220 / 6–20 kV. Regulation is performed in most cases in the stabilization voltage mode rarely counter regulation is applied.

Control actions for the both methods occur after the actual changing of mode settings that do not provide an optimal effect. In modern works is proposed to operate branches of tap changers of transformers using intelligent algorithms. In the Fig. 1 is shown example of working of such algorithm [1]. In the Fig. 1 are presented consumers with different load graphs and with different requirements to the power quality. As it can be clearly seen, all the consumers' requirements for power quality is full-filled, but there is a problem with huge amount of tap-changings, which is lined by blue stepped line (up to 10 in one day). The resource of these devices is limited and it leads to the need to solve the optimization problem of distribution of control operations in time to ensure the best mode of all consumers. To solve this problem and provide an active-adaptive regulation is necessary to build short-term forecasts (from several minutes to a day) based on archive measurement and ambient parameters (temperature, light, day of the week).

N.T. Nguyen et al. (Eds.): ICCCI 2017, Part I, LNAI 10448, pp. 433–441, 2017.
DOI: 10.1007/978-3-319-67074-4_42

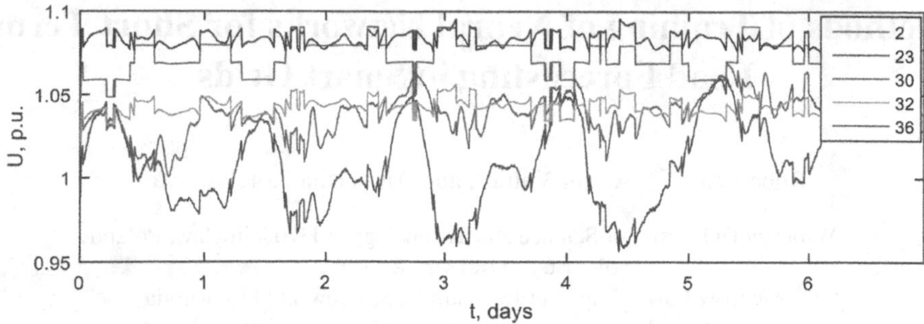

Fig. 1. The results of the work of active-adaptive control algorithm [5] (Color figure online)

Thus, for the successful work of the active-adaptive algorithms, based on results of the known methods of load forecasting, it is necessary to choose the most effective method of forecasting and apply it to the available data [2].

The aim of this paper is to build the neural network load-forecasting model as the part of the active-adaptive algorithm of regulating of voltage for specific Moscow network based on real load data. In recent years, the new training approaches for neural networks achieved and it is needed to compare its performance and to choose the best solution for particular network.

2 Methods of Load Forecasting

In general, forecasting techniques are divided into intuitive and formal methods. The group of formalized methods is divided into statistical methods and artificial intelligence techniques, so-called heuristics. All the methods listed can be applied for the load forecasting, but the selection of specific technique should be done based on particular problem and understanding for which purpose the chosen technique is used, which advantages or disadvantages it have and could it provide adequate results in this case or not [3, 8]. In this paper for forecasting the neural networks approach is implemented, because the data set is not big enough for using statistical methods, the fuzzy logic and SVM techniques gives big error in forecasting and requires an expert for setting the rules and building the model. Further some advantages and disadvantages for methods listed is given. In addition, recommendations and possible cases in which following techniques could be implemented is done.

2.1 Statistical Methods

Usually statistical methods may accurately predict the daily schedule load on ordinary days, but they lack the ability to analyze the load on holidays or other days, due to the absence of flexibility of their structure [9]. Statistical methods include multiple linear and non-linear regression, stochastic time series, the general exponential smoothing, the methods of state space, and others.

The apparent advantage of statistical methods is their "transparency", i.e. when solving problems of forecasting there is a known equation, based on which the problem is solved one way or another.

The main disadvantage of statistical methods are their computational complexity, long duration of computing and the need for large amounts of archived data. Statistical methods are difficult to apply to a model in which there are non-linear dependence, rapid load changes or missing data.

2.2 Heuristic Methods

Heuristic methods include artificial neural networks, fuzzy logic systems, SVM-methods. An important advantage of these methods is their adaptability, i.e. the ability dynamically adapt to changing conditions. This property is one of the key for applications of the method as part of an intelligent control system for distribution networks.

Heuristics are well suited for the prediction for the following reasons: first, they are capable numerically approximate any continuous function with a given accuracy. Secondly, to predict there is no need to build an accurate model of the system [4].

In addition, this methods are able to provide the required prediction accuracy in low-quality initial data, the presence in the archive gaps and abnormal deviations.

The disadvantages include the complexity of their initial design and the fact that most of them is "black boxes".

2.3 Artificial Neural Networks

A detailed comparison of classical and heuristic methods is given in [5]. Particular attention is paid to neural networks, since they are well suited for forecasting loads. They are able to generate a forecast for the load schedule of any complexity based on previous experience. In addition, there is no need to build a mathematical model of the network under consideration. For predictions with reasonable accuracy, it is sufficient to have retrospective data of the measured load values [7, 10].

To solve the load prediction problem, the following types of neural networks are mainly used: multi-layer perceptron, radial-basis functions and linear networks. For each practical task, the quality of prediction in these models is estimated by the MAPE (Mean Absolute Percentage Error). In most cases, the best results are obtained by a multilayer perceptron.

The quality of neural network prediction is influenced by various factors, such as adjusting the model parameters and training algorithms.

The training of a multilayer perceptron is most often carried out by the method of back propagation of the error (BPE). Its advantages include ease of implementation and speed, disadvantages - the ability to find a locally optimal solution instead of global, the sensitivity to the order of training examples.

In addition, evolutionary algorithms, such as the genetic algorithm (GA) and the particle swarm method (PSO), have recently been used to train the network. Particle swarm algorithm quickly converge to the best solution, the method is easy to implement and is very effective. However, with incorrect selection of optimization parameters, the training time of the network increases [6].

3 The Used Model of Neural Network

Today a multilayer perceptron is one of the most widely used neural network models because of its ability to reflect complex nonlinear relationships between input and output parameters. The network consists of several layers of neurons and weight coefficients, reflecting the connections between them. The transmission of information is carried out based on direct dissemination. The model of the neural network implemented in this paper is shown in Fig. 2.

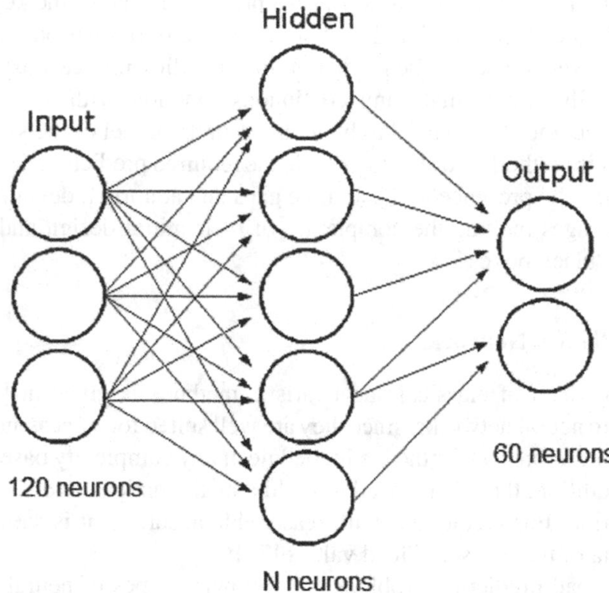

Fig. 2. Structure of used model

To predict the electrical load one hour ahead, a model with the following parameters was used:

- the number of layers – 3 (input, hidden and output), because this configuration is the most common in solving forecasting problems;
- the number of neurons in the input layer – 120 (120 values of the actual power chart in the previous 2 h, each point represents one minute measurement);
- the number of neurons in the hidden layer will be defined in 4th Chapter;
- the number of output neurons – 60 (forecasting done for 1 h ahead, each point represents one minute measurement);
- the function of activation of the hidden layer is tangential, since it has the property of amplifying weak signals better than strong ones, and prevents saturation from

strong signals, hence this function solves the dilemma of noise saturation of the Grossberg;

- training algorithm (a) back propagation of the error (b) swarm of particles.

The main settings of the back propagation error algorithm are learning rate, the number of iterations, momentum.

The main parameters of particle swarm algorithm is the number of particles in the swarm, the number of iterations, the self-acceleration coefficients and the value of velocity changes.

These parameters for both of algorithms was setted in the Chap. 4.

In this study, the mean absolute error in percentage (MAPE) is used to evaluate the training and test samples.

4 Results

In this study, the results of load measurement in the distribution networks of the Moscow region in 2008–2010 with the averaging interval of 1 min were used. To model the mode, the MATLAB2017a environment was used.

As far as the aim of this paper to create the neural network, which will adequate forecast the load curve for particular system, the network was trained on archived load data, which were provided by PAO «MOESK». After that, it was tested on a sample from data that was not used in the training samples.

Based on the lowest MAPE the number of hidden neurons for training method back propagation of the error is set to 38 using simple increasing of the hidden neurons

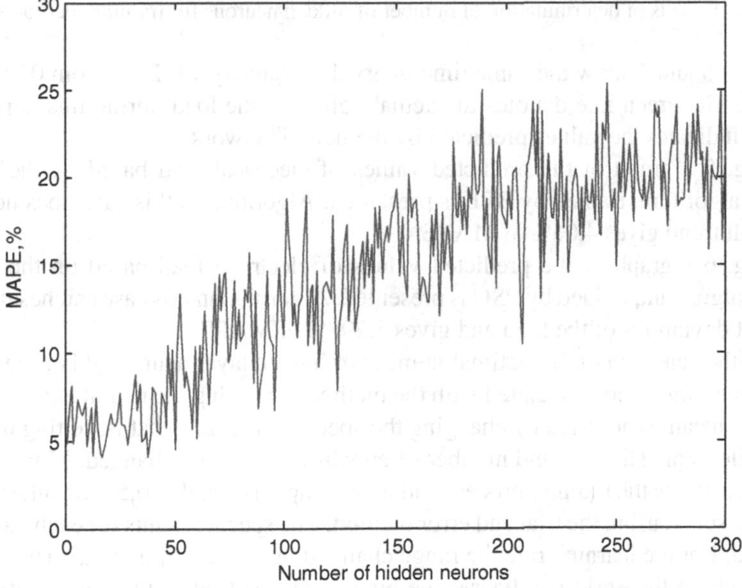

Fig. 3. Results of determination of number of hidden neurons for training method BPE

number. In each case network was trained and the resulting value of error was evaluated. MAPE in this case is equal 4.06%. The results of the experiment is shown on the Fig. 3.

Based on the lowest MAPE the number of hidden neurons for training method particle swarm is set to 27. Test was made in the same mode as for BPE algorithm. MAPE in this case is equal 1.298%. The results of the experiment is shown on the Fig. 4.

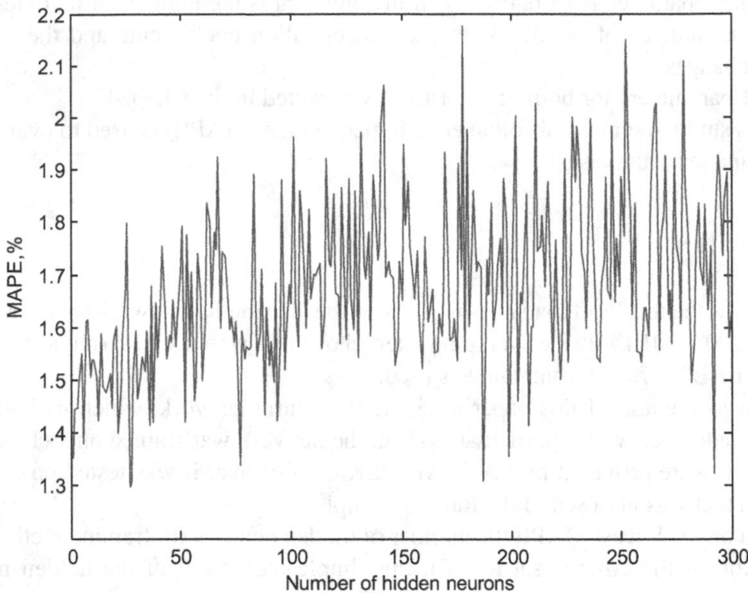

Fig. 4. Results of determination of number of hidden neurons for training method PSO

Figures 5 and 6 show the same time interval on January 13, 2010 from 01:00:00 to 02:00:00. The green line denotes the actual values of the load during this period; the blue line indicates the values predicted by the neural network.

In Fig. 5 a graph of the predicted values of electrical load based on the created software algorithm, trained by BPE is presented. Algorithm in this case does not catch the overall trend gives 4.06% of MAPE.

In Fig. 6 a graph of the predicted values of electrical load based on the created software algorithm, trained by PSO is presented. Algorithm in this case catches the trend and small deviations of the load and gives 1.298% of MAPE.

After the selection of the optimal number of hidden layer neurons it is necessary to select other parameters associated with the method of training the neural network. Back propagation can be adjusted by changing the speed of learning, that is, setting up a step gradient descent. The date and number of epochs also may be changed. Momentum is used to accelerate the training process and in this paper is equal to 0,5. The other parameters were setted using the trial and error method. The specific points for each parameter were used. For the learning rate the range changed in [0,01; 0,5] interval. The smallest values leads to the «grid paralysis», the biggest values lead to the non-ability of the network to find the global minima. The number of epochs insignificantly influence on

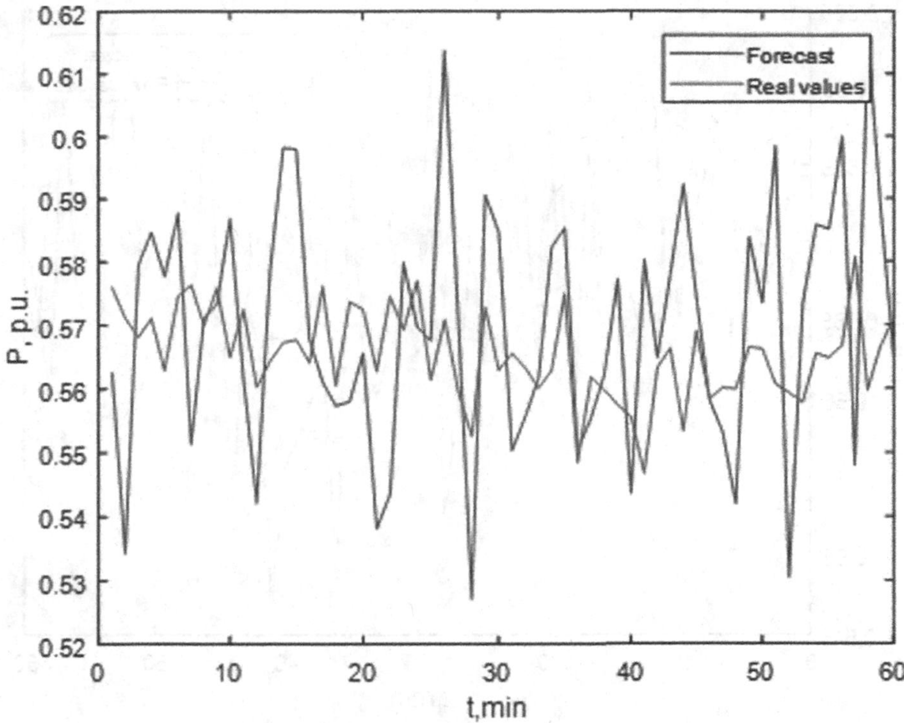

Fig. 5. The results of forecasting for the ANN trained by BPE (Color figure online)

the training time and was changed in the range [50; 500]. The selection can be done using special algorithms such as self-organizing maps, genetic algorithms, but it is not the purpose of this paper and could be done for improving of the performance in next studies. The best results of setting the parameters is shown in Table 1.

Table 1. The results of forecasting.

Name of training method	MAPE, %	Note
ANN (BPE)	1.79–6.68	Tested on the data, which were not used for configuring (Number of hidden neurons = 38, learning rate = 0.1, number of epochs = 300)
ANN (PSO)	1.61–4.68	Tested on the data, which were not used for configuring (Number of hidden neurons = 27, population size = 20, number of iterations = 75)
ANN (BPE)	1.69	The best variant during configuring (Number of hidden neurons = 38, learning rate = 0.1, number of epochs = 300)
ANN (PSO)	1.44	The best variant during configuring (Number of hidden neurons = 27, population size = 20, number of iterations = 75)

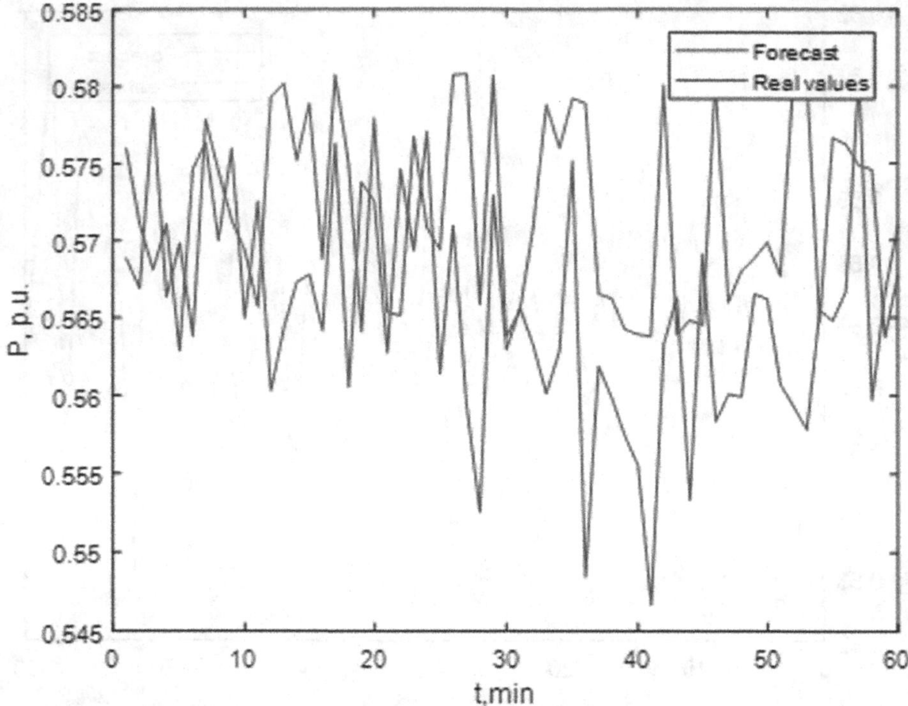

Fig. 6. The results of forecasting for the ANN trained by PSO (Color figure online)

Selection of the optimal parameters for changing the velocity of the particles is the subject of separate studies. In this paper are accepted parameters c1 = 1,49445, c2 = 1,3, found by the other authors and providing the best results for this method of training for the other problems such as pattern recognition. To this training algorithm, the population size is important, it directly affects the speed of finding the optimal solution, because the small size of the population make the search for the optimum longer. The population size were changed in the range of [5; 150]. The iterations number influence slightly on the training time and was setted in the range [50; 500]. The selection was done using trial and error method. The best variant of setting the parameters is shown in Table 1.

5 Conclusions

- Figures in the table show that, using a "particle swarm" training algorithm neural network is able to reduce the average prediction error by 0,18–2% and to reach a desired value.
- Forecast result in this case not only shows the general trend, but also reflects the changes in the local load curve. In addition, this method of training has a great speed, which is important for operating in real time.

- Back propagation algorithm for training the neural networks is simple to implement and configure, but for complex load profiles is inapplicable because it does not allow to achieve the required accuracy of prediction (1,79–6,68%).
- The algorithm "swarm of particles" for training the neural networks is slightly more difficult to implement and configure, but gives an acceptable error value (1,61–4,68%).
- Comparative analysis of training algorithms showed that for complex, non-linear load curves for neural network training is necessary to use evolutionary algorithms, as they give the best results of the absolute values of errors and training time.
- The prospects for further development include finding methods to reduce training time, and reducing the MAPE error, using of intelligent techniques for defining optimal parameters of the network (self-organizing, setting using genetic algorithms).

References

1. Vanin, A., Aleshin, S., Nasirov, R., Novikov, D., Tulsky, V.: Investigation of voltage control at consumers connection points based on smart approach. MDPI Inf. **7**(3), 42 (2016)
2. Rutkovskaja, D.: Neural Networks, Genetic Algorithms and Fuzzy Systems. Gorjachaja linija - Telekom, Moscow (2006)
3. Voronov, I.V.: Improving the efficiency of operation of the enterprises of power supply systems through the integrated use of smart grid and neural networks. Vestnik Kuzbasskogo gosudarstvennogo tehnicheskogo universiteta **2**(90), 63–66 (2012)
4. Lis, R.A.: Application of data mining techniques to identify critical voltage control areas in power system. In: Jędrzejowicz, P., Nguyen, N.T., Hoang, K. (eds.) ICCCI 2011. LNCS, vol. 6922, pp. 152–162. Springer, Heidelberg (2011). doi:10.1007/978-3-642-23935-9_15. ISBN 978-3-642-23934-2
5. Badar, I.: Comparison of conventional and modern load forecasting techniques based on artificial intelligence and expert systems. Int. J. Comput. Sci. Issues **8**(5), 504–513 (2011)
6. Roy, A.: Training artificial neural network using particle swarm optimization algorithm. Int. J. Adv. Res. Comput. Sci. Softw. Eng. **3**(3), 430–434 (2013)
7. Haykin, S.: Neural Networks: A Comprehensive Course, 2nd edn. Viliams, Moscow (2008)
8. Buhari, M.: Short-term load forecasting using artificial neural network. In: Proceedings of the International Multi Conference of Engineers and Computer Scientists 2012, vol. 1, 14–16 March, Hong Kong (2012)
9. Kennedy, J., Eberhart, R.: Particle swarm optimization. In: Proceedings of IEEE International Conference on Neural Networks (Perth, Australia). IEEE Service Center, Piscataway (1995)
10. Khothanzad, A., Hwang, R.C., Abaye, A., Maratukulam, D.: An adaptive modular artificial neural network hourly load forecaster and its implementation at electric utilities. IEEE Trans. Power Syst. **10**(3), 1716–1722 (1995)

Scheduling Sensors Activity in Wireless Sensor Networks

Antonina Tretyakova[1], Franciszek Seredynski[1], and Frederic Guinand[1,2](✉)

[1] Department of Mathematics and Natural Sciences, Cardinal Stefan Wyszynski University in Warsaw, Warsaw, Poland
a.tretyakova@uksw.edu.pl, frederic.guinand@univ-lehavre.fr
[2] LITIS Laboratory, University of Le Havre, Le Havre, France

Abstract. In this paper we consider Maximal Lifetime Coverage Problem in Wireless Sensor Networks which is formulated as a scheduling problem related to activity of sensors equipped at battery units and monitoring a two-dimensional space in time. The problem is known as an NP-hard and to solve it we propose two heuristics which use specific knowledge about the problem. The first one is proposed by us stochastic greedy algorithm and the second one is metaheuristic known as Simulated Annealing. The performance of both algorithms is verified by a number of numerical experiments. Comparison of the results show that while both algorithms provide results of similar quality, but greedy algorithm is slightly better in the sense of computational time complexity.

Keywords: Maximum lifetime coverage problem · Metaheuristics · Energy-efficient coverage preserving protocol

1 Introduction

Wireless Sensor Networks (WSNs) are one of the faster developing computer-communication technologies currently involving in many spheres of human activities, like healthcare, agriculture, industry environment, military (see, e.g. [1,2]), etc. WSN is a set of a huge number of small devices, called sensors or sensor nodes, enabling to monitor surroundings, gather information about environment and perform many other tasks. For many missions, sensors are randomly distributed over the monitoring area in environments, where human access is limited or impossible. Therefore, batteries of sensors cannot be usually rechargeable or renewable. Such scenarios of WSN can be executed in deserts, forests, wilderness, mountain terrains and etc. Exhaustion of battery charge implies the change in topology of the WSN, quality of its work and reduction of its lifetime. In WSNs energy-efficient management is intrinsically important task.

One of the main tasks posed before wireless sensor networks is an area monitoring. According to the area applications a WSN should perform different functions, among which are sensing environmental characteristics, gathering data, transmission data to a sink, etc. Due to their tiny construction WSN nodes

N.T. Nguyen et al. (Eds.): ICCCI 2017, Part I, LNAI 10448, pp. 442–451, 2017.
DOI: 10.1007/978-3-319-67074-4_43

have limitations on energy power, computing power, sensing range, transmission distance and bandwidth. These restrictions lead to a number of optimization problems, goals of which are to maximize lifetime of the system via effective managing the capabilities of the network. Limited energy sources of sensor devices demand to equip them by energy - efficient coverage preserving protocol. Such a kind of protocol: centralized or localized has to solve a variant of maximum lifetime coverage problem (MLCP) in WNSs. MLCP can be considered as a specific deterministic scheduling problem, where it is necessary to schedule sensors activity in time in a such way to maximize lifetime of the network maintaing at the same time some quality parameter like e.g. amount of covered by active sensors monitored area.

MLCP is known to be NP-hard problem [3,4], therefore, one can relay on delivering rather approximate solutions instead of exact ones. Recently, a number of nature-inspired algorithms applied to optimization problems in WSNs have appeared in the literature. Among them are genetic algorithms [5], evolution strategies [6], particle-swarm optimization [7], etc. As we already mentioned, the quality of solutions and computational complexity are not satisfactory. Coverage problems were considered under different scenarios and types of WSNs, namely wireless multimedia sensor networks in dynamic environment [6], directional sensor networks [8]. These assumptions lead to different problems statement and, therefore, each approach should be modified to enable to solve coverage problem in another type of WSNs. Our research on a direct applying nature inspired metaheuristics of general purpose to solve MLCP [9] shows that they are not enough efficient. We believe that further improving the quality of approximate solutions and computational time complexity can be achieved by incorporation into a searching engine of an algorithm of a specific knowledge about the problem. In this paper we propose two knowledge based algorithms to solve a variant of MLCP. It is a greedy heuristic and an algorithm based on Simulated Annealing (SA).

The rest of the paper is organized as follows. In the next section the problem is stated. The two following sections present our greedy algorithm and SA-based algorithm. Sections 5 and 6 contain results of simulation experiments and conclusion remarks.

2 Problem Statement

Let us consider a homogeneous sensor network $S = \{s_1, \ldots, s_N\}$ consisting of N sensor nodes randomly distributed over a given *target field* F, a two-dimensional rectangular area of $W \times H$ m^2. The target field F is uniformly divided on points of interest (POIs) with a step g, sensors are responsible for detection of an intruder (a target point) and sending an alarm message to the sink node. A sensor s_j is defined as a point of coordinates (x_j, y_j) in two-dimensional area, sensing range R_s and battery capacity b. An example of a sensor network randomly deployed over the target field is depicted in Fig. 1. It is assumed that each sensor can work in two modes: *active mode* and *sleeping mode*. In active mode a sensor

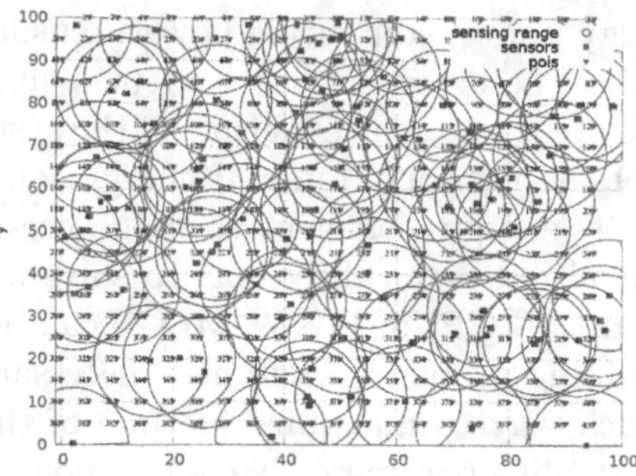

Fig. 1. An example of sensor network deployed over the target field

observes a circle area within its sensing range and can transmit or receive a signal. Let us denote the mode of i-th sensor during j-th time interval as $state(s_i, t_j)$, where $state(s_i, t_j) \in$ {ON, OFF}. The value of $state(s_i, t_j)$ equals ON means that i-th sensor s_i during j-th time interval is in active mode, otherwise, $state(s_i, t_j) =$ OFF.

Below we give a number of definitions concerning the problem statement.

Definition 1. *A sensor* $s_i(x_i, y_i)$ *covers a POI* p(x, y) *iff the Euclidean distance* d(s_i,p) *between them is less than the sensing range* R_s.

Let us denote a set of POIs covered by i-th sensor s_i as $POIs_{obs}(s_i)$ and call as *coverage area* of i-th sensor. All POIs covered by an active network during j-th time interval is denoted as $POIs_{obs}(t_j)$, i.e.

$$POIs_{obs}(t_j) = \cup_{i=1}^{N} POIs_{obs}(s_i)|_{state(s_i,t_j)=ON} \tag{1}$$

Definition 2. *Coverage of a target field F at j-th time period* t_j *denoted as* $cov(t_j)$ *is a real number equal to a ratio of a number of POIs covered by an active network during j-th time interval* t_j *to all* $POIs$, *i.e.*

$$cov(t_j) = \frac{|POIs|_{obs}(t_j)}{|POIs|} \tag{2}$$

Let us denote a number of POIs covered by i-th sensor as $cov(s_i)$.

A sensor is assumed to consume energy for monitoring area and it depends on its sensing range R_s. Consider a homogeneous sensor network, where all sensors have the same sensing range, the energy consumption per time interval is constant. A potential solution is a schedule of a network S deployed over a target field prescribing states of activity for all sensors in the network during the whole period of time of the network operation.

Definition 3. *A schedule of a WSN is a binary* $T_{max} \times N$ *matrix denoted as* Sol, *i.e.*

$$Sol(S) = \{state_i^j\}, where\ i = 1, ..., N\ and\ j = 1, ..., T_{max}, \tag{3}$$

where $state_i^j \in [0,1]$ *is a state of i-th sensor during j-th time interval, 0 corresponds to OFF state and 1 is related to ON state.*

Each row of the matrix is related to one of the sensors and represents its schedule of activity over all period of network operation from t_1 till $t_{T_{max}}$. Let us assume a sensor to spend one unit of energy during one time unit of its activity.

Definition 4. *A schedule* $Sol(S)$ *is a feasible solution if the following equality is met:*

$$(\forall i)_{i=1,\dots,N} | \sum_{j=1}^{T_{max}} state_i^j = b \qquad (4)$$

Definition 5. *Coverage string is a set of real values, each of which corresponds to the coverage of a target field F during each time interval of the WSN operation, i.e.*

$$coverage\ string = \{cov(t_1), cov(t_2), ..., cov(t_{T_{max}})\} \qquad (5)$$

Definition 6. *Lifetime of the WSN denoted as* $Lifetime(q)$ *is defined as a sum of time intervals, during which the coverage requirement is met,*

$$Lifetime(q) = \sum_{i=1}^{T_{max}} i|_{cov(i)\geq q\}} \qquad (6)$$

Maximal time of network performance is restricted by the characteristics of the network such as a number of sensors and their distribution, a distribution of POIs over the target field, sensing range, battery capacity value and the level of coverage required. The parameter T_{max} is a predefined number and should be set greater than the performance time $Lifetime(q)$ and less than the upper bound of network operation denoted as $Lifetime^{Up}$.

We consider Maximum Lifetime Coverage Problem (MLCP) as a scheduling problem applied to a WSN solving the area coverage problem in the discrete two-dimensional space. MLCP has as an objective to prolong lifetime of a WSN by minimizing a number of redundant sensors during each time interval in order to minimize energy consumption. The function $Lifetime(q)$ is maximized over the space of all feasible solutions. Coverage requirement is given by a coverage ratio q, which means that at least q-th part with small declination δ of all targets is covered by at least one sensor.

MLCP - specific knowledge. A searching process conducted by both a greedy heuristic and SA-based algorithm (see, below) incorporates the MLCP specific knowledge and is based on the following classification of columns in a schedule. All columns of the schedule solution are divided on three groups called three subsequences: *Redundant Subsequence (RS)*, *Excellent Subsequence (ES)*, *Unsatisfactory Subsequence (US)*. Each subsequence groups time intervals such that a network of active sensors covers the target area with certain *coverage ratio*, i.e.

$$t_i \in RS,\ if\ cov(t_i) > q + \delta, \qquad (7)$$

$$t_i \in ES,\ if\ |cov(t_i) - q| \le \delta \tag{8}$$

$$t_i \in US,\ if\ cov(t_i) < q - \delta \tag{9}$$

Let us denote a number of elements in RS, ES and US as N_R, N_E and N_U respectively.

3 A Greedy Heuristic to Solve MLCP

In this section, we present an iterative knowledge-based stochastic greedy heuristic to solve MLCP. The algorithm is based on constructing a tree of solutions. A root of the tree is a randomly created solution. In each iteration a solution called a predecessor is changed under two steps described below to form one more solution called a successor. The next iteration continues from the node corresponding to the best solution between a predecessor and its successor from the previous iteration. The pseudocode of the algorithm is presented in Algorithm 1. At each iteration a schedule is a subject of two types procedures, pseudo codes of which are sketched in Procedure 1 (Algorithm 1, lines 6–12) and Procedure 2 (Algorithm 1, lines 18–21). The aim of procedure 1 is to improve a current solution via joining active subnetworks during time intervals when necessary coverage is not achieved. This purpose can be obtained by multiple shifting several columns from US toward ES or RS. A schedule is changed under the Procedure 1 as follows. Two new columns are generated by applying boolean-valued functions OR and AND to a pair of two values from same row from US columns. The new first column contains the values resulted of OR operator, and the second column contains the results of AND operator applied. As the algorithm proceeds it may happen that a new solution is not improved in the sense of *Lifetime(q)*. In that case a parameter k (a number of use Procedure 1) is increased by 1. Initially, k is equal to 1. The solution obtained by the first modification (Procedure 1) is next changed by the Procedure 2. The aim of this stage is to reduce redundant consumption of energy, i.e. a randomly chosen sensor in active state from RS time interval is switched off and, next, is switched on during US time interval.

The Procedure 2 is executed on a current solution and consists of the two steps. Firstly, from the i-th RS column a cell is randomly selected with probability p_i, where n_1 is a number of "1" cells in the column. Let us denote a row of the selected cell as j. Second, the "0" cell in the in first US column in j-th row is changed on "1". Therefore, the first selected cell is equal to 1. The second selected cell is taken as the first "0" cell from US and from the same row as previous cell was. The selected cells swap their values. If there is not a cell with the value "0" in US, the predecessor solution coincides with its successor. The above mentioned two steps are repeated consequently N_R times for each RS column.

The predecessor schedule and its successor are evaluated by *Lifetime(q)* metric and the best one is saved as a current schedule for the next iteration to be applied. These steps are repeated until stop condition is met. The last saved schedule is a result of the algorithm.

Algorithm 1. Pseudocode - Greedy algorithm for random initial solution.

```
Input : WSN, Target field, N_I, q, δ, T_max
initialize random sol_cur(N, T_max)
k = 1
for i ← 1 to N_I do
  compute US,
  for i ← 1 to N_U - k + 1 do
    for j ← 1 to k do
      modify i and i + j columns from US,
      j = j + 1
    end for
    i = i +1
  end for
  compute Lifetime(q)
  if Lifetime(q) of the predecessor > Lifetime(q) of the successor then
    k = k + 1
  end if
  compute RS, US for the successor,
  for i ← 1 to N_R do
    modify the i − th column in RS in the solution,
    i = i +1
  end for
  compute Lifetime(q) for the successor,
  keep the best from the predecessor and its successor,
  i = i+1
end for
return sol
```

4 Simulated Annealing Algorithm to Solve MLCP

Simulated Annealing (SA) is one of the nature inspired metaheuristics based on the physical annealing process observed in glass manufacturing process and metallurgy.

The performance of SA depends on construction of a given solution neighbourhood. Generating the sequential solutions is based on a swap of a pair of opposite values in one row. A neighbouring solution is differed from the given solution by a number of bits changing comparatively with the given solution. Let us call this characteristic defined by a number of changing pairs of bits, as *neighbourhood size* and denoted as k_{neigh} - neighbourhood. In such a way, in case of changing two random cells, we obtain a solution in 1-neighbourhood from a given solution. The random neighbour is generated as follows: k_{neigh} times two opposite values chosen at random from the same row are swapped (see, Algorithm 2, lines 8–16). The additional information about solution is computed such as coverage of each time slot according which the time line is divided on RS, ES and US. The idea of knowledge-wise neighbourhood generating procedure is to switch off an active sensor from redundant subsequence in order to reduce a number of redundantly covered POIs and to switch it on in the first unsatisfactory subsequence with the aim of increasing coverage at the additional time interval. These steps may increase lifetime of the generated solution. The pseudo code of SA solving MLCP is presented in Algorithm 2. SA algorithm works until termination condition meets performing cycles of searching solutions in the neighbourhood of the initial solution. Each cycle consists of several iterations characterized by the temperature. During one iteration a random solution within the current solution's neighbourhood is created, Lifetime(q) for the current solution and its neighbour is computed, the current solution is compared with the created neighbour. The better new solution always replaces the old one, while in case of worse neighbour it replaces the current solution with probability $\exp\frac{\Delta}{T}$ so that the probability of acceptance of the new solution depends on the temperature value and difference in values of evaluation function between two solutions. At the end of an iteration temperature level is increased according with the cooling scheme.

Algorithm 2. Pseudocode of SA algorithm

```
Input : WSN, Target field, N_I, q, δ, T_max
Initialize unit Sol(N, T_max)
Maximal temperature L
k
while termination condition is not fulfilled do
    for i ← 1 to L do
        life = compute Lifetime(q) of Sol
        for i ← 1 to k_neigh do
            for j ← 1 to RS.size() do
                choose a random cell of "1" value from i − th RS column, the row of the
                gene let us denote as l
                find the first "0" gene from US and l − th row
                swap the values of the chosen pair
                j = j + 1
            end for
            i = i + 1
        end for
        lifeN = compute Lifetime(q) of SolN
        Δ = life - lifeN
        if Δ ≤ 0 then
            Sol = SolN
        else
            Sol = SolN with probability exp(Δ/T)
        end if
        decrease T(i)
    end for
    update termination values
end while
Sol
```

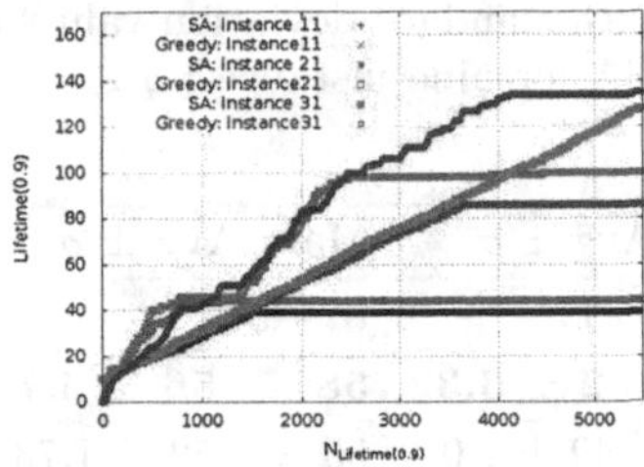

Fig. 2. An example of a typical run of two algorithms: *greedy* and SA for instance11, $R_s = 20$, $b = 10$.

5 Experimental Results

We consider WSN consisting of a number N of sensors equal to 100, 200 and 300, respectively. For each value of N, we created 3 instances, which differ by random allocation of sensors, so 9 instances were used in experimental study. Each instance is described as Instance{*indicator of network size*}{*order number of WSN instance*}, where N is equal to *indicator of network size* times 100. To give an example, Instance23 represents the third instance of WSN consisting of 200 sensors. The algorithm's parameters should be chosen as the set of the best values for each of the algorithms: greedy heuristic and SA. SA is defined by the following values. Temperature is cooled according to the logarithmic scheme with initial temperature 50, length of the temperature cycle 25, the frozen level 10 and maximal number of iterations 100. The termination condition is as follows: exceeding maximal number of iterations or achieving the frozen temperature level. Greedy heuristic needs to set a number of iterations equal to 150, after which the algorithm stops. The main component of computational cost of both algorithms is related to calculation of $Lifetime(q)$ function. An example of a typical run of these two algorithms is presented in Fig. 2, which presents the dynamics of $Lifetime(0.9)$ obtained by greedy and SA for three instances consisting of 100, 200 and 300 nodes as a function of a number of computation the $Lifetime(q)$ function. From the figure one can observe that *greedy* converges quicker for all instances with a slightly better quality than SA. However, when the size of an instance is relatively large (instance 31) the *greedy algorithm* achieves a local optimum, while SA continuously improves quality of a solution.

Let us finally discuss the overall results on the MLCP input data. In order to present a broader view, WSN instances with different properties will be used. For testing purposes we will consider nine WSN instances with three types of densities and sensing coverage range 20 deployed over the same target field <100, 100, 5>. Coverage requirement is represented by three values: 0.85, 0.9 and 0.95. Battery capacity is equal to 10.

Table 1 presents results of systematic study of both algorithms conducted for representatives of three types of instances, which support our previous observations. The table contains maximal, average and standard deviation of the goal function values obtained by *greedy* and SA based on averaging of 10 runs.

Table 1. Maximal, average with standard deviation values of Lifetime(q) obtained by two algorithms: *greedy* and SA for nine instances; $q \in = \{0.85, 0.9, 0.95\}$, $R_s = 20$, $b = 10$.

q	*algorithm*	Max	Avg ± σ	Max	Avg ± σ	Max	Avg ± σ
		*instance*11		*instance*12		*instance*13	
0.85	*greedy*	**54**	**52 ± 3.32**	**58**	**56 ± 4.7**	**52**	**49 ± 4.36**
	SA	51	49 ± 1.0	55	52 ± 1.73	49	47 ± 1.0
0.9	*greedy*	**44**	**42 ± 3.88**	**47**	**44 ± 4.36**	**41**	**39 ± 3.88**
	SA	42	38 ± 2.0	46	43 ± 1.41	38	36 ± 1.0
0.95	*greedy*	**32**	**30 ± 3.47**	**35**	**33 ± 3.32**	**29**	**27 ± 4.36**
	SA	29	26 ± 1.0	33	29 ± 1.73	25	23 ± 1.0
		*instance*21		*instance*22		*instance*23	
0.85	*greedy*	109	107 ± 4.36	108	105 ± 5.48	109	107 ± 4.48
	SA	**110**	106 ± 1.41	**108**	104 ± 2.23	**110**	**107 ± 1.41**
0.9	*greedy*	**93**	**90 ± 4.7**	**93**	**88 ± 7.82**	**93**	**91 ± 4.48**
	SA	88	85 ± 1.73	87	84 ± 1.41	90	86 ± 1.73
0.95	*greedy*	65	60 ± 6.71	60	58 ± 3.75	63	61 ± 4.48
	SA	64	**62 ± 0.0**	**61**	**59 ± 1.0**	**66**	**63 ± 1.41**
		*instance*31		*instance*32		*instance*33	
0.85	*greedy*	166	162 ± 6.49	158	156 ± 4.8	165	160 ± 9.44
	SA	165	162 ± 1.73	**163**	**158 ± 2.0**	164	161 ± 1.41
0.9	*greedy*	**139**	**135 ± 6.33**	**136**	**129 ± 9.28**	**139**	**136 ± 4.36**
	SA	137	130 ± 2.44	130	126 ± 2.23	135	131 ± 2.44
0.95	*greedy*	97	92 ± 9.8	88	85 ± 5.39	98	93 ± 8.84
	SA	**98**	**94 ± 1.73**	**94**	**89 ± 2.82**	**98**	**95 ± 1.0**

The remaining parameters are as follows: q-requirement is based on the set {0.85, 0.9, 0.95}, nine instances of WSN with $R_s = 20$, and $b = 10$. One can see from the table that the average and the maximal values of $Lifetime(q)$ are differed slightly. In the most cases, it is evident that *greedy* provides better solutions than SA, the results concerning instances of WSN consisting of 100 nodes can serve as an example. Meanwhile, with growth of the problem complexity, when N is equal to 200 or 300, SA finds better solutions, for instance, see q equal to 0.85 or 0.95. It should be notice that standard deviation computed for SA results in all cases are better than σ-values computed for the results provided by *greedy algorithm*. This indicates, that SA is a more stable than *greedy algorithm*. In such a way we can assume that in the case of bigger problem instance SA enables to provide better solutions than *greedy*. To summarize aforementioned discussion it is shown that there are two different approaches solving MLCP providing good different solutions of the problem.

6 Conclusion

In this paper the problem of lifetime maximization in WSNs stated as MLCP with assumption of not full coverage defining by a coverage ratio requirement q was considered. The problem belongs to a class of NP-hard problems characterized by high computational complexity, what motivates to use algorithms provided approximate solutions.

To solve the problem we proposed and study two centralized knowledge-based algorithms: stochastic greedy heuristic and simulated annealing algorithm. All of the algorithms were studied on the same testbed and under the same assumptions. Results of experimental study of the algorithms shows that the greedy algorithm is efficient in both a quality of solutions and time complexity for a medium sizes of the problem instances. When the size of the problems becomes relatively large simulated annealing provides better quality of solutions, however it is achieved by increasing computational time of the algorithm.

References

1. Nesamony, S., Vairamuthu, M.K., Orlowska, M.E., Sadiq, S.W.: On sensor network segmentation for urban water distribution monitoring. In: Zhou, X., Li, J., Shen, H.T., Kitsuregawa, M., Zhang, Y. (eds.) APWeb 2006. LNCS, vol. 3841, pp. 974–985. Springer, Heidelberg (2006). doi:10.1007/11610113_104
2. Pierce, F.J., Elliott, T.V.: Regional and on-farm wireless sensor networks for agricultural systems in Eastern Washington. Comput. Electron. Agric. **61**(1), 32–43 (2008)
3. Cardei, M., Wu, J.: Energy-efficient coverage problems in wireless ad-hoc sensor networks. J. Comput. Commun. Arch. **29**, 413–420 (2006)
4. Garey, M.R., Johnson, D.S.: Computers and Intractability: A Guide to the Theory of NP-Completeness. W.H. Freeman and Co., New York (1979)
5. Sahoo, B., Ravu, V., Patel, P.: Observation on using genetic algorithm for extending the lifetime of wireless sensor networks. In: IJCA Special Issue on 2nd National Conference- Computing, Communication and Sensor Network, pp. 9–13 (2011)
6. Fayyazi, H., Sabokrou, M., Hosseini, M., Sabokrou, A.: Solving heterogeneous coverage problem in Wireless Multimedia Sensor Networks in a dynamic environment using Evolutionary Strategies. In: ICCKE2011, Mashhad, Iran, 13–14 October 2011
7. Abbasi, M., Abd Latiff, M.S., Modirkhazeni, A., Anisi, M.H.: Optimization of wireless sensor network coverage based on evolutionary algorithm. IJCCN **1**(1), 104 (2011)
8. Gil, J.M., Han, Y.H.: A target coverage scheduling scheme based on genetic algorithms in directional sensor networks. Sensors **11**(2), 1888–1906 (2011)
9. Tretyakova, A., Seredynski, F.: Application of evolutionary algorithms to maximum lifetime coverage problem in wireless sensor networks. In: 27-th IEEE IPDPS, Boston, USA (2013)

Application of Smart Multidimensional Navigation in Web-Based Systems

Ivan Soukal[1] and Aneta Bartuskova[2(✉)]

[1] Department of Economics, Faculty of Informatics and Management, University of Hradec Kralove, Rokitanskeho 62, 500 03 Hradec Kralove, Czech Republic
ivan.soukal@uhk.cz

[2] Faculty of Informatics and Management, Center for Basic and Applied Research, University of Hradec Kralove, Rokitanskeho 62, 500 03 Hradec Kralove, Czech Republic
aneta.bartuskova@uhk.cz

Abstract. In this paper we further develop a new method for implementing multidimensional navigation. This technique is based on advantages of web-based techniques such as site maps, vertical menus and tag clouds. It combines high informational density of tags with structural quality of traditional hierarchical navigation. The result is organization by all org. schemes at once, increased information density and reduced interaction and attention-switching cost. Multidimensional navigation is expected to enhance efficiency of information retrieval especially on websites and web-based systems. It can be however applied also on information systems such as knowledge management or educational systems.

Keywords: Multidimensional navigation · Web-based systems · Navigational design · Organization schemes · Web interaction

1 Introduction

This paper follows our previous research, published under the title "The Novel Approach to Organization and Navigation by Using All Organization Schemes Simultaneously" by Springer [1]. We have proposed a method for implementing multidimensional navigation in order to make information retrieval more efficient. It is based on web design principles and interfaces, however it can be applied on any information system, knowledge management system, educational system or other system, which has advanced requirements on organization [1].

Motivation for research on this topic was mostly based on organizational and navigational issues of web-based systems. These are especially disorientation of users, subjectivity of navigating, unused descriptive potential of navigation and high interaction and attention-switching cost. Even though usability is the basic presumption of successfully using any system, many systems have poor usability which discourages potential users. In our research, organization schemes are used for enhancing the usability of navigation.

N.T. Nguyen et al. (Eds.): ICCCI 2017, Part I, LNAI 10448, pp. 452–461, 2017.
DOI: 10.1007/978-3-319-67074-4_44

2 Theoretical Background

Theoretical background will be presented only briefly, as our previous introductory paper covered this area in detail [1].

2.1 Organization Schemes

We can describe organization schemes as constructs by which the information is organized. Basic division of schemes can be regarded as (1) objective schemes, which divide information into well-defined and exclusive sections, and (2) subjective schemes, dividing information into categories without exact definition [2].

Lidwell et al. [3] suggested five ways of organizing information. The approach is well-known as the LATCH (Location, Alphabet, Time, Category and Hierarchy). Organizing by time equals an organization by chronological sequence. Organization by location is by geographical or by spatial reference. Hierarchy (or also continuum) refers to organization by magnitude, e.g., highest to lowest, best to worst etc. [3]. Category can be used for organizing by similarity within the information.

2.2 Usability of Navigation

Navigation is part of usability, which is a complex construct, depending on many design decisions. Towards effective navigation contribute among others structure, organization, labelling, browsing and searching [2]. As was already mentioned in the introduction, usability is the basic presumption of successfully using any system. However commonly implemented navigation often suffers from many shortcomings.

The confusing and disorganized navigation structure is one of the main contributors to the problem of disorientation [4]. While it is known that a common solution to the disorientation of web users is the presentation of a site map or other overview of the site structure, websites usually contain only limited subset of the standard navigation aids [5].

The other issue is the overall subjectivity of navigation means. While creating navigation according to subjective organization scheme, we can come across issues such as: broad vague categories, poor grouping of categories and poor organization of menu options [6]. In the subjective organization scheme, someone other than the user has made an intellectual decision to group items together [2].

Navigation has some descriptive potential, which can be formed e.g. by spatial arrangement or visual cues. However navigational design is mostly subjected to overall interface appearance and related graphic design decisions and as such bears only limited information value. It was however researched that users appreciated descriptive characteristics of tag clouds [7]. Consequently, we assume that higher informational density of navigation could be valued by users.

There are two major types of cost in web environment: interaction cost and attention-switching cost [8]. Leuthold et al. confirmed in their experiment, that vertical menus (which reveal all navigation items at once) outperform dynamic menus (which display only one level of navigation and reveal lower levels upon mouse interaction) [9]. Moreover, vertical menus were subjectively preferred by users [9].

3 Conceptual Proposal

This section will again briefly summarize our conceptual proposal, which is in detail described in [1]. The saved space will allow us to focus more on the actual implementation of the proposal.

3.1 Fundamental Starting Points

The proposed method can be characterized as a layering technique, defined by Lidwell et al. as the process of organizing information into related groupings in order to manage complexity and reinforce relationships in the information [3]. Layer of organization is then formed by each org. scheme. All layers can be seen by users simultaneously as the information-rich navigational area. Presenting more information in one place is also presumed to reduce interaction and attention-switching cost and increase usability. Another benefit is a decreased subjectivity in construction of navigation by implementing several exact means of organization.

Tag clouds emerged as a usable technique in presenting more information through navigation while occupying a relatively small space. The concept of tags is very inspiring as it offers organization by three organization schemes simultaneously. Our intention is to implement more than three organization schemes in one arrangement.

Objectives of the proposed solution are based on identified navigational issues. These objectives are: reduced subjectivity, reduced disorientation of users, reduced interaction and attention-switching cost and increased information density [1].

3.2 Differentiation by Spatial and Visual Aspects

Golombisky and Hagen presented the seven elements of design as basic units of visual communication - space, line, shape, size, pattern, texture and value [10]. Fowler and Stanwick suggested using visual cues such as color, font and size to signify group organization [11]. In this part of paper we will present selected design items as means for arrangement by different org. schemes (Table 1).

Table 1. Selected design aspects to differentiate individual items

Type	Design aspects
Visual	Color (or pattern), font, size
Spatial	Starting position (x,y)

We can use color in design to group elements and suggest meaning [3]. Size is a design element, which can be used for signifying magnitude or/and importance. Tag clouds use this technique - font size of tags reflects the number of matching instances for individual tag [12]. As for space, in the two-dimensional display like a computer monitor, the element's position is defined by horizontal and vertical value. These were usable possibilities for aspects applicable inclusively, i.e. on navigation items themselves, or in fact on their text labels. In addition we can use separate design elements for the arrangement by remaining org. schemes. These additional elements (i.e. shapes) can signify differentiation by some of the already mentioned techniques.

3.3 Creating the Arrangement by All Organization Schemes

The proposed solution of navigation is capable of supporting all organization schemes, which are: alphabet, time, category (and/or tags), continuum and location. The differentiation of each arrangement from others is ensured by combined use of textual, spatial and visual techniques. The inclusive representation is preferable in order to avoid cluttering of the navigation area, however in some cases its implementation reveals usability issues. We tried to find balance between advantages of inclusive representation (expressed by differences in color, size or position) and necessity of adding new design elements. Individual methods based on these design variations were consequently assigned to organization schemes (see Table 2).

Table 2. Overview of the identified issues and their solutions, adopted from [1]

Org. scheme	Method	Explanation
Alphabet	Vertical arrangement	Entries are sorted traditionally by alphabet
Time	Horizontal arrangement	Entries are arranged by time by difference in horizontal starting position
Category	Square with different color/pattern	Entries are visually associated with a color/pattern, distinct for each individual category
Subcategory	Text label	Separate entries are grouped in subcategories, which are represented by displayed text labels
Continuum	Bar rating with different length	Entries are associated visually with a bar of length, which reflects value
Location	Associative link	Entries are linked to place in the adjacent map

We can represent category by the principle of similarity or grouping. Alphabetical organization can be made either horizontally, through sorting entries as inline elements, or vertically such as block elements, which should be better for readability. Continuum and time - continuous org. schemes - can be represented by size differences (width, height) or position (vertical, horizontal). Vertical organization was already assigned to alphabet, so for these variables could be used horizontal arrangement. Representation by text size difference would compromise usability, therefore we propose for the second technique adding a new element, which will specify value by difference in its width. The color of this element would correspond with the item's category to further support grouping effect. Finally, we can represent location by an associative link to the respective map.

3.4 Spatial Design of the Solution

Spatial design of the proposed solution was the last concluding section in our previous paper [1]. It combines conclusions from previous sections regarding (1) conceptual proposal and argumentation, (2) use of design techniques for the defined objectives, and (3) arrangement by all organization schemes using these techniques (Fig. 1).

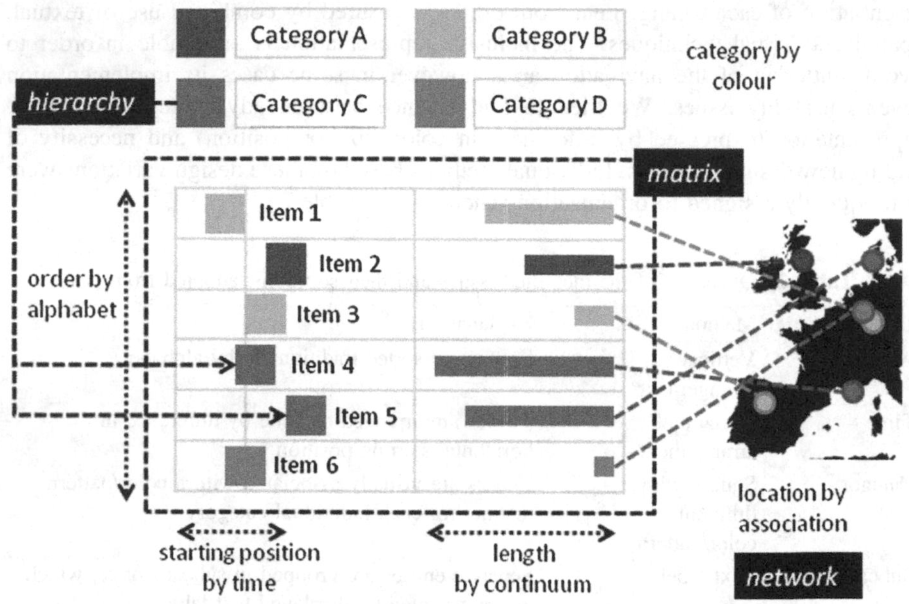

Fig. 1. The summarized schema of all schemes and structures [1]

3.5 From Schemes to Actual Variables

Until now we have discussed organization schemes in general without actual variables. While concept of alphabet or category is apparent, implementation of time and continuum should be further specified. What organization scheme is possible to use and what will be useful at the same time, depends on particular system and its content. As for time, we can use specific data such as when the item was added or last edited and organize items accordingly. We can also use some relative time measures, e.g. the time order in which the user should proceed, which would be especially useful in sites with learning or instructional purposes. Organization by time when the item was added (i.e. by newest items) would be very usable in process-oriented systems.

While organization by time was mostly exact, organization by continuum can easily facilitate social navigation. As a continuum we can use e.g. popularity, implemented as number of views, number of comments (in case of blog articles) or value of rating (if the rating system is implemented). It is known that recommendations based on popularity are usually appreciated by users.

In the case of using subcategories, continuum can represent a number of individual items in the category. By implementing this organization scheme, the user can identify right away, which topics (represented by subcategories) are the most frequent in this system. This is a very useful information to receive at the first sight and would take much longer to realize by using traditional navigation.

4 Scope and Technical Implementation

As was already mentioned, the proposed solution is suitable for various information systems, not only web-based ones. The proposed approach can enhance any navigation from a list of items to a dynamically updated overview of all content in the system. Nowadays when we are overwhelmed with all available content, such consistent knowledge visualization can save us a lot of time and effort. With the presented method, any meaningful list of items can be turned into this arrangement and can facilitate efficient orientation in presented choices.

The visual representation schema is presented in the following figure (Fig. 2). It is important to note that individual aspects of implementation can vary for various websites according to their purpose, content and audience. Incorporating each organization scheme into the navigation should be both feasible and useful for users.

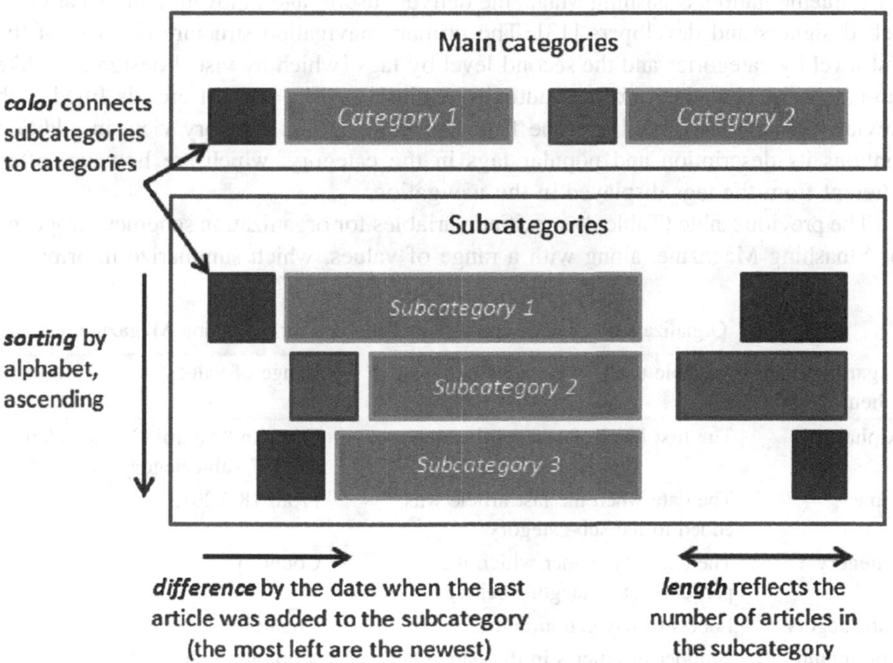

Fig. 2. The schema of visual representation

The following table specifies technical solution for implementing each organization scheme into the final arrangement (Table 3). Location is omitted from implementation, because it is not so common in web-based systems that navigation items are connected to location.

Table 3. Implementation with a use of technologies for web development

Org. scheme	Technical implementation
Alphabet	Order of items in code editor
Time	Value of items' left margins according to their differences in time variable, in the demonstration implemented as mono-spaced
Category	Square visual elements adjacent to individual items and categories, associated by the same background color
Continuum	Visual element adjacent to individual items with a width according to their value in continuum variable, implemented as proportional

5 Implementing Navigation to the Existing System

For the demonstration of our proposal's usage, we have selected web-based system due to its accessibility for anyone, who would like to compare the original navigation and our implementation. Smashing Magazine delivers useful and innovative information to web designers and developers [13]. The primary navigation structure consists of the first level by categories and the second level by tags (which by visual design look like sub-categories, but they are not mutually exclusive). The content area is filled with previews of articles, sorted by time from the newest. The category view in addition contains its description and popular tags in the category, which are however often different from the tags displayed in the navigation.

The previous table (Table 4) specified variables for organization schemes, proposed for Smashing Magazine, along with a range of values, which summarize information

Table 4. Organization schemes applied on variables for Smashing Magazine

Organization scheme	Variable used by the scheme	Range of values[a]
Alphabet	The first letter of every sub-category	From an "Android" to a "Web design" sub-category
Time	The date when the last article was added to the sub-category	From 18.3.2013
Category	The category, under which the particular sub-category belongs	Count: 6
Subcategory	Labels of navigation items	Count: 23
Continuum	Number of articles in the particular sub-category	Minimum value: 22 Maximum value: 270
Location	Not applicable	Not applicable

a At the time of writing this paper

gathered by manually clicking and searching. For the purposes of this study, we dealt with tags as with sub-categories and only with those included in the primary navigation. The following figure depicts implementation of the method on Smashing Magazine, as it was displayed in a web browser (Fig. 3).

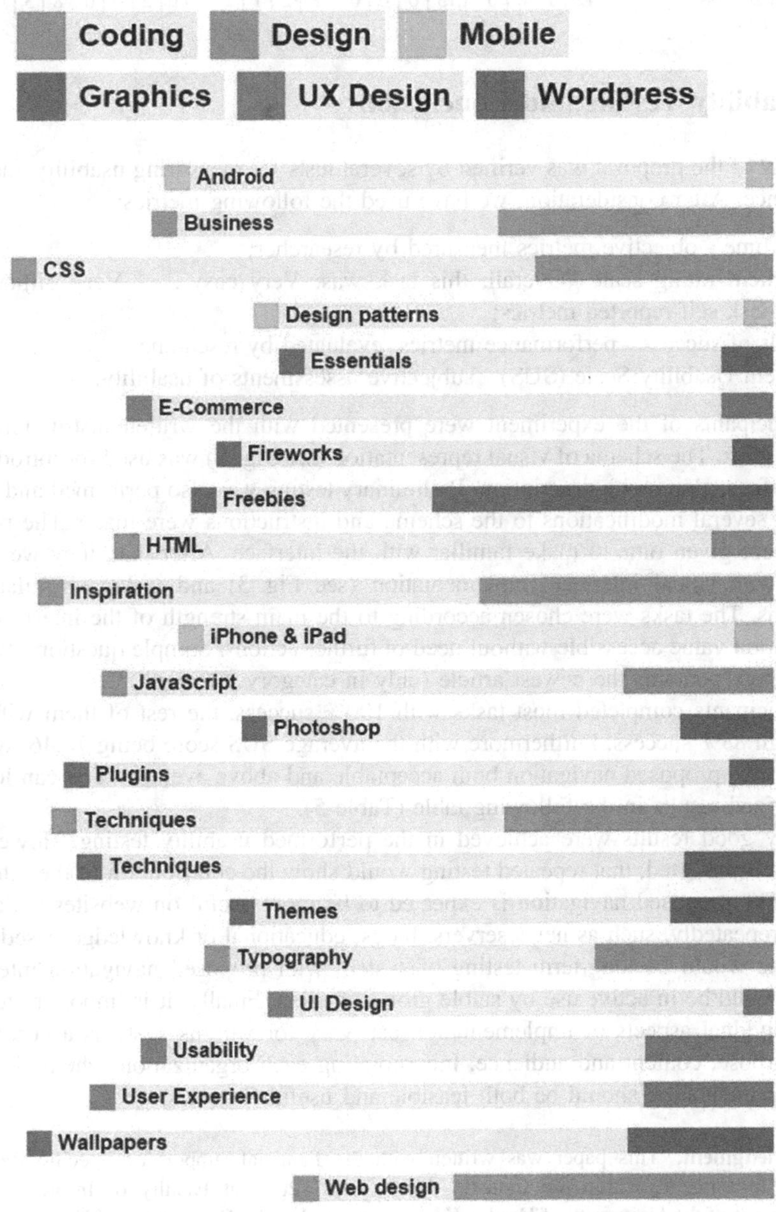

Fig. 3. Implementation of the proposed method for Smashing Magazine

Table 5. The results - summarization of collected data for each question

	Q1	Q2	Q3	Q4	Q5
Frequencies	1 | 2 | 3 | 4 | 5	1 | 2 | 3 | 4 | 5	1 | 2 | 3 | 4 | 5	1 | 2 | 3 | 4 | 5	1 | 2 | 3 | 4 | 5
Correct answers (%)	100%	100%	85%	100%	85%
Avg. time (s)	7, 9	7, 5	14, 3	13, 4	19, 8
Inquiries (*frequencies*)	10 | 3 | 0 | 0 | 0	10 | 3 | 0 | 0 | 0	2 | 7 | 3 | 1 | 0	3 | 6 | 3 | 1 | 0	0 | 5 | 4 | 3 | 1
Levels of success (*frequencies*)	12 | 1 | 0 | 0	13 | 0 | 0 | 0	8 | 2 | 1 | 2	11 | 2 | 0 | 0	8 | 3 | 0 | 2

6 Usability Testing and Conclusions

Usability of the proposal was verified by several tests for measuring usability and user experience. After consideration, we have used the following metrics:

1. task time - objective metrics measured by researcher;
2. one-item rating scale "Overall, this task was: Very easy…… Very difficult." - post-task self-reported metrics;
3. levels of success - performance metrics, evaluated by researcher;
4. System Usability Scale (SUS) - subjective assessments of usability.

Participants of the experiment were presented with the written instructions and answer sheet. The schema of visual representation (see Fig. 2) was used for introducing the interface, along with description. Preliminary testing was also performed and based on that, several modifications to the schema and instructions were made. The participants were given time to make familiar with the interface. After that, they were presented with actual interface implementation (see Fig. 3) and tasks, formulated as questions. The tasks were chosen according to the main strength of the interface - an information value accessible without need of further actions. Sample question: "Which subcategory contains the newest article (only in category "Design")?".

Participants completed most tasks with 100% success, the rest of them with still very high 85% success. Furthermore with the average SUS score being 78.46, we can consider the proposed navigation both acceptable and above average. You can look at summarized results in the following table (Table 5).

Very good results were achieved in the performed usability testing. However it needs to be asserted, that repeated testing would show the contribution of the interface better. The proposed navigation is expected to be more useful on websites which are visited repeatedly, such as news servers, blogs, educational or knowledge-based sites. Desirable would be long-term testing of system with proposed navigation interface, which would be in active use by stable group of users. Finally, it is important to note that individual aspects of implementation can vary for various systems according to their purpose, content and audience. Incorporating each organization scheme into the resulting navigation should be both feasible and useful for users.

Acknowledgment. This paper was written with the financial support of Specific Research Project "Investments within the Industry 4.0 concept" 2017 at Faculty of Informatics and Management of the University of Hradec Kralove, granted to the Department of Economics. We thank Martin Kral for help with organizing usability testing.

References

1. Bartuskova, A., Soukal, I.: The novel approach to organization and navigation by using all organization schemes simultaneously. In: Řepa, V., Bruckner, T. (eds.) BIR 2016. LNBIP, vol. 261, pp. 99–106. Springer, Cham (2016). doi:10.1007/978-3-319-45321-7_7
2. Rosenfeld, L., Morville, P., Arango, J.: Information Architecture: For the Web and Beyond, 4th edn. O'Reilly Media, Newton (2015)
3. Lidwell, W., Holden, K., Butler, J.: Universal Principles of Design, Revised and Updated: 125 Ways to Enhance Usability, Influence Perception, Increase Appeal, Make Better Design Decisions, and Teach through Design. Rockport, Beverly (2010)
4. Fang, X., Holsapple, C.W.: An empirical study of web site navigation structures' impacts on web site usability. Decis. Support Syst. **43**(2), 476–491 (2007)
5. Danielson, D.R.: Web navigation and the behavioral effects of constantly visible site maps. Interact. Comput. **14**(5), 601–618 (2002)
6. Kalbach, J.: Designing Web Navigation. O'Reilly Media, Newton (2007)
7. Walhout, J., Brand-Gruwel, S., Jarodzka, H., van Dijk, M., de Groot, R., Kirschner, P.A.: Learning and navigating in hypertext: navigational support by hierarchical menu or tag cloud? Comput. Hum. Behav. **46**, 218–227 (2015)
8. Hong, L., Chi, E.H., Budiu, R., Pirolli, P., Nelson, L.: SparTag.us: a low cost tagging system for foraging of web content. In: Proceedings of the Working Conference on Advanced Visual Interfaces, AVI 2008, pp. 65–72 (2008)
9. Leuthold, S., et al.: Vertical versus dynamic menus on the world wide web: Eye tracking study measuring the influence of menu design and task complexity on user performance and subjective preference. Comput. Hum. Behav. **27**(1), 459–472 (2011)
10. Golombisky, K., Hagen, R.: White Space is Not Your Enemy: A Beginner's Guide to Communicating Visually through Graphic, Web and Multimedia Design, 2nd edn. Focal Press, Waltham (2013)
11. Fowler, S., Stanwick, V.: Web Application Design Handbook: Best Practices for Web-Based Software. Morgan Kaufmann, San Francisco (2004)
12. Zhang, X., Song, D., Priya, S., Daniels, Z., Reynolds, K., Heflin, J.: Exploring linked data with contextual tag clouds. Web Semant. Sci. Serv. Agents World Wide Web **24**, 33–39 (2014)
13. Smashing Magazine (2016). Retrieved from: http://www.smashingmagazine.com/about/. Accessed 2016

WINE: Web Integrated Navigation Extension; Conceptual Design, Model and Interface

Ivan Soukal[1] and Aneta Bartuskova[2(✉)]

[1] Department of Economics, Faculty of Informatics and Management, University of Hradec Kralove, Rokitanskeho 62, 500 03 Hradec Králové, Czech Republic
ivan.soukal@uhk.cz

[2] Faculty of Informatics and Management, Center for Basic and Applied Research, University of Hradec Kralove, Rokitanskeho 62, 500 03 Hradec Králové, Czech Republic
aneta.bartuskova@uhk.cz

Abstract. Limitations in the area of web navigation and organization usually lead to poor usability, which further leads to user disorientation and dissatisfaction. In this paper we present WINE (Web Integrated Navigation Extension) as the novel alternative solution for efficient information retrieval. With combining website's data and local user data, WINE can offer useful navigation options in a consistent stable environment. In the case of missing support, partial solution is offered. One of the main goals of this interface is reducing interaction cost and user disorientation and provide means of personalization in the scope of each website or system.

Keywords: Web navigation · Information retrieval · Usability · Web interaction · Navigation schemes · Interaction cost

1 Introduction

The internet presents many challenges in the area of navigation, organization and consistency. Limitations in this area usually lead to poor usability of information-rich websites and web information systems. Poor usability reflects on user experience, satisfaction and consequently user's willingness to stay on the particular website. In this paper we propose a novel approach to information retrieval, called WINE (Web Integrated Navigation Extension). We expect that this functionality in a form of browser extension would make information retrieval more efficient.

WINE is based on the idea of integrated navigational interface, placed under the address bar of the internet browser. This interface would provide the most usable navigation links of any website in logical navigation groups in the consistent environment. This would create an alternate access to website's content.

Technically, fully supported WINE would integrate website's data and custom user data. Website's data would be stored in the respective database and delivered to WINE either by JSON, RSS or other format for data interchange. Custom user data would be managed by local storage. Partial WINE implementation (in the case of missing support

N.T. Nguyen et al. (Eds.): ICCCI 2017, Part I, LNAI 10448, pp. 462–472, 2017.
DOI: 10.1007/978-3-319-67074-4_45

from the side of individual websites and systems) would be reduced to local user data, i.e. personalization possibilities.

The conceptual design of WINE presents many advantages for users. It would primarily reduce dealing with each website's individual structure and navigation schema, which presents high interaction and attention-switching cost. Next, users could access the most usable navigation links from each website in one consistent environment. They could also manage individual pages as favourites grouped under the respective domain or track changes on websites and one's own progress.

2 State of the Art

2.1 Disorientation on Web Interfaces

Web users commonly experience disorientation while browsing, which has a negative effect on their performance [1]. According to Amadieu et al. [2], disorientation may be structural (related to the physical space of hypertexts) or conceptual (related to the conceptual space of hypertexts). Structural disorientation reflects a cognitive load linked to the processing of physical space (such as location of the position in the physical space) and conceptual disorientation concerns the users' difficulties to meaningfully link the different concepts conveyed by a hypertext [2]. One of the main contributors to this problem is confusing and disorganized navigation structure [3].

This can originate from the overall subjectivity, as website navigation consists primarily of subjective schemes [4]. Subjective organization schemes divide information into categories that defy exact definition, they are difficult to design and maintain and they can be difficult to use [5]. Problems with creating useful navigation according to subjective organization scheme include: broad vague categories, poor organization of menu options and poor grouping of categories [4].

2.2 Usability of Web Interface

Fang and Holsapple [3] defined two prominent approaches regarding to knowledge acquisition from a web site: search (e.g. via keywords) and navigation (e.g. via links). Users are often discouraged to go through the navigation structure and use search form whenever possible. However according to Lynch and Horton [6], even though web search is powerful, it is no substitute for coherent site architecture, carefully expressed in the page design and navigation. Users often do not know what they should look for or they express it in a way which will not find the information.

Navigation is an important part of usability of any website [7] and affects site credibility (e.g. [4]). Usability can be defined as a quality attribute that assesses how easy user interfaces are to use [8]. It was confirmed that providing web users with a usable environment can lead to significant savings and improved performances [9, 10].

However there is no standard in model-based user interface development environment yet and graphical user interface is still being created in an ad hoc manner [11]. There is no universal web interface, where everyone knows where to expect which objects. Such interface would save time needed for adapting to various interfaces and structural organizations. Consistent approach to layout and navigation allows users to adapt quickly to design and to predict the location of information [6].

2.3 Cost of Web Interactions

Hong et al. [12] identified two types of cost: interaction cost (mouse clicks, button presses, typing) and attention-switching cost (moving attention from one window to another). Activating particular navigation link contains both costs - interaction cost of mouse click and attention-switching cost of window reload. The return to previous state (e.g. by Back button) or moving to another page includes again both costs. Users use the Back button to return to a landmark page or hub, which indicates that the user has either extracted all he wanted from the current page, or that the page does not contain desired information [13].

Ware presented an idea of cognitive cost, which originates from moving through space [14]. He reviewed the basic costs of some common modes of information access, where internal pattern comparison is much more efficient than mouse hovering, selecting or clicking [14]. Leuthold et al. [15] confirmed the hypothesis that opening the dynamic menu needs an additional mouse movement and is thus more costly than just scanning the navigation items. Leavitt and Shneiderman [16] also stated that content should be formatted to facilitate scanning, and to enable quick understanding. This applies to all interactive systems with graphical user interface, including websites, web applications or web information systems.

3 Problem Definition

Potential issues related to websites and web information systems were already indicated in the previous section. They were focused primarily in the area of organization and navigation. In this section, we would like to discuss those limitations which are most relevant to our proposed solution.

3.1 Inefficient Navigation Schemes

Apart from technical shortcomings, such as poorly implemented interaction design or readability issues, disorientation belongs to the biggest navigation problems, as was discussed in Sect. 2.1. This concerns primarily larger information-rich websites and systems, where creation of navigation schema presents especially difficult task, prone to weaknesses such as:

- broad, vague or artificial categories
- poor organization of navigation links
- deep multi-level nesting
- many navigation areas displayed at once
- poor grouping of categories
- duplicities of navigation links
- vocabulary (web developer x user)

All of these issues increase interaction cost, time of interaction, number of errors and wrongly chosen pages, followed by increase of attention-switching cost as well. As a consequence, interaction with information-rich website is often ineffective and frustrating for users, even more so if the website or system is not optimized for performance. There are many alternate navigation possibilities, which do not suffer from the same limitations as subjective schemas. The intention is not to replace these navigation schemas, but to provide alternate means of navigation. The general idea is to let users easily access the important content which they would use more likely than the rest. This is in accordance with generally applicable Pareto principle.

Commonly used navigation sometimes underestimates the division of content according to user roles or audience type. However every user comes on the particular website in some specific role, therefore this division would be clear to him, unlike other subjective organization schemes. Of course the possible application vary with the type of website or system. E.g. university could apply roles "future student", "student", "graduate", "teacher" and "press". Determining target audience is important for creating efficient navigation, as well as distinguishing new and returning users.

Another easily implemented but often neglected navigation concept is easy access to the most popular content or the most frequent tasks (depending on context) on the website/system. In the case of a university, these would be e.g. "contacts", "entrance exams", "fields of study", "timetable" or "events". Finally, metadata such as "lastly edited" or "count of views" can also create useful navigation options. The most common strategies for implementing these concepts and thus differentiating special content are summarized in Table 1.

Table 1. Common strategies for differentiating special content or user groups on websites

Type	Group	Common strategies
Roles	New users	Homepage; landing pages accessed by sponsored links through search services
	Returning users	Session/login-related client area
	Users in a specific role	Contextual navigation
Users (all)	The most popular content/frequent tasks	Navigation links duplicated in a separate area (sidebar, footer, …)
User (individual)	Favourite content/repeatedly performed tasks	Session/login-related client area
	Recently viewed content	
Changes	Recently added content	Navigation links duplicated in a separate area (sidebar, footer, …)
	Recently edited content	

These alternate navigation strategies are often implemented as an afterthought more than as a result of careful planning. Duplicating navigation links this way can lead to user's confusion and disorientation as well. For example, if you look for information about entrance exams on our university's website, you can find the relevant link in several places. Being a popular content, its link is also duplicated in footer, however user has to scroll to it first and find it among over thirty other links, which is hardly a prominent place for one of the most searched content.

3.2 High Interaction Cost

Traditional navigation schemas often suffer from high interaction cost. We have analysed various task-oriented interactions on several information-rich websites to support this claim. As an example, we present a full cost analysis of searching for "sample tests on FIM (Faculty of Informatics and Management)" on our university's website. From this analysis, depicted in the following figure (Fig. 1), we can see more ways of accessing the same content.

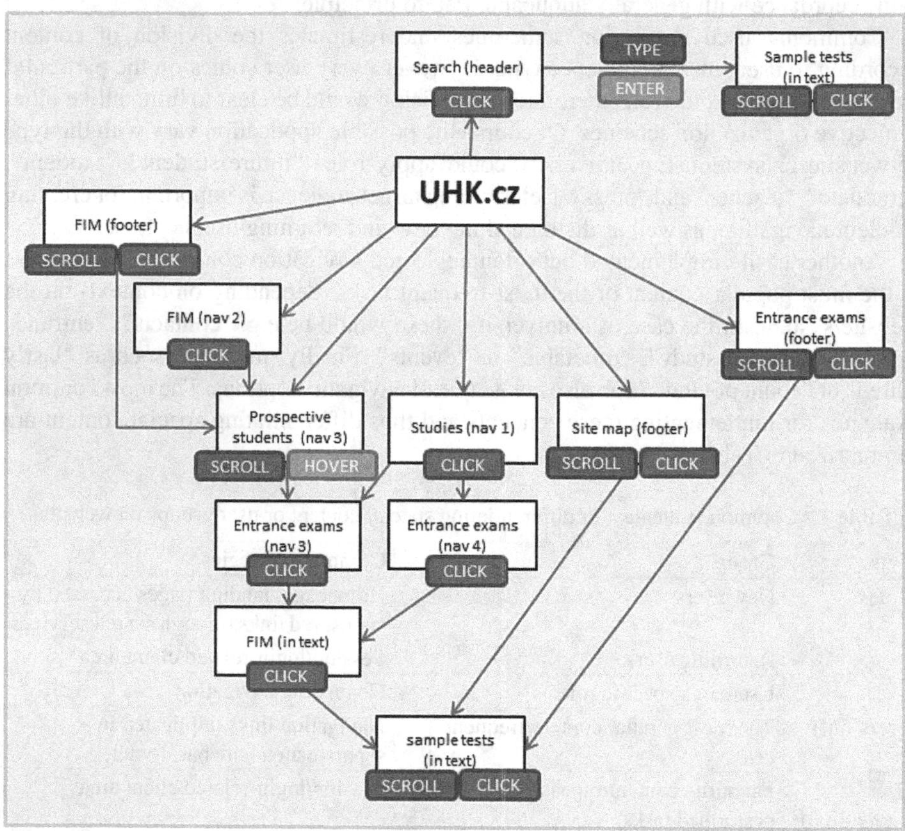

Fig. 1. Cost analysis of accessing the content "sample tests". nav 1 = uppermost navigation bar in text = main content area, nav 2 = bar of departments under nav 1 header = above nav 1, nav 3 = quick links under image banner footer = under main content area, nav 4 = navigation in left column. Labels were translated into English for better readability, however original content was accessed in Czech, since English version of the website contains far less information.

Each box represents a navigation link to a new page or view (in case of hover action), parentheses contain information about placement of the particular link and arrows signify direction of user browsing process. Smaller oval boxes represent necessary action(s). These actions were identified as: click, scroll, hover, type and enter (submitting search form either by clicking or pressing key). Each page loading requires switching attention, thus increasing attention-switching cost. Actions, on the other hand, increase interaction cost, by clicking, typing or pressing buttons. Scrolling - which was not mentioned in the analysis by Hong et al. [12] - could pertain to both groups, as it requires both interacting and attention-switching.

The performed cost analysis does not explicitly depict two actions: "switch attention to new page" and "look up the right navigation link". Switching attention to new page always follows after clicking navigation link and pressing enter in search form, since AJAX is not implemented on tested website and new page is loaded every time. Looking up the right navigation link follows after loading a new page, if search is not used instead, and can be performed simultaneously with scrolling. Looking up the search form is not considered as action with cost in the following calculations, because it is placed where users expect it to be (see [17]).

According to the performed analysis, the straightest path to desired content is by using search, with cost of 6 actions (click, type, enter, switch, scroll, click). Of course we assume that the search query is adequate, in our case e.g. "sample tests fim" or just "sample tests". The next identified efficient path is by using links in footer, either through the link "Site map" or "Entrance exams", with cost of 7 actions (look-up/scroll, click, switch, look-up/scroll, click).

However it is important to note several issues of such cost analysis. First, we need to consider how time-consuming each of these actions is. E.g. scrolling down to the bottom of the page and looking up the right link (2 actions) is usually not quicker than clicking into search field, typing keyword or two and pressing enter (3 actions). Even two actions of the same type are not of the same cost. E.g. looking up something in site map ought to be more time-consuming than looking up the same thing in smaller topical section. Similarly, there can be huge differences between cost for returning user (who already knows where to click) and new users (who needs to read at least several labels before they can confidently click on anything).

To conclude, a new user cannot know (and returning user probably won't remember) which path is more straightforward. In this manner, spreading more links across the website seems like a reasonable idea, however we believe that there is a more efficient solution. If the desired navigation link was present in primary or secondary navigation, the cost could be only 2 actions (look-up, click). This is mostly not possible in the case of information-rich websites. The most popular and useful links could be however prioritized, as is the main idea behind our proposal.

4 Proposed Approach

Previous sections indicated a room for improvement especially in the area of navigation, which relates also to organization, consistency and interaction design [18]. In this paper we propose a novel approach to navigation of information-rich websites and web information systems. We will address our proposal as WINE (Web Integrated Navigation Extension) for brevity in the rest of this paper. Users can either look up a certain web page for the particular piece of information or just browse the internet without any particular goal of their browsing. Our approach is intended for the first use case, with the goal to make information retrieval more efficient.

4.1 Conceptual Design

Our solution is based on the idea of standardized interface, which would integrate the most usable navigation links of any website in the consistent environment. WINE functionality could be technically available as a browser extension, with interface placed between address bar and inner window of the web browser, where the particular website is displayed. This extension would create an additional navigation layer above any website or web-based system, providing alternate access to website's content and a novel approach to information retrieval on the internet. The content of WINE would change according to the particular website during new page load.

WINE's purpose is to integrate website's data and custom user data. Website's data would be stored in the respective database and delivered to WINE either by JSON, RSS or other format for data interchange. Custom user data would be stored locally on front-end. This arrangement offers many advantages for users:

- reducing dealing with each website's individual structure and navigation schema, which presents high interaction and attention-switching cost
- accessing the most usable navigation links from each website in one consistent environment
- adding individual pages to favourites in the same stable environment, possibly creating custom groups of pages, grouped under relevant domain
- tracking changes on websites and one's own progress
- possibilities of personalization and customization

WINE uses exported data from dynamic website's database and at the same time data from local storage. Model is created on the fly for the combination of particular user and particular website. Controller includes logic of the WINE solution, in consistency with MVC (Model-View-Controller).

4.2 Structure of WINE Model

The purpose of WINE is to offer generally applicable navigation objects, which can be used by any website/system, and configurable navigation objects. These additional navigation objects provide navigation patterns specific to the particular site and to the particular user. The following table (Table 2) presents a list of all three identified types. Main groups of these objects were marked as "Views", consisting of "Roles", "Global view" and "Local view". Each of these groups is further divided to general and site-specific navigation objects. The term "Navigation groups" means set of related navigation links (individual pages). Support in terms of data export by individual websites is needed for the majority of WINE navigation objects, otherwise only "Local view" facilitated by local storage would be available for these websites.

Table 2. Proposed navigation objects for WINE

View	Type	Navigation groups	Support needed
Roles (site data)	General	New users	JSON
		Returning users	JSON
	Site-specific	Site-specific groups	JSON
Global view (site data)	General	Most popular	JSON
		Recently added	JSON/RSS
		Recently edited	JSON/RSS
	Site-specific	Site-specific groups	JSON
Local view (user data)	General	Favourites	Local storage
		Recently viewed	Local storage
		To view later	Local storage
	User-specific	User-specific groups	Local storage

The amount of navigation links in navigation groups depends on data received from website and locally stored user data and preferences. Especially dynamically created groups such as "recently added" should be limited by number of entries.

4.3 Arrangement of WINE Interface

This section presents WINE model transformed into the graphical user interface, which will be placed under the address bar. This interface can be optionally opened, minimalised or closed, depending on the particular website and user preferences. The next figure (Fig. 2) presents a structural layout of the proposed interface:

Fig. 2. Structural layout of WINE interface

Default behaviour of the interface on the particular website should be:

- opened, if the website fully supports WINE
- minimalised, if the website does not support WINE, but user data exists
- closed, if the website does not support WINE and there are no user data yet

Visual design of proposed interface could be either as indicated in structural design, i.e. by expandable select boxes. This arrangement is less space-demanding, however as was already mentioned in the previous sections, there is strong argumentation against dynamic navigation. Ware stated that internal pattern comparison is much more efficient than mouse hovering, selecting or clicking [14]. Leuthold et al. [15] confirmed that opening the dynamic menu needs an additional mouse movement and Leavitt and Shneiderman [16] also stated that content should be formatted to facilitate scanning. Therefore individual navigation objects should be presented in easy-scannable lists. However because of space requirements, this arrangement should not be the default one. It would be available as an option in locally stored preferences. In the case of missing support on the particular website, WINE interface would be limited to "Local view" (Fig. 3).

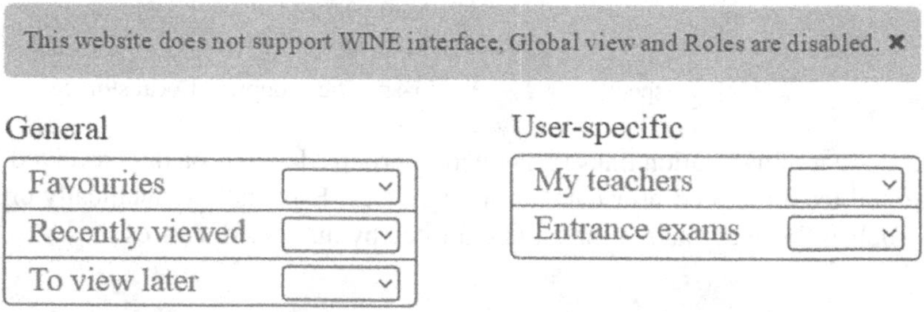

This website does not support WINE interface. Global view and Roles are disabled. ✖

General

Favourites

Recently viewed

To view later

User-specific

My teachers

Entrance exams

Fig. 3. WINE interface with missing support

5 Conclusions and Discussion

In this paper, we have proposed a novel approach to information retrieval on the internet; WINE - Web Integrated Navigation Extension. This extension would provide an interface, integrating the most usable navigation links of any website in the consistent environment. This functionality would be implemented as a browser extension, placed under the address bar.

This idea however presents many difficulties and limitations. Because of the use of local storage, user data would have to be associated with the particular browser. Also the functionality itself would be bound to the browser as the browser extension. To fully utilize power of WINE, websites would have to support it by delivering relevant data. Without this support, WINE is limited only to custom user data, which are created by individual users and stored locally.

We also have to take into account that any website can deviate from commonly used content structures and as such cannot fully benefit from WINE. It is also important to note that WINE integrates many concepts already present in web browsing, yet with varying implementations in different browsers and to be found in different places. E.g. in Google Chrome there are bookmarks (as Favourites) in the top bar or e.g. history (as Recently Viewed) accessible in settings. The main goal of WINE is to unify these features into one consistent environment.

Acknowledgement. This paper was written with the financial support of Specific Research Project "Investments within the Industry 4.0 concept" 2017 at Faculty of Informatics and Management of the University of Hradec Kralove, granted to the Department of Economics. We thank Jan Hruska for help with conducting cost analysis.

References

1. McDonald, S., Stevenson, R.J.: Navigation in hyperspace: an evaluation of the effects of navigational tools and subject matter expertise on browsing and information retrieval in hypertext. Interact. Comput. **10**(2), 129–142 (1998)
2. Amadieu, F., Tricot, A., Mariné, C.: Prior knowledge in learning from a non-linear electronic document: disorientation and coherence of the reading sequences. Comput. Hum. Behav. **25**(2), 381–388 (2009)
3. Fang, X., Holsapple, C.W.: An empirical study of web site navigation structures' impacts on web site usability. Decis. Support Syst. **43**(2), 476–491 (2007)
4. Kalbach, J.: Designing web navigation. O'Reilly Media, Sebastopol (2007). ISBN 978-0596528102
5. Rosenfeld, L., Morville, P., Arango, J.: Information Architecture. For the Web and Beyond, 4th edn. O'Reilly Media, Inc., Newton (2015)
6. Lynch, P.J., Horton, S.: Web Style Guide: Basic Design Principles for Creating Web Sites, 3rd edn. Yale University Press, New Haven (2009). ISBN: 978-0300137378
7. Bartuskova, A., Krejcar, O.: Design requirements of usability and aesthetics for e-learning purposes. In: Sobecki, J., Boonjing, V., Chittayasothorn, S. (eds.) Advanced Approaches to Intelligent Information and Database Systems, pp. 235–245. Springer, Cham (2014). doi: 10.1007/978-3-319-05503-9_23
8. Nielsen, J.: Usability 101. http://www.nngroup.com/articles/usability-101-introduction-to-usability/. Accessed 20 Mar 2016
9. Unal, Z., Unal, A.: Evaluating and comparing the usability of web-based course management systems. J. Inf. Technol. Educ. **10**, 1019–1038 (2011)
10. Kibaru, F., Dickson-Deane, C.: Model for training usability evaluators of e-learning. In: Proceedings of World Conference on E-Learning in Corporate, Government, Healthcare, and Higher Education, pp. 517–522 (2010)
11. Mustakerov, I., Borissova, D.: A conceptual approach for development of educational Web-based e-testing system. Expert Syst. Appl. **38**, 14060–14064 (2011)
12. Hong, L., Chi, E.H., Budiu, R., Pirolli, P., Nelson, L.: SparTag.us: a low cost tagging system for foraging of web content. In: Proceedings of the Working Conference on Advanced Visual Interfaces, AVI 2008, pp. 65–72 (2008)
13. Danielson, D.R.: Web navigation and the behavioral effects of constantly visible site maps. Interact. Comput. **14**(5), 601–618 (2002)
14. Ware, C.: Visual Thinking for Design. Morgan Kaufmann Publishers Inc., Burlington (2008)
15. Leuthold, S., et al.: Vertical versus dynamic menus on the world wide web: eye tracking study measuring the influence of menu design and task complexity on user performance and subjective preference. Comput. Hum. Behav. **27**(1), 459–472 (2011)
16. Leavitt, M., Shneiderman, B.: Research Based Web Design and Usability Guidelines. Government Printing Office, Washington (2006)

17. Roth, S.P., Schmutz, P., Pauwels, S.L., Bargas-Avila, J.A., Opwis, K.: Mental models for web objects: where do users expect to find the most frequent objects in online shops, news portals and company web pages? Interact. Comput. **22**(2), 140–152 (2010)
18. Bartuskova, A., Krejcar, O., Soukal, I.: Framework of design requirements for e-learning applied on blackboard learning system. In: Núñez, M., Nguyen, N.T., Camacho, D., Trawiński, B. (eds.) ICCCI 2015. LNCS, vol. 9330, pp. 471–480. Springer, Cham (2015). doi:10.1007/978-3-319-24306-1_46

Real-Life Validation of Methods for Detecting Locations, Transition Periods and Travel Modes Using Phone-Based GPS and Activity Tracker Data

Adnan Manzoor(✉), Julia S. Mollee, Aart T. van Halteren, and Michel C.A. Klein

Behavioural Informatics Group, Department of Computer Science,
Vrije Universiteit Amsterdam, Amsterdam, The Netherlands
{a.manzoorrajper,j.s.mollee,a.t.van.halteren,michel.klein}@vu.nl

Abstract. Insufficient physical activity is a major health concern. Choosing for active transport, such as cycling and walking, can contribute to an increase in activity. Fostering a change in behavior that prefers active transport could start with automated self-monitoring of travel choices. This paper describes an experiment to validate existing algorithms for detecting significant locations, transition periods and travel modes using smartphone-based GPS data and an off-the-shelf activity tracker. A real-life pilot study was conducted to evaluate the feasibility of the approach in the daily life of young adults. A clustering algorithm is used to locate people's important places and an analysis of the sensitivity of the different parameters used in the algorithm is provided. Our findings show that the algorithms can be used to determine whether a user travels actively or passively based on smartphone-based GPS speed data, and that a slightly higher accuracy is achieved when it is combined with activity tracker data.

Keywords: Intelligent applications · Data analytics · Health support systems · Physical activity · Clustering

1 Introduction

Physical inactivity is a major health concern: according to the WHO, every year around three million people die because of physical inactivity [1]. One of the causes of physical inactivity is that people are more inclined to passive modes of transportation. Active travelling options such as biking and walking provide ample opportunities to improve physical activity [2]. Studies have shown that people who frequently use public transport are more physically active than those who use other types of inactive transport [3, 4], as walking to and from a public transport can also lead to a substantial increase in physical activity levels. Self-monitoring is a well-known and often used behavior change technique for supporting people in improving their physical activity [5]. Fostering a change in behavior that prefers active transport over inactive transport could start with self-monitoring of travel choices. In order to provide support in behavior change or self-monitoring, automatic measuring of (active/inactive) travel behavior is inevitable.

N.T. Nguyen et al. (Eds.): ICCCI 2017, Part I, LNAI 10448, pp. 473–483, 2017.
DOI: 10.1007/978-3-319-67074-4_46

This paper describes and evaluates an approach that can be used within an online mobile coaching system for stimulating physical activity [6]. The approach exploits existing techniques for determining frequently visited locations, transition periods (periods of traveling between two locations) and travel modes (active vs. inactive transport) of individuals based on GPS and accelerometer data. Unsupervised learning methods (density based clustering methods) on GPS data are used to determine frequent locations. These locations and the other GPS readings are subsequently used to derive transition periods. Finally, GPS speed measurements and accelerometer data are combined to obtain a more precise result about travel modes and activity levels of individuals. Most of the techniques that we use have been developed and tested in lab-settings or controlled environments. We evaluate the approach in a real-life pilot study. In this study, people kept an online diary of their travelling behavior, which was compared with the results of the automatic approach. The main research question investigated in this paper is whether this combination of existing techniques can reliably be used to detect important locations, transitions, and transport modes in a real-life context with smartphone-based GPS data and an off-the-shelf activity tracker. Our hypothesis is that the combination of both (GPS and activity tracker) will provide better results compared to only using on one of the sources.

This study was conducted in the context of the development of an online mobile coaching system [6]. Although in the current implementation of that system transportation options were asked in the form of user input, the objective is to automate the process of travel mode detection from raw GPS data and activity tracker data. The ultimate aim of such an integrated system is to provide personalized support to individuals based on their traveling context and physical environment.

2 Background

This section discusses the state-of-the-art in terms of the detection of locations, transitions, and travel modes.

2.1 Location Detection

Literature suggests various techniques for identifying important places within a list of visited locations. An important place could be home, work, college and/or office. Clustering is a popular approach to perform this task [7, 8]. One class of clustering methods that is of particular relevance for clustering geospatial data is based on density. Density is defined as the number of points within a given radius [9]. In [7], K-means clustering is compared with the density-based technique DJ-cluster (a variation of DBSCAN). The conclusion is that density-based clustering provides better results for finding important places. A disadvantage of the K-means approach is that it needs to know the number of clusters in advance.

Besides clustering, other techniques for detecting important locations are also applied convincingly on GPS data. For example, in [10] a kernel-based algorithm is applied to synthetic GPS data; this gives better results compared to traditional approaches. However, the authors admit that one drawback of their study is that they use synthetic data, which are uninterrupted, while real world data is usually interrupted because of various reasons (e.g., inside a building, underground metro station).

2.2 Identification of Transitions

One of the difficult steps in the process of detecting travel behavior is to find transitions [11] from one location to another. A transition has two aspects: the travel period and the start and end locations. A transition occurs between two locations (e.g., between home and study). A travel period is usually detected by separating the periods with a significant speed from periods in which the speed is almost zero; in addition, the change in GPS location itself can be used.

The detection approaches presented in the literature are usually based on dedicated GPS trackers in combination with geographical information systems (GIS software) [12, 13]. The problem with approaches that depend on external sources such as GIS is that the transition can only be detected once the data is loaded into a GIS based application, which means that it cannot always be executed on devices with low computing power and bandwidth (e.g., smartphones). When smartphone-based GPS data is used, the experiments are usually conducted in extremely controlled lab settings [11, 14]. For example, in the study conducted by Reddy et al., participants were asked to attach six phones to different parts of their body [14].

2.3 Travel Mode Detection

In the literature, mainly two approaches are used for the identification of the *mode* of travel: one is time dependent and the other is trip dependent. In a time dependent approach, a transport mode is assigned to every time frame of travel data (usually one minute [15] or one second [16]), while in trip level approaches, a mode is assigned to the whole trip. Zheng, Liu et al. visualize participants' GPS log data in a prototype application named GeoLife [17, 18]. This application shows the user a visualized version of his/her own GPS data. Besides plainly visualizing the path the user has taken, several other features are available as well. An example of an additional feature is the possibility to see the distinction between different travel modes (like traveling by foot or car) in terms of color-differences. In GeoLife this is done using a supervised learning algorithm which is based on the raw GPS data. This algorithm divides each GPS track in different trips and each trip in different segments. A segment consists of a (part of a) trip using one single travel mode.

Since the current study is conducted in the context of a coaching system (to increase physical activity levels of individuals by encouraging active transport), we are more concerned with the activity level of participants. Therefore, our aim is to differentiate active and inactive travel modes by means of raw GPS and activity tracker data; detecting the precise means of transportation is of less concern.

3 Our Approach

An integrated approach is required that starts with collecting raw data and is finally able to suggest active transport options to the participants. This means that our approach has to integrate different analyses of GPS data: the extraction of frequently visited locations, the detection of transition periods and finally determination of the travel modes.

3.1 Location Detection

Raw GPS data are processed with a clustering algorithm to find the frequently visited locations. We use the OPTICS [19] clustering method in this step, which is a density based clustering technique and requires two parameters: the minimum number of points within a cluster (*MinPts*) and the maximum distance between two points (*Eps*). We have chosen this algorithm because it does not require to specify the number of clusters in advance. OPTICS is an extension of the DBSCAN algorithm. DBSCAN is able to find arbitrarily shaped clusters, but is not effective when it comes to finding clusters with varying densities because it is sensitive to the particular settings of its parameters. The OPTICS algorithm computes an augmented ordering of points. To extract the actual clusters, another algorithm is used, which is known as OPTICSXI and is also suggested in [19]. In our approach, we have used the implementation of the algorithm in the ELKI [20] framework.

Additionally, selecting an appropriate distance metric is important for the clustering process. Since we are dealing with spatial data, the "great circle distance" is used, which is a shortest distance between two latitude and longitude points. The Haversine formula [21] is used to calculate the "great circle distance". Several experiments were conducted to see which value of parameters (*Eps* and *MinPts*) provides better results.

3.2 Transitions

Based on the results of the cluster analysis, the travel behaviors are extracted in the form of transitions (travel periods between locations). This step involves separating periods of transition from periods at locations. A transition is detected by combining different factors, according to the following algorithm. First, the periods are identified in which an increase in average GPS speed coincides with a change in clustered location. Different thresholds were tested i.e. between 0.4 and 1.5 m/s inclusive (with a step size of 0.1 m/s) which are close to the average walking speed (between 2 and 6.6 km/h [22]). As a second step, transitions are merged when the time difference between two or more consecutive transitions is less than some threshold. Different time intervals were checked, namely 3, 4, 5, 6 and 7 min. These transitions can be the result of, for instance, waiting at a train station, bus stop or perhaps erroneous GPS readings. There are various possible causes for such errors, for example being inside a building, in an underground metro station, or because of connectivity problems.

3.3 Travelling Modes

As the main goal of the envisioned system is to find physical activity opportunities it is relevant to detect the mode of travels. To find these modes, combinations of both kinds of data sources are used: accelerometer and GPS. Travel modes are classified in terms of active (biking and walking) or inactive (train, tram, metro, etc.). We do not differentiate the transportation modes at a more detailed level; rather, the focus is on active versus inactive options.

We investigate three different ways to detect travel modes. First, a threshold value is used for the GPS speed parameter to decide on the mode of travelling. Another method is using accelerometer data for determining the travel modes. In this approach, we used the average number of steps per minute during a transition period. Various thresholds were compared for average speed (5, 6, 7, 8, 9, 10 m/s) and for average steps per minute (10, 15, 20, 25, 30, 35) to see which works better. In the third method, we used the combination of GPS speed and average steps per minute to see whether that performs better than one of the individual forms.

4 Experimental Setup

A pilot study was conducted between January 21 and March 18, 2015. A total of 26 persons participated in this study. Individuals were recruited via personal networks of the researchers. All participants had an Android based smartphone with GPS logger software installed on it. We did not impose any restriction on the model of the smartphone, apart from running on Android. Activity data (floors, steps, calories) were collected by means of a Fitbit One activity tracker. The GPS logger records the data for each participant every minute. The average age of participants was 22 years, ranging between 18–26 years, 15 of them were male. Twenty of the participants were university students. Two participants dropped out due to technical problems. During this study, each individual filled in an intake questionnaire and replied to daily questionnaires. The intake questionnaire includes questions about demographics, location information and travel options. Daily questionnaires include questions about travel log to work/study/sports location and sports activity log. These questionnaires provide the ground truth for each participant's important places and their travel behavior and are used to validate the travel behavior and mode of transport. Due to problems with the daily questionnaires, the questions about the travel modes were not asked in the beginning of the experiment. Therefore, the participants were asked to report their travel modes in the first weeks of the experiment in hindsight together with a confidence level.

To facilitate the validation process, travel periods in the ground truth file are labeled as active/inactive depending on the minutes of active travel during the travel period. When the minutes of active travel exceeds the minutes of inactive for a travelling period, the whole transition is labeled as active. Travel mode data in the survey questionnaire are reported in the following form: "*walk: 7, metro: 8, walk: 5, train: 35, walk: 5*". This example is an instance of mixed mode travel, and since the active travel minutes are 17 and the passive travel minutes are 43, the travel mode is labeled as "inactive". Another part of the ground truth consists of a set of latitude and longitude values corresponding to each participant's reported significant places. This file is used to validate significant locations.

5 Results

In this section, we validate the described approach by comparing the results of the algorithms with the ground truth as provided in the travel logs.

Before explaining the results, we first provide an illustration of the collected data in Fig. 1. This figure shows a map of two locations which are marked as home and work, and a transition path between them. The transition path is generated based on the heuristics we used to find transitions (see Sect. 3.2) in the location data. The high spikes in Fig. 1(b) show that the person uses an active mode of transport – in this case walking. Similarly, the high spikes in Fig. 1(c) illustrate that the person uses an inactive mode of transport with a relatively high speed. In this example, the user started his/her travel to the work location by taking 9 to 10 min of walk to a nearby public transport facility. This is visible in Fig. 1(b), which shows the steps taken during time period 7:34 and 7:43. After that, an inactive mode of transport is used to travel further, as the speed graph (Fig. 1(c)) shows that the speed is quite high between 7:44 and 7:53.

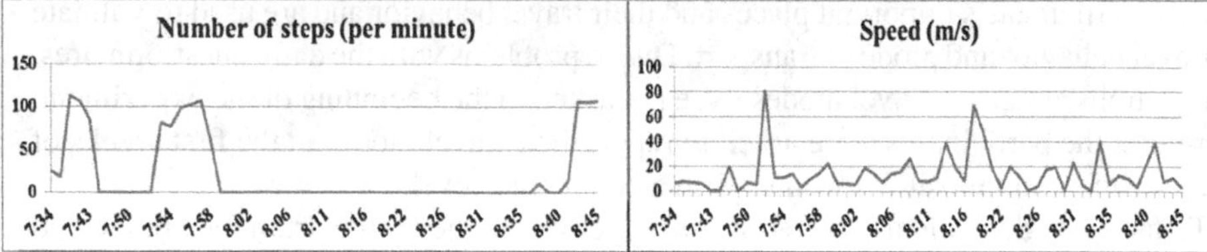

Fig. 1. An example transition period between two locations that includes active and inactive travelling modes. (a) refers to map, (b) to activity tracker data (steps per minute) and (c) to GPS speed (meter per second).

5.1 Locations

We evaluate the performance of the OPTICS clustering algorithm based on the recall score, as we would like to find most of the significant locations listed by the participants, but it is not a problem if other frequently visited locations are found as well. An instance of the clustering result is considered to be a true positive when it matches with one of the user locations given in the intake survey, a false negative is a location instance for which we do not find any cluster, and a false positive is a cluster instance which does not match to any location specified in the survey. A higher number of true positives means that a higher number of significant locations is returned that could also lead to better chances of extracting more transitions and travel modes. Figure 2 gives an overview of the recall scores obtained when *Eps* varies between 90–330 and *MinPts* between 50–100 To find the optimal parameters for the remaining analysis, we first select the *Eps* values with a high recall, i.e. 270, 290, 310, 330, which all have a similar score. From these, we choose a value based on the lowest number of false positives. The choice for *Eps* is reduced to 290 and 330 with *MinPts* 40 or 50. Finally, we selected *Eps* = 290 and *MinPts* = 50 because of the smallest *Eps* distance.

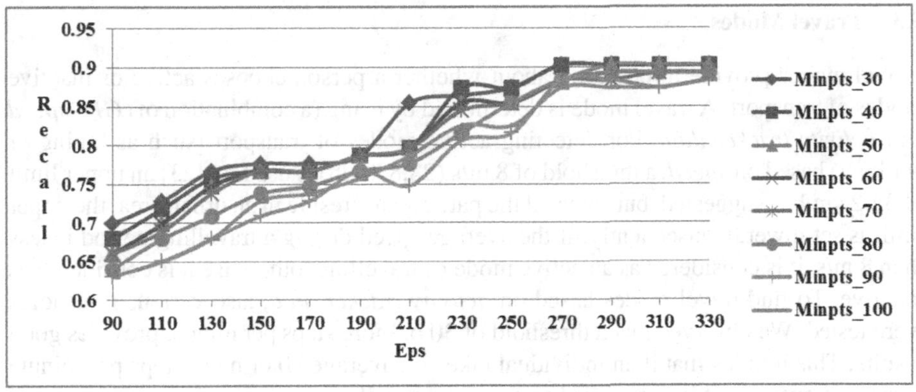

Fig. 2. An overview of the recall scores, with *Eps* between 90–330 and *MinPts* between 50–100.

5.2 Transitions

As described in the Sect. 3.2, two parameters are used to detect transitions. When checking all combinations of average speed and interval length in minutes, the recall score ranged between 0.4 and 0.44. The best recall was achieved when the average speed was set to a threshold of 1.2 m/s and minutes interval to 5 min. We use two different methods to evaluate the algorithm: (1) check the travel period only, i.e. start and end time of a transition, and (2) in combination with travel locations. For evaluation method (1) a travel period is considered a true positive when it has overlapping minutes with an actual travel period reported by the participants. For evaluation method (2), in addition to the previous criterion, also the (start and end) locations should match. False positives are those travel periods that are not found in the user's survey and which are probably results of shorter or longer trips to unimportant locations that users did not mention in the surveys. False negatives are the travel periods which are not found by the algorithm. We compare the results of the algorithm using different time windows for what we consider to be a match between the time according to the algorithm and the time reported in the survey: half an hour, 45 min and one hour. It is apparent that increasing the time window also increases the recall. A threshold of one hour gives a recall of 64% while a 54% recall is obtained when the time window is half an hour. Table 1 provides the detailed results. It can be seen that all values in the third column are smaller than the respective values in the second column.

Table 1. Recall scores for detecting transitions for two different methods

Time window	Detecting travel periods (w/o filtering on locations)	Detecting travel periods (taking locations into account)
30 min	0.54	0.44
45 min	0.59	0.45
60 min	0.64	0.47

5.3 Travel Modes

Travel modes provide information about whether a person chooses active or inactive modes of transport. A travel mode is determined by using (a combination of) *GPS speed* and *activity tracker data*. For detecting active modes of transport (such as biking or walking) based on *speed*, a threshold of 8 m/s (28,8 km/h) is used. In [23] an upper limit of 32.2 km/h is suggested, but since all the participants reside in an urban area, the upper limit is set lower. Consequently, if the average speed during a travelling period is less than 8 m/s, it is considered as an active mode of travelling, otherwise it is considered as inactive. To find travel modes based on *activity tracker data*, also various thresholds were tested. We observed that a threshold of 30 or more steps per minute provides good results. This implies that if an individual takes on average 30 or more steps per minute during a transition, it is considered as an active travel.

Table 2. Travel modes detected with speed and/or activity tracker

	Total reported travels	Correctly detected	Recall
Speed	162	146	0.90
Activity tracker	162	115	0.70
Speed and tracker	162	158	0.97

In our experiment, travel modes are detected for the transitions found in the previous step using a time window of half an hour and taking locations into account. Table 2 illustrates the results for the different approaches. The first row shows that a recall of 0.9 is obtained when modes are detected based on GPS speed only. If only the activity tracker is used, a score of 0.7 is obtained. When both parameters are combined, then a recall of 0.97 is achieved. We can conclude that the activity tracker data on itself is not a very accurate method, but that it can be used to improve the results of the method based on GPS speed.

6 Discussion and Conclusion

The ultimate aim of our work is to create a mechanism that can be employed in a system that motivates and encourages individuals to engage more in physical activity by exploiting their physical context and using this to find opportunities for active ways of travelling. In this study, existing algorithms for finding important locations and detecting transitions and travel modes were validated using a smartphone-based GPS and an off-the-shelf activity tracker. It turned out that in this setting, the clustering algorithm (OPTICS) for detecting locations provides a good recall with a setting of 290 for Eps and 50 for MinPts. The high number of false positives could be explained by the fact that participants were only asked to report the most important locations of type home, work, study, sports location, but that it was not required to mention every significant location. Of course, in daily life we do visit more locations than the ones listed above.

There are a few limitations of our study that we would like to mention. First, the participants reported about their travel modes in hindsight. Due to technical issues the participants provided the travel information for the first period only after two weeks, for the remaining period information was provided on the day after the travel took place. For this reason, a confidence level was associated with each trip in the travel log. It is possible that people underestimated or overestimated their travel periods which could affect the reliability of data and hence leads to less performance. However, in our validation we did not take the confidence level into account, all reported travel modes were considered equally important.

There are a number of possible explanations for the low number of transitions that were found. First of all, it is known that the young population that we have in our pilot study tends to underreport or misreport travels [24]. Another possible reason is that the detection process becomes more complex when different types of smartphones are used [25]. Since in this study no restriction was imposed to use a particular brand (apart from the operating system), we had a large variety of devices. Furthermore, it is also known that a dedicated GPS receiver is more accurate than the GPS sensor in a smartphone [25].

Our attempt to detect locations, trips, and mode was made in the context of a mobile coaching app. We therefore abstained from using GIS, since the goal was to automatically detect in real time. However, as can be seen in the literature, using GIS data provides more reliable trip information as compared to only relying on GPS and accelerometer data [12, 13].

We can conclude that it is difficult to find transition periods based on GPS locations only. Our hypothesis is that this is caused by the fact that there are many gaps in the GPS logs. Several factors could have contributed to this problem, for example: the smartphone is not charged, an accurate location cannot be obtained – especially when inside a building, or a transition is not reported correctly. However, for the transition periods that were correctly identified, we have shown that travel modes can be very well detected based on GPS speed. Adding accelerometer data to GPS speed further improves detection performance. The results of travel mode detection are comparable to the literature. For example Ellis et al. [15] reported an overall accuracy between 89.8% and 91.9% based on different methods when combining GPS and accelerometer. For the trip level detections Gong et al. reported 82.6% accuracy [26].

Acknowledgments. This research is supported by Philips and Technology Foundation STW, Nationaal Initiatief Hersenen en Cognitie NIHC under the partnership program Healthy Lifestyle Solutions. The authors would like to thank Lars Rouvoet and David Rip for their contribution to the data collection and their help in conducting the study.

References

1. WHO | Physical activity. http://www.who.int/topics/physical_activity/en/
2. Sahlqvist, S., Song, Y., Ogilvie, D.: Is active travel associated with greater physical activity? The contribution of commuting and non-commuting active travel to total physical activity in adults. Prev. Med. **55**, 206–211 (2012)
3. Rissel, C., Curac, N., Greenaway, M., Bauman, A.: Physical activity associated with public transport use—a review and modelling of potential benefits. Int. J. Environ. Res. Public Health **9**, 2454–2478 (2012)
4. Saelens, B.E., Moudon, A.V., Kang, B., Hurvitz, P.M., Zhou, C.: Relation between higher physical activity and public transit use. Am. J. Public Health **104**, 854–859 (2014)
5. Sanders, J.P., Loveday, A., Pearson, N., Edwardson, C., Yates, T., Biddle, S.J., Esliger, D.W.: Devices for self-monitoring sedentary time or physical activity: a scoping review. J. Med. Internet Res. **18**(5), e90 (2016). http://www.jmir.org/2016/5/e90/
6. Klein, M.C., Manzoor, A., Middelweerd, A., Mollee, J.S., te Velde, S.J.: Encouraging physical activity via a personalized mobile system. IEEE Internet Comput. **19**, 20–27 (2015)
7. Zheng, Y., Zhang, L., Xie, X., Ma, W.-Y.: Mining interesting locations and travel sequences from GPS trajectories. In: Proceedings of the 18th International Conference on World Wide Web, pp. 791–800. ACM (2009)
8. Zhou, C., Bhatnagar, N., Shekhar, S., Terveen, L.: Mining personally important places from GPS tracks. In: 2007 IEEE 23rd International Conference on Data Engineering Workshop, pp. 517–526. IEEE (2007)
9. Ester, M., Kriegel, H.-P., Sander, J., Xu, X.: A density-based algorithm for discovering clusters in large spatial databases with noise. In: Proceedings of the Second International Conference on Knowledge Discovery and Data Mining KDD 1996, pp. 226–231 (1996)
10. Thierry, B., Chaix, B., Kestens, Y.: Detecting activity locations from raw GPS data: a novel kernel-based algorithm. Int. J. Health Geogr. **12**, 14 (2013)
11. Fan, Y., Chen, Q., Liao, C.-F., Douma, F.: UbiActive: a smartphone-based tool for trip detection and travel-related physical activity assessment. In: TRB 92nd Annual Meeting Compendium of Papers, Transportation Research Board, TRB 2013 Annual Meeting (2013)
12. Bohte, W., Maat, K.: Deriving and validating trip purposes and travel modes for multi-day GPS-based travel surveys: a large-scale application in the Netherlands. Transp. Res. Part C Emerg. Technol. **17**, 285–297 (2009)
13. Chung, E.-H., Shalaby, A.: A trip reconstruction tool for GPS-based personal travel surveys. Transp. Plan. Technol. **28**, 381–401 (2005)
14. Reddy, S., Mun, M., Burke, J., Estrin, D., Hansen, M., Srivastava, M.: Using mobile phones to determine transportation modes. ACM Trans. Sens. Netw. TOSN **6**, 13 (2010)

15. Ellis, K., Godbole, S., Marshall, S., Lanckriet, G., Staudenmayer, J., Kerr, J.: Identifying active travel behaviors in challenging environments using GPS, accelerometers, and machine learning algorithms. Public Health **2**, 39–46 (2014)
16. Feng, T., Timmermans, H.J.: Transportation mode recognition using GPS and accelerometer data. Transp. Res. Part C Emerg. Technol. **37**, 118–130 (2013)
17. Zheng, Y., Wang, L., Zhang, R., Xie, X., Ma, W.-Y.: GeoLife: managing and understanding your past life over maps. In: The Ninth International Conference on Mobile Data Management (MDM 2008), pp. 211–212. IEEE (2008)
18. Zheng, Y., Liu, L., Wang, L., Xie, X.: Learning transportation mode from raw GPS data for geographic applications on the web. In: Proceedings of the 17th International Conference on World Wide Web, pp. 247–256. ACM (2008)
19. Ankerst, M., Breunig, M.M., Kriegel, H.-P., Sander, J.: OPTICS: ordering points to identify the clustering structure. ACM Sigmod Rec. **28**, 49–60 (1999). ACM
20. Schubert, E., Koos, A., Emrich, T., Züfle, A., Schmid, K.A., Zimek, A.: A framework for clustering uncertain data. Proc. VLDB Endow. **8**, 1976–1979 (2015)
21. Shumaker, B.P., Sinnott, R.W.: Astronomical computing: 1. Computing under the open sky. 2. Virtues of the haversine. Sky Telesc. **68**, 158–159 (1984)
22. Transportation Research Board: Special Report 209 (1994)
23. Taylor, D., Mahmassani, H.: Coordinating traffic signals for bicycle progression. Transp. Res. Rec. J. Transp. Res. Board. **1705**, 85–92 (2000)
24. Nustats, J.Z., Geostats, J.W., Zmud, J.: Identifying the Correlates of Trip Misreporting—Results from the California Statewide Household Travel Survey GPS Study (2003)
25. Montini, L., Prost, S., Schrammel, J., Rieser-Schüssler, N., Axhausen, K.W.: Comparison of travel diaries generated from smartphone data and dedicated GPS devices. Transp. Res. Procedia **11**, 227–241 (2015)
26. Gong, H., Chen, C., Bialostozky, E., Lawson, C.T.: A GPS/GIS method for travel mode detection in New York City. Comput. Environ. Urban Syst. **36**, 131–139 (2012)

Adaptive Runtime Middleware: Everything as a Service

Achilleas P. Achilleos[1(✉)], Kyriaki Georgiou[2], Christos Markides[1], Andreas Konstantinidis[1], and George A. Papadopoulos[2]

[1] Frederick University, 7 Y. Frederickou Str., Nicosia, Cyprus
{com.aa,com.mc,com.ca}@frederick.ac.cy
[2] University of Cyprus, 1 University Avenue, Nicosia, Cyprus
{kgeorg04,george}@cs.ucy.ac.cy

Abstract. The Internet of Things applies and has a large impact on a multitude of application domains, such as assistive technologies and smart transportation, by bringing together the physical and virtual worlds. Due to the large scale, the extreme heterogeneity and the dynamics of the IoT there are huge challenges for leveraging the IoT within software applications. The management of devices and the interactions with software services poses, if not, the greatest challenge in IoT, so as to support the development of distributed applications. This paper addresses this challenge by applying the service-oriented architecture paradigm for the dynamic management of IoT devices and for supporting the development of distributed applications. A service-oriented approach is a natural fit for both communication and management of IoT devices, and can be combined logically with software services, since it is currently the paradigm that excels and dominates the virtual domain. Building on our past and ongoing work on middleware platforms, this work reviews middleware solutions and proposes a service-oriented middleware platform to face IoT heterogeneity, the interactive functionality of IoT and promote modular-based development to scale as well as provide flexibility in the development of IoT-based distributed applications.

Keywords: Middleware · IoT · Services · Mobile devices · Distributed applications

1 Introduction

During the last decade, key trends have been observed in the world of embedded devices, which refer mainly to miniaturization, increased computation, cheaper hardware and the shift of software approaches towards service-oriented integration in the Internet of Things (IoT). On the basis of the stated-by-many vision for the IoT, the majority of the devices will soon have communication and computation capabilities that they will use to connect, interact, and cooperate with their surrounding environment [1, 2], including other devices and services. Business-oriented complex distributed applications are being developed on the basis of composition and collaboration among diverse services, in many cases across different vendors.

The Internet of Services (IoS) vision [3] assumes this on a large scale, where services reside in different layers of the enterprise, IT networks, or even running directly on

N.T. Nguyen et al. (Eds.): ICCCI 2017, Part I, LNAI 10448, pp. 484–494, 2017.
DOI: 10.1007/978-3-319-67074-4_47

devices and machines of the company [4]. As the Internet proved its merit both for content and services, we are now facing a trend where service-based information systems blur the border between the physical and virtual worlds, offering a fertile ground for a new breed of real-world aware distributed applications. Therefore, for the success of the IoT, future research, vision and business ecosystems require a merge between cloud computing and the IoT, by enabling a model of "Everything as a Service".

Such a model will deliver an IoT paradigm, where applications rely on the cooperation between heterogeneous devices and software applications, all of which are offered as dynamic web services. Thus, functions such as dynamic discovery, query, selection, and provisioning of web services will be needed for facilitating access and interaction between real-world objects (i.e. devices) and virtual objects (i.e. software services) [4]. The future Internet will provide the capability for embedded heterogeneous real-world entities, similar to virtual entities, to offer their functionalities (e.g. provisioning of sensor data) as RESTful/Web APIs [6]. This will enable virtual entities (i.e. enterprise services) to interact with real-world entities since both will be offered as services in a realisation of the "Everything as a Service" model.

The added value brought by real-world services, (i.e. services) provided by embedded systems that are linked to the physical world, is the increased efficiency of the decision making process due to the fact that they offer real-time data about the world. Thus, the critical issue about such a model is that embedded heterogeneous devices will be able to offer their functionality as web services, which can be used by applications, other services, or even other devices. In this case, device drivers will not be needed anymore and a new level of efficiency will be achieved as web service clients can be generated dynamically at runtime [4]. This will result in a mashup of services where horizontal collaboration between devices will be possible, as well as vertical collaboration of devices with software services and enterprise applications that provide correspondingly interaction capabilities with people [5].

In related work [4, 7], the key challenges continue to be open and need to be addressed [8], such as providing topology dynamics, high scalability and overcoming heterogeneity in such a dynamic IoT environment. In fact, such a highly dynamic environment is also further augmented by the fact that peoples' needs evolve over time, so a scalable and reliable IoT environment needs to be designed with inherent built-in modularity, flexibility and a variety of components in order to meet diverse individual situations and to remain attractive to end-users over time. The following list outlines the key unresolved IoT challenges [8]:

- **Heterogeneity:** Sensors and actuators are the main actors in an IoT environment, where due to the highly heterogeneous nature of IoT devices, enabling interoperability is a complex task.
- **Scalability:** To accurately represent the real world, a sensing/actuating task will more often require the cooperation and coordination of numerous things.
- **Flexibility:** Different configurations may be required for different situations.

A service-oriented approach that follows the concepts of the Internet can provide a solution to the above mentioned challenges. Web technologies provide the base on which a service-oriented approach for the IoT can be formulated to properly address these restrictions. This enables different devices and software components to work together by ex-posing their functionalities to others as web services. Services are the entities that enable users to access the capabilities through pre-defined interfaces in accordance with the policies and constraints, which are part of the description of that service [10]. Web services are platform-independent and can be accessed through the Internet. The original contribution of the web was as a content-provisioning medium. Today, the key role of the web is to act as a facilitator in service outsourcing [2]. This role enables businesses to collaborate dynamically, thus reducing overheads. Therefore, the service deployment model can be applied to any component, physical or virtual, so as to make it available as a service [2, 9].

In the IoT, there is huge heterogeneity in both the communication technologies, and the system level technologies. Therefore, apart from service-oriented computing, a middleware system can support heterogeneity of both communication and system-level technologies that are diverse and many in the world of the IoT. In general, a middleware abstracts the complexities of the system or hardware, allowing the application developer to focus all his effort on the task to be solved [10]. In fact, a middleware system offers a software layer between applications, the operating system and the network communications layers. Based on the above, it is evident that a middleware system can offer an abstraction layer for handling the complexities of the IoT and addressing the challenges faced in developing such applications.

This work aims to support the development of dynamic, flexible and distributed IoT applications, by combining service-oriented computing and middleware technologies. The proposed service-oriented system, coined Adaptive Runtime Middleware (ARM), allows addressing heterogeneity, scalability and flexibility issues. The RESTful architectural pattern is adopted, which enables upon deployment of smart devices (e.g. sensors, actuators), to discover these devices, detect their capabilities, regenerate and redeploy the middleware injecting new RESTful service interfaces (i.e. APIs).

ARM's key capability is the annotation-driven runtime code generation that drives dynamic injection of device capabilities in the form of new service interfaces. The ARM's self-adaptive nature enables interoperability amongst heterogeneous devices, automates device discovery and management, and exposes these devices as services. This offers simplicity for the developer, without introducing additional technologies that increase the already profound complexity in the IoT. In fact, the developer will only need to base its client application implementation on the generated documentation, which describes the generated services that enable management of the IoT devices.

2 Related Work

2.1 Context-Aware Middleware

Several approaches and research work has been performed for the development of middleware systems in different research domains. The Cooltown project [11] supports wireless mobile devices in interacting with a web-enabled environment by assigning URLs to devices, people and things as a web-presence identifier – providing therefore a "rich" interface to the entity. Middleware systems include the Gaia [12] that aims to provide a distributed functionality similar to an operating system, the MiddleWhere [13], which provides enhanced and enriched location information to applications by utilizing a number of location sensing techniques based on a location model, and the MobiPADS [14] that targets mobile environments and its services are provided through various migrated entities from different MobiPADS environments.

A context-aware middleware is also developed in the MUSIC EU project [15], which is a comprehensive open-source software development framework. MUSIC is an ubiquitous OSGi-based context-management middleware system for developing adaptive applications and services for ubiquitous environments.

2.2 Middleware for the Internet of Things

Several middleware IoT architectures and frameworks have been proposed, aiming for a more usable connection among, often, complex and already existing applications that were not originally designed to be connected. The essence of the IoT is making it possible for just about anything to be connected and communicate data over a network, where the middleware framework is part of the architecture thus enabling that connectivity among heterogeneous devices and software services.

An example of a scalable and modular architecture that integrates various components and technologies is openHAB [16]. OpenHAB is an open-source, agnostic automation software with an active community, which encompasses different home automation systems and technologies under the same umbrella of a single solution, enabling the user to define the interaction of systems and devices through automation rules and uniform user interfaces. It is also OSGi-based, and provides APIs for integration with other systems, where REST API is used for remote communication.

OpenIoT is an open-source middleware for connecting cloud sensors and collecting information, extending the IoT solution and exploring efficient ways to use and manage cloud environments [17, 18]. Through an adaptive middleware framework, which is deployed on the basis of one or more distributed nodes, data are collected, filtered, combined and semantically annotated from virtual sensors or physical devices. The proposed middleware does not support though access via service interfaces.

2.3 Service-Oriented Middleware

The service-oriented design paradigm deals with the implementation of software or applications in the form of services by following the concepts and ideas of service-oriented computing (SOC). SOC benefits, such as technology neutrality, loose coupling, service reusability, service composability, and service discoverability [19], can be also beneficial to IoT applications. However, IoT's heterogeneity, scalability and flexibility make service discovery, deployment and composition challenging.

The Hydra EU research project set out to develop a middleware for Networked Embedded Systems. The Hydra middleware allows developers to incorporate heterogeneous physical devices into their applications by offering easy-to-use web service interfaces for controlling any type of physical device irrespective of its network technology such as Bluetooth, RF, ZigBee, RFID, WiFi, etc. As stated in [20], the software middleware is based on Service-Oriented Architecture (SOA), which means that the communication occurs transparently between the lower layers. The aim of the middleware, coined LinkSmart, was to support diverse and heterogeneous connected devices, which enable developers to implement applications that depend on and adapt to context information [21]. *Services are defined statically* in the proposed middleware.

CHOReOS [22] is a service-oriented middleware that enables large scale choreographies of adaptable and heterogeneous services in IoT. It aims to address scalability, interoperability, and adaptability issues via *static service interfaces*. The SenseWrap service-oriented middleware combines Zeroconf protocols with hardware abstraction using virtual sensors [23]. A virtual sensor provides transparent discovery of resources, through the use of Zeroconf protocols, which applications can use to discover sensor-hosted services. SenseWrap also provides a standardized communication interface to hide the sensor-specific details from the applications.

3 The Adaptive Runtime Middleware (ARM)

3.1 Our Contribution

This paper builds on our research work on middleware systems and in research projects such as MUSIC, AsTeRICS and Prosperity4All, so as to design and develop an adaptive middleware system for the IoT. Such a system will have the ability when a smart module (e.g., sensor, actuator) is installed, to discover it, detect its capabilities, regenerate and re-configure the middleware. The middleware supports annotation-driven runtime code generation of device capabilities in the form of *dynamic services*. The key aspect is simplicity for the developer, without introducing additional technologies, IDEs and platforms. The developer can use the generated service interfaces to manage devices, and thus create cross-platform distributed applications (e.g. Android, iOS, HTML5).

3.2 ARM Architecture

The proposed middleware takes advantage of the principles of RESTful architectural pattern and exposes devices as services. The functionalities of each device (e.g., Smart Light – turn light on, dim light) are implemented as annotated Java functions available within each device-specific OSGi component. The key idea is that OSGi components correspond to IoT devices, which can be accessed and managed using RESTful interfaces. In addition, there are two main services implemented as OSGi components, which refer to the middleware core functionality and the REST server that hosts the device resources. The middleware core functionality detects the device capabilities via the annotations in the OSGi component that is installed and allows generating the service interfaces that the REST server component exposes, which correspond to the functionalities of the installed device.

The smallest components in the architecture are the individual OSGi bundles. Each one of these bundles implements the device capabilities, which could be as simple as turning a light on/off, or could be as complex as the interactions between multiple actuators and sensors. The middleware core functionality detects the capabilities of newly installed bundles, thus generating the RESTful interfaces that expose and enable access to these capabilities.

The communication between the components of the proposed adaptive runtime middleware system is illustrated in Fig. 1, where the architecture of the middleware system. The middleware bundle contains the application server that enables communication with the installed bundles via the generated service interfaces. Furthermore, the REST-based architecture enables to access devices over the network in distributed end-user locations (e.g., home, office). The developer is able to develop client applications that make use and even allow interaction between devices in distributed locations, since the service interfaces can be accessed seamlessly via the middleware system available in these locations.

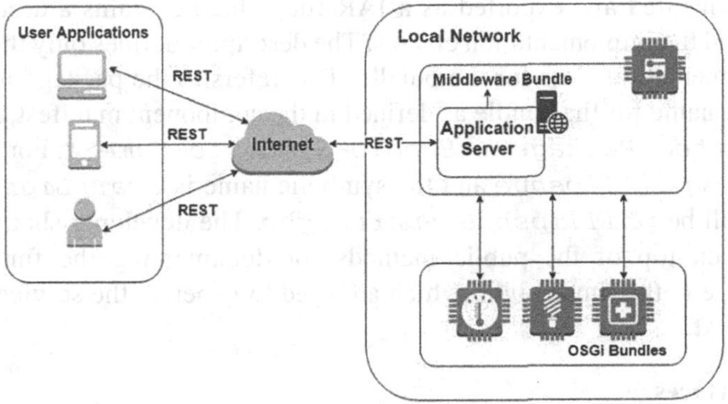

Fig. 1. ARM middleware architecture.

3.3 ARM Implementation

The dynamic middleware is realized as an OSGi bundle, which utilizes the benefits of the OSGi specification for enabling the modularity and scalability of the system. The implemented middleware is built on top of the OSGi Equinox framework, used also in the Eclipse IDE, which is actually an implementation of the OSGi specification. In addition, the REST architecture satisfies and offers solutions in terms of the flexibility needed in an IoT environment. For the implementation of the REST OSGi bundle, the Java API for RESTful Web Services (JAX-RS) specification and its analogous Jersey implementation were used. The OSGi-JAX-RS Connector (i.e., Staudacher) was used, since it packages the Jersey implementation in the form of a bundle and thus integrating consistently the Jersey and OSGi frameworks.

Apart from the core bundles, the middleware and REST server, each device or software service can be defined in a separate OSGi bundle. For instance, a WiFi smart socket can be implemented as an OSGi bundle. This approach offers many advantages since it enables above all flexibility, heterogeneity and scalability as new devices and software services can be supported. The requirement is that developers create a new bundle that enables communication to the device or software service.

Java annotations are syntactic metadata that can be added in the code. Hence, when a new bundle is installed, the middleware detects and starts the component, parses the annotations of public methods and generates the service interfaces that enable direct access to the new resource. The middleware will also generate the documentation for the service, based on the annotations of each public method defined in the bundle. These annotations define the functionality of the bundle, the signature of public methods including the input and output parameters of the method. This mechanism is exploited to enable runtime code generation of the service interfaces.

Descriptor and Annotations

The developer of each IoT device bundle needs to follow a set of guidelines, in order to utilize the adaptive runtime functionality of the middleware. Each bundle can be implemented and exported as a JAR file, which contains a descriptor (i.e. XML file) and the implementation classes. The descriptor defines only the full name of the implementation class for the bundle. This refers to the package followed by the symbolic name for the bundle as defined in the component manifest, in the form of: "`<Exported-Package>.<Bundle-Symbolic-Name>`". For instance, if the package is `phillipshue` and the symbolic name is `SmartLight`, then the full name will be `phillipshue.SmartLight`. The developer should use Java annotations on top of the public methods for documenting the functionalities provided by, e.g. the SmartLight, which are used to generate the service interfaces as defined next.

Service Interfaces

The generated service interfaces need to be consistent and adhere to a simple resource path definition logic, which enables developers of client applications to easily access, and learn how to invoke and thus make use of device functionalities. Table 1 presents the generic definitions for the service interfaces that provide access to device or software services. These refer to the resource paths automatically generated by the middleware.

Table 1. Middleware path hierarchy for generated service interfaces.

Path	Description
<baseURL>	Lists information on the available bundles, including description and interface definition. The *baseURL* represents the service interface of the middleware
<baseURL>/<BundleName>	Provides information on a specific bundle. *BundleName* is the middleware name for the bundle. The *BundleName* can be retrieved by invoking the *baseURL*
<baseURL>/<BundleName>/<MethodName>	Invokes functionality by the *MethodName*, which is appended after the *BundleName* (no parameters)
<baseURL>/<BundleName>/<MethodName>/<Parameter-Data>	Invokes functionality by the *MethodName*, which is appended after the *BundleName* (accepts parameters)
<baseURL>/<BundleName>/<MethodName>/def	Presents information related to the functionality offered by this method of the specified bundle

4 Smart Light Use Case Demonstrator

The use case demonstrator introduced in this section presents the installation of the bundle, as well as accessing, communicating and controlling the Philips Hue smart light. In this use case scenario, an HTML5 client is implemented and used for demonstrating the middleware capabilities. The bundle implementation offers access to four device capabilities: (1) turn light on, (2) turn light off, (3) dim light and (4) set light level. First the bundle descriptor needs to be defined by the bundle developer as follows:

```
<bundle-definition>
  <fullname>
    osgi_SmartPhilipsHUELight.SmartPhilipsHUELight
  </fullname>
</bundle-definition>
```

Figure 2, presents a fraction of the code that showcases how annotations are defined for the "turn light on" method implementation of the Phillips Hue. The next step involves exporting the bundle. The middleware will then parse the descriptor containing the bundle's full name. If the bundle is not already installed, the middleware will automatically install and start it. Using reflection, the middleware detects all device capabilities, re-configures the middleware via runtime code generation and injects/publishes the discovered functionalities as RESTful service interfaces.

```
package osgi_SmartPhilipsHUELight;

import osgi_annotations.interfaces.*;

@ClassDescription(value = "SmartPhilipsHUELight is an OSGi bundle for communicating with a Philips HUE Light. "
        + "This OSGi bundle implements four basic functionalities for interacting with the smart light: "
        + "\t* Turn Light On\t* Turn Light Off\t* Dim Light\t* Set Light Level")

public class SmartPhilipsHUELight {

    @MethodDescription(value = "TurnLightOn method turns the light On. "
        + "Returns true if the light is turned On, otherwise false.")

    public boolean TurnLightOn() {
        boolean lightIsOn = InternalOperationTurnLightOn();
        return lightIsOn;
    }
}
```

Fig. 2. Code snippet of the Phillips Hue implementation class.

Figure 3 showcases the resource paths for invoking the device capabilities. The developer of the client application is now able to invoke the base URL, which will return the description of the bundles currently installed and the paths for retrieving details on how to invoke each device capabilities. Figure 4 presents the currently installed Phillips

Hue bundle and exposes the description and paths for invoking the implemented functionalities of the device. An HTML5 client has been implemented, which allows showcasing the use of the dynamically generated service interfaces that enable accessing and controlling the Phillips Hue Smart Light device (demo video[1]).

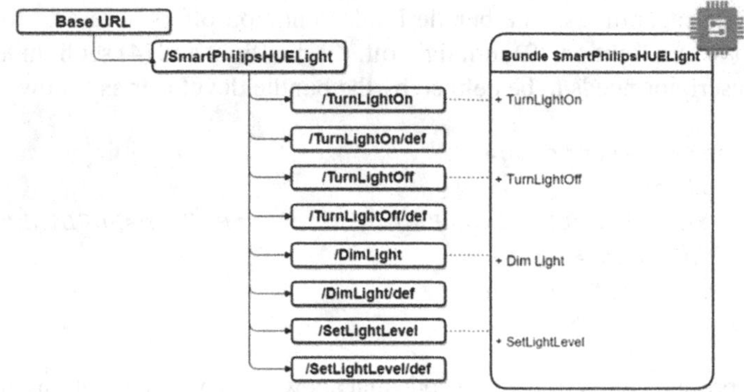

Fig. 3. Generated RESTful service interfaces for the Phillips Hue Smart Light.

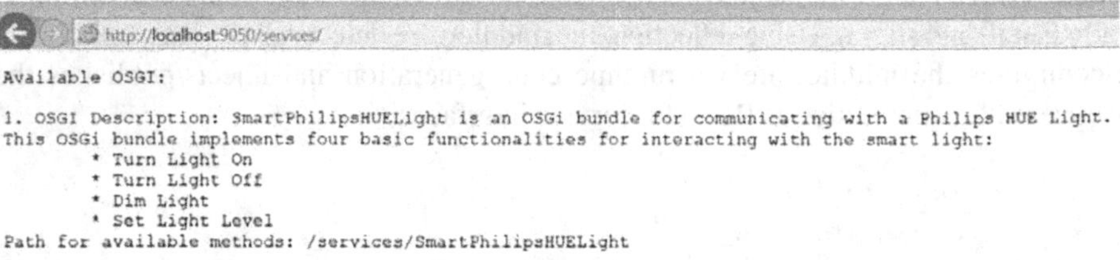

Fig. 4. Generated RESTful service interfaces for the Phillips Hue Smart Light.

5 Conclusions and Future Work

The research work presented in this paper aims to provide an Adaptive Runtime Middleware (ARM). The proposed middleware allows utilizing the benefits of the OSGi framework, the Java reflection mechanism and the RESTful architectural pattern in order to provide solutions to the IoT challenges of scalability, heterogeneity and flexibility. The presented use case scenario demonstrates the adaptive capabilities of the proposed ARM. The architecture of ARM enables to address the aforesaid IoT issues, since for each IoT device or Cloud service a corresponding bundle can be developed following the guidelines presented in this work. The bundle can be then exported and the ARM - can install, start and parse the bundle so as to generate at runtime the required service interfaces. Future research work aims to extend the middleware capabilities so as enable dynamic generation of Server-Sent Events (SSE) for handling sensor devices data as soon as they become available.

[1] Available at: https://www.youtube.com/watch?v=NQ0tzv5Ob48&sns=em.

Finally, a rules engine will be implemented for defining dependencies between device and/or software services, for example motion detected → turn on camera.

References

1. Fleisch, E., Mattern F.: Das Internet der Dinge: Ubiquitous Computing und RFID in der Praxis: VTAH. Springer, Berlin (2005)
2. Zeshan, F., et al.: Service discovery framework for distributed embedded real-time systems. In: Ghani, I., Kadir, W., Ahmad, M. (eds.) Handbook of Research on Emerging Advancements and Technologies in Software Engineering, pp. 126–147. IGI Global, Hershey (2014). doi:10.4018/978-1-4666-6026-7.ch007
3. Lizcano, D., Jiménez, M., Soriano, J., Cantera, J.M., Reyes, M., Hierro, J.J., Garijo, F., Tsouroulas, N.: Leveraging the upcoming internet of services through an open user-service front-end framework. In: Mähönen, P., Pohl, K., Priol, T. (eds.) ServiceWave 2008. LNCS, vol. 5377, pp. 147–158. Springer, Heidelberg (2008). doi:10.1007/978-3-540-89897-9_13
4. Guinard, D., et al.: Interacting with the SOA-based internet of things: discovery, query, selection, and on-demand provisioning of web services. IEEE Trans. Serv. Comput. **3**(3), 223–235 (2010)
5. Issarny, V., Bouloukakis, G., Georgantas, N., Billet, B.: Revisiting service-oriented architecture for the IoT: a middleware perspective. In: Sheng, Q.Z., Stroulia, E., Tata, S., Bhiri, S. (eds.) ICSOC 2016. LNCS, vol. 9936, pp. 3–17. Springer, Cham (2016). doi: 10.1007/978-3-319-46295-0_1
6. Guinard, D., Trifa, V.: Towards the web of things: web mashups for embedded devices. In: Proceedings of Workshop Mashups, Enterprise Mashups and Lightweight Composition on the Web (MEM 2009) (2009)
7. Teixeira, T., Hachem, S., Issarny, V., Georgantas, N.: Service oriented middleware for the internet of things: a perspective. In: Abramowicz, W., Llorente, I.M., Surridge, M., Zisman, A., Vayssière, J. (eds.) ServiceWave 2011. LNCS, vol. 6994, pp. 220–229. Springer, Heidelberg (2011). doi:10.1007/978-3-642-24755-2_21
8. Paganelli, F., et al.: A DHT-based discovery service for the Internet of Things. J. Comput. Netw. Commun. **2012**, 11 (2012). doi:10.1155/2012/107041. Article ID 107041
9. Zhu, Y., Xiao-hua, M.: A Framework for service discovery in pervasive computing. In: 2nd International Conference on Information Engineering and Computer Science (ICIECS) (2010)
10. Neely, S., et al.: Adaptive middleware for autonomic systems. Ann. Técommun. **61**(9–10), 1099–1118 (2006)
11. Barton, J., Kindberg, T.: The Cooltown user experience (2001)
12. Román, M., et al.: Gaia: a middleware infrastructure to enable active spaces. IEEE Pervasive Comput. **1**(4), 74–83 (2002)
13. Ranganathan, A., Al-Muhtadi, J., Chetan, S., Campbell, R., Mickunas, M.D.: MiddleWhere: a middleware for location awareness in ubiquitous computing applications. In: Jacobsen, H.-A. (ed.) Middleware 2004. LNCS, vol. 3231, pp. 397–416. Springer, Heidelberg (2004). doi: 10.1007/978-3-540-30229-2_21
14. Chan, A., Chuang, S.-N.: Mobipads: a reflective middleware for context-aware mobile computing. IEEE Trans. Softw. Eng. **29**(12), 1072–1085 (2003/12)
15. Paspallis, N., et al.: Developing self-adaptive mobile applications and services with separation-of-concerns'. In: At Your Service: Service-Oriented Computing from an EU Perspective, chap. 6, pp. 129–158. MIT Press (2009)

16. openHAB (2017). http://www.openhab.org/. Accessed 24 Jan 2017
17. Soldatos, J., et al.: OpenIoT: open source internet-of-things in the cloud. In: Podnar Žarko, I., Pripužić, K., Serrano, M. (eds.) Interoperability and Open-Source Solutions for the Internet of Things. LNCS, vol. 9001, pp. 13–25. Springer, Cham (2015). doi: 10.1007/978-3-319-16546-2_3
18. OpenIoT Consortium: OpenIoT—Open Source cloud solution for the Internet of Things (2016). http://www.openiot.eu/. Accessed 24 Jan 2017
19. Papazoglou, M.: Service-oriented computing: concepts, characteristics and directions. In: Proceedings of 4th International Conference on Web Information System Engineering (WISE 2003), pp. 3–12 (2003)
20. Sarnovsky, M., et al.: First demonstrator of hydra middleware architecture for building automation. In: Snášel, V. (ed.) Proceedings of Znalosti Conference, pp. 204–214. FIIT STU Bratislava, Slovakia (2008)
21. Souza, M.C., et al.: A novel smart home application using an Internet of Things middleware. In: Smart SysTech 2013, Erlangen/Nuremberg, Germany, pp. 1–7 (2013)
22. Hamida, A.B., Kon, F., et al.: Integrated CHOReOS middleware—enabling large-scale, QoS-aware adaptive choreographies (2013)
23. Evensen, P., Meling, H.: SenseWrap: a service oriented middleware with sensor virtualization and self-configuration. In: International Conference on Intelligent Sensors, Sensor Networks and Information Processing (ISSNIP 2009), Melbourne, VIC, pp. 261–266 (2009). doi: 10.1109/ISSNIP.2009.5416827

Decision Support & Control Systems

Adaptive Neuro Integral Sliding Mode Control on Synchronization of Two Robot Manipulators

Parvaneh Esmaili[1(✉)] and Habibollah Haron[2]

[1] Department of Computer Engineering, Faculty of Engineering, Girne American University, Kyrenia, Cyprus
p.esmaili1984@gmail.com
[2] Department of Computer Science, Faculty of Computing, Universiti Teknologi Malaysia (UTM), 81310 Skudai, Johor, Malaysia
habib@utm.my

Abstract. Designing a new adaptive synchronization controller for multiple robot manipulators is main purpose of this study. But, this synchronization between robots are considered without having direct communication between robots. The adaptive synchronization method is consisted of the integral sliding mode controller improved with adaptive neural network controller. In order to analyze the performance of the proposed method, four different situations are considered. Also, the result are compared with the ANFIS method. The proposed method is guaranteed by Lyapunov stability method.

Keywords: Synchronization controller · Adaptive neural network controller · Integral sliding mode control

1 Introduction

Nowadays, the efficiency and quality of production processes such as assembly, transportation and welding tasks are relied on complex, nonlinear and integrated systems. As for increasing maneuverability, flexibility and manipulability of production processes, multiple robot manipulator systems are replaced with single robot systems. Practically, the satisfactory efficiency is based on the constraints in the multiple robot manipulators system and the environment, coordination and cooperation between them are one of the most important challenges which are studied by researchers.

In coordination of multi robot manipulators, synchronization methods are presented as an alternative. In [6] synchronization strategy is only based on the kinematic and/or dynamic constraints and direct force controller is not investigated in the method.

As far as robustness in the presence of uncertainties in real world is concerned, synchronization method with sliding mode controller (SMC) is used which SMC is a strong method to overcome with uncertainties [11]. In 2011 [9], a low pass filter based on sliding mode control was designed. Approximation ability of neural network turns it out as a significant method to help the systems in the presence of uncertainties in the real world. These uncertainties can be structured and un-structured uncertainties. In [2],

N.T. Nguyen et al. (Eds.): ICCCI 2017, Part I, LNAI 10448, pp. 497–506, 2017.
DOI: 10.1007/978-3-319-67074-4_48

adaptive coordination method is proposed on neural network by ideal back propagation. That can be face with friction and external effect. Siqueira [5] is developed a hybrid method in fully actuator or under actuated coordination. This method is combination of H∞ and adaptive neural network but just can solve actuator nominal dynamics uncertainties. To deal with follower uncertainty and leader's acceleration in the leader - follower method is used an RBF neural network based synchronization algorithm [10]. The neural network tries to solve follower's calibration problems but still have problem in communication failures. In the multiple robotic systems, direct and un-direct communication topologies are proposed. In the direct communication robots are connect physically but in un-direct robots are not connect and used connection ways like wireless networks to communicate between them. But in un-direct topologies the system would be faced with failure connection problem. So, in this work a combination of direct and un-direct communication is used as implicit communication. In this system, robots are not connected each other. They are just connected to the object and the object is used as a communication point. Each robot tries to control its trajectory based on the object (especially middle of the object). If each robot can be balanced with the object, it leads to the balancing with the other robots. In addition, during grasping due to finger contacts may be object be moved un-predictably. This caused catastrophic failures such as dropping object, different end-effectors configuration and change in the originally planned grasp (called as a set point in this work) which caused changes in the position and orientation of the object being grasped [4]. In this paper, in addition to compensate lumped uncertainties, grasping uncertainties also considered to overcome.

This paper is presented into five sections as follows: After description on problem definition in coordination multiple robot manipulators in the introduction part, in Sect. 1 is described problems formulation. Section 2 is described the proposed controller for two robot manipulators which the stability analysis is guaranteed by Lyapunov method. The simulation results reveal the performance of proposed method in compare with ANFIS controller [3] in Sect. 3. The conclusion is given in Sect. 4.

2 Problems Formulation

2.1 Communication Topology in the Multiple Robot Manipulators

In this cooperative system, the end-effectors of each robot are considered as a set points which are shown in Fig. 1. According to this Figure, the end-effectors are fixed on the beam. The middle of the beam is called reference in this work. The position of the set points is related to the reference. Therefore, each end-effector tries to compensate grasping uncertainties of its end-effector with respected to the reference. Design a decentralized synchronization approach by using implicit communication is the purpose of this work. It should be noted that the contact forces between beam and end-effectors are not measured but shadow of forces is considered. For example, when the end-effectors are not completely fixed on the object, contact forces exerted on the robot and the beam. Any change of each end-effector with respected to the beam is considered as grasping uncertainties and the controller aimed to compensate it. Besides,

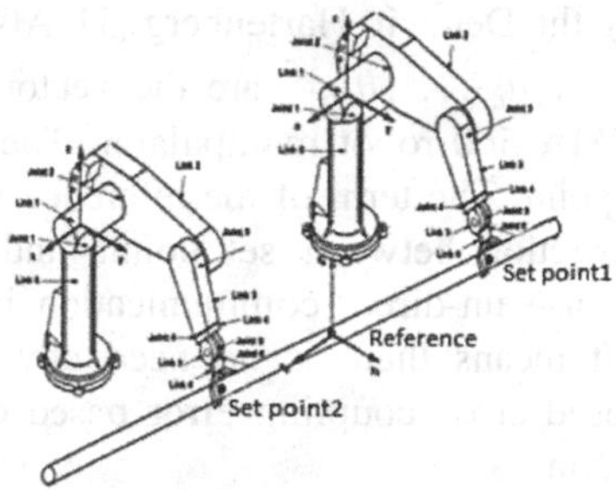

Fig. 1. Two PUMA560 robot manipulators handle a lightweight beam

the controller is faced with parameter uncertainties and external disturbances. The position of the middle of the beam (x_c, y_c, z_c) is played a role as a reference that each robot manipulators (x_i, y_i, z_i), i = 1, 2, are adopted by reference. The grasping uncertainties between robot and reference, which is approximated by $\alpha_i^x, \alpha_i^y, \alpha_i^z$ generates a reaction force f_i to robot i.

$$\alpha_i^x = x_i - x_c, \alpha_i^y = y_i - y_c, \alpha_i^z = z_i - z_c \tag{1}$$

It assume that the reaction force f_i is a positive-definite function $P_i(\alpha_i^x, \alpha_i^y, \alpha_i^z)$. Note that if $\alpha_i^x = \alpha_i^y = \alpha_i^z = 0$, there is no grasping uncertainty between end-effectors and payload which implies that $P_i(\alpha_i^x, \alpha_i^y, \alpha_i^z)$ satisfies the following:

$$f_i = P_i\left(\alpha_i^x, \alpha_i^y, \alpha_i^z\right) = 0 \Leftrightarrow \alpha_i^x = \alpha_i^y = \alpha_i^z = 0 \tag{2}$$

To solve grasping uncertainties problem, by using inverse kinematics matrix, joint angle of robot i and reference are calculated, then grasping uncertainties function $D_{i,j}$ is defined on the joint angle.

$$\begin{aligned} \text{IK} = (x_i, y_i, z_i) = \theta_{i,j=1,2,\ldots 6},\ \text{IK} = (x_c, y_c, z_c) = \theta_{c,j=1,2,\ldots 6}, D_{i,j=1,2,\ldots 6} \\ = \theta_{i,j=1,2,\ldots 6} - \theta_{c,j=1,2,\ldots 6} \end{aligned} \tag{3}$$

2.2 Dynamic Model of the System

Dynamic equation of two PUMA560 robot manipulators by investigating the Newton-Euler method is described as follows.

$$M_i(\theta_i)\ddot{\theta}_i + B_i\left(\theta_i, \dot{\theta}_i\right) + C_i\left(\theta_i, \dot{\theta}_i\right) + G_i(\theta_i) = \tau_i + d_i(t), i = 1, 2 \tag{4}$$

where, M_i is a [6 × 6] mass matrix, B_i is the Corioli-Coefficient which is [6 × 15], C_i is Centrifugal-Coefficient which is [6 × 6] matrix, G_i is Gravity term [6 × 1] matrix, τ_i is Torque [6 × 1] matrix and d_i is lumped system uncertainty caused by friction, backlash, modeling error and external disturbance, etc. The measured parameters of

PUMA 560 are driven using the Denavit–Hartenberg [1]. Also, $\theta_i = [\theta_{i,1}, \theta_{i,2}, \ldots, \theta_{i,6}]$, $\dot{\theta}_i[\dot{\theta}_{i,1}, \dot{\theta}_{i,2}, \ldots, \dot{\theta}_{i,6}]^T$, $\ddot{\theta}_i = [\ddot{\theta}_{i,1}, \ddot{\theta}_{i,2}, \ldots, \ddot{\theta}_{i,6}]^T$ are the vector of the position, velocity and acceleration of the PUMA 560 robot manipulator. The grasping position on the object is considered as set point. The term of the reference is related to the middle of the beam. The direct connection between set points and reference considered as communication which is cause un-direct communication between robots (called as implicit communication). It means there is no need any communication topology between robots. The proposed cross coupling error based on the implicit communication is presented in equation.

$$E_i^*(t) = C_i e_i + \beta_i \int_0^t Se_i^R(t) d\zeta, \quad Se_i^R(t) = D_{i,j=1,2,\ldots 6} \tag{5}$$

where, e_i is the position tracking error of ith robot and e_R is the ideal error refer to the reference $\ddot{\theta}_i \in R^{mn}$, $\ddot{\theta}_i^d \in R^{mn}$ are second derivative of position tracking, desired position tracking vectors, Se_i^R is synchronization errors, respectively. The proposed cross coupling error is designed for tuning parameters of controller with respected to the changes of environment.

2.3 Controller Design

The main objective of this work is to design a new synchronization controller by combination of neural network and integral sliding mode control to guarantee position tracking error $e_i(t)$ and synchronization error $Se_i(t)$ simultaneously in the presence of any uncertainties. This controller called as Adaptive neuro integral sliding mode control (ANNISMC). The structure of the proposed controlling system is shown in Fig. 2. As shown in Fig. 2, the proposed controller consists of two sections. The first section is to compensate lumped uncertainties using integral sliding mode control. To improve the performance of the integral sliding mode control is added a two layers feed forward neural network. To compensate grasping uncertainties is defined an adaptive law based on the proposed cross coupling error.

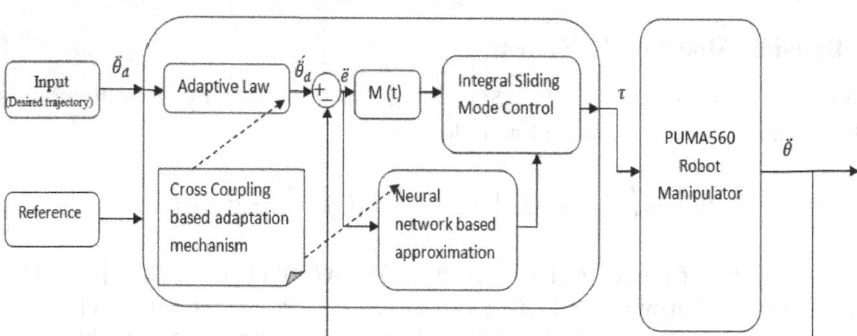

Fig. 2. The structure of the proposed controller

2.3.1 Improved Integral Sliding Mode Control (MISMC)

Integral sliding mode control is a class of sliding mode control that is robust in the entire response of system [7]. To improve the performance of integral sliding mode control, a two layer feed forward neural network is proposed on nonlinear dynamic *f*(*x*) function.

2.3.1.1 Integral Based Sliding Mode Control

In the Integral sliding mode control [7], the control input of dynamic model is defined as follows.

$$u = u_0 + u_1 \tag{6}$$

where, u_0 is the nominal control and u_1 is discontinuous part of the controller to overcome uncertainties. By considering dynamic model of robot manipulator from Eq. (6) and using well-known computed torque method in the ideal situation without uncertainties, the joint torque of robot manipulator turns to u_0 by using joint position.

$$u_0 = \tau_0 = \hat{M}\left(\ddot{\theta}\right)\left[\ddot{\theta}_d + k_v\dot{E}^* + k_pE^*\right] + f(\ddot{\theta}) \tag{7}$$

where, $\hat{M}$ is known mass matrix. By integrating u_0 into the dynamic model of robot manipulator, the error of the dynamic system is defined as

$$\ddot{E}^* + k_v\dot{E}^* + k_pE^* = \hat{M}^{-1}f(\ddot{\theta}) \tag{8}$$

So, by considering Eq. (16), it is obvious even by set the best values of k_v, k_p, the error not only will not be stable but also will not converge to zero. Hence, the integral sliding surface will be integrated with error and we have:

$$s = \dot{E}^* + k_vE^* + k_p\int_0^t E^*(\zeta)d\zeta - k_vE^*(0) - \dot{E}^*(0) \tag{9}$$

By enforcing sufficient joint toque, the $\dot{s}$ will be properly converge to zero. The second part of control input u_1 is discontinuous part which is defined as follows.

$$u_1 = -k_c sat(s) \tag{10}$$

where, $k_c \in R^{n\times n}$ is a positive matrix. Then the control input will be appearing as Eq. (18).

$$u = \hat{M}\left(\ddot{\theta}\right)\left[\ddot{\theta}_d + k_v\dot{E}^* + k_pE^*\right] + f\left(\ddot{\theta}\right) - k_c sat(s) \tag{11}$$

So, the overall structure of ISMC is presented in Fig. 3.

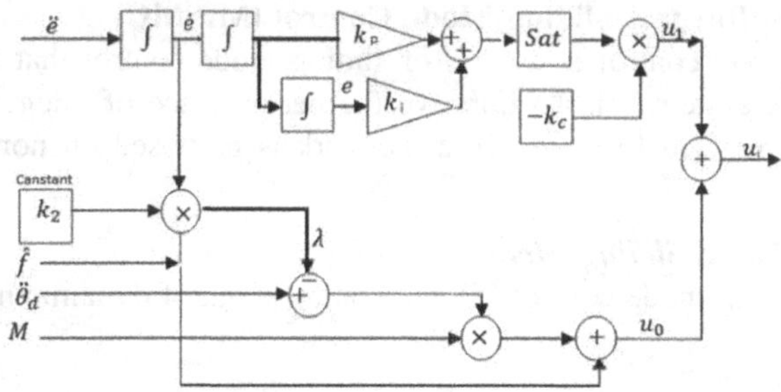

Fig. 3. The structure of integral sliding mode control

2.3.1.2 Feed Forward Neural Network Control on Integral Sliding Mode Control

As mentioned before a two layer feed forward neural network model is proposed to compensate uncertainties of nonlinear $f(\theta_i)$ of dynamic model. A two-layer feed-forward neural network model with 6 inputs neurons, 6 output neurons and 10 hidden nodes in hidden layer is shown in Fig. 4. The 6 inputs of neural network are related joint angles of the 6-DOF PUMA 560 robot manipulator and the 6 outputs are related to estimation of $f(\theta_i)$ of dynamic model. The output of the neural network is considered as follows.

$$z_i = \sum\nolimits_{j=1}^{10} \left[w_{ij}\sigma\left(\sum\nolimits_{k=1}^{6} v_{ij}y_k + \theta_{vj}\right) + \theta_{wj}\right], i = 1, 2, \ldots, 6. \tag{12}$$

where, $\sigma(\cdot)$ is activation function. The interconnection weights between hidden layer to output layer is denoted by w_{ij} and the interconnection weights between input layer to hidden layer is denoted by v_{ij}. θ_{vj} and θ_{wj} are denoted as bias weights. The collecting form of the neural network can write such as Eq. (21). The values of V^T and W^T are related to v_{ij} and w_{ij}. The activation and pureline functions are used as activation function. The back propagation algorithm is considered as learning algorithm in the neural network model.

$$Z = W^T\sigma(V^Ty) \tag{13}$$

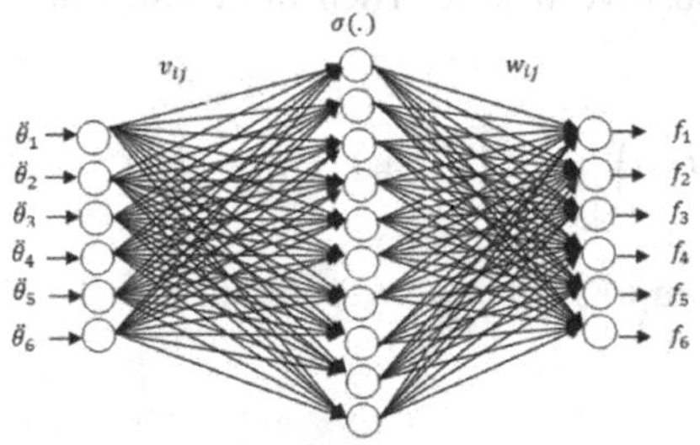

Fig. 4. The structure of neural network model

For any $x \in R^n$ the vector of activation function is defined by $\sigma(x) = [\sigma(x_1), \sigma(x_2), \ldots, \sigma(x_n)]^T$. The bias weights is consisted the first column of weights matrices and need to be argument by replacing 1 as the vectors y and $\sigma(\cdot)$ for example $y = [1 y_1 y_2 \ldots y_n]^T$. Based on the approximation property of neural network, a continue function of n variables is denoted by h(y).

$$h(y) = W^T \sigma\left(V^T y\right) + \in (y) \tag{14}$$

where, $\in$ denote as neural network approximation error. This error is satisfying by $\| \in (y) \| < \in_N$.

$$\hat{f}(\theta_i) = h(y) \tag{15}$$

Based on the (19) and (23) the control input turn as (24).

$$u_{smc} = M(\theta_i)\left(\ddot{\theta}_{d_i} - K\dot{E}^*(t)\right) + h(y) - k_c sat(s(t)) \tag{16}$$

2.3.1.3 Adaptive Synchronous Law

To overcome with grasping uncertainties improved integral sliding mode is not completely enough. So, an adaptive law based on the proposed cross coupling error is added to the controlling system to achieve synchronization manner between robot manipulators. The output of the adaptive law as ideal desired input is shown as $\ddot{\theta}'_{d_l}$.

$$\ddot{\theta}'_{d_l} = \ddot{\theta}_{di} \times £, \quad £ = E_i^*(t) \tag{17}$$

Based on the (24) and (25) the control input turn as (26).

$$u_{ANNISMC} = M\left(\theta'_l\right)\left(\ddot{\theta}'_{d_l} - K\dot{E}^*(t)\right) + h(y) - k_c sat(s(t)) \tag{18}$$

So, the output of the controller is shown at (27).

$$\tilde{\theta}_i(t) = M(\theta_i)\left(u_{ANNISMC} + \hat{f}(\theta_i)\right) \tag{19}$$

By considering Lyapunov stability method, bounds of parameters of sliding mode controller to stabilize the proposed system can be determined. Positive definite Lyapunov function is defined as follows. In terms of stability, the Lyapunov function V(x) should be positive and the derivative of the Lyapunov function should be negative.

$$V(x) = \frac{1}{2} s(t)^T s(t), \quad \dot{V} = s\dot{s} < 0 \tag{20}$$

$$s = \dot{E}^* + k_v E^* + k_p \int_0^t E^*(\zeta) d\zeta - k_v E^*(0) - \dot{E}^*(0) \tag{21}$$

$$u_{ANNISMC} = M(\theta_l')\left(\ddot{\theta}_{d_l}' - K\dot{E}^*(t)\right) + h(y) - k_c sat(s(t)) \quad (22)$$

$$\dot{V} = s(-M^{-1}\hat{f} - M^{-1}k_c\text{sat}(s(t))) \quad (23)$$

As mentioned before, $\text{f} = \hat{f} + \text{h.f}$, we have:

$$\text{f} - \hat{f} = \text{h.f} > 0,\ \hat{f} - \text{f} < 0,\ \dot{V} = \text{s}(\hat{\text{f}} - \text{f} - \text{K}_\text{c}\ \text{sat}(\text{s}(\text{t})) < -\text{K}|\text{s}| < 0 \quad (24)$$

Therefore, the stability of system is guaranteed by Lyapunov stability method.

3 Result and Discussion

The proposed synchronization scheme is performed to solve uncertainty on the trajectory of set point in compared with the trajectory of grasping object. The proposed scheme tries to compensate for the difference between trajectories of set point with respect to the trajectory of the grasping object.

Dynamic parameters of PUMA 560 robot manipulator are derived by [1]. The object is a rigid lightweight beam. The initial pose of joints of robot manipulator is θ_i = [0, 0, 0, 0, 0, 0] and the final pose is θ_i = [0.7854, 1.5708, −1.5708, 0, 0, 0]. To compensate the nonlinear dynamic f function in two robot manipulator, a feed forward neural network model is investigated which the features of neural network are considered in Table 1. The parameters and results of neural network model are shown in Table 2.

Table 1. Results of neural network model

Features in neural model	Value
Number of input nodes	6
Number of hidden nodes	10
Number of output nodes	6
Number of epochs	1000

Table 2. Results of neural network model

Sample no.	Training sample no.	Validation sample no.	Epoch no.	Mean squared error (mse)	Gradient	Regression
474	71	71	9	1.9641e−11	1e−11	9.99999e−1

In the PI-type sliding surface, the k_p and k_i parameters are set to 2 and 0.0001, respectively. To analyze the proposed approach two 6 DOF PUMA560 robot manipulators are modeled by [1]. In this section, several cases are performed to validate the proposed approach on several cases.

Case 1: Without uncertainties
Case 2: With Grasping uncertainties
Case 3: With lumped uncertainties
Case 4: With both uncertainties

Both methods are compared in four cases by T-Test analysis as shown as in Table 3. In each case, three measurement factors are investigated named as Coupled error (CE), Position tracking error (PE) and Synchronization error (SE). In Table 3, P's value is shown that comparison between two methods is significantly different. The T-test value is shown the value of difference between two methods. The minus T-test value means that second parameter of analysis have better difference. In the average, the proposed method has 3% better performance in compare with ANFIS method for handling object by two robot manipulators.

Table 3. Comparison results between two methods by T-Test analysis

ANFIS, ANNISMC		Mean	Variance	Pearson correlation	T-Test	P (T ≤ t) two tail
Case 1	CE	0.008571941	0.001118439	0.998158852	−3.893600744	0.000293346
	PE	0.008571941	0.001118439	0.998158852	−30.68769608	4.22121E−34
	SE	2.55492E−14	1.25827E−23	0.960649108	−1.057870232	0.2951946E−4
Case 2	CE	0.001082149	0.001094133	0.993674988	−1.079891077	0.2852358 E−4
	PE	−0.00882168	0.00109470	0.993682744	−1.100280192	0.2769245 E−4
	SE	0.003768297	0.000157102	0.940622938	−0.544449844	0.5885998 E−4
Case 3	CE	0.024471564	0.013261823	0.999614105	−0.654899661	0.5155018 E−4
	PE	0.014487355	0.013262095	0.999612519	−0.674642388	0.5030653 E−4
	SE	0.001375477	4.36781E−05	0.969481172	−2.730848481	0.008291
Case 4	CE	0.019334449	0.017465581	0.999591466	−1.01838298	0.313392 E−4
	PE	0.00945537	0.01746755	0.999591513	−1.03090541	0.3075458 E−5
	SE	−0.0061414	0.004295757	0.996515413	−1.111072978	0.2718546 E−4

4 Conclusion

Achieving synchronous manner between multiple robot manipulators with minimum communication between robots is an important issue in this study. This manner obtained by a robust, decentralized and low cost intelligent based controlling method which is guaranteed by Lyapunov stability method. In the future work the focus is on optimal adaptation mechanism in the intelligent part to optimize the performance of the system.

References

1. Armstrong, B., Khatib, O., Burdick, J.:. The explicit dynamic model and inertial parameters of the PUMA 566 arm. In: IEEE, pp. 510–518 (1986)
2. Bouteraa, Y.: Adaptive backstepping synchronization for networked Lagrangian systems. Int. J. Comput. Appl. **42**(12), 1–8 (2012)

3. Esmaili, P., Haron, H.: Intelligent synchronization tool using ANFIS for multi robot manipulators. In: Fujita, H. (ed.) New Trends in Software Methodologies, Tools and Techniques, pp. 37–50. IOS Press, Amsterdam (2014). doi:10.3233/978-1-61499-434-3-37s
4. Kim, J., Iwamoto, K., Kuffner, J.J., Ota, Y., Pollard, N.S.: Physically based grasp quality evaluation under pose uncertainity. IEEE Trans. Rob. **29**, 1424–1439 (2013)
5. Siqueira, A.A.G., Terra, M.H.: Neural network-based control for fully actuated and underactuated cooperative manipulators. Control Eng. Pract. **17**(3), 418–425 (2009)
6. Sun, D.: Synchronization and Control of Multiagent Systems. CRC Press, Boca Raton (2010)
7. Utkin, V., Shi, J.S.J.: Integral sliding mode in systems operating under uncertainty conditions. In: Proceedings of 35th IEEE Conference on Decision and Control, vol. 4, pp. 1–6 (1996)
8. Zeinali, M., Notash, L.: Adaptive sliding mode control with uncertainty estimator for robot manipulators. Mech. Mach. Theory **45**(1), 80–90 (2010)
9. Zhao, D., Li, C., Zhu, Q.: Low-pass-filter-based position synchronization sliding mode control for multiple robotic manipulator systems. Proc. Inst. Mech. Eng. Part I J. Syst. Control Eng. **225**(8), 1136–1148 (2011)
10. Zhao, D., Zhu, Q.: Position synchronized control of multiple robotic manipulators based on integral sliding mode. Int. J. Syst. Sci. **45**(3), 556–570 (2014)
11. Zhao, D., Zhu, Q., Li, N., Li, S.: Synchronized control with neuro-agents for leader–follower based multiple robotic manipulators. Neurocomputing **124**, 149–161 (2014)

Ant-Inspired, Invisible-Hand-Controlled Robotic System to Support Rescue Works After Earthquake

Tadeusz Szuba(✉)

Department of Applied Computer Science, AGH University, Cracow, Poland
szuba@agh.edu.pl

Abstract. A bridge/scaffolding system, based on ants-like robots is proposed. Collective Intelligence of the system is derived from Adam Smith's Invisible Hand phenomena (ASIH). Such a bridge system will be air/truck delivered to a particular earthquake location. Next, the bridge-robots, initialized by trained supervisors, will collectively assemble a bridge by means of transporting a chain or scaffolding structure similar to that of bridges made by ants joining their bodies together. Additionally, bridge-robots, task-controlled by supervisors, should also provide the functions of light crane or conveyor belt systems to manipulate heavy debris. Different rescue scenarios have been presented with the help of 3D graphics.

Keywords: Bridges made of bodies of ants · Invisible hand · Collective intelligence · Distributed algorithms · Robots like ants · Self-constructing bridges for emergency applications

1 Introduction

What is critical, is an immediate (in term of hours) bidirectional access of well (i.e. optimally) resourced rescue teams to the location where victims are expected to be trapped. In most cases, heavy equipment is not available and/or applicable because of streets blocked (see Fig. 1) by collapsed buildings, unstable buildings or mountainous environment. Moreover, specialized heavy equipment is required, different to that available on a typical/nearby construction site.

The proposed intelligent, self-assembling bridges should not only perform the transportation functions (see Fig. 7), but also act as a self-assembling scaffolding (ladder) enabling access to locations on upper floors of unstable and/or partially collapsed buildings, where some victims may be trapped (see Fig. 8). The proposed ant-like robots can, to some extent, replace heavy equipment. This is due to their robotic structure, consisting of a strong body with leg-actuators, which allow lifting heavy debris and to extricating victims from debris (see Fig. 9).

Applying the theory of Adam Smith's Invisible Hand (ASIH) [8,9] it is possible to derive a hypothetical structure of ants' social algorithm to build astonishing bridges made of their joined bodies, for example see Fig. 2. They build such

N.T. Nguyen et al. (Eds.): ICCCI 2017, Part I, LNAI 10448, pp. 507–517, 2017.
DOI: 10.1007/978-3-319-67074-4_49

Fig. 1. Post earthquake street. Amatrice, Italy, 2016.

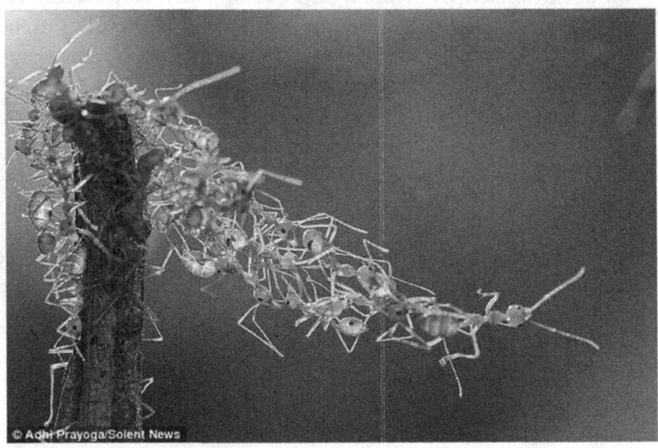

Fig. 2. A scaffolding and the start of an ant bridge construction. This is a great and inspiring example of how scaffolding and nonstandard transportation structure for e.g. a ruined building, can be designed and implemented.

bridges almost spontaneously, without any planning and strength calculations. ASIH theory also explains how this distributed algorithm is calculated on the platform of their primitive brains. This social algorithm is basis for a proposed, ants-inspired, robotic system to support rescue works after earthquake.

2 Controlling a Team of Rescue Robots with the Help of Adam Smith's Invisible Hand Paradigm (ASIH)

Ant Colony Optimization algorithms (ACO) [2] demonstrate that computational problems which humans solve using digital computers and algorithms designed for such a computational platform, can be alternatively solved by quite different algorithms based on another computational platform, i.e. that of social behavior. This computational platform based on interaction of ants' brains should be considered as a kind of analog computer [11]. Similarly, a distributed bridge building algorithm (able to be hosted by ants' brains) based on Adam Smith's Invisible Hand paradigm, can be proposed. In such an algorithm, strength calculations and construction problems are solved on the basis of proper social behavioral rules.

The Invisible Hand is a term used by Adam Smith (Scottish moral philosopher, pioneer of political economy) to describe the **unintended social benefits of individual actions**. The phrase was employed by Smith with respect to income distribution (1759) and production (1776). The exact phrase is used just three times in Smith's writings [8], but has come to capture his notion that **individuals' efforts to pursue their own interest may frequently benefit society more than if their actions were directly intending to benefit society**.

The society is highly divided about the existence and nature of the Invisible Hand [5]. For example such minds as the Nobel Prize-winning economist Joseph E. Stiglitz and Noam Chomsky (an American linguist, philosopher, cognitive scientist, historian, logician, social critic, and political activist) are counted among the critics of this paradigm. John Keynes was generally against it[1].

However, there are also supporters who consider the Invisible Hand not as a myth but as a serious hypothesis with mathematical arguments supporting it [1]. Another Nobel Prize-winning economist Friedrich Hayek supported and interpreted it as a *spontaneous order*. He provided so many pro arguments [4].

We are currently developing a kernel algorithm for a single ant to pursue its own (present) interest for future benefit of the social structure (e.g. the good of the bridge and welfare of the anthill). If a single ant analyzing the present bridge structure in terms of improvement, will pursue it's own interest and will respect interest of other ants in the bridge structure (not to overload them), the emerging bridge will be safe from a strength point of view, and will provide safety limits comparable to those we can find in engineering manuals. We rely in algorithm on Adam Smith's ideas, and works of F. Hayek [4], who developed Smith's thoughts.

A simple algorithm emerges, based on only 6 behavioral rules necessary for the bridge construction algorithm. Means-Ends Analysis[2], from the advent of AI or meta-heuristics [3] are perfectly applicable here. It is a simple problem solving algorithm, perfectly fitted to be hosted by simple ant's brains. The problem of the bridge construction can be easily expressed in terms of various *differences* removal, between the current construction state of the bridge and a future bridge shape, which is still unknown. Local, temporary bridge structure overloading, can

[1] Just before his death in 1946, Keynes said: "*I find myself more and more relying for a solution of our problems on the invisible hand which I tried to eject from economic thinking twenty years ago.*".

[2] Proposed in 1972 by Allen Newell and Herbert Simon (Nobel Prize in Economics) and published in the book entitled "Human Problem Solving". The description of Means-Ends Analysis (MEA) is available in all AI textbooks.

be also be considered as a specific *difference* to be removed. A simple decision table, based on behavioral patterns can easily be defined for the removal of such *differences.* The problem of evacuating victims form a partially collapsed building, and problem of lifting heavy debris, can be considered in the same way.

Basic behavioral rules for bridge construction require following mental states of single ant (considered as automate):

1. An ant is busy transporting food to the anthill, or tired if it is just released from the bridge structure, e.g. to rest. Such an ant will not respond to the bridge initiative;
2. An ant is heading towards the food deposit and is assumed to be rested;
3. A rested ant when subordinated by the foreman's pheromone will change state. From now on, it will participate in the bridge construction process;
4. An ant feels exhausted and wants to leave its position in the structure of the bridge;
5. An ant feels overloaded because the tension on some parts of its body exceeds its personally assumed limits (reflecting the age, condition, etc.);
6. A given ant as a component of the bridge structure feels OK. Thus, its body can be used for further bridge development or via its body other ants can travel transporting the food cargo.

The use of pheromones is a natural way to communicate[3] the state of a given ant to its neighbors. Additionally, such pheromones will easily identify the location of, for example, an overloaded or exhausted ant. Thus, only 7 pheromones, including the foreman's pheromone, are necessary for the bridge algorithm to

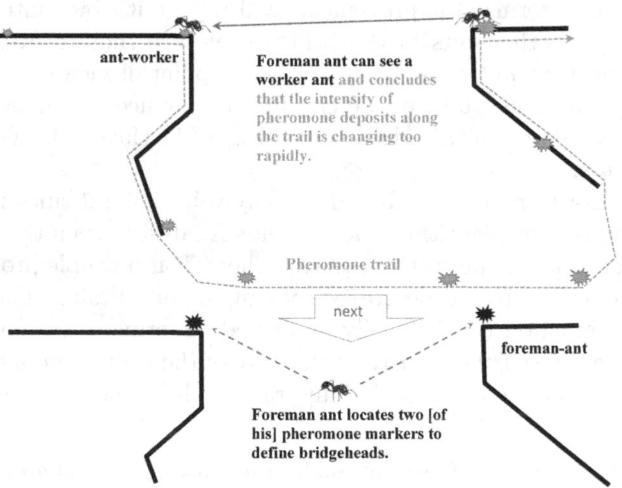

Fig. 3. A situation which will inspire ants to build a bridge.

[3] Alternatively, acoustic signals can be used.

work. This pheromone system can be easily mapped into the bridge-robots communication system, e.g. WiFi-based.

Probably, the simplest possible situation pattern for executing the bridge initiative is given in Fig. 3. In the world of real ants, most probably there are more powerful algorithms, e.g. aimed to compare candidate bridge locations in order to find the best one. As mentioned before, the bridgeheads defined by the foreman-ant pheromone will also provide a social mechanism which will trigger a switch from the previous activity to the bridge works. Such mobilization must slightly disturb the current anthill economy. Therefore, it is proposed that this pheromone will locally affect only the fresh, rested ants heading for the food deposit, and not the tired ant, with a cargo of food, heading towards the anthill. Successful bridge construction will immediately speed up food transportation for the ants arriving with a food cargo to the bridge location, which should compensate for the lack of ants engaged in the bridge structure.

3 Ant-Like Rescue Robot

Although this rescue system is inspired by ant bridges made up of their joined bodies, quite a new system of ant-like, intelligent, autonomous robot has been designed, because the functionality of ants is different from the functionality of robotized bridge elements.

The bridge-robot nicknamed AIBE (Ant-Inspired Bridge Element) will not have any articulated joints (like an ant), allowing to shift bending/twisting stress. Articulated joints optimize walking and running, but weaken the truss strength. The price of this restructuring is the need for additional DOFs[4] at shoulders and wrists and more complicated algorithms to perform movements required at the moment.

Bridge-robots are assumed to travel, for example, on a truck in a folded configuration and upon the arrival at the rescue scene, must be able to self-unfold, descend from the truck and move to the required destination. The assumed speed of the AIBE is no more than 5 km/h (the speed of a walking human). The danger results from the environment of ruins, not predators.

There are 10 handles on the thorax and the abdomen, where one bridge-robot can grasp another bridge-robot. They are rotary and U-shaped. The general proposal of the AIBE is presented in Fig. 4.

The AIBE has to precisely perceive the shape and pose of other bridge robots for the sake of collision calculations in the thicket of the arms, thoraxes and abdomens.

This can be solved with the help of perceiving markers, allowing to infer the precise current shape of the robot-partner on the basis of the precise knowledge of the AIBE shape stored in their memory. An example location of markers allowing the nearby partner-robot to infer the actual pose of a given AIBE is given in Fig. 5.

[4] Degrees of freedom.

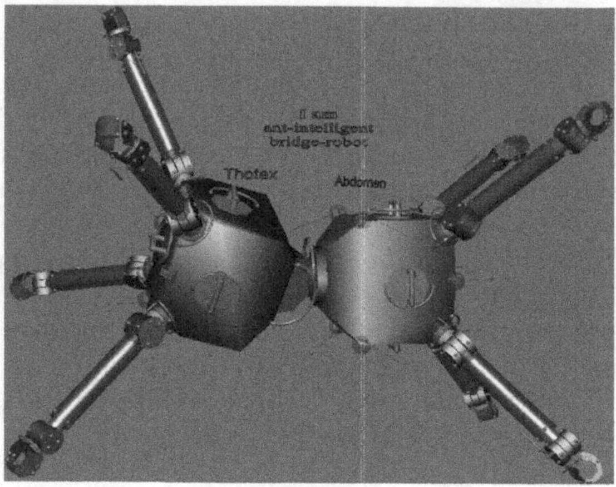

Fig. 4. A possible structure of the bridge-robot from robotics point of view.

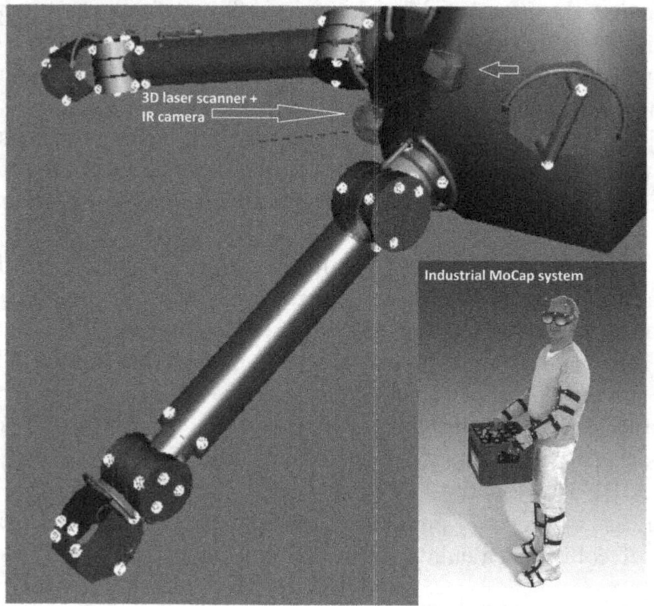

Fig. 5. The idea of the perception of partner robots by a bridge-robot is based on industrial MoCap systems, as given in the sub-photo in in the bottom right corner. The glass domes of two compound eyes are pointed to by white arrows. They can rotate in the eye sockets. Such an eye dome hosts laser 3D scanners (to scan ruins under and around the robot) and an IR camera for markers triangulation.

Most probably, human-supervisors and other participating rescue workers (for safety reasons) will be forced to wear rescue suits with such markers to allow bridge-robots to easily sense their presence and location.

It is assumed (ants inspired) that the whole transport along the bridge or the scaffolding structure made of AIBEs will be done only with the help of cargo containers. The lack of a deck surface (with safety barriers) will make the bridge/scaffolding much easier to be built. In the case of the scaffolding, the shape of the deck is a great problem. This will also exclude heavy vehicles from entering. The arms of the AIBEs will immediately detect a container which is too heavy. Such containers will be transported by an intelligent bridge structure, like the larva body. Our proposed form of a universal container is a rectangular (made of steel pipes) structure, without walls, capable of holding up to 200 kg of load. For example, it can be a toboggan transported towards the action site, with a rescue team member + some equipment. On return, 2 victims will be transported back. The tubular structure of the container allows easy and robust grasping by the AIBE arms (see Fig. 6).

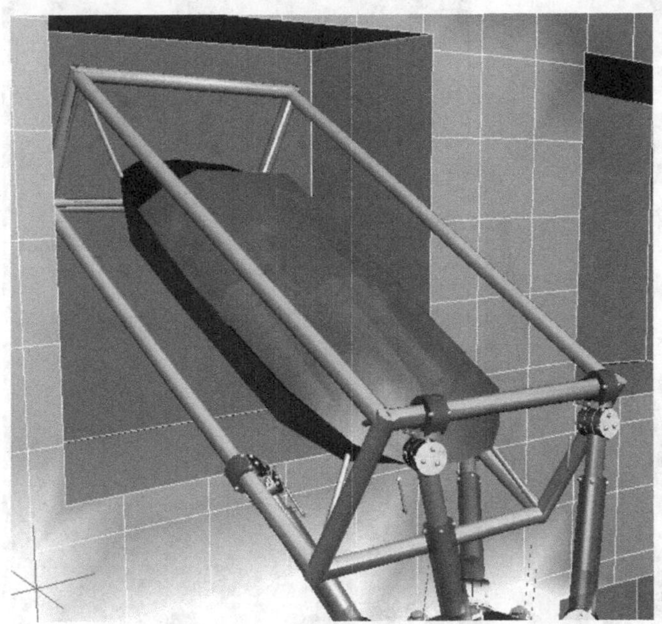

Fig. 6. A tubular container with 2 earthquake victims inside a rescue toboggan being evacuated from a ruined building.

4 Rescue Supported by Robots - Scenarios

Below are depicted (Figs. 7, 8 and 9) three most important cases which rescue teams face.

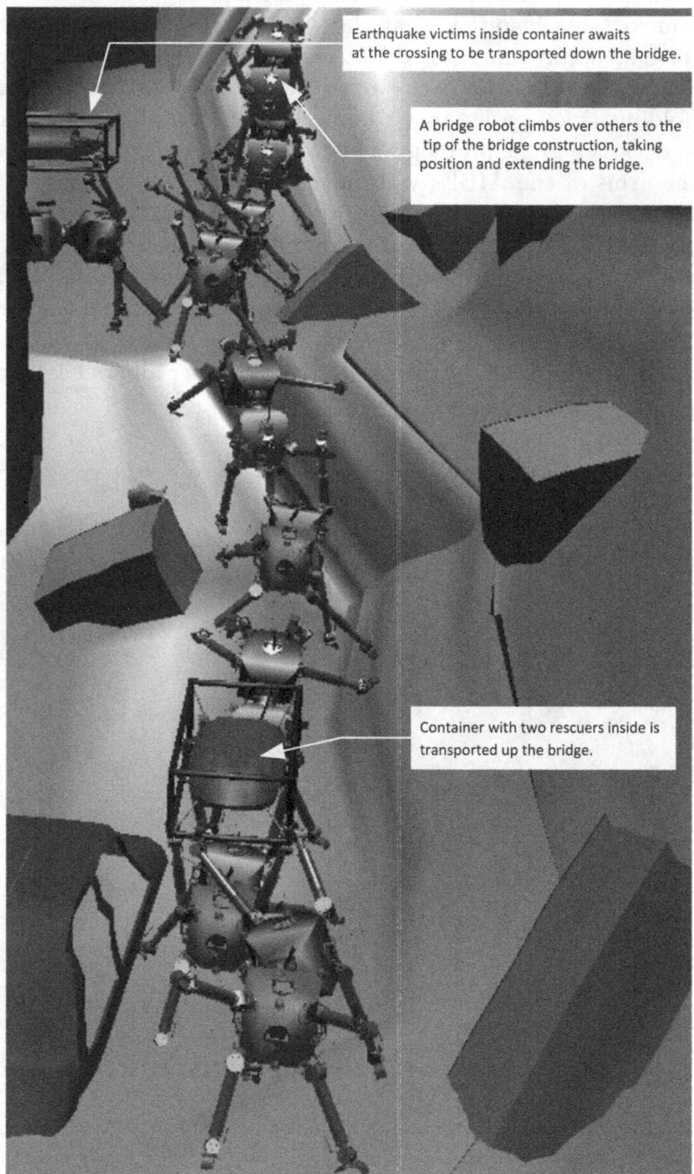

Fig. 7. Transportation functions provided by ant-like robotized bridge. Please note the ability to create intersections and bidirectional transfer.

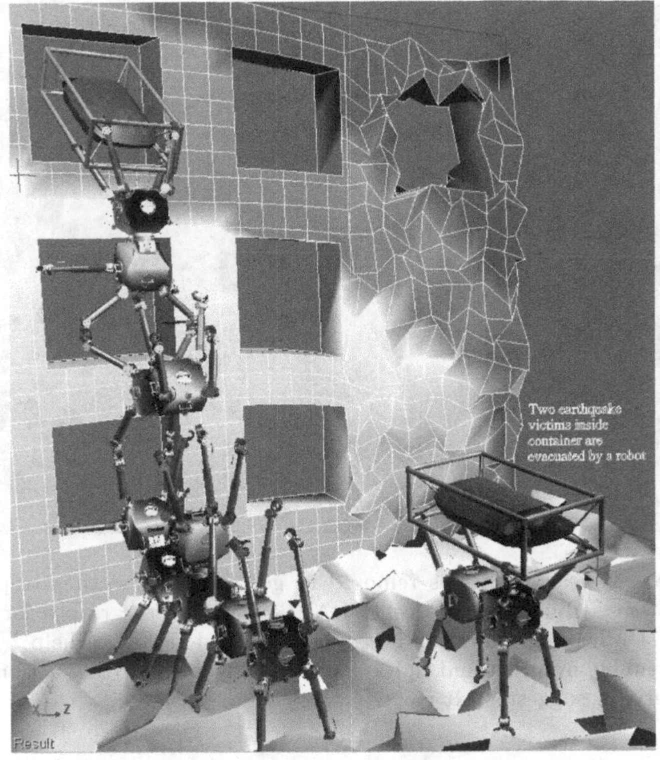

Fig. 8. An evacuation scaffolding.

5 Conclusion and Future Plans

Many politicians, macro-economists, managers, etc. consider ASIH as a harmful myth [5], which threatens their freedom to control or manipulate markets and social structures. **Demonstrating that ASIH, even if considered as a myth, allows to build a completely new control paradigm in AI-robotics is a powerful blow to this belief.**

Now our efforts focus on two goals:

- To build a consortium (EU countries that suffer frequent earthquakes) and to get funds in terms of EU Grant. We target Horizon2020 "Secure societies Protecting freedom and security of Europe and its citizens";
- Social rules providing spontaneous bridge construction, seems to be ready - however, they require more tests and development. Social rules to implement behavior depicted in Figs. 8 and 9 are under construction.

Fig. 9. A piece of heavy debris is removed to get access to victims trapped inside a partially collapsed building. The central AIBE is trying to grasp a large piece of rubble, while 3 others are supporting its abdomen and trying to help by means of grasping this rubble from left and right. This is a difficult and risky action, therefore a human-supervisor suggests next moves.

References

1. Blaug, M.: The fundamental theorems of modern welfare economics, historically contemplated. Hist. Polit. Econ. **39**(2), 185–207 (2007). doi:10.1215/00182702-2007-001
2. Dorigo, M., Gambardella, L.M.: Ant colony system: a cooperative learning approach to the traveling salesman problem. IEEE Trans. Evol. Comput. **1**, 53–66 (1997)
3. Dudek-Dyduch, E.: Algebraic logical meta-model of decision processes - new metaheuristics. In: Rutkowski, L., Korytkowski, M., Scherer, R., Tadeusiewicz, R., Zadeh, L.A., Zurada, J.M. (eds.) ICAISC 2015. LNCS, vol. 9119, pp. 541–554. Springer, Cham (2015). doi:10.1007/978-3-319-19324-3_48
4. Hayek, F.A.: Individualism and economic order (1948). Pdf available at books.google.com
5. Kennedy, G.: The myth of the Invisible Hand – A view from the trenches. Social Science Research Network (SSRN) (2012). Available at: SSRN: http://ssrn.com/abstract=2143277 or http://dx.doi.org/10.2139/ssrn.2143277
6. Nozick, R.: Invisible-hand explanations. Am. Econ. Rev. **84**, 314–318 (1991)
7. Reida, C.R., Lutzb, J., Powellc, S., Kaod, A.B., Couzine, I.D., Garniera, S.: Army ants dynamically adjust living bridges in response to a cost-benefit trade-off. Proc. Natl. Acad. Sci. **112**, 15113–15118 (2015)
8. Smith, A.: An Inquiry into the Nature and Causes of the Wealth of Nations. W. Strahan and T. Cadell, London (1776). http://www2.hn.psu.edu/faculty/jmanis/adam-smith/wealth-nations.pdf

9. Szuba, T., Szydlo, S., Skrzynski, P.: Formal and computational model for Adam Smith invisible hand paradigm. In: Computational Collective Intelligence: 7th international conference, ICCCI 2015, Madrid, vol. 1 (2015)
10. Szuba, T.: Computational Collective Intelligence. Wiley Series on Parallel and Distributed Computing. Wiley, New York (2001). Hardcover, 410 pages, Monograph
11. Ulmann, B.: Analog Computing. De Gruyter Oldenbourg Pub. (2013). Hardcover, 319 pages

Estimation of Delays for Individual Trams to Monitor Issues in Public Transport Infrastructure

Marcin Luckner(✉) and Jan Karwowski

Faculty of Mathematics and Information Science, Warsaw University of Technology, ul. Koszykowa 75, 00–662 Warszawa, Poland
{mluckner,jan.karwowski}@mini.pw.edu.pl

Abstract. Open stream data on public transport published by cities can be used by third party developers such as Google to create a real–time travel planner. However, even a real data based system examines a current situation on roads. We have used open stream data with current trams' localisations and timetables to estimate current delays of individual trams. On that base, we calculate a global coefficient that can be used as a measure to monitor a current situation in a public transport network. We present an use case from the city of Warsaw that shows how a critical situation for a public transport network can be detected before the peak points of cumulative delays

1 Introduction

The City of Warsaw is a good example of a city that opened urban data for third–party developers. One of published data sources is a stream with locations of public transport vehicles. Such data can be used to create a real–time travel planner. However, directly used localisation data is not enough to plan a long term or even mid term travel. The planner needs an additional prediction of the situation in the public transport network.

We decided to use a conglomeration of trams' localisation data with their schedules to create a projection of current situation in the city based on delays of individual trams. The trams are the most stable elements of public transport with a relatively low interaction with another form of transportation. Therefore, a collective information on tram delays should be stable in similar hours of the day. Moreover, a significant deviation can be a good alert for upcoming public transport issue.

We have prepared an architecture based on Apache Hadoop components to collect, process, and present public transport data. The data come from separate sources. The first source is a localisation stream, the second data source is timetables. Merged data allow us to estimate delays for individual trams.

N.T. Nguyen et al. (Eds.): ICCCI 2017, Part I, LNAI 10448, pp. 518–527, 2017.
DOI: 10.1007/978-3-319-67074-4_50

Using an aggregation of individual trams' delays, we created measures to monitor a global public transport status. The aggregation is done separately for different degrees of exemption from a planned schedule. We defined a condition based on three sigma rule to alarm in case of a growing total delay in the city.

We tested our tool on data from the City of Warsaw. The tool was tested both on data from days without any serious disturbances in public transport and on data from a day with a critical situation in public transport. The tool did not create false alarms and alarmed about the critical situation in advance.

The remaining part of the paper is organised as follows: Sect. 2 summarises related works. Section 3 presents technical details on the architecture for gathering and processing data. Section 4 describes estimation of delays for separate trams and calculation of global coefficients for prediction of stoppages in trams networks. An example of a predicted gridlock is given in Sect. 5. The work is concluded in Sect. 6.

2 Related Work

Our system is a dynamic solution that informs about threats for public traffic capacity. Such dynamic information can be used in dynamic route planning system and in managing of public transport systems.

A need of dynamic information was stressed in work [9]. The authors noticed that only 11% of models available in the public transport business were dynamic models that include real-time information for passengers.

According to work [5] three main subsystems of public transport information systems are: a timetable information subsystem, a route information subsystem, and a payment subsystem. In this categorisation information about delays for the routes are key for the route information subsystem and global information on delays may be used to update the real timetable information subsystem.

The collected information can be used in a majority of the public transport models [1,7,10]. The model should recalculate routes touched by the delays. Our system can convey all the necessary information.

An alternative architecture for event detection of stream data processing was presented in [8]. The architecture was created for SCATS and Twitter data. Therefore, both architectures cannot be compared directly.

Event detection in public transport data is a subject of several works. In our work GPS from trams were used for to localise vehicles. An alternative solution is a crowd-sending based localisation [2,4]. When a GPS signal taken from a tram's device is connected directly to the vehicle, a crowd-sending method enables a verification of GPS localisation quality.

Another approach to event detection is based on social media mostly Twitter [3,6]. The method, similar to ours, can only detect major accidents that caused long delays and disruptions. The main problem with that approach is a limited number of localised messages.

3 Gathering and Processing Data

We have developed a system running on Apache Hadoop. The system gathers and processes the data from Warsaw public transit buses and trams.

We gathered data from the following sources:

- timetables of buses and trams which can be downloaded,
- current GPS positions of vehicles (trams and buses) provided as an on-line stream of position updates for each vehicle,
- event counts in mobile phone cells in Warsaw.

The timetables and public transport vehicles locations were obtained from the City of Warsaw open data portal (https://api.um.warszawa.pl). The timetables can change daily when the locations change every 10 s.

The statistical mobile data were made available by a cellular network operator Orange for an internal usage in the VaVeL project[1], which aims to use urban data in applications that can identify and address citizen needs and improve urban life. The data are used to estimate the *a priori* probability distribution of demand for public transport.

Figure 1 presents an architecture of the system. The system consists of several modules based on Apache components.

The first module of our system – labelled (1) in Fig. 1 – is responsible for merging data from timetables and vehicle GPS into a stream containing information about the location and delay of each vehicle. The module downloads the timetables for all vehicles in daily intervals and stores them in our system. Then during the day the system subscribes to the stream of vehicle GPS updates. After merging with timetables the module provides Kafka stream of entries containing vehicle coordinates with a timestamp and a delay. The generated stream is then used in two ways: stored on a hard drive and used by on-line delay visualisation.

The second module – labelled (2a) and (2b) in Fig. 1 – prepares data for visualisation, processes delay data generated by the previous module. The data processing is done with Apache Spark and yields two vectors of data: unique vehicle positions in each minute and counts of trams with respectively no delay, small, medium and large delay.

Our third component – labelled (3) in Fig. 1 – is a visualisation module and is built around Apache Zeppelin platform. The module is a javascript browser application that visualises positions of vehicles accompanied with heat-map of delays drawn on the city map. Besides the map, the plot of the delay trends is presented. Delay trends are values of vehicle delays assigned to different classes. This module is bound to the second module using Zeppelin data binding.

The third module can be used both in an off-line fashion, displaying stored archive data and on-line fashion, directly displaying data from the Kafka stream. The on-line stream is very responsive and displays data with the delay of approximately 1 min. Additionally, the off-line version of the visualisation can be supported by information of count of events in the cellular network in given region (which is an estimate of a number of people being currently in the region).

[1] http://www.vavel-project.eu.

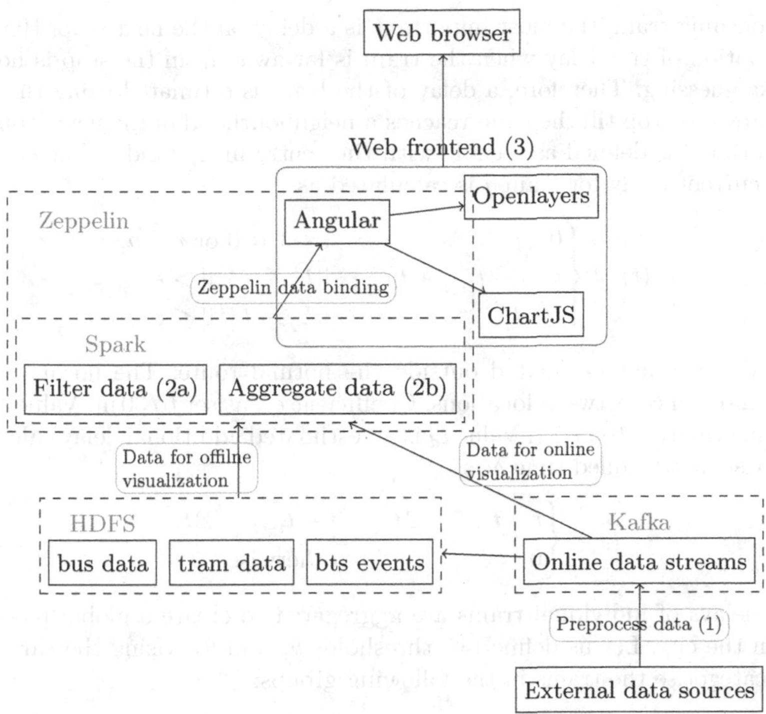

Fig. 1: Architecture of the system used to gather and analyse tram delay data.

4 Estimation and Aggregation of Delays

Let us define a route of tram j as r^j. The route contains fixed number of stops n know before the tram starts the route labelled as $r^j_0, \ldots, r^j_n$. The first and the last stop are technical stops. Let us define a normal route of a tram as a route without the technical stops $r^j_1, \ldots, r^j_{n-1}$.

The stops on the route are defined by their localisation $L_i^{r^j}$ and scheduled time $t_i^{r^j}$, where i is a number of the stop on the route.

A tram on the route is defined by its estimated localisation $\hat{L}_j$. At every moment of the normal route the tram is between two stops r^j_{i-1} and r^j_i. Stop r^j_{i-1} is the last stop where the tram had stopped. We assume that the distance between the two stops will be covered at the scheduled time:

$$\Delta t = t_i^{r^j} - t_{i-1}^{r^j}. \tag{1}$$

Our task now is to estimate a delay for a single tram. The delay can be calculated according to the previous and the next stop. For the passengers waiting for an approaching tram, the most important is a delay on the next stop. However, An estimation of the delay when the tram is far away from the stop is nothing more like guessing. Therefore, a delay of the tram is estimated using the delay on the previous stop till the tram reaches a neighbourhood of the next stop. The neighbourhood is defined as a circle with the centre in $L_i^{r^j}$ and radius ϵ.

The current delay for tram j is calculated as

$$d_j(t) = \begin{cases} 0 & i-1=0 \text{ or } i=n, \\ t_{i-1} - t_{i-1}^{r^j} + t_d & ||\hat{L}_j - L_i^{r^j}|| > \epsilon \\ t - t_i^{r^j} & ||\hat{L}_j - L_i^{r^j}|| \leq \epsilon. \end{cases} \quad (2)$$

The delay is not calculated outside the normal route. The norm $||.||$ is a Euclidean distance between locations. Coefficient ϵ was set to 30 m. Value t_{i-1} is an arriving time at stop r_{i-1}^j. Value t_d is an estimated additional delay calculated on the base of scheduled time δt as

$$t_d = \begin{cases} t - t_{i-1} - \Delta t & t - t_{i-1} > \Delta t, \\ 0 & \text{otherwise.} \end{cases} \quad (3)$$

The delays of individual trams are aggregated to create a global picture of delays in the city. Let us define two thresholds θ_e and θ_d. Using the thresholds we can categorise the trams in the following groups:

Regular trams. Trams without delays. Estimated delay $d_j(t)$ is in range $[\theta_e, \theta_d]$.
Early trams. Trams without delays. Estimated delay $d_j(t)$ is less than θ_e.
Low delayed trams. Trams with delays. Estimated delay $d_j(t)$ is in range $(\theta_d, 2\theta_d]$.
High delayed trams. Trams with delays. Estimated delay $d_j(t)$ is in range $(2\theta_d, 3\theta_d]$.
Huge delayed trams. Trams with delays. Estimated delay $d_j(t)$ is in greater than $3\theta_d$.

For all defined types of trams we can define a membership function $\delta^j(t)_{[\alpha,\beta]}$ that returns 1 if and only if the delay of a tram is in a defined range $[\alpha, \beta]$.

The ratio of delayed trams from the given category to all n trams can be now defined as

$$R(t)_{[\alpha,\beta]} = \frac{1}{n} \sum_j^n \delta^j(t)_{[\alpha,\beta]} \quad (4)$$

We are going to compare ratio from delayed trams from two categories $R_{(\theta_d,2\theta_d]}$ and $R_{(3\theta_d,\infty)}$ for clarity we mark them as $R(t)_{\theta_d}$ and $R(t)_{3\theta_d}$ respectively. We assume that critical issues in the public transport infrastructure can be detected when the following formula is true

$$\frac{R(t)_{3\theta_d}}{R(t)_{\theta_d}} > 1. \quad (5)$$

The fulfilled formula means that the total number of trams with huge delays is greater than the total number of trams with the low delays at the given moment t.

5 Use Case - Trams Monitoring in City of Warsaw

We analysed a public transport data in the City of Warsaw. A Detailed description reports from periodic updates of the location of the vehicles managed by Public Transport Authority in the City of Warsaw are published at their web page[2].

The data contain information from 750 trams. The frequency of updates is 10 s. The size of a single update is 0.2 MB every 10 s, which is 72 MB per hour approximately.

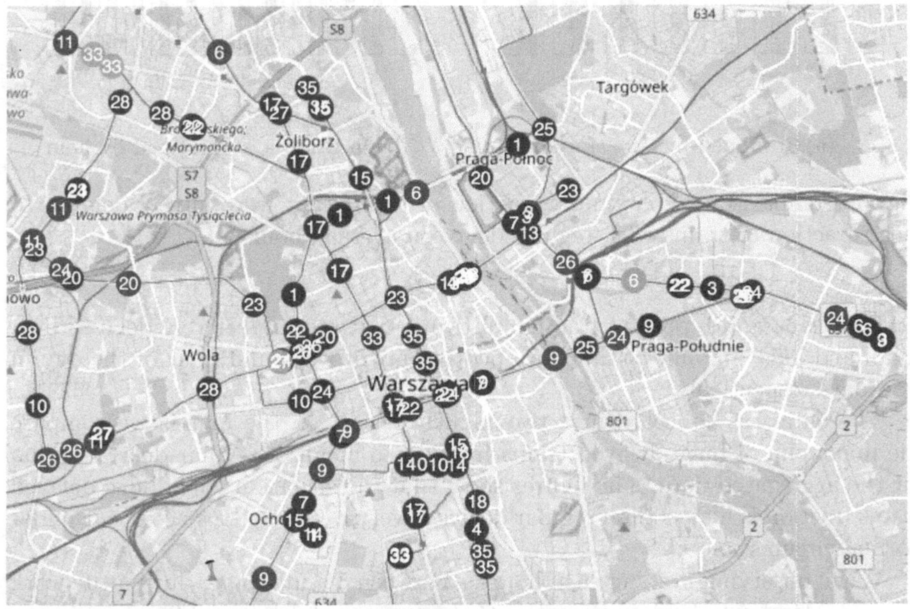

Fig. 2: Delays for individual trams. Trams are labelled with a line number. The colours symbolised delays: a blue is an early tram, a green is a tram on schedule, and an orange is a delayed tram. (Color figure online)

The created system presents all positions of trams with estimated delays on-line. Figure 2 presents delays of individual trams. The trams that are on the schedule are marked as green, early trams are marked as blue, and delayed trams are marked as orange. All trams are labelled with a line number. According to

[2] https://api.um.warszawa.pl.

public transport authorities in the City of Warsaw θ_e equals −1 min and θ_d equals 3 min.

For verification of out hypothesis that a ratio of different type of delays can be used to detect anomaly situations in a public transport network we collected data on Warsaw's trams by two months (from February to March in 2017).

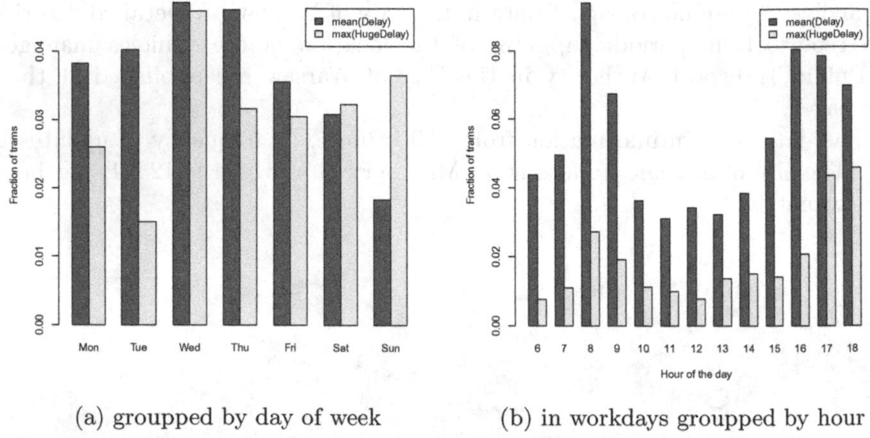

(a) groupped by day of week (b) in workdays groupped by hour

Fig. 3: Comparison of maximum fraction of trams that were hugely delayed to mean fraction of trams with regular delay on average hour delay

Figures 3a and b present fraction of trams that are respectively delayed by a small and huge amount of time grouped by day of week and hour of the day in workdays respectively.

The data was collected in a two–month period. In the figures we compare the maximum of huge delayed trams fraction to mean of regular delay fraction for the aggregated data. The figures show that the situation when huge delayed trams fraction is higher than regular delays is very rare and does not occur under normal circumstances.

When observing evening rush hours – between 5 and 6 pm – during the work day (i.e. Monday to Friday) we notice that approximately 80 % of trams arrive on time. Figure 4a presents a situation on 07–03–2017. Excluding some fluctuations the ratio of trams in each group stays stable. What is important, the groups are mostly well separated. A plot for the huge delayed trams did not cross a plot for the low delayed trams. Therefore, condition (5) was never fulfilled.

The situation is diametrically different for the evening rush hours observed on 08–03–2017. On this day a manifestation of women took place in the centre of Warsaw. Figure 4b shows how the ratio of the huge delayed trams grows among time. The peak point of the ratio is obtained at 17:57. However, condition (5) was fulfilled at 16:57.

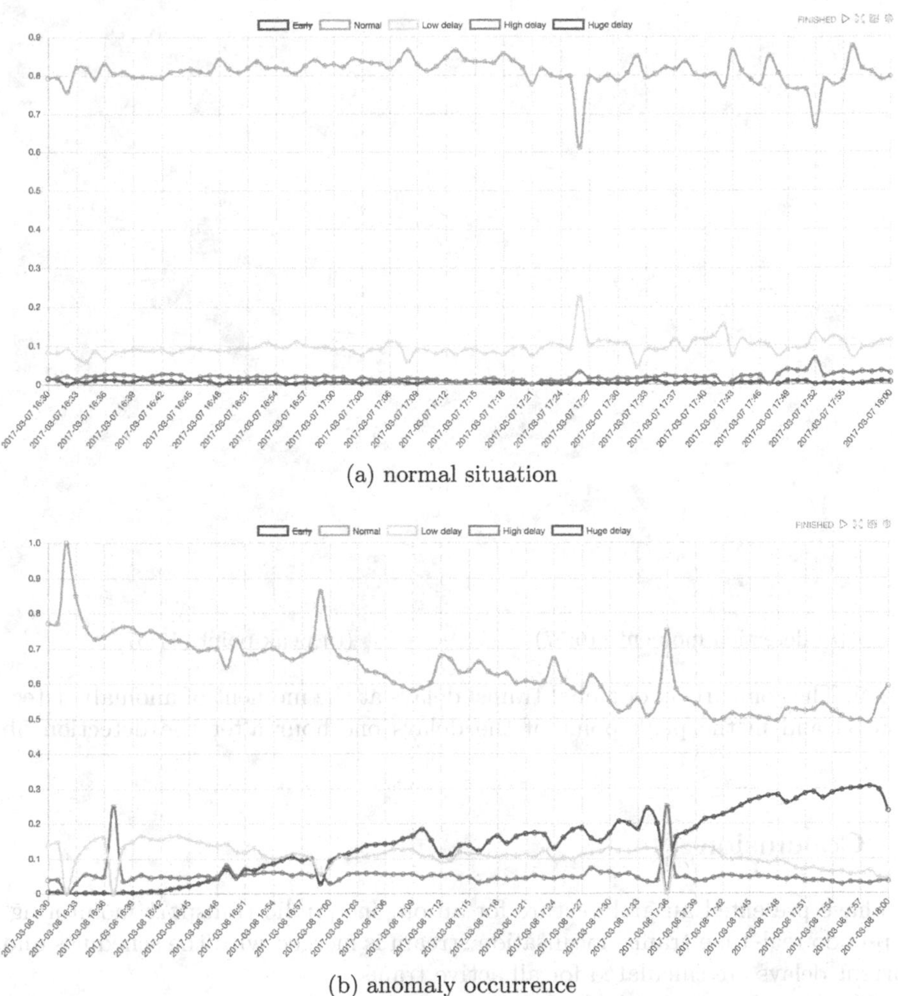

(a) normal situation

(b) anomaly occurrence

Fig. 4: The comparison of a typical distribution of delays among trams and an anomaly situation registered during manifestations. Distributions present a ratio of normal and delayed trams. Both distributions were registered during evening rush hours.

Figure 5 compares a situation in the city at the moment when the condition was fulfilled and one hour later when the peak point was obtained. In the first situation presented on Fig. 5a local delays are visible on one tram line. However, after one hour the whole city is paralysed as it is shown on Fig. 5b.

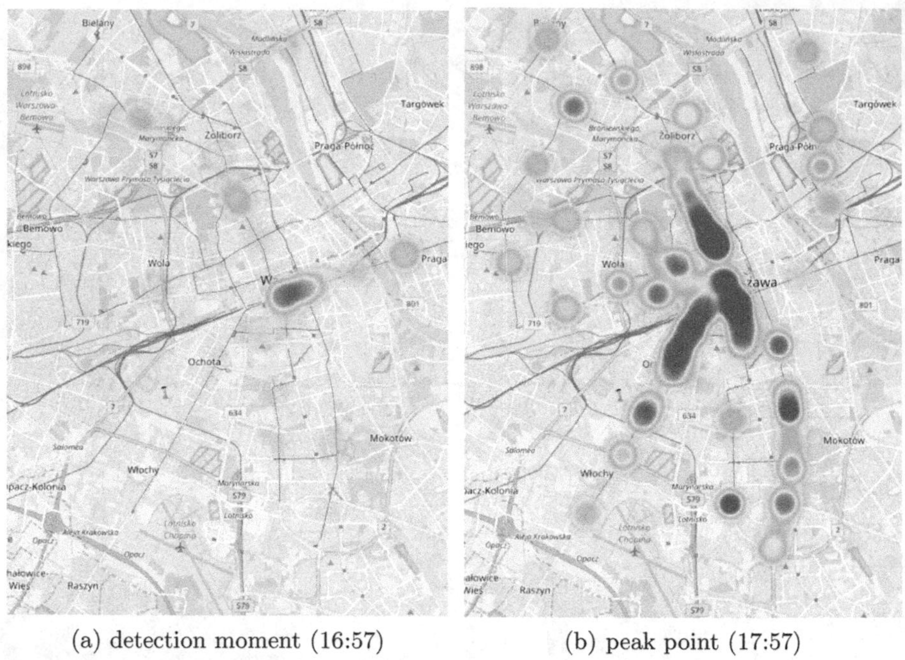

(a) detection moment (16:57) (b) peak point (17:57)

Fig. 5: The comparison of global trams' delays at the moment of anomaly detection 5a and in the peak point of the delays one hour after the detection 5b

6 Conclusion

We have presented an architecture for an on–line public transport monitoring. A pseudo real–time trams' localisation stream is merged with the schedules and current delays are calculated for all active trams.

The data from individual trams were aggregated to propose a measure for anomaly events detection. The tests on real data proved that the proposed condition is fulfilled in exceptional situations only.

The obtained results can be used to update a real–time travel planning system and to predict issues in a public transport monitoring system.

Our future work aims at a statistical analysis of the proposed measure. The model can be also updated by a prediction of travel time [11] and a bias from such factors as the weather and the number of passengers estimated by cellular network clients concentration.

Recently, the City of Warsaw has opened data with bus locations and the analysis can be repeated for that data. Also, data from other cities can be used if only the thresholds θ_e and θ_d can be estimated.

Acknowledgements. This research has been supported by the European Union's Horizon 2020 research and innovation programme under grant agreement No. 688380 *VaVeL: Variety, Veracity, VaLue: Handling the Multiplicity of Urban Sensors.*

References

1. Álvarez, A., Casado, S., González Velarde, J.L., Pacheco, J.: A computational tool for optimizing the urban public transport: a real application. J. Comput. Syst. Sci. Int. **49**(2), 244–252 (2010). http://dx.doi.org/10.1134/S1064230710020103
2. Boutsis, I., Kalogeraki, V., Guno, D.: Reliable crowdsourced event detection in smartcities. In: 2016 1st International Workshop on Science of Smart City Operations and Platforms Engineering (SCOPE) in partnership with Global City Teams Challenge (GCTC) (SCOPE - GCTC), pp. 1–6 (2016)
3. Cui, L., Zhang, X., Zhou, X., Salim, F.: Topical Event Detection on Twitter, pp. 257–268. Springer International Publishing, Cham (2016). http://dx.doi.org/10.1007/978-3-319-46922-5_20
4. Farkas, K., Feher, G., Benczur, A., Sidlo, C.: Crowdsending based public transport information service in smart cities. IEEE Commun. Mag. **53**(8), 158–165 (2015)
5. García, C.R., Pérez, R., Lorenz, Á., Alayón, F., Padrón, G.: Supporting Information Services for Travellers of Public Transport by Road. In: Moreno-Díaz, R., Pichler, F., Quesada-Arencibia, A. (eds.) EUROCAST 2009. LNCS, vol. 5717, pp. 406–412. Springer, Heidelberg (2009). doi:10.1007/978-3-642-04772-5_53
6. Panagiotou, N., Katakis, I., Gunopulos, D.: Detecting Events in Online Social Networks: Definitions, Trends and Challenges, pp. 42–84. Springer International Publishing, Cham (2016). http://dx.doi.org/10.1007/978-3-319-41706-6_2
7. Rodrigues, F., Borysov, S., Ribeiro, B., Pereira, F.: A Bayesian additive model for understanding public transport usage in special events. IEEE Trans. Pattern Anal. Mach. Intell. **99**, 1–1 (2016)
8. Souto, G., Liebig, T.: On Event Detection from Spatial Time Series for Urban Traffic Applications, pp. 221–233. Springer International Publishing, Cham (2016). http://dx.doi.org/10.1007/978-3-319-41706-6_11
9. Tyrinopoulos, Y.: A complete conceptual model for the integrated management of the transportation work. J. Public Transp. **7**(4), 101–121 (2004)
10. Zheng, P., Wang, W., Ge, H.: The influence of bus stop on traffic flow with velocity-difference-separation model. Int. J. Mod. Phys. C **27**(11), 1650135 (2016). http://www.worldscientific.com/doi/abs/10.1142/S0129183116501357
11. Zychowski, A., Junosza-Szaniawski, K., Kosicki, A.: Travel time prediction for trams in warsaw. In: Kurzynski, M., Wozniak, M., Burduk, R. (eds.) CORES 2017. AISC, vol. 578, pp. 53–62. Springer, Cham (2018). doi:10.1007/978-3-319-59162-9_6

Novel Effective Algorithm for Synchronization Problem in Directed Graph

Richard Cimler[1(✉)], Dalibor Cimr[1], Jitka Kuhnova[2], and Hana Tomaskova[1]

[1] Faculty of Informatics and Management, University of Hradec Kralove, Hradec Kralove, Czech Republic
richard.cimler@uhk.cz

[2] Faculty of Science, University of Hradec Kralove, Hradec Kralove, Czech Republic

Abstract. An effective algorithm for solving synchronization problem in directed graph is presented. The system is composed of vertices and edges. Entities are going through the system by given paths and can leave the vertex if all other entities which are going through this vertex have already arrived. The aim of this research is to create an algorithm for finding an optimal input vector of starting times of entities which gives minimal waiting time of entities in vertices and thus in a whole system. Asymptotic complexity of a given solution and using of brute-force method is discussed and compared. This algorithm is shown on an example from a field of train timetable problem.

Keywords: Synchronization · Oriented graph · Optimization · Brute-force

1 Introduction

Different real life problems can be converted to graph tasks that can help us find the right solution. Those problems are often approximated with varying degrees of abstraction. This article deals with scheduling of system entities in a directed graph. The problem is close to the Train Timetable Problem (TTP) where departure and arrival times for trains on their lines are set. Detailed surveys and more about this problematic can be found in [1–3,6].

There are several assumptions in studied system. The system entities enter into a graph by initial vertices. Those entities go through previously specified paths (graph edges) to the final vertices. The example of this system is described in the Fig. 1, where P_1, P_2, P_3 and P_4 are initial vertices for four entities. All entities have known respective paths from the beginning. These paths are indicated by a line style. Paths between two vertices can be the same for more entities, but each path can have a different length.

In the following part of the paper, the problem is described in a detail. The second part deals with a proposed methodology for solving the problem containing informations about our algorithm and possible brute-force solution. In the third part of the paper, complexity of our solution and brute-force is discussed.

N.T. Nguyen et al. (Eds.): ICCCI 2017, Part I, LNAI 10448, pp. 528–537, 2017.
DOI: 10.1007/978-3-319-67074-4_51

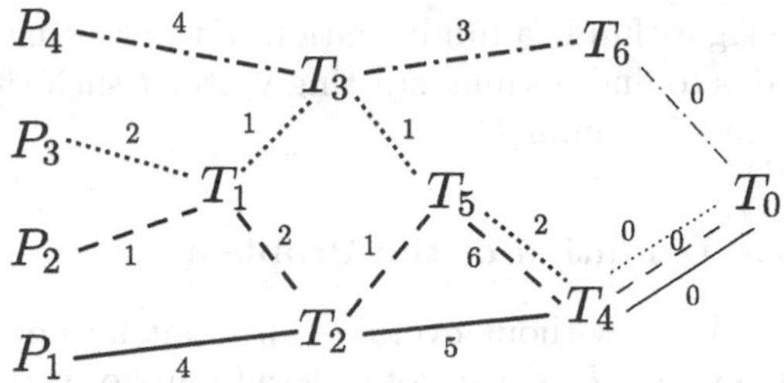

Fig. 1. Example of a studied system

1.1 Description of Problematics

This research has been inspired by an issue solved in [4] by extremal algebras, where authors are focused on a search for allowable solutions in analogical synchronization problem. Some of the assumptions are different, therefore solutions are impossible to compare. Before the problem itself, the definition of terms needs to be described.

Entity: an object entering the system and going through vertices
Vertices: Points the entities are going through
initial vertex – point where entity begin its instance (Vertices P_1, P_2, P_3, P_4 in the Fig. 1)
final vertex – point where entities end their instance (Vertices T_4, T_5 in the Fig. 1)
synchronization vertex – point which has to be leaved by entities in the same time (Vertices T_1, T_2, T_3, T_5 in the Fig. 1)
imaginary final vertex – auxiliary point for final synchronization of entities. If there is not single final vertex in the system, it is necessary to add imaginary vertex with zero distance from final vertices of all entities (vertex T_0 in the Fig. 1).
Edge: direct weighted path between vertices
Loop: pair of vertices, which is common for a set of entities and there is no other common vertex between this pair for this set (pairs (T_2, T_4), (T_1, T_5), (T_5, T_4))
Layer: number of steps between an actual vertex and a final (or imaginary final) vertex. If there are more paths between the vertex and the final vertex, a layer is the maximal number of edges on these paths. (layer 1: T_4, T_6, layer 2: T_5, layer 3: T_2, T_3, layer 4: T_1)

We assume n initial vertices $(P_1, P_2, \ldots, P_n)^T$ for entering of n entities. Starting time $t_i, i = 1, \ldots, n$, of an entity can be different for all entities and forms an initial time vector $\boldsymbol{t}$. Paths of entities are described by lists of edges. For each entity, there is only one path between each pair of vertices. Entities can leave a given synchronization vertex only at the same time, therefore each must wait for others, which creates a waiting time in a system.

Objective of the algorithm is a minimization of the waiting time in synchronization vertices and is to find optimal starting vector $\boldsymbol{t}$ such that the total time of entities in the system is minimal.

1.2 Mathematical Definition of the Problem

Oriented graph $G = (V, E)$ without cycles is an input into our algorithm. The set V is a set of all vertices, E is a set of ordered pairs of vertices called edges. There is a set of starting vertices $P_i \in V$, where $i \in I = \{1, \ldots, n\}$, and sequence of edges which are interconnected in a path $\boldsymbol{p}_i$.

Lets assume that each edge-weighted path $\boldsymbol{p}_i = (e_{i1}, e_{i2}, \ldots, e_{is_i})$ starts from one starting point P_i. Weight of an edge e_{ij} is $w_{ij} > 0$ for $i \in I$, $j = 1, \ldots, s_i$. A count of sequence elements $|\boldsymbol{p}_i| = s_i$ gives a length of the path $\boldsymbol{p}_i$, a weight of such path is $w(\boldsymbol{p}_i) = \sum_{j \in J} w_{ij}$.

Lets assume that there is a moving entity g_i with a starting time t_i on each path $\boldsymbol{p}_i$. Movement of entities is ruled by a condition C, which causes waiting in the transit vertices. Starting time from a starting vertex is t_i, waiting time of entity g_i in the j-th vertex of the path is δ_{ij}.

Starting time t_i in a sum with a time which takes to travel between two vertices (edge weight) w_{ij} gives a total time of a travel.

Aim of the algorithm is to find such starting times $t_1, \ldots, t_n$, in order to get minimal total waiting time

$$z = \sum_{i \in I} \sum_{j \in J_i} \delta_{ij}$$

in the system where synchronization condition C is applied.

Synchronization condition $\boldsymbol{C}$. Lets have an entity g_i on a path $\boldsymbol{p}_i$ going through vertex T:

$$\boldsymbol{p}_i = (e_{i1}, e_{i2}, \ldots, e_{ij_T}, e_{i(j_T+1)}, \ldots, e_{is_i}).$$

Time to get to vertex T, where it is necessary to wait for other entities, is given by:

$$t_{ij_T} = t_i + \sum_{k=1}^{j_T} w_{ik} + \sum_{k=1}^{j_T - 1} \delta_{ik}.$$

It is a sum of a starting time, time spent travelling between vertices and total waiting time in all previous vertices.

Departure time of an entity g_i from a vertex T is:

$$t_{ij_T} + \delta_{ij_T}.$$

An inequality

$$\max\{t_{1j_T}, t_{2j_T}, \ldots, t_{nj_T}\} \leq \min\{t_{1(j_T+1)}, t_{2(j_T+1)}, \ldots, t_{n(j_T+1)}\}$$

must be fulfilled in order to get non-negative waiting times δ_{ij_T}.

2 Proposed Solution

In this section, two types of solutions are described. The first one is our proposed method. It solves the problem in a specific asymptotic time which is described in the Sect. 3. The second solution is a brute-force method for verification of our results because it tests all possible solutions. The complexity of these two solutions is discussed.

2.1 Data Preparation

We use a matrix D containing distances of the synchronization vertices from the beginning of the path as an input.

$$D = \begin{pmatrix} d_{11} & d_{t2} & \cdots & d_{1m} \\ d_{21} & d_{22} & \cdots & d_{2m} \\ \vdots & \vdots & \ddots & \vdots \\ d_{n1} & d_{n2} & \cdots & d_{nm} \end{pmatrix}.$$

Columns represent paths of entities and rows represent vertices. If the path does not include some vertex, it is described by a zero on a given position.

Following this step, matrix W with information about distance between directed pair of vertices is created. At the same time, matrix O with information about order of path vertices is created as well. (e.g. if there is a vector for an entity g_i in a matrix D such as $\boldsymbol{d}_i = (2,0,0,4,9,5)^T$, then corresponding vector in a matrix W is $\boldsymbol{w}_i = (2,0,0,2,4,1)^T$ and in a matrix O is $\boldsymbol{o}_i = (1,4,6,5,0,0)^T$).

During this step, matrix

$$A = \begin{pmatrix} a_{11} & a_{t2} & \cdots & a_{1n} \\ a_{21} & a_{22} & \cdots & a_{2n} \\ \vdots & \vdots & \ddots & \vdots \\ a_{n1} & a_{n2} & \cdots & a_{nn} \end{pmatrix},$$

is filled; a_{ij} contains the value of how many loops is on an edge of an entity g_j to a vertex T_i. Simultaneously, a matrix with number of solved loops is also created. Next part of the solution is an identification of layers. Last part of data preparation is a creation of a list of loops.

2.2 Synchronization Part of an Algorithm

The goal of a synchronization is that a set of entities which go through the same transit vertex must leave this vertex at the same time. In case that transit times of entities are not the same, entities with lower arrival time are paused. The waiting times are saved into the matrix. Sum of elements of this matrix represents a total waiting time. This total waiting time needs to be minimized.

The methodology is managed by following main rules in the given order. In each step of an algorithm one of the loop waiting time is solved or partly solved.

There are three options how to find the optimal waiting time. All three options are mutually exclusive. For finishing off a given option, all sub-steps have to be completed. The default is the first option, if the conditions are not fulfilled, second option is applied, if the condition of the second option is not fulfilled, the third option is used.

1. Sequential testing of edges of the entity g in the loop l_i
   (a) The edge must belongs also into another loop l_j.
   (b) An entity g has to be in both loops l_i and l_j.
   (c) Transit time of the entity g in the loop l_j has to be lower than transit time of at least one other entity in l_j.
2. Finding of the edge which does not belong into another loop.
3. Finding the edge which is in the fewest number of other loops.
   (a) Edges in the loops that have not been solved yet are proffered.

More Detailed Description of Steps. First step is to find the edges in the loop l_i, which belongs into other loop l_j. If we denote v_1 and v_2 the layer of a starting and ending point of a loop l_j respectively, and u_1 and u_2 as the layer of a starting and ending vertex of the edge respectively, following conditions must be fulfilled

$$v_1 \geq u_1 \quad \wedge \quad v_2 \leq u_2.$$

Then we know that the edge belongs into the loop l_j. After that we add the maximal possible value of a difference in a loop l_j, which means the difference between the transit time of an entity g and maximal transit time of another entity in the loop l_j. In case the sufficient value is not added, we continue in looking for another edge to add remaining value.

First option – thanks to increasing of a waiting time in one of vertices of another loop, there is partially or completely solved loop where such vertex belongs. Such vertex has not been already solved in another step of an algorithm. If the first option fails and any such vertex is not found, it is necessary to go to the second option. The task of the second option is to find one of the edges which does not affects any another loop. If any such vertex does not exist, third option has to be used.

The third option solves the loop l_i, but increases the difference between entities in another loop which includes an entity o too. For that reason, it is important to choose an edge, which extends the lowest number of other loops. If there are several such edges, a loop is found that is not solved yet. If there is again more than one, it means more than one possible solutions.

After successful completion, Δ matrix, which represents system waiting time on vertices, is increased by necessary value to solve a loop and the value a_{ij} of matrix A is raised by one (number of already solved loops on a_{ij}).

At the end of this part of a problem solution, we have waiting times of all entities in a system. Final part is to count up a starting vector $\boldsymbol{t}$. Thanks to synchronization of entities, it means to find out an entity with the highest value of a total time (sum of waiting and transit time) in a system and subtract all other values from this value.

2.3 Brute-Force

Common brute-force solution means sequential testing of all options. By this method, we are looking for a solution which matches a consistent state of the system. This is achieved by aggregating of transit times, waiting times and starting times for all entities. For each entity, the value must be the same and means that waiting times in a matrix Δ is a feasible solution but may not be the best.

For searching of the optimal solution, we have to generate all possible matrices Δ ordered by a sum of its elements. If we want to generate all options with matrix size $a \cdot b$ with all variation with repetition in the range r, we have $r^{(a \cdot b)}$ matrices. Entities may not have a dependency on all vertices, it is possible to generate only a number of feasible positions. Because matrices are generated in order, first successfully found result is the best one.

3 Complexity of Solutions

In this section, we describe asymptotic complexity of both methods. For calculating of the complexity we will use the following notation:

p – number of entities
t – number of vertices

3.1 The Proposed Algorithm

The first step is transferring of a matrix D to the matrix W and creation of a matrix O. It means to find the distance between vertices for entities and ordering of vertices by the distance from the initial vertex. Complexity of it corresponds to the formula

$$p \cdot t + t^2,$$

where $p \cdot t$ is a complexity of a creation of W and t^2 complexity of ordering of vertices. The algorithm has been implemented in Matlab, which use quicksort method.

The second step is identification of layers. Complexity of this step is:

$$t^2 + t + t^2 - \frac{t(t+1)}{2} = \frac{3t^2 + t}{2},$$

where $t^2 + t$ corresponds to a complexity of an evaluation of a matrix A for finding vertices with already determined layer, t^2 is removing of already resolved vertices and $\frac{t(t+1)}{2}$ means reduction of a complexity by this removing.

Vector containing a number of loops in which are each points presented has to be created. Its complexity is:

$$p \cdot t + t + p,$$

where $p \cdot t$ corresponds to an evaluation of an input matrix determining the number of edges. Besides that, it is necessary to subtract $t + p$ of ones because of synchronization vertices.

In a next step, loops of a system are found. It means comparing pairs of vertices which are the same for a set of entities. At the same time, absence of other common vertices is tested. Complexity of this step is

$$\frac{(pt)^2 - pt}{2} + \frac{pt \cdot (pt+1) \cdot (pt+2)}{6} = \frac{pt((pt)^2 + 6pt - 1)}{6},$$

The first part of the formula is finding of feasible loops and the second part is theirs verification. The complexity of the second part is also represented by tetrahedral (or triangular pyramidal) numbers [7].

Complete algorithm has a polynomial complexity $\Theta(n^k)$, where n is equal to a number of vertices and $k = 5$. This complexity is given by cubic complexity $\Theta(n^3)$ of a number of edges which are extended in loops. For this edges, steps 1, 2 and 3 (described in Sect. 2.2) are executed and complexity of this part is $\Theta(n^2)$, so overall complexity is:

$$\Theta(n^2) \cdot \Theta(n^3) = \Theta(n^5),$$

3.2 Brute-Force

Brute-force complexity includes the complexity of matrices generation and their evaluation. We are generating all possible solutions of a problem. Starting with a matrices with lowest sum of waiting times. Numbers are not generated on all positions of a matrix only on a matrix positions, where might be some waiting. The value of a maximal waiting time is not limited, so there has to be set a maximal sum of waiting times.

For brute-force algorithm, we have a number of verticies t and a sum of waiting time i. Firstly, we generate a matrix composed of lower triangular matrices of numbers 1,..., i. The complexity of a creating of such a matrix is not greater than $i \cdot t$.

Next step is to combine selected rows of the matrix to get a vector with a sum of elements equal to i. In the worst case, we have to combine vectors with only one nonzero element with ascending order. Hence we have vectors with numbers 1, 2,..., a, such that their sum is i. Then $\frac{a(a+1)}{2} = i$ must be satisfied. In order to get only positive number of steps, the maximal number of steps is a floor of a, where $a = \frac{\sqrt{8i+1}-1}{2}$. We have to find all combinations of rows. Number of all combinations is $\frac{(i \cdot t)!}{j!(i \cdot t - j)!}$, where $j = 1, \ldots, a$; we can use the worst possibility where j is the ceil of $\frac{i \cdot t}{2}$. Hence, the number of all combinations is $m = \dfrac{(i \cdot t)!}{\lceil \frac{i \cdot t}{2} \rceil ! \, (i \cdot t - \lceil \frac{i \cdot t}{2} \rceil)!}$. This number increases with an increasing product of $i \cdot t$ geometrically. The ratio of two consecutive numbers is either 2 or $\frac{2(i \cdot t+1)}{i \cdot t+2}$ (which converges to 2) depending on a parity of these numbers. Therefore, the complexity of a finding all combinations of rows is $t \cdot 2^{i \cdot t}$. After that, it is necessary to reshape m rows (complexity $\lceil \frac{i \cdot t}{2} \rceil \cdot t$) into one vector. All of this is done for a combinations of rows.

After that we have some combinations duplicated, therefore we have to find only unique combinations of rows (complexity $\frac{a(a+1)}{2}$).

Finally, we have to find all permutations of these unique combinations. The number of unique combinations are limited by Fibonacci sequence, therefore for a sum i, there are $F(i+1) = \frac{1}{\sqrt{5}}(\varphi^{i+1} - (1-\varphi)^{i+1})$ number of combinations. Complexity of a creating of permutations is linear and depends only on a number of verticies t.

If we put all this complexities together, we get

$$i \cdot t + a\left(t \cdot 2^{i \cdot t} + m \left\lceil \frac{i \cdot t}{2} \right\rceil \cdot t\right) + \frac{a(a+1)}{2} + t \cdot F(i+1).$$

The number of a maximal generated matrices corresponds with the previous algorithm and is equal to $\binom{i+t-1}{i}$. Which increases in the worst case with 2^{i+t-1}.

The overall complexity of determining whether the matrix is the optimal result can be described by formula

$$\left(p \cdot \frac{t^2+t}{2}\right) + (p \cdot t + p) + (p \cdot t) = \frac{p \cdot (t^2 + 5t + 2)}{2},$$

where $\left(p \cdot \frac{t^2+t}{2}\right)$ stands for an adding of a waiting time to all following vertices, $(p \cdot t + p)$ is generating and adding of a start vector and $(p \cdot t)$ means validation of solution.

In a worst case scenario the complexity of brute-force algorithm is

$$\Theta(2^{i \cdot t}).$$

4 Testing

The algorithm has been tested on many systems setups. One of the possible system layouts and its testing is shown on the following example. Layout of the system has been the same (see Fig. 1) and in each example length of edges has been changed in order to get systems with final solutions (sum of waiting times in a system) from 0 to 11 (what corresponds to indexes of matrices D_0–D_{11}).

$$D_0 = \begin{bmatrix} 0 & 1 & 2 & 0 \\ 4 & 3 & 0 & 0 \\ 0 & 0 & 4 & 4 \\ 11 & 10 & 11 & 0 \\ 0 & 4 & 5 & 0 \\ 0 & 0 & 0 & 11 \end{bmatrix} \quad D_1 = \begin{bmatrix} 0 & 1 & 2 & 0 \\ 4 & 3 & 0 & 0 \\ 0 & 0 & 4 & 4 \\ 11 & 10 & 11 & 0 \\ 0 & 4 & 5 & 0 \\ 0 & 0 & 0 & 10 \end{bmatrix} \quad D_2 = \begin{bmatrix} 0 & 1 & 2 & 0 \\ 4 & 3 & 0 & 0 \\ 0 & 0 & 4 & 4 \\ 11 & 10 & 11 & 0 \\ 0 & 4 & 5 & 0 \\ 0 & 0 & 0 & 9 \end{bmatrix} \quad D_3 = \begin{bmatrix} 0 & 1 & 2 & 0 \\ 4 & 3 & 0 & 0 \\ 0 & 0 & 4 & 4 \\ 11 & 10 & 11 & 0 \\ 0 & 4 & 5 & 0 \\ 0 & 0 & 0 & 8 \end{bmatrix}$$

$$D_4 = \begin{bmatrix} 0 & 1 & 2 & 0 \\ 4 & 3 & 0 & 0 \\ 0 & 0 & 4 & 4 \\ 11 & 10 & 11 & 0 \\ 0 & 4 & 5 & 0 \\ 0 & 0 & 0 & 7 \end{bmatrix} \quad D_5 = \begin{bmatrix} 0 & 1 & 2 & 0 \\ 4 & 3 & 0 & 0 \\ 0 & 0 & 4 & 4 \\ 11 & 10 & 10 & 0 \\ 0 & 4 & 5 & 0 \\ 0 & 0 & 0 & 7 \end{bmatrix} \quad D_6 = \begin{bmatrix} 0 & 1 & 2 & 0 \\ 4 & 3 & 0 & 0 \\ 0 & 0 & 4 & 4 \\ 11 & 10 & 9 & 0 \\ 0 & 4 & 5 & 0 \\ 0 & 0 & 0 & 7 \end{bmatrix} \quad D_7 = \begin{bmatrix} 0 & 1 & 2 & 0 \\ 4 & 3 & 0 & 0 \\ 0 & 0 & 4 & 4 \\ 11 & 10 & 8 & 0 \\ 0 & 4 & 5 & 0 \\ 0 & 0 & 0 & 7 \end{bmatrix}$$

$$D_8 = \begin{bmatrix} 0 & 1 & 2 & 0 \\ 4 & 3 & 0 & 0 \\ 0 & 0 & 4 & 4 \\ 11 & 10 & 7 & 0 \\ 0 & 4 & 5 & 0 \\ 0 & 0 & 0 & 7 \end{bmatrix} \quad D_9 = \begin{bmatrix} 0 & 1 & 2 & 0 \\ 4 & 3 & 0 & 0 \\ 0 & 0 & 4 & 4 \\ 10 & 10 & 7 & 0 \\ 0 & 4 & 5 & 0 \\ 0 & 0 & 0 & 7 \end{bmatrix} \quad D_{10} = \begin{bmatrix} 0 & 1 & 2 & 0 \\ 4 & 3 & 0 & 0 \\ 0 & 0 & 4 & 4 \\ 9 & 10 & 7 & 0 \\ 0 & 4 & 5 & 0 \\ 0 & 0 & 0 & 7 \end{bmatrix} \quad D_{11} = \begin{bmatrix} 0 & 1 & 2 & 0 \\ 4 & 3 & 0 & 0 \\ 0 & 0 & 3 & 4 \\ 9 & 10 & 6 & 0 \\ 0 & 4 & 4 & 0 \\ 0 & 0 & 0 & 7 \end{bmatrix}$$

The average run time of ten repetitions on each setup of brute-force and our proposed algorithm was compared (see Table 1). Results show a large increase of brute-force solution time consumption. It shows that more complicated system is impossible to be solved by brute-force method in a short time.

Table 1. The average time of brute-force and our methodology in seconds.

Matrix	Brute force	Our solution	Matrix	Brute force	Our solution
D_0	0.002	0.026	D_6	3.7028	0.0396
D_1	0.0324	0.0326	D_7	11.8098	0.0408
D_2	0.0522	0.034	D_8	26.2068	0.0434
D_3	0.13	0.0338	D_9	76.0954	0.042
D_4	0.426	0.0328	D_{10}	165.2368	0.0584
D_5	1.2612	0.0394	D_{11}	360.9798	0.0574

5 Discussion

In a future research we would like to focus on a possibility of converting this solution for a CPM (Critical path method). CMP is a project modeling method, which has been introduced in the fifties of the last century [5]. This method is used in a project management. Given tasks of a project and their length are showed in a directed graph. This methods find the longest path in a given graph. It finds critical tasks and the earliest and the latest time that each activity can start and finish without making a delays in a project.

CPM computation time consumption is based on a number of tasks and their dependencies on previous tasks. In the conversion of our problem into CPM, tasks refers to edges. In a case that every vertices are in a different layer and their number is constant, we assume that the computation complexity of our solution is linear but in CPM complexity is quadratic with every new entity occurrence. We assume that CPM method might be more effective for solving problems with more vertices.

6 Conclusion

In this paper, we introduced an algorithm for a solution of a synchronization problem in an oriented graph. Our algorithm has asymptotic complexity $\Theta(n^5)$ while the complexity of a brute-force solution is $\Theta(2^{i \cdot n})$, where n is a number of vertices and i is a total waiting time in a system. This method might help to solve for example specific train timetable problems. In the future work, the proposed solution will be tested against critical path method. We aim to find an effective solution for solving systems where a big amount of entities travel through a given system in which synchronization is needed.

Acknowledgment. The support of the Specific Research Project at FIM UHK is gratefully acknowledged.

References

1. Arenas, D., Chevrier, R., Hanafi, S., Rodriguez, J.: Solving the train timetabling problem, a mathematical model and a genetic algorithm solution approach. In: 6th International Conference on Railway Operations Modelling and Analysis (2015)
2. Binder, S., Maknoon, Y., Bierlaire, M.: The multi-objective railway timetable rescheduling problem. Transp. Res. Part C: Emerg. Technol. **78**, 78–94 (2017)
3. Cacchiani, V., Toth, P.: Nominal and robust train timetabling problems. Eur. J. Oper. Res. **219**(3), 727–737 (2012)
4. Gavalec, M., Ponce, D., Karel, Z.: The two-sided (max/min, plus) problem is NP-complete. Submitted to: Fuzzy sets and systems (2016)
5. Kelley, Jr., J.E., Walker, M.R.: Critical-path planning and scheduling. In: Papers presented at the December 1–3, 1959, eastern joint IRE-AIEE-ACM Computer Conference, pp. 160–173. ACM (1959)
6. Siebert, M., Goerigk, M.: An experimental comparison of periodic timetabling models. Comput. Oper. Res. **40**(10), 2251–2259 (2013)
7. Sloane, N.J.A.: https://oeis.org/A000292

Bimodal Biometric Method Fusing Hand Shape and Palmprint Modalities at Rank Level

Nesrine Charfi(✉), Hanene Trichili, and Basel Solaiman

Head Image and Information Processing (iTi) Dept., IMT Atlantique, Plouzané, France
nesrine.charfi@ieee.org, hanene.trichili@telecom-bretagne.eu, basel.solaiman@imt-atlantique.fr

Abstract. Person identification becomes increasingly an important task to guarantee the security of persons with the possible fraud attacks, in our life. In this paper, we propose a bimodal biometric system based on hand shape and palmprint modalities for person identification. For each modality, the SIFT descriptors (Scale Invariant Feature Transform) are extracted thanks to their advantages based on the invariance of features to possible rotation, translation, scale and illumination changes in images. These descriptors are then represented sparsely using sparse representation method. The fusion step is carried out at rank level after the classification step using SVM (Support Vector Machines) classifier, in which matching scores are transformed into probability measures. The experimentation is performed on the IITD hand database and results demonstrate encouraging performances achieving IR = 99.34% which are competitive to methods fusing hand shape and palmprint modalities existing in the literature.

Keywords: Biometry · SIFT descriptors · Hand shape · Palmprint · Fusion · Classification

1 Introduction

With the growing of fraud attacks in our society, the person identification has become an important task in order to control the identity in public spaces. Biometry is the safest technology which allows identity recognition. In fact, this technology is based on physical modalities such as fingerprint [HG], palmprint [RM], hand shape [NC1] and behavioral modalities such as voice [ZD] and signature [VA]. In this paper, we focus on bimodal biometric system fusing hand shape and palmprint modalities. In the literature, several studies have been developed in this topic [NC2, NC3, NC4, NC6]. Concerning hand shape modality, Guo et al. [JG] obtained a CIR = 96.23% by the extraction of 34 geometrical features of the hand from 6000 hand images. Moreover, Luque-Baena et al. [LB] achieved an EER = 4.51% using the IITD hand database and an EER = 4.64% using the CASIA hand database, by extracting 403 geometric features. Regarding palmprint modality, Hong et al. [DH] extracted Fast Vese-Osher decomposition

N.T. Nguyen et al. (Eds.): ICCCI 2017, Part I, LNAI 10448, pp. 538–547, 2017.
DOI: 10.1007/978-3-319-67074-4_52

model from palmprint images and achieved EER = 0.92% using IITD hand database. On the other hand, Guo et al. [XG] investigated a palmprint recognition method based on HEBD (Horizontally Expanded Blanket Dimension). Experimental results evaluated on PolyU and CASIA palmprint databases showed a high recognition rate. Other researchers focused on the fusion of palmprint and hand shape modalities. In fact, the fusion scheme may be performed in five different levels: data level based on the fusion of different acquisition data, features level based on the fusion of different modality features, matching score level based on the fusion of different scores, rank level based on the combination of different ranks provided from different sources and decision level which is based on the fusion of different decisions generated from different biometric sources. For example, Wang et al. [CW] combined contour features of hand shape modality and wavelet features of palmprint modality at feature level. On the other hand, Ferrer et al. [MF] fused 15 geometrical features from hand geometry and Gaussian filter from palmprint modality at score level.

In this paper, we propose to fuse hand shape and palmprint modalities at rank level for person identification. The aim is to represent sparsely invariant descriptors which are SIFT descriptors for these two modalities and to transform scores generated after classification step, into probabilities for fusion at rank level. Hence, the proposed method contains the following steps: the segmentation step for both hand and palmprint modalities, the SIFT feature extraction and sparse representation steps, the classification step and the fusion step at rank level to further person identification. Experimental results showed promising performances by the fusion at rank level, by obtaining IR = 99.34%. The general flowchart of the proposed method is given in Fig. 1.

This paper is organized as follows. Section 2 presents the preprocessing step. Section 3 describes the sparse feature representation step for the two modalities. Whereas Sect. 4, it presents the classification and fusion steps adopted in our work. Finally, Sect. 5 shows performances of the proposed approach.

2 Preprocessing Step

2.1 Hand Segmentation

This step consists on separating the hand from background. In fact, the morphological operators are firstly applied to remove separated debris of foreground. Then, in order to eliminate shape artifacts (rings), an algorithm of removal cavities is adopted [EY]. After that, hand images are rotated to normalize hands into a unique orientation, which is then adopted for palmprint segmentation step.

2.2 Palmprint ROI Localization

The palmprint region of interest (ROI) is localized using the following steps. Initially, the centroid of the hand is detected in order to align the hand depending on the vertical axis. Then, the tip of the middle finger is detected in order to

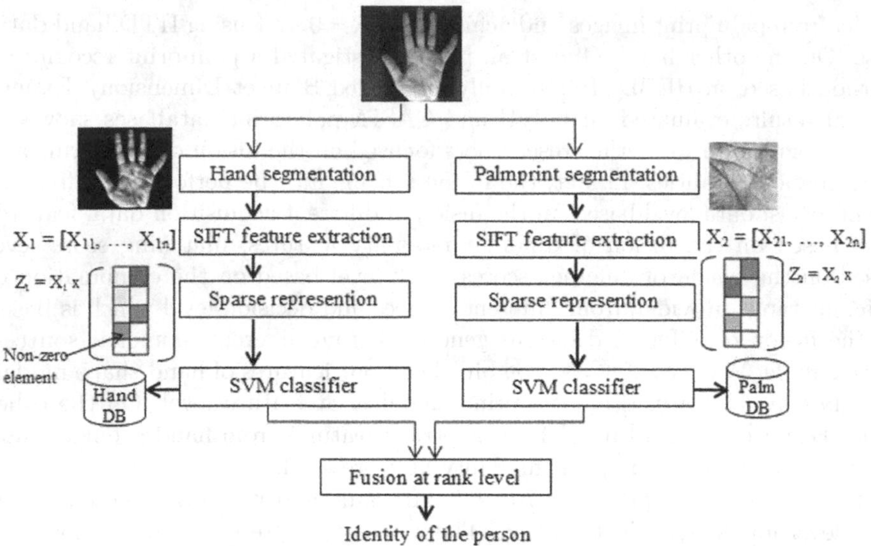

Fig. 1. Flowchart of the proposed identification method.

rotate the position of the hand according to the vertical axis. After, the valleys of the 5 fingers are localized with the new position of the hand image. The valleys points localized around the middle finger, presented by V_2 and V_3 in Fig. 2, are related by a reference line to intersect the hand contour. This intersection point is related to the valley point V_2 in order to localize the center point M_2 as presented in Fig. 2. The last step is repeated to localize the midpoint M_2 through the valleys points V_3 and V_4 (see Fig. 2). Finally, the two midpoints M_1 and M_2 are linked as the side of the square which represents the palmprint ROI.

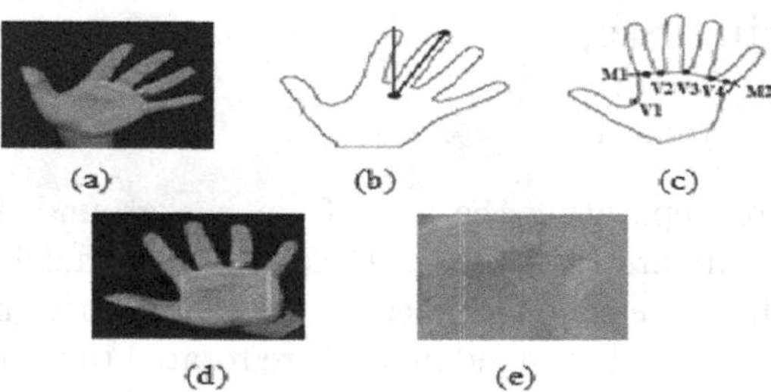

Fig. 2. Palmprint preprocessing: (a) Original palmprint image; (b) Rotation depending on the vertical axis; (c) Localization of two midpoints M_1 and M_2; (d) Localization of palmprint ROI; (e) Extraction of the palmprint image.

3 Sparse Feature Representation Step

In our work, the Scale Invariant Feature Transform (SIFT) is adopted for hand shape and palmprint modalities. In fact, this descriptor has been firstly proposed by Lowe [DL2] and widely emerged in computer vision as discriminant extracted features from images. SIFT method may be summarized as follows:

3.1 Scale Space Detection

In order to build the Gaussian scale space, the palmprint image is convolved using a subset of Gaussian kernels in various scales, via the following equation:

$$I(x, y, \sigma) = S(x, y, \sigma) * P(x, y) \tag{1}$$

where $I(x,y,\sigma)$ is the Gaussian kernel in scale σ, $P(x,y)$ is the segmented palmprint image and $S(x,y,\sigma)$ is a variable-scale function represented by:

$$S(x, y, \sigma) = \frac{1}{2\pi\sigma^2} e^{-}(x^2 + y^2)/2\sigma^2 \tag{2}$$

3.2 Keypoint Detection

The Key point is detected using the difference of Gaussians (DoG), represented by the Gaussian scale transform. The DoG is obtained by subtracting two neighbor scales separated by the scale factor f [DL1], as defined in Eq. (3).

$$D(x, y, \sigma) = I(x, y, f\sigma) - I(x, y, \sigma) \tag{3}$$

3.3 Keypoint Description

The key point is described depending on the orientation and the gradient magnitude of each key point. Each key point orientation is computed using the histogram of 36 orientations. The peak orientation in the histogram is regarded as the principal one for the current key point. To achieve rotation invariance, axis are rotated in accordance with the considered key point orientation. The description of the current key point is made using 16 sub-blocks (4×4) around. For each sub-block, the gradient orientation histogram is calculated with 8 orientations. Therefore, the feature vector for each key point is formed by $4 \times 4 \times 8 = 128$ values [DL1]. In this work, the shape of the hand is described by keypoints detected on the contour of the hand. In fact, the standard algorithm of SIFT descriptors localize keypoints inside the hand region rather than its external boundary. So, to detect keypoints on the hand contour, we firstly extracted the whole contour from the hand region. Next, we eliminated redundant keypoints of the contour. After that, the silhouette is sampled on the external boundary, with uniform spacing.

3.4 Sparse Representation

Each image of hand shape or palmprint modality is described by a matrix of SIFT description. So, the idea is to combine SIFT vectors of the matrix in order to describe discriminately each image using a unique vector. Sparse Representation is adopted in order to solve this idea. In fact, this method is based on the representation of training features approximately, by combining linearly basis vectors. Thus, the final feature vector of each image is represented as a signal $f \in \mathbb{R}^a$ using a dictionary Φ (basis vectors) and the sparse vector which contains weights (coefficients), as follows:

$$f = \Phi\alpha_0 \tag{4}$$

where $\Phi = [\Phi_1, \Phi_2, \ldots, \Phi_z] \in \mathbb{R}^{D\times a}$ $(a \ll D)$, is the matrix which contains the t training set related to z classes, $\alpha_0 = [0, \ldots, 0, \alpha_{p,1}, \alpha_{p,2}, \ldots, \alpha_{p,a_z}, 0, 0]^\top \in \mathbb{R}^s$ is the vector of coefficients where non zero values are concerned solely with the k-th class (person) and D represents the feature vector dimension of training images. In the case of large number of classes z, α_0 will be sparse. To learn an overcomplete dictionary, we adopted in this work the Lagrange dual technique [HL] which consists in searching the dictionary leading to the best possible representation. Thus, the dictionary is provided according to this optimization problem, defined as follows:

$$\min_{\alpha_i} \frac{1}{2}\|f_i - \Phi\alpha_i\|_2^2 + \lambda\|\alpha_i\|_1 \tag{5}$$

where α represents the sparse representation and $\|\alpha\|_1$ is the $l_1 - norm$ of α.

4 Classification and Fusion Steps

The classification step consists on comparing the test feature vector to the whole vectors stored in database, to form a one-to-many comparison, in order to find the identity of the person. In our work, the number of classes (persons) is higher than 100, so we adopted the multi-class linear SVM (Support Vector Machines) classifier thanks to its efficiency and performance. In fact, this method uses optimization to form a separating hyperplane for dissimilar classes and extend the safety margin between this hyperplane and the closest vectors of each class. For a training set $\{(u_i, v_i)\}_{i=1}^2$, $q_i \in$ Q = {1, ..., L}, the idea of a linear SVM classifier is to train L linear functions $\{p_s^\top u | c \in Q\}$. In fact, for a test sample u, we define the label of the predicted class as:

$$v = \max_{c\in Q} p_c^\top u \tag{6}$$

This aim allows solving the unconstrained convex optimization problem, as follows:

$$\min_{p_c}\{J(p_c) = \|p_c\|^2 + C\sum_{t=1}^{n} l(p_c; v_t^c, u_t)\} \tag{7}$$

where $v_t^c = 1$ if $v_t = c$, else $v_t^c = -1$, and $l(p_c; v_t^c, u_t)$ is the hinge loss function. Thus, the training step may be generated by employing a method of gradient-based optimization.

After the classification step, the fusion of hand shape and palmprint modalities is presented at rank level. In fact, this fusion is performed using scores provided from SVM classification method, which are affected to each sample of each person. These scores are considered as the belonging degree of each sample to the whole classes or persons stored in database. So, the idea is to convert the SVM scores to probability values, in order to determine the probability in which the sample s is belonging to the person x. The transformation of SVM scores to probability values is performed via the following steps.

4.1 Basis of Probability Knowledge

This basis is built from the score values space provided using SVM classifier. In our work, each sample is described sparsely and a feature vector is generated for each image. Secondly, the feature vectors of all images are fed to the SVM classifier which provides weights of belonging of each feature vector to each class (or person) as: $\omega[v_{k1}, v_{k2}, ..., v_{kM},]$. So, a probability distribution $Pr_{kn}(v_{kn})$ is estimated for each score value v_{kn}, where $k \in \{1, 2, ..., K\}$; K is the number of persons; $m \in \{1, 2, ..., M\}$ and Ω_{kM} represents the definition space related to v_{kn}.

4.2 Probability Distributions

To build the probability distribution, a function allowing the transformation of SVM scores into probability values is defined. In fact, for each score a triangular probability function is defined using the following steps: [HG].

- For each test image, the standard deviation of the triangular probability function is computed as follows:

$$D = \sqrt{\sum_{j=1}^{n} (s_j - p)^2} \tag{8}$$

  where n represents the total number of persons; s_j is the score of each person $j \in [1...n]$; $p = (\sum_{j=1}^{n}(s_j))/n)$.
- Build the triangular probability function according to the following steps:
  - Define the coordinates of lower and upper limits (l and u) of the triangular function. These coordinates are defined depending on the deviation D and the coordinate c of the peak location (when probability measure = 1). $(x_l^j, y_l^j) = (s_j - D, 0)$ and $(x_u^j, y_u^j) = (s_j + D, 0)$

- The triangular distribution is performed according to the probability density function, expressed as:

$$f(x) = \begin{cases} \dfrac{2\,(x - x_l)}{(x_u - x_l)(x_c - x_l)}\,; x_l \le x \le x_c \\ \dfrac{2\,(x_u - x)}{(x_u - x_l)(x_u - x_c)}\,; x_c \le x \le x_u \\ 0\,; x < x_l, x > x_u \end{cases} \quad (9)$$

- The probability measures determined from the function $f(x)$ are maintained for each sample of test.
- The previous steps are repeated for each sample.

In this work, we aim to discriminate between several persons. However, some ambiguities still exist for some persons which leads to a misclassification. Thus, we aim to compute the Ambiguity Ratio (AR) of each class, based on the ratio of probabilities of the second and first pertinent classes, expressed in Bounhas et al.'s work [MB] as:

$$AR(i, cl_1, ..., cl_m) = \frac{Pr(cl_2|i)}{Pr(cl_1|i)} \quad (10)$$

where cl_1 and cl_2 represents the first and second pertinent classes, respectively, of the sample i considered for classification.

The identification step is performed depending on the AR measure, calculated from our two biometric modalities (hand shape and palmprint traits). In fact, the decision about the final identity is achieved from the modality having the minimum value of AR.

5 Experimental Results

Our experiments are assessed using a publicly touch-less hand database called IITD hand database. In fact, 1150 hand images were acquired by a CMOS camera. These images were captured using a touch-less device, in bitmap format. In fact, left hand images have been collected from 230 persons, whose their ages vary between 14 and 56 years old, and five samples have been captured from each person.

The identification step consists in comparing a test sample to all templates stored in database. So, it is based on one-to-many comparisons. To assess our identification method, the correct identification rate (CIR) is computed in our experiments. In our work, SIFT descriptors are extracted from hand shape and palmprint modalities, and represented then sparsely via sparse representation method. Our experiments are performed on 230 persons using 3 and 4 training images per person. Regarding the hand shape modality, 300 keypoints detected on the external boundary of the hand and are adopted for SIFT description. The feature vector of each keypoint is formed by 128 features. For sparse representation step, we learnt a dictionary of SIFT patches (16×16) with a size of 1024.

Promising results are achieved with CIR = 96.15% for 3 training images and CIR = 96.96% for 4 training images, as seen in Table 1. Concerning palmprint modality, the standard SIFT algorithm is employed for sparse representation method. The size of dictionary is 1024 which is learnt from 16 × 16 SIFT patches. Our experiments demonstrate encouraging results by obtaining CIR = 96.73% using 3 training images and CIR = 97.83% using 4 training images. The fusion of hand shape and palmprint modalities was performed at rank level using the probability distribution functions. Our fusion method at rank level shows the efficiency of the proposed system (Table 1) for 3 training images by achieving CIR = 99.34% as well as for 4 training images by achieving CIR = 100% which are competitive rates in hand biometric field. Figure 3 illustrates the CMC (Cumulative matching characteristic) curve of the performance of proposed method for each modality alone and the fusion at rank level using 3 training images.

Table 1. Identification performances

Modality	Train	Test	CIR(%)
Hand shape	3	2	96.15
-	4	1	96.96
Palmprint	3	2	96.73
-	4	1	97.83
Hand shape + palmprint at rank level	6	4	99.34
-	8	2	100

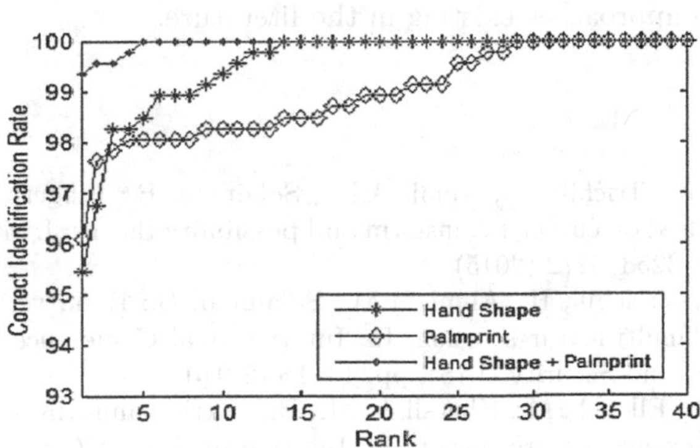

Fig. 3. Cumulative matching characteristic (CMC) curve for the proposed approach.

A comparison of performances between the proposed approach and other existing approaches on the literature is presented in Table 2. This comparison is performed using the same database (IITD hand database) and fusing the same biometric modalities. The proposed method presents better performances (CIR = 99.34%) compared to the method of Ferrer et al. [MF] and Charfi et al. [NC5]. In fact, Ferrer et al. [MF] fused geometric features of the hand and wavelet filter of the palmprint modality, at feature level and they obtained a CIR = 99.21%. On the other hand, Charfi et al. [NC5] proposed a bimodal method fusing SIFT descriptors from hand and palmprint modalities at score level and a CIR = 97.82% was achieved.

Table 2. Performance comparison to other existing methods using IITD hand DB

Method	Features	Fusion level	IR(%)
Ferrer et al. 2011 [MF]	Geometric features and Wavelet filter	Feature	99.21
Charfi et al. 2014 [NC5]	SIFT descriptors	Score	97.82
Proposed method	SIFT sparse features	Rank	99.34

6 Conclusion

The proposed identification method fusing hand shape and palmprint modalities is described, in this paper. These two modalities are represented using SIFT sparse representation method. The fusion is performed at rank level by transforming scores generated by SVM classifier to probability measures using triangular probability distributions. This approach has proven its efficiency by achieving promising performances CIR = 99.34% which are competitive to other hand recognition approaches existing in the literature.

References

[HG] Guesmi, H., Trichili, H., Alimi, A.M., Solaiman, B.: Fingerprint verification system based on curvelet transform and possibility theory. J. Multimedia Tools Appl. **74**, 3253–3272 (2015)

[NC1] Charfi, N., Trichili, H., Alimi, A.M., Solaiman, B.: Hand verification system based on multi-features fusion. In: International Conference on Information Assurance and Security (IAS), pp. 13–18 (2015)

[RM] Mokni, R., Elleuch, M., Kherallah, M.: Biometric palmprint identification via efficient texture features fusion. In: International Joint Conference on Neural Networks (IJCNN), pp. 4857–4864 (2016)

[ZD] Darojah, Z., Ningrum, E.S.: The extended Kalman filter algorithm for improving neural network performance in voice recognition classification. In: Intelligent Technology and Its Applications, pp. 225–230 (2016)

[VA] Anikin, I.V., Anisimova, E.S.: Handwritten signature recognition method based on fuzzy logic. In: Dynamics of Systems, Mechanisms and Machines, pp. 1–5 (2016)

[JG] Guo, J.M., Hsia, C.H., Liu, Y.F., Yu, J.C., Chu, M.H., Le, T.N.: Contact-free hand geometry-based identification system. Expert Syst. Appl. **39**, 11728–117361 (2012)

[LB] Luque-Baena, R.M., Elizondo, D., Ezequiel, L.R., Palomo, E.J., Watson, T.: Assessment of geometric features for individual identification and verification in biometric hand systems. Expert Syst. Appl. **40**, 3580–3594 (2013)

[DH] Hong, D., Liu, W., Wu, X., Pan, Z., Su, J.: Robust palmprint recognition based on the fast variation Vese-Osher model. Neurocomputing **174**, 999–1012 (2016)

[XG] Guo, X., Zhou, W., Wang, Y.: Palmprint recognition algorithm with horizontally expanded blanket dimension. Neurocomputing **40**, 152–160 (2013)

[CW] Wang, W.C., Chen, W.S., Shih, S.W.: Biometric Recognition by Fusing Palmprint and Hand-geometry Based on Morphology. In: International Conference on Acoustics, Speech and Signal Processing, pp. 893–896 (2009)

[NC2] Charfi, N., Trichili, H., Alimi, A.M., Solaiman, B.: Bimodal biometric system for hand shape and palmprint recognition based on SIFT sparse representation. Multimedia Tools Appl. 1–26 (2016)

[NC3] Charfi, N., Trichili, H., Alimi, A.M., Solaiman, B.: Local invariant representation for multi-instance toucheless palmprint identification. In: IEEE International Conference on Systems, Man, and Cybernetics (SMC), pp. 3522–3527 (2016)

[MF] Ferrer, M.A., Vargas, F., Morales, A.: BiSpectral contactless hand based biometric system. In: Conference on Telecommunications (CONATEL), pp. 1–6 (2011)

[DL1] Lowe, D.G.: Distinctive image features from scale-invariant keypoints. Int. J. Comput. Vis. **60**, 91–110 (2004)

[DL2] Lowe, D.G.: Object recognition from local scale-invariant features. In: International Conference on Computer Vision, pp. 1150–1157 (1999)

[MB] Bounhas, M., Mellouli, K., Prade, H., Serrurier, M.: Possibilistic classifiers for numerical data. Soft. Comput. **17**, 733–751 (2012)

[NC4] Charfi, N., Trichili, H., Alimi, A.M., Solaiman, B.: Personal recognition system using hand modality based on local features. In: International Conference on Intelligent Systems Design and Applications (ISDA), pp. 189–194 (2015)

[EY] Yoruk, E., Konukoglu, E., Sankur, B., Darbon, J.: Shape-based hand recognition. IEEE Trans. Image Process. **15**, 1803–1815 (2006)

[HL] Lee, H., Battle, A., Raina, R., Ng, A.Y.: Efficient sparse coding algorithms. In: NIPS, pp. 801–808 (2007)

[NC5] Charfi, N., Trichili, H., Alimi, A.M., Solaiman, B.: Bimodal biometric system based on SIFT descriptors of hand images. In: IEEE International Conference on Systems, Man, and Cybernetics (SMC), pp. 4141–4145 (2014)

[NC6] Charfi, N., Trichili, H., Alimi, A.M., Solaiman, B.: Personal verification system using hand modalities. J. Inf. Assur. Secur. **11**, 157–168 (2016)

Adaptation to Market Development Through Price Setting Strategies in Agent-Based Artificial Economic Model

Petr Tucnik(✉), Petr Blecha, and Jaroslav Kovarnik

University of Hradec Kralove, Hradec Kralove, Czech Republic
{petr.tucnik,petr.blecha,jaroslav.kovarnik}@uhk.cz

Abstract. The paper is focused on the incorporation of costs in relation to the selection of the price setting strategies which, in general, represent crucial part of economic agents' decision making. Study and research of efficient decision making related to the price setting are important especially in agent-based economic systems which are intended application area for obtained results. This paper provides detailed description of various cost types used in traditional economic analysis and an effort has been made to identify constant (i.e. stable) and/or dynamic factors, typically related to the volume of production) in the cost calculation. Simulation part supports provided discussion about these design questions of artificial economic models in several scenarios.

Keywords: Agent · Cost calculation · Virtual economy · Agent-based economic model · Price setting

1 Introduction

The selection of the proper price making strategy is crucial component for establishment of functional trading relations. The price is primary (and in some cases sole) indicator of potential partner's suitability when considering trade. While this paper is focused primarily on the segment of industrial production (since it is most frequently used for application of research results), mechanics for establishment of trading relationships with partners are similar for all participants in the distribution chain (such as service providers, distributors, end customers, raw materials suppliers, etc.). The persistent goal of any economic entity is to secure sustainable sales, supply, cashflow, and its existence in general. Because the market environment is highly dynamic and individual agents representing companies can adjust their behaviour to improve their situation, it is important to focus on study of possible approaches how to do it effectively to maximize performance of the entire system.

The motivation for this line of research comes as a reflection of a today's trend of increasing automation of industrial production. With the concept of Industry 4.0 [1], autonomy of a plant control grew continually from the operational to tactical level, and in future years this trend may possibly impact even strategic decision making. With application of the agent-based simulation approach as a decision support tool, adjustment of individual approach has potential to be beneficial for company decision makers or for optimization of company's internal processes. Positive contribution can be even

N.T. Nguyen et al. (Eds.): ICCCI 2017, Part I, LNAI 10448, pp. 548–557, 2017.
DOI: 10.1007/978-3-319-67074-4_53

more significant in cases of multi-plant production [2] because geographical distribution of facilities and variability of local specifics could be difficult to grasp or simply too time demanding when quick reaction is needed.

The model used for experimental purposes works under following constraints: (i) it is a closed economy with a single currency, (ii) financial markets and institutions are not taken into consideration, (iii) services are not included (industrial products only). More information about our model may be found in [3]. Although this study uses only virtual "lab environment" for agents to interact within, establishment of such artificial economic system allows us to research important questions related to such system`s construction and functioning.

2 Relevant Works

The primary interest of this paper is decision making related to the price setting. Although the output of this decision-making process is quite straightforward (value of market price for goods), it cannot be separated from the internal processes of the company and various aspects of company operation (supplies, human resources management, maintenance, production volume, technology of production etc.) which are to be considered as well. We may consider final product price to be a major factor contributing to the (economic) performance of the company. Although the company can dynamically adjust only a part of factors contributing to the market price of products (see Sect. 3), such as profit margin, there are various strategies which can be implemented to influence the overall performance of the company. The key aspects related to out topic could be divided into two primary areas: (i) adaptation and (ii) supply chain management/logistics.

The adaptation of the company can be achieved in several ways. The idea of self-adaptive manufacturing is thoroughly discussed by Busaferri et al. [2]. Busaferri states that the future factory automation systems should be able to promptly react to changing exogeneous conditions, like consumer expectations, market dynamics, design innovation, new materials, and components integration. Some of these aspects are related to manufacturing, but there are obvious market factors strongly related to establishment of the flexible (reconfigurable) manufacturing systems.

Another example of adaptive economic model is the adaptive production network described by Hamichi et al. [4]. There is a strong similarity with our model of production chain because the Hamichi's concept of a "network" is a grid consisting of layers with nodes representing firms, where layering begins with input of natural resources and ends in supermarket layer for end customers. Hamichi also describes mechanism for profit and production capacity management. There is certain abstraction involved because profits are initially placed in liquid assets and can be used for the modification of production if needed and the change projects into next time iteration of the model. The diverse types of costs are not recognized in detail.

Other studies focus on different aspects of adaptive systems, apart from manufacturing. A dynamic adaptation is studied from the perspective of scheduling systems on virtual market in real time by Skobelev et al. [5], or through genetic algorithms [6]. More general theoretical approach towards self-adaptiveness using model-based reasoning offers Steibauer and Wotawa [7]. Incorporation of cognitive elements into decision making processes is implemented in the concept of so-called "cognitive factory" [8].

With gradual paradigm shift from simple operations management research to the research and study of more complex concept of automated factories, the attention can be quite naturally extended to the area of autonomous logistics. The logistics is an integral part of internal business processes and is crucial for efficient supply chain management and smooth business operation. The motives for autonomous control and handling of logistics tasks summarizes Gehrke [9] who also presents various views on autonomous control of these processes.

3 Costs and Price Setting

3.1 Cost Oriented Price

The process of setting a price is crucial process for every company. Different approaches can be used for the setting of the final price, but the most frequently used are cost-oriented price, competition-oriented price, or sell-oriented price. Some of these approaches are described for example in [10] or [11]. The goal of every agent is to maximize its satisfaction in general, which can be represented namely by profit, living conditions, food supplies level, etc. Since the trade is primary topic here, it will be considered a singular objective of agents within this paper. For the maximization of profit, SA (Store Agent – it represents stores companies) agents usually apply strategy of profit margin manipulation to increase/decrease price of product on the market to weaken the competition and become more dominant on the market, which might result in higher turnover, etc.

As far as competition and sell-oriented prices are concerned, if somebody wants to use these methods, the knowledge of the other subjects on the market is essential. It is almost impossible to use these methods if you do not know the prices of your competitors or the needs of your potential customers. Moreover, the price established based on these methods may not lead to profit, or more precisely a net income. Net income can be calculated as a difference between the revenues and expenditures, while total revenues are the number of units sold multiplied by the price. It is obvious that the price established as the competition or sell-oriented does not take into account the expenditures of the company, therefore such price can be very low and it can lead into situation, where the expenditures are higher and company has a net loss.

Third possible approach solves this potential problem. Cost-oriented price is calculated directly based on the expenditures of the company. In case that company sells all its units, such price will certainly lead into net income.

For agent-based systems is this cost-oriented price probably the best possible solution. Costs are divided into two important groups in this method, namely direct costs, and indirect costs, based on whether they can be calculated to the cost object, or not. Cost object may be a product, a department, a service, a project, etc.

Direct costs can be defined as costs which can be accurately traced to a cost object with little effort. Indirect cost (in other words so called overhead) cannot be attributed to specific cost objects. These typically conclude multiple cost objects and it is impracticable to accurately trace them to individual product. However, the classification of any cost either as direct or indirect is done by taking the cost object into perspective. Therefore, a particular cost may be direct cost for one cost object and indirect cost for another cost object.

Because of the indirect costs are not set on individual product, these amounts are usually significantly higher than the amounts of direct costs. Therefore, there exist a huge amount of different methods how to split indirect costs per one unit as well. The simplest method is called simple calculation of dividing, while this method is probably the best option for agent-based simulation. The principle of this method is to divide the indirect costs by number of expected units. The disadvantage of this method is obvious. Direct costs are traced to particular product or service; therefore, the total amount of these costs grow together with the amount of created products and vice versa. In other words, the direct costs per one unit remain same no matter how many products have been created. The indirect costs are usually fixed and therefore the amount per unit will be different according to created units, but it is not possible to know this number at the moment of calculation, it is only estimated. Therefore, companies always prepare one calculation before the manufacturing process with expected number of units, and another calculation after the manufacturing process with real values both of costs (direct and indirect ones) and of units.

In accounting system of the Czech Republic is usually used so called general calculation formula, where direct and indirect costs are divided into several groups, namely: (1) direct material, (2) direct salaries, (3) other direct costs, (4) manufacturing overheads, (a) own production costs, (5) administrative overheads, (b) own serving costs, (6) sales overheads, (c) total own serving costs, (7) profit margin, and (d) Total Price. Items (a), (b), (c), and (d) present sum of above mentioned items, precisely a) is sum of 1 to 4, (b) is (a) plus 5 (or costs 1 to 5), etc. Profit margin can be established as a fix amount, but more frequently as percent from previous sum, usually from total own serving costs. Final price subsequently present sum of all direct and indirect costs per one unit plus profit margin. Therefore, if the company sells all expected units (where this amount has been used as the basis for dividing of indirect costs), not only all costs will be covered, but also expected profit will be achieved.

3.2 Different Strategies for Price Establishment

If we use the traditional terminology used for description of agent-oriented system, the agent can be perceived as an entity functioning in an infinite sense-think-act cycle formed between agent and its environment (with all other agents). The strategies are basically actions which agent carries out to influence the environment and market response (demand) as a feedback which can be used to adjust its behavior. It is a matter of the elementary rationality that the general goal is to maintain prices of inputs as low and prices of outputs as high as it is possible. However, all suppliers and customers do the same. Moreover, in a competitive environment, actions of every competitor have impact as well. This all contributes to market dynamics and crucial for the rational decision-making is market feedback. Key for the survival of the fittest is fast adaptation and this is reason why the underlying goal of our research is self-adaptation through strategy selection.

As was mentioned above, the cost-based price seems to be the best choice for agent-based system. However, it is possible to analyse different strategies (or kind of behaviour) of every agent in terms of price establishment. The microeconomic theory shows that the profit can be calculated as follows (from [12]):

$$\P = TR - TC, \quad (1)$$

where TR means total revenues and TC means total costs. These indicators can be subsequently calculated as follows:

$$TR = P * Q, \quad (2)$$

$$TC = FC + VC = FC + (Q * VC \text{ per unit}), \quad (3)$$

where P means price, Q is number of produced (sold) units, FC are fixed costs, and VC are variable costs, where these costs can be calculated as number of produced units multiplied by the variable costs per one unit. Fix costs are payments which remain same no matter what, for example, rent, salaries of managers, advertisement, etc. Moreover, these amounts are paid even if there is no production at all. On the other hand, variable costs are directly traced to the production, such as material, salaries of workers, energy, etc. These amounts are zero in case when nothing is produced and grow together with the number of produced unit.

However, question remains whether the company stop its production in case of loss or not. The answer is not as easy as it may appear. In every company exists so called point of closure, which is a point where the total revenues are lower than variable costs. It is obvious that company starts with loss at the very beginning of its existence because it has to pay fix costs, even if there is no production yet. However, when the first unit is produced, total revenues are created (in the amount of price, because Q = 1, and therefore TR = P * 1), but also variable costs are created. If the price is higher than these variable costs, the production is realised, because the price can cover whole variable costs and at least part of fix cost. That means that because of this production, the loss is lower. However, in case that this price should be lower than variable costs, company will not produce anything. Final loss will be lower (only in the amount of fix costs), while in case of production the price cannot cover the variable costs and the loss would be higher (as a sum of fix costs and not covered part of variable costs).

Probably the most important point in existence of every company is so called point of profit maximization. As was already described, the profit can be calculated as a difference between total revenues and total costs, therefore the point of profit maximization is when marginal revenues are equal to marginal costs.

The behaviour of agents in agent-based system can be, according to above explained theories, relatively different.

Agents Oriented on the Quantity

There can be such agent, which either does not have unique product or operates in the environment close to perfect competition. This agent will take the price as a constant established on the market, calculated by other competitors and accepted by the customers.

After that, this agent compares the price with its costs and different situations can occur depending on the size of different types of costs:

- If the price is lower than its variable costs, the company will not produce anything and will create the loss in the amount of fix costs (point of closure).
- If the price is higher than its variable costs, the company will start the production. It will create loss at the beginning, but this loss is lower than in the first case because price manages to cover not only variable costs but part of the fix costs as well.

Agents Oriented on the Price
There can be such agent, which used above described cost-based price, where some particular amount of production is used as a base for dividing of overhead costs per one unit. Final price is consequently calculated as a sum of total costs (both direct and overhead) and profit margin, where this agent deals with the limitation whether it is possible to sell all calculated units with such price (whether the market accepts this calculated price or not).

The behaviour of such agent can be described as follows:

- The agent tries to sell its production with calculated price.
- If the price is too high, but the agent wants to sell whole production, the price has to be reduced. The profit margin will decrease.
- If the price is still decreasing, there will be no profit, but the loss occur.
- If the price is lower than variable costs (direct ones), agent will stop the production (point of closure).

4 Simulation Scenarios

4.1 Simulation Set-up

Each of the three simulations described in this section begins with the same initial parameters. The key performance indicator in the model is the capital measured in days of model runtime. Each company begins with random amount of the capital (500 k–1.000 k (of unspecific units of measurement)). Storage level is at the zero level, i.e. no goods are stored beforehand. Companies have to obtain the semi-products from the market first in order to start production and compute final price of products.

The timeframe for each simulation is always set to 120 days of model runtime (which were empirically evaluated as sufficient time range). There are five production companies with the same type of products in a single city. These companies have initial profit level set to 25%, after first thirty days of model runtime, and each time a company with the least capital is selected and one of the new strategies is applied to it. Important role plays amount of stored goods because implemented model has mechanism for modifying price of the contracts according to the used storage capacity. When the storehouses are full, price is lower, and vice versa. This mechanism allows and supports more dynamic adjustment to the changes in market price.

4.2 Simulation #1 – Double Profit Strategy

The Simulation #1 measures impact of increase in profit level to double rate (25% → 50%) in the least efficient company (labelled Factory0 in the Fig. 1). Even with (randomized) good initial position with capital of 890 k, this company was unable to compete in highly competitive environment during the first month and earned significantly less profit than other companies. After thirty days of model runtime, the company made change in policy and adopted the "Double Profit Strategy".

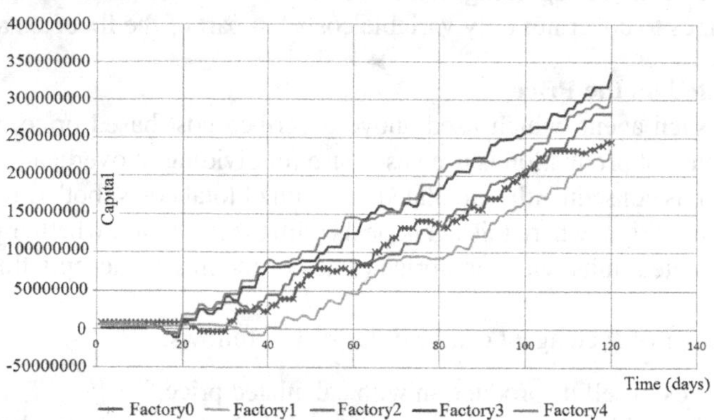

Fig. 1. Simulation #1 – Double profit strategy (Factory 0).

As it is shown in the Fig. 1, the new profit level leads to almost immediate increase in sales. This is result of the favourable market situation, where demand for the product is high and non-saturated by the competition. After two months, there is a change in the trend and Factory0 becomes third in the profit ranking. This can be considered the trend breakpoint and profit is lowered after this period.

4.3 Simulation #2 – Lower Profit Strategy

The Simulation #2 was used to test impact of decrease of the profit margin with two of the least profitable companies (Factory0, Factory4) within the city region after one month. With the initial capital of 880 k and 200 k, Factory0 was in the best and Factory4 in the worst starting position, respectively.

Results of the Simulation #2 are shown in the Fig. 2. Both companies had sustained significant loss after the first month of the model runtime. Adopted lower profit strategy leads to decline of the profit rate from 25% to 10%. In the subsequent month, both companies experienced significant increase in capital, especially Factory0 (with higher initial capital, thus more favourable starting position) became able to compete against other companies with higher initial profits. The reason is lower purchase price of the product which results in the increase of sales volume. Both companies (Factory0, Factory4) became profitable (Fig. 3).

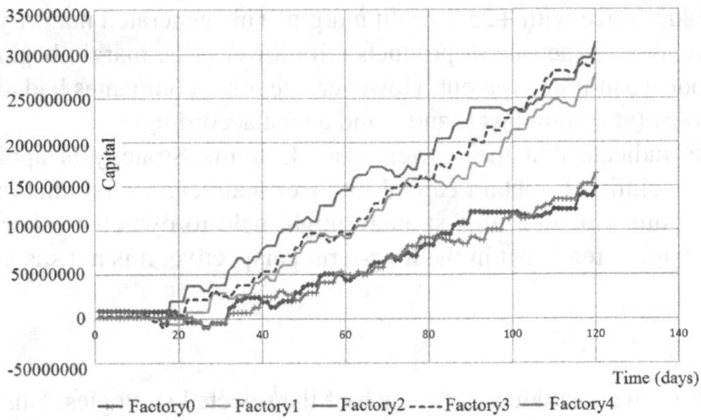

Fig. 2. Simulation #2 – Lower profit strategy (Factory 0, Factory 4).

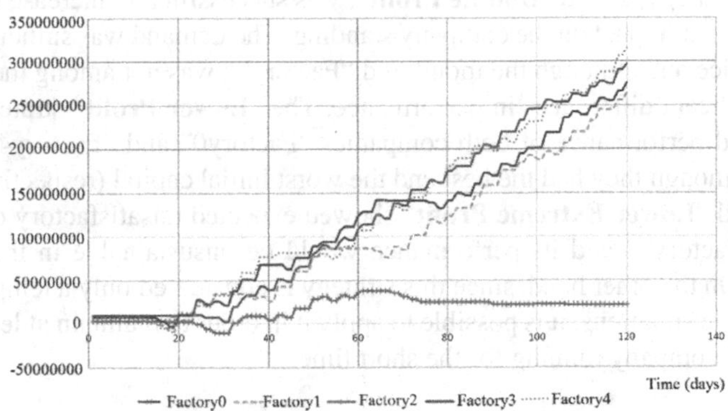

Fig. 3. Simulation #3 – Lower profit extreme strategy (Factory 2).

In the long run, this strategy has not lead to better overall performance. Stored products accumulated during a low-profit period were sold and other companies were able to afford lower price margin as well, which allowed them to maintain their market share. This slowed down improvement of the both followed companies and market participants adjusted to the adopted strategy.

4.4 Simulation #3 – Lower Profit Extreme Strategy

The last strategy used extreme reduction of the profit margin even when facing negative values – specifically, products are being sold at lower than production price. This approach might be valid in certain situations, e.g. when company has large quantities of products for sale and not enough finances to cover maintenance costs.

According to the obtained data, Factory2 had the worst performance during the first month. Radical change of strategy (decrease in profit from 25% to −10%) was expected to improve financial balance. However, the profit in the time period immediately following the first month was generated mostly through the sales of the remaining contracts for older price with +25% profit margin. This generated majority of profit in this period. As soon as the sale of products with lower price margin began, there was perceivable short-term improvement. However, the other companies had also products for sale in storage (at certain level) and could adjust accordingly.

The results indicate that the Lower Profit Extreme Strategy is applicable only as a short time solution to obtain capital to cover maintenance/production costs. For companies in a dire situation, this strategy might help to overcome short periods of unfavourable market trends but in the long-term perspective, it is not sustainable.

4.5 Results

Three simulations were conducted for each of the selected strategies. During all three simulations, there is an initialization period which lasts approximately for the first 20 days of the model runtime. After initialization, the following results were obtained:

The first strategy, called "**Double Profit**", was successful and increase of the profit had no significant impact on the company standing. The demand was sufficient to cover increase in price and although the monitored "Factory0" was not among the best, there were no significant differences in performance. The "**Lower Profit**" strategy was less successful and performance of both companies "Factory0" and "Factory4" were less successful, although they had the best and the worst initial capital (respectively). Third strategy called "**Lower Extreme Profit**" showed expected unsatisfactory outcome for monitored "Factory2" and its performance would be unsustainable in the long-term perspective. On the other hand, since this strategy is considered only a temporary solution for extreme situations, it is possible to apply it in order to maintain at least minimal cash flow and company running for the short time.

5 Conclusion and Future Work

It has been shown that with a limited number of participants, simulating small, closed, geographically localized economy, there is a strong feedback loop between all participating agents on the market, and intensive mutual influence resulting from their decisions. Presented simulations showed potential for testing more complex market behaviour patterns in such virtual environments. All three simulations were focused on a single product market, which is, obviously, far from the real-world situation. However, future proceeding in simulation will be focused on involvement of several layers of production chain simultaneously, with gradually increasing complexity up to the involvement of all distribution chain participants, thus creating system which is dynamically complete. Together with involvement of more market participants, this is intended for the future work.

Acknowledgements. The Financial support of the Specific Research Project "Autonomous Socio-Economics Systems" of FIM UHK is gratefully acknowledged.

References

1. Premm, M., Kirn, S.: A multiagent systems perspective on industry 4.0 supply networks. In: Müller, P.J., Ketter, W., Kaminka, G., Wagner, G., Bulling, N. (eds.) MATES 2015. LNCS, vol. 9433, pp. 101–118. Springer, Cham (2015). doi:10.1007/978-3-319-27343-3_6
2. Brusaferri, A., Ballarino, A., Carpanzano, E.: Distributed intelligent automation solutions for self-adaptive manufacturing plants. In: Ortiz, Á., Franco, R.D., Gasquet, P.G. (eds.) BASYS 2010. IAICT, vol. 322, pp. 205–213. Springer, Heidelberg (2010). doi: 10.1007/978-3-642-14341-0_24
3. Bureš, V., Tučník, P.: Complex agent-based models: application of a constructivism in the economic research. Ekon. Manage. **17**, 152–168 (2014)
4. Hamichi, S., Brée, D., Guessoum, Z., Mangalagiu, D.: A multi-agent system for adaptive production networks. In: Di Tosto, G., Van Dyke Parunak, H. (eds.) MABS 2009. LNCS, vol. 5683, pp. 49–60. Springer, Heidelberg (2010). doi:10.1007/978-3-642-13553-8_5
5. Skobelev, P., Budaev, D., Laruhin, V., Levin, E., Mayorov, I.: Practical approach and multi-agent platform for designing real time adaptive scheduling systems. In: Corchado, J.M., et al. (eds.) PAAMS 2014. CCIS, vol. 430, pp. 1–12. Springer, Cham (2014). doi: 10.1007/978-3-319-07767-3_1
6. Chen, P., Zhu, L., Li, X.: Multi-resource balanced scheduling optimization based on self-adaptive genetic algorithm. In: Cai, Z., Tong, H., Kang, Z., Liu, Y. (eds.) ISICA 2010. CCIS, vol. 107, pp. 19–28. Springer, Heidelberg (2010). doi:10.1007/978-3-642-16388-3_3
7. Steinbauer, G., Wotawa, F.: Model-based reasoning for self-adaptive systems – theory and practice. In: Cámara, J., de Lemos, R., Ghezzi, C., Lopes, A. (eds.) Assurances for Self-Adaptive Systems. LNCS, vol. 7740, pp. 187–213. Springer, Heidelberg (2013). doi: 10.1007/978-3-642-36249-1_7
8. Zaeh, M.F., Ostgathe, M., Geiger, F., Reinhart, G.: Adaptive job control in the cognitive factory. In: ElMaraghy, H.A. (ed.) Enabling Manufacturing Competitiveness and Economic Sustainability, pp. 10–17. Springer, Heidelberg (2012). doi:10.1007/978-3-642-23860-4_2
9. Gehrke, J., Herzog, O., Langer, H., Malaka, R., Porzel, R., Warden, T.: An agent-based approach to autonomous logistic processes. Künstl. Intell. **24**, 137–141 (2010)
10. Walker, J.: Chapter 9 - The analysis of cost. In: Accounting in a Nutshell, 3rd edn., pp. 227–253. CIMA Publishing, Oxford (2009)
11. Mahanty, A.K.: Chapter Nine - Theory of costs. In: Intermediate Microeconomics with Applications. Academic Press, New York, pp. 211–239 (1980)
12. Weller, P.: 15 - Economic theory 2: Microeconomics A2 - Fletcher, John. In: Information Sources, 2nd edn., pp. 228–233. Butterworth-Heinemann, Oxford (1984)

Efficacy and Planning in Ophthalmic Surgery – A Vision of Logical Programming

Nuno Maia[1], Manuel Mariano[2], Goreti Marreiros[3], Henrique Vicente[4,5], and José Neves[5(✉)]

[1] Departamento de Informática, Universidade do Minho, Braga, Portugal
nuno.maia@mundiservicos.pt
[2] Centro Hospitalar Baixo Vouga, EPE, Aveiro, Portugal
msmmariano@gmail.com
[3] GECAD – Grupo de Engenharia do Conhecimento e Apoio à Decisão, Departamento de Engenharia Informática, Instituto Superior de Engenharia do Porto, Porto, Portugal
goreti@dei.isep.ipp.pt
[4] Departamento de Química, Escola de Ciências e Tecnologia, Universidade de Évora, Évora, Portugal
hvicente@uevora.pt
[5] Centro Algoritmi, Universidade do Minho, Braga, Portugal
jneves@di.uminho.pt

Abstract. Different variables should be considered in order to identify the critical aspects that influence ophthalmologic surgery and, in particular, the patient's conditions that can become the key factor in this process, i.e., in situations that can influence the stability and surgery of the patient. Protocol of ophthalmologic surgery has as main concern *Glycemic Index*, *Maximum Blood Pressure*, *Abnormal Cardiac Index*, and *Cardiac-Respiratory Insufficiency*. Such variables will be used to construct a dynamic virtual world of complex and interacting entities that map real cases of surgical planning situations, understood here as the terms that make the extensions of mathematical logic functions that compete against one another in a rigorous selection regime in which fitness is judged by one criterion alone, its *Quality-of-Information*. Indeed, one focus is on the development of an *Evolutionary Clinical Decision Support System* to evaluate patient stability and assist the physicians in the decision of doing or postponing surgery, once cataract is the leading cause of blindness in the world.

Keywords: Ophthalmologic surgery · Logic programming · Evolutionary case based reasoning · Knowledge representation and reasoning

1 Introduction

Health, which is considered as the state of being free from disease or injury, is one of the key aspects in the competitive capacity of societies. This concern is reflected in the *European Commission's report Towards Social Investment for Growth and Cohesion* [1], which sets the role of health as a strategic vector in the *Europe 2020 Program*. Indeed, the current trend of demographic evolution in most developed countries lies in

N.T. Nguyen et al. (Eds.): ICCCI 2017, Part I, LNAI 10448, pp. 558–568, 2017.
DOI: 10.1007/978-3-319-67074-4_54

the phenomenon of population aging [2]. This puts pressure on governments, since national authorities must maintain the quality of health and lifestyle of their citizens. Thus, health fund management has become a priority in the allocation of available resources, where criteria of rigor and fairness are the fundamental principles for ensuring sustainability. In fact, efficiency and planning are closely linked. The former one can be defined as the result of an optimization process where the inputs are the resources available with the objective of guaranteeing the best allocation, as well as their profitable use [3].

The latter, on the other hand, is a technique designed to identify the critical activities of a particular process and to prioritize the allocation of the resources needed to achieve the required objectives. Planning is linked to the coordination of all activities to ensure that resources (e.g., technical, technological, human) are available and operational for the development of tasks that lead to desired outcomes. In other words, planning is a systematization of the activities associated with the process that guarantees an improvement of efficiency.

When comparing the actual performance of the *Health Units* (*HCUs*), a paradox arises, i.e., they obtain different results despite the similarities in organization and resources. This fact highlights the difficulty in establishing and implementing standard solutions to guarantee the improvement, sustainability and greater efficiency of *HCUs* [4]. To meet this challenge, it is necessary to have consistent methodologies for problem solving as well as computational techniques to deal with unknown, incomplete or even self-contradictory data, information or knowledge.

Thus, the development of *Evolutionary Decision Support Systems* to evaluate patient stability can be an asset to health professionals, in particular to assist them in the decision of doing or postponing surgery. Thus, this work aims to specify an Evolutionary Clinical Decision Support System to plan Cataract Surgery based on a set of historical data, under an *Evolutionary Case-Based Reasoning* (*ECBR*) approach to problem solving [5–11].

This paper encompasses five sections. Thus, the former one includes an introduction to the problem. Then the proposed approach to *Knowledge Representation* and *Reasoning* and an *ECBR* view to computing are introduced. In the third and fourth sections it is set a case study and presented a solution to a patient's illness state. Finally, in the last section we look at the outcome of this work and future work is delineated.

2 Related Work

2.1 Surgical Efficiency and Planning

Surgical efficiency aims to minimize the costs of a surgical procedure. However, the planning process that leads to the improvement of surgical efficiency depends on a set of interdependent factors [12–14], being influenced by:

- The reliability of the diagnosis that makes the screening of the pathology and determines the more appropriate surgical procedures;
- The material resources available for surgery (e.g., operating rooms, materials, and equipment); and
- The availability of human resources both in number, and in skills.

Different techniques were tested to ensure maximum efficiency in the planning of the surgical procedures, like the *Health in All Policies* (*HiAP*) method [12]. However, in many situations, practice shows that discrepancies between the estimated and achieved results are significant. Indeed, in addition to the previously described factors, there are personal ones, which differ from patient to patient, that are crucial in order to minimize delays and/or postponements. The patient's stability is one of them. Although it depends on a set of physiological variables (evaluated through standard procedures, according to the type of surgery and the psychological and emotional level of the patient), patient's stability cannot be assessed in a deterministic way [15]. Indeed, even following all the procedures of the surgical protocol, there is a set of psychological and emotional variables that are uncontrollable. These variables depend on the relationship with the patient – physician (or medical team), where respect, trust, loyalty, friendliness, and psycho-emotional moods and feelings influence the patient's stability [16]. Another critical aspect is related with the choice of the anaesthesia type (e.g., local, topical, topical with sedation). This choice should weigh, on one hand, the patient's stability and, on the other hand, the ability to regulate the patient's anxiety. On the other hand, the use of less aggressive anaesthesia type, which will have benefits in terms of effects and costs, but is only possible when the patient is calm [17]. Indeed, their results indicate that Surgical Efficiency and Planning involves interaction with different human decision makers, namely in social, organizational, or economic settings. Undeniably, this is not surprising given the fact that qualities like autonomy, communication or planning are characteristic of human beings, and therefore such factors that will be considered in the work that follows.

2.2 Knowledge Representation and Reasoning

Regarding the computational paradigm it were considered extended logic programs with two kinds of negation, classical negation, ¬, and default negation, *not* [5, 6]. An *Extended Logic Program* is a finite set of clauses as shown in Program 1.

$\{$

$\neg p \leftarrow not\ p, not\ exception_p$

$p \leftarrow p_1 \wedge \cdots \wedge p_n \wedge not\ q_1 \wedge \cdots \wedge not\ q_m$

$?\ (p_1 \wedge \cdots \wedge p_n \wedge not\ q_1 \wedge \cdots \wedge not\ q_m)(n, m \geq 0)$

$exception_{p_1}$

$\cdots$

$exception_{p_j}\ (0 \leq j \leq k)$, *being k an integer*

$\} :: scoring_{value}$

Program 1. The Archetype of a Generic Extended Logic Program.

where the first clause of Program 1 depict the predicate's closure, "$\wedge$" denotes "*logical and*", while "*?*" is a domain atom denoting "*falsity*". The "p_i, q_j, and p" are classical ground literals, i.e., either positive atoms or atoms preceded by the classical negation sign "$\neg$" [5]. Indeed, "$\neg$" stands for a "*strong declaration*" that speaks for itself, and "*not*" denotes "*negation-by-failure*", i.e., a flop in proving a given statement, once it was not declared explicitly. According to this formalism, every program is associated with a set of "*abducibles*" [18, 19], given here in the form of "*exceptions*" to the extensions of the predicates that make the program. In order to model the universe of discourse in a changing environment, the breeding and executable computer programs will be ordered in terms of the *Quality-of-Information* (*QoI*) and *Degree-of-Confidence* (*DoC*) that stems out of them, when subject to a process of conceptual blending [20]. In blending, the structure or extension of two or more predicates is projected to a separate blended space, which inherits a partial structure from the inputs, and has an emergent structure of its own. Meaning is not compositional in the usual sense, and blending operates to produce understandings of composite functions or predicates, the conceptual domain, i.e., a basic structure of entities and relations at a high level of generality (e.g., the conceptual domain for journey has roles for traveler, path, origin, destination). Here it will be followed the normal view of conceptual metaphor, i.e., the system will carry structure from one conceptual domain (the source) to another (the target) directly. Being i ($i \in \{1, \ldots, m\}$) the predicates whose extensions compose an extended logic program or theory that model the universe of discourse, and j ($j \in \{1, \ldots, n\}$) the attributes of the mentioned predicates. Let $x_j \in [min_j, max_j]$ be a value for attribute j. To each predicate it is also coupled a scoring function $V_j^i[min_j, max_j] \rightarrow 0 \cdots 1$, that gives the score predicate i assigns to a value of attribute j taking into account its domain (for the sake of simplicity, scores are kept in the interval 0…1), given in terms of *all* (*attribute exception list*, *sub expression*, *invariants*) productions. The former predicate generates a list of all possible value combinations (e.g., pairs, triples) as a list of sets defined by the domain size plus the invariants. The second predicate recourses through this list, and makes a call to the third predicate for each exception combination. The third predicate denotes *sub expression* and is constructed in the same form. Thus, the *QoI* with respect to a generic predicate K is given by $QoI_K = 1/Card$, where *Card* stands for the cardinality of the exception set for K, if the exception set is not disjoint. Conversely, if the exception set is disjoint, the *QoI* is given by:

$$QoI_K = \frac{1}{C_1^{Card} + \cdots + C_{Card}^{Card}} \tag{1}$$

where C_{Card}^{Card} is a card-combination subset, with *Card* elements.

The next element to be considered stands for the relative importance that a predicate assigns to each of its attributes. w_j^i stands for the relevance of attribute j for predicate i (assuming that the weights of all predicates are normalized, i.e., $\sum_{1 \leq j \leq n} w_j^i = 1$). Thus, it is possible to define a predicate's scoring function, in the multi-dimensional space defined by the attributes domains, given by:

$$V^i(x) = \sum_{1 \leq j \leq n} w_j^i V_j^i(x_j) \tag{2}$$

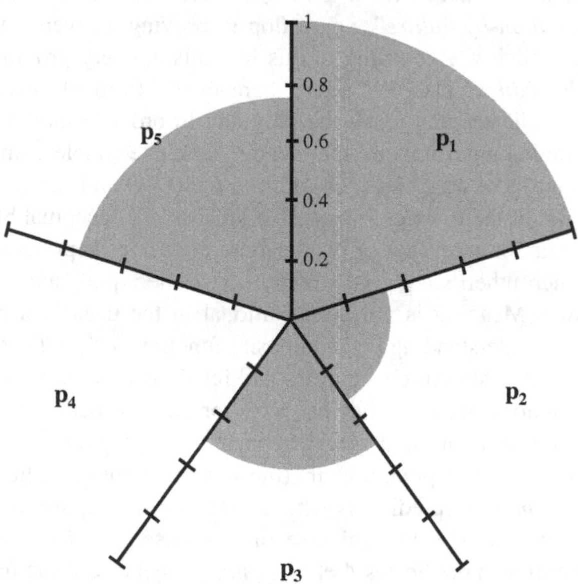

Fig. 1. A measure of *QoI* for the logic program or theory *P*

In order to measure the *QoI* that stems from a logic program or theory the $V^i(x)$ values are posting into a multi-dimensional space. The axes denote the logic program or theory, with a numbering ranging from 0 (at the center) to 1. Figure 1, shows an example of an extended logic program or theory *P*, built on the extension of 5 (five) predicates, $p_1 \ldots p_5$, where the dashed area stands for the respective *QoI.*

Regarding the *DoC*, it is a measure of one's confidence that the argument values of the terms that make the extension of a given predicate, with relation to their domains, fit into a given interval. The *DoC* is computed using $DoC = \sqrt{1 - \Delta l^2}$, where Δl denotes the length of the argument interval, which was set to the interval [0, 1] (Fig. 2).

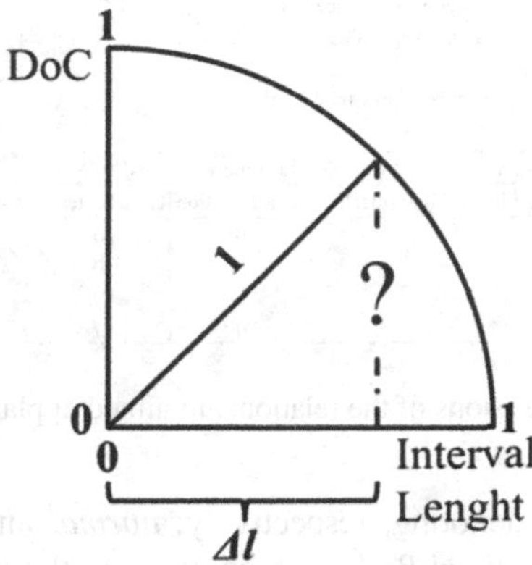

Fig. 2. *DoC's* evaluation

3 Methods

In order to develop an intelligent system aiming at the planning process of *Cataract Surgery* a database, based on the health records of patients at the *Ophthalmology Surgery Service* at the *Hospital of Baixo Vouga, Aveiro, Portugal*, was set. This section depicted, succinctly, how the information is processed.

3.1 Data Processing

The extensions of the relationships illustrated in in Fig. 3, refer to information management aiming at *Cataract Surgery Planning*. Incomplete and/or unknown information are present in the database. For instance, in case 1 the *Cardiac Respiratory Failure* is unknown, which is depicted by the symbol $\perp$, while the *Infectious Pathology* ranges in the interval [0, 1].

In *Infectious Pathology* table 0 (zero) and 1 (one) denote, respectively, *absence* and *presence* of the correspondent pathology. To set the information present in the *Patient Collaboration* table, the procedures described in [21] were followed.

The columns *Glycaemia Index*, *Maximum Blood Pressure*, *Cardiac Index*, and *Cardiac Respiratory Failure* of the *Planning of Cataract Surgery* table are populated

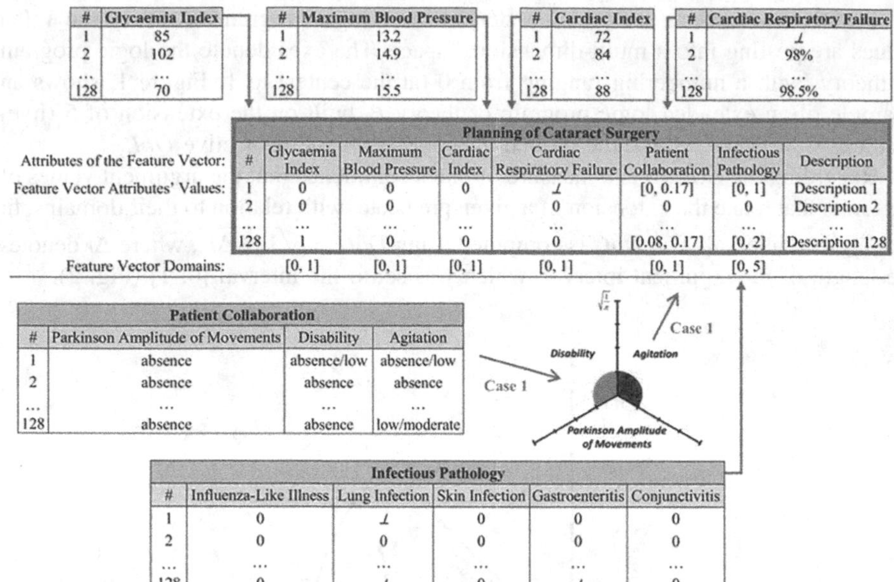

#	Glycaemia Index
1	85
2	102
...	...
128	70

#	Maximum Blood Pressure
1	13.2
2	14.9
...	...
128	15.5

#	Cardiac Index
1	72
2	84
...	...
128	88

#	Cardiac Respiratory Failure
1	⊥
2	98%
...	...
128	98.5%

Planning of Cataract Surgery

	#	Glycaemia Index	Maximum Blood Pressure	Cardiac Index	Cardiac Respiratory Failure	Patient Collaboration	Infectious Pathology	Description
Attributes of the Feature Vector:								
Feature Vector Attributes' Values:	1	0	0	0	⊥	[0, 0.17]	[0, 1]	Description 1
	2	0	0	0	0	0	0	Description 2
	...	...	...	...	...	...	...	...
	128	1	0	0	0	[0.08, 0.17]	[0, 2]	Description 128
Feature Vector Domains:		[0, 1]	[0, 1]	[0, 1]	[0, 1]	[0, 1]	[0, 5]	

Patient Collaboration

#	Parkinson Amplitude of Movements	Disability	Agitation
1	absence	absence/low	absence/low
2	absence	absence	absence
...	...	...	...
128	absence	absence	low/moderate

Infectious Pathology

#	Influenza-Like Illness	Lung Infection	Skin Infection	Gastroenteritis	Conjunctivitis
1	0	⊥	0	0	0
2	0	0	0	0	0
...	...	...	...	...	...
128	0	⊥	0	⊥	0

Fig. 3. A fragment of the extensions of the relationship aimed at planning the *Cataract Surgery*.

with 0 (zero) and 1 (one) denoting, respectively, *normal* and *abnormal* values. The values presented in the *Infectious Pathology* column are the sum of the values present in the correspondent table. The *Description* column includes text (free or structured) that describes the relevant features of the patient.

Applying the algorithm presented in [22] to the fields that make up the knowledge base for *Cataract Surgery Planning* (Fig. 3), and excluding from this process the *Descriptions* that will not be object of attention, it is possible to set the arguments of the state of the predicate ***cataract_surgery_planning*** (cs_{plan}) referred to below, whose extensions denote the objective function regarding the problem under analyse.

$$cs_{plan}: G_{lycaemia}I_{ndex}, M_{aximum}B_{lood}P_{ressure}, C_{ardiac}I_{ndex}, C_{ardiac} \\ R_{espiratory}F_{ailure}, P_{atient}C_{ollaboration}, I_{nfectious}P_{athology} \rightarrow \{0, 1\} \quad (3)$$

where 0 (zero) and 1 (one) denote, respectively, the truth values *false* and *true*.

As an example consider the Program 2, regarding a term (whose variable values represent that of a particular patient) that presents the feature vector $G_{lycaemia}\ I_{ndex} = 0$, $M_{aximum}\ B_{lood}\ P_{ressure} = 0$, $C_{ardiac}\ I_{ndex} = 0$ $C_{ardiac}\ R_{espiratory}\ \mathrm{F}_{ailure} = \bot$ $P_{atient}\ C_{ollaboration} = 1$, $I_{nfectious}\ P_{athology} = [0, 2]$:

{

$$\neg cs_{plan}\left(((A_{GI}, B_{GI})(QoI_{GI}, DoC_{GI})), \cdots, ((A_{IP}, B_{IP})(QoI_{IP}, DoC_{IP}))\right)$$

$$\leftarrow cs_{plan}\left(((A_{GI}, B_{GI})(QoI_{GI}, DoC_{GI})), \cdots, ((A_{IP}, B_{IP})(QoI_{IP}, DoC_{IP}))\right)$$

$$cs_{plan}\underbrace{\left(((0,0)(1,1)), \cdots, ((0, 0.4)(1, 0.92))\right)}_{\text{attribute's values ranges once normalized and respective QoI and DoC values}} :: 1 :: 0.82$$

$$\underbrace{[0,1] \quad \cdots \quad [0,1]}_{\text{attribute's domains once normalized}}$$

}:: 1

Program 2. The *Extended Logic Program* for the patient characteristics vector referred to above.

4 Results and Discussion

The extensions of the relationships illustrated in Fig. 3 refer to one hundred and forty-eight patients aged 35–88 years, mean of 69 ± 14 years. The gender distribution was 59.2.3% and 40.8% for women and men, respectively.

The tables presented in Fig. 3 show how information is gathered. In this section, a soft computing approach has been defined to model the discourse universe, where the computational part is based on an *ECBR* approach to computing. In contrast to other problem solving methodologies (e.g., *Decision Trees* or *Artificial Neural Networks*), in the *ECBR* relatively little work is done offline [9]. Undeniably, in almost all situations the work is done at the time of the consultation. The main difference between the new method [23, 24] and the typical *CBR* [7, 8] is based on the fact that not only all cases have their arguments defined in the interval [0, 1], but also incomplete, unknown or even self-contradictory data or knowledge may be handled in a uniform way. Thus, the *CB* terms are given in terms of the following pattern:

$$Case = \{ < Raw_{data}, Normalized_{data}, Description_{data} > \} \tag{4}$$

where Raw_{data} and $Normalized_{case}$ stand for themselves, and $Description_{data}$ is made on a set of strings or even in free text.

The algorithm presented in [22] is now applied to a new case, with feature vector $G_{lycaemia}\ I_{ndex} = 0$, $M_{aximum}\ B_{lood}\ P_{ressure} = 0$, $C_{ardiac}\ I_{ndex} = 0$ $C_{ardiac}\ R_{espiratory}$ $\mathrm{F}_{ail\text{-}ure} = 0$ $_{Patient}\ C_{ollaboration} = [1, 3]$, $I_{nfectious}\ P_{athology} = \perp$. Then, the computational process may be continued, with the upshot:

$$cs_{plan_{newcase}} \underbrace{(((0,0)(1,1)),\cdots,((0,1)(1,0))) :: 1 :: 0.82}_{\substack{cattribute's\ values\ ranges\ once\ normalized \\ and\ respective\ QoI\ and\ DoC\ values}} \tag{5}$$

The *new case* is now compared with every *retrieved case* using a similarity function *sim*, given in terms of the average of the modulus of the arithmetic difference between the arguments of each retrieved case and those of the *new case*, i.e., [25]:

$$\begin{array}{l} retrieved_{case_1}(((0,0)(1,1)),\cdots,((0.2,0.8)(1,0.8))) :: 1 :: 0.97 \\ \vdots \\ retrieved_{case_n}\underbrace{(((0,0)(1,1)),\cdots,((0.2,1)(1,0.6)))}_{normalized\ cases\ that\ make\ the\ retrieved\ ones} :: 1 :: 0.93 \end{array} \tag{6}$$

where *n* speaks for itself. Assuming that every attribute has equal weight, for the sake of presentation, the *dis(imilarity)* between *new* $_{case}$ and the $retrieved_{case1}$, i.e., *new* $_{case \longrightarrow 1}$, may be computed as follows:

$$dis^{DoC}_{newcase \to 1} = \frac{\|1-1\| + \cdots + \|0-0.8\|}{6} = 0.14 \tag{7}$$

Thus, the *sim(ilarity)* for $sim^{DoC}_{newcase \to 1}$ is set as 1 – 0.14 = 0.86. Regarding *QoI* the procedure is similar, returning $sim^{QoI}_{newcase \to 1} = 1$. Thus, one may have:

$$sim^{QoI,DoC}_{newcase \to 1} = 1 \times 0.86 = 0.86 \tag{8}$$

The proposed model was tested on a real data set with 128 examples, which was divided into unique subsets by cross-validation of ten times [26]. In the execution of the respective division procedures, ten executions were carried out for each of them. To ensure the statistical significance of the results obtained, 30 (thirty) experiments were applied in all tests. The accuracy of the model was 88.3% (i.e., 113 postures correctly classified in 128). Thus, the predictions made by the *ECBR* model are satisfactory, reaching accuracies close to 90%. The sensitivity and specificity of the model were 91.4% and 84.5%, while the *Positive* and *Negative Predictive Values* were 87.7% and 89.1%, respectively.

5 Conclusions

The purpose of this study was to specify and develop an *Evolutionary Clinical Decision Support System* to plan *Cataract Surgery* with a focus on the evolution of the decision making process, once cataract is the leading cause of blindness in the world. The proposed model is based on a formal framework grounded on *Logical Programming* for *Knowledge Representation and Reasoning*, complemented with an *ECBR* approach to computing. Taking into account that the outcome measures of cataract surgery (e.g., visual acuity, accuracy of refractive correction, occurrence of significant operative and postoperative complications) has to be recorded routinely, being the data made available to care providers and commissioners, in future work may emerge some developments on how to assist in resource planning based on optimization and simulation techniques.

Acknowledgments. This work has been supported by COMPETE: POCI-01-0145-FEDER-007043 and FCT – Fundação para a Ciência e Tecnologia within the Project Scope: UID/CEC/00319/2013.

References

1. European Commission: Investing in Health: Towards Social Investment for Growth and Cohesion – Including Implementing the European Social Fund 2014–2020. The European Commission, Brussels (2013)
2. Commission of the European Communities: Together for Health: A Strategic Approach for the EU 2008–2013. The Commission of the European Communities, Brussels (2007)
3. Mossialos, E., Dixon, A.: Funding health care: an introduction. In: Mossialos, E., Dixon, A., Figueras, J., Kutzin, J. (eds.) Funding Health Care: Options for Europe, pp. 1–30. Open University Press, Buckingham (2002)
4. Directorate-General of Health: National Program for Health of Elderly People. Directorate-General of Health, Lisbon (2004)
5. Neves, J.: A logic interpreter to handle time and negation in logic databases. In: Muller, R., Pottmyer, J. (eds.) Proceedings of the 1984 Annual Conference of the ACM on the 5th Generation Challenge. Association for Computing Machinery, New York, pp. 50–54 (1984)
6. Neves, J., Machado, J., Analide, C., Abelha, A., Brito, L.: The halt condition in genetic programming. In: Neves, J., Santos, M.F., Machado, J. (eds.) Progress in Artificial Intelligence. LNAI, vol. 4874, pp. 160–169. Springer, Berlin (2007)
7. Aamodt, A., Plaza, E.: Case-based reasoning: foundational issues, methodological variations, and system approaches. AI Commun. **7**, 39–59 (1994)
8. Richter, M.M., Weber, R.O.: Case-Based Reasoning: A Textbook. Springer, Heidelberg (2013). doi:10.1007/978-3-642-40167-1
9. Carneiro, D., Novais, P., Andrade, F., Zeleznikow, J., Neves, J.: Using case-based reasoning and principled negotiation to provide decision support for dispute resolution. Knowl. Inf. Syst. **36**, 789–826 (2013)
10. Janssen, R., Spronck, P., Arntz, A.: Case-based reasoning for predicting the success of therapy. Expert Syst. **32**, 165–177 (2014)
11. Ying, S., Joël, C., Arnelle, J., Kai, L.: Emerging medical informatics with case-based reasoning for aiding clinical decision in multi-agent system. J. Biomed. Inf. **56**, 307–317 (2015)
12. World Health Organization: Health in All Policies (HiAP) Framework for Country Action. http://www.who.int/cardiovascular_diseases/140120HPRHiAPFramework.pdf
13. Fraunfelder, F.: Game plan for a winning pratice. Ophthalmol. Manag. **16**, 16–22 (2012)
14. Dexter, F., Macario, A., Penning, D., Chung, P.: Development of an appropriate list of surgical procedures of a specified maximum anesthetic complexity to be performed at a new ambulatory surgery facility. Anesth. Analg. **95**, 78–82 (2002)
15. Dexter, F., Traub, R.D.: Statistical method for predicting when patients should be ready on the day of surgery. Anesthesiology **93**, 1107–1114 (2000)

16. American College of Eye Surgeons: Guidelines for Cataract Practice. www.aces-abes.org/guidelines_for_cataract_practice.htm
17. Dexter, F., Epstein, R., Traub, R., Xiao, Y.: Making management decisions on the day of surgery based on operating room efficiency and patient waiting times. Anesthesiology **101**, 1444–1453 (2004)
18. Kakas, A., Kowalski, R., Toni, F.: The role of abduction in logic programming. In: Gabbay, D., Hogger, C., Robinson, I. (eds.) Handbook of Logic in Artificial Intelligence and Logic Programming, vol. 5, pp. 235–324. Oxford University Press, Oxford (1998)
19. Pereira, L., Anh, H.: Evolution prospection. In: Nakamatsu, K. (ed.) New Advances in Intelligent Decision Technologies—Results of the First KES International Symposium IDT 2009. Studies in Computational Intelligence, vol. 199, pp. 51–64. Springer, Heidelberg (2009). doi:10.1007/978-3-642-00909-9_6
20. Turner, M., Fauconnier, G.: Conceptual integration and formal expression. J. Metaphor Symb. Act. **10**, 183–204 (1995)
21. Fernandes, A., Vicente, H., Figueiredo, M., Neves, M., Neves, J.: An evaluative model to assess the organizational efficiency in training corporations. In: Dang, T.K., Wagner, R., Küng, J., Thoai, N., Takizawa, M., Neuhold, E. (eds.) FDSE 2016. LNCS, vol. 10018, pp. 415–428. Springer, Cham (2016). doi:10.1007/978-3-319-48057-2_29
22. Fernandes, F., Vicente, H., Abelha, A., Machado, J., Novais, P., Neves J.: Artificial neural networks in diabetes control. In Proceedings of the 2015 Science and Information Conference (SAI 2015), pp. 362–370, IEEE Edition (2015)
23. Quintas, A., Vicente, H., Novais, P., Abelha, A., Santos, M.F., Machado, J., Neves, J.: A case based approach to assess waiting time prediction at an intensive care unity. In: Arezes, P. (ed.) Advances in Safety Management and Human Factors. AISC, vol. 491, pp. 29–39. Springer, Cham (2016). doi:10.1007/978-3-319-41929-9_4
24. Silva, A., Vicente, H., Abelha, A., Santos, M.F., Machado, J., Neves, J., Neves, J.: Length of stay in intensive care units—a case base evaluation. In: Fujita, H., Papadopoulos, G.A. (eds.) New Trends in Software Methodologies, Tools and Techniques, Frontiers in Artificial Intelligence and Applications, vol. 286, pp. 191–202. IOS Press, Amsterdam (2016)
25. Figueiredo, M., Esteves, L., Neves, J., Vicente, H.: A data mining approach to study the impact of the methodology followed in chemistry lab classes on the weight attributed by the students to the lab work on learning and motivation. Chem. Educ. Res. Pract. **17**, 156–171 (2016)
26. Haykin, S.: Neural Networks and Learning Machines. Pearson Education, Upper Saddle River (2009)

A Methodological Approach Towards Crisis Simulations: Qualifying CI-Enabled Information Systems

Chrysostomi Maria Diakou(✉), Angelika I. Kokkinaki, and Styliani Kleanthous

University of Nicosia, Makedonitissas Ave. 46, 1700 Nicosia, Cyprus
cmdiakou@gmail.com, kokkinaki.a@unic.ac.cy,
styliani.kleanthous@gmail.com

Abstract. Low probability high impact events (LoPHIEs) disrupt organizations' processes severely. Existing methods used for the anticipation and management of such events, suffer from common limitations resulting in a huge impact to the quantification of probability, uncertainty and risk. Continues studies in the field of Crisis Informatics, present an opportunity for the development of a framework that fits the uncertainty related properties of LoPHIEs.

The paper identifies the need for the development and conduction of a series of experiments, aiming to address the factors that qualify Collective Intelligence-enabled Information Systems with respect to their applicability towards support for LoPHIEs; and aims to propose an experiment framework as a methodology for scenario design in LoPHIEs settings.

Keywords: Collective Intelligence · Low probability · High impact · Emergencies

1 Introduction

Low Probability High Impact Events (LoPHIEs), also known as black swan events, which include crises, disasters or emergencies, have significant implications upon the operation of involved and affected organizations [1]. The recent wildfires in Troodos Mountains in Cyprus, in 2016, the 7.8 magnitude earthquake that struck Nepal in 2015, as well as the sequence of disasters resulting from the earthquake in Japan in 2011, are few examples of LoPHIEs. Within such contexts, consequences are disproportionate and difficult to contain or predict [2].

The aforementioned catastrophes that led to technological disasters, economic disruptions and life losses, highlight the complexity of dealing with global risks in an effective manner [2] and imply that as the identification of direct causality between risks becomes progressively problematic, traditional risk management needs to be reinforced with new concepts designed to fit an environment with uncertain qualities.

Appropriate preparations to such events may make the difference between a major disruption of operations in the affected organizations or their resilience and survival [3, 4]. Organizations' preparedness to LoPHIEs is usually distinguished into three phases, namely methods that prepare the organization BEFORE the event; methods that are initiated DURING the event to limit damage and methods that examine the aftermaths

N.T. Nguyen et al. (Eds.): ICCCI 2017, Part I, LNAI 10448, pp. 569–578, 2017.
DOI: 10.1007/978-3-319-67074-4_55

[3, 5]. Such approaches exhibit some fundamental limitations [6]. The most common being the biases raised in judgment or decision making that usually, have a huge impact in the quantification of probability, uncertainty and risk [e.g., 7–13]. Sustainable management for LoPHIEs requires real-time diagnostics, based on both a pragmatic and prepared approach as well as an engagement in effective deliberations on a global scale [3, 5, 14–16]. The research on Crisis Informatics has become more prominent in the recent years with many researchers and practitioners actively exploring different aspects of the use of Collective Intelligence and crowdsourcing processes for crisis management [e.g., 4, 17–20].

Collective Intelligence (CI), through the outreach of many subjective views (which may carry biases individually), results into a diversity of assumptions, solutions and beliefs, that collectively are found to mitigate human biases and lead to more objective decision outcomes [21–27]. Recent studies suggest CIs successful integration in resilience, capacity building and mitigation, for the management of adverse events [28, 29] and present an opportunity for the development of a framework that fits the uncertainty related properties of our modern world.

A main question arising, concerns the factors that qualify CI-related Information Systems with respect to their applicability towards support for LoPHIEs; leading to the hypothesis and main consideration of *whether the proposed framework can provide more comprehensive results as compared to other existing methods/approaches that do not incorporate CI aspects and are deployed for addressing LoPHIEs*. In order to test the hypothesis and meet the following research objectives, the need for the development and conduction of a series of simulated-based experiments is critical:

RO1-Explore indicators related to LoPHIEs management in the presence of CI decision making;
RO2-Explore the relation of CI-based Information Systems for the support of CI-enabled management of LoPHIEs.

In this context, the contribution of this work lies in the development of an experiment framework as a methodology for scenario design in LoPHIEs settings. The application and evaluation of this methodology is part of the immediate future work.

The remainder of the paper is structured as follows: Sect. 2 examines the literature on Crisis Simulations, covering aspects of crisis management scenario design, scenario modeling and construction as well as methodologies for simulation conceptualization and acts as the basis of the proposed methodology. Section 3, using the *Context leads to Scenario* methodology for simulation conceptualization, examines a geopolitical disaster aiming to form the foundations of a crisis scenario for simulated-based experiments within the defined context. Finally, the content of the paper is summarized in a concluding section, providing future work direction.

2 Crisis Simulations

Crisis simulations can be applied to a broad variety of LoPHIEs such as natural disasters, international and political-military conflicts and confrontations. Two types of simulations appear particularly predominant in practice. The first type, concerns simulations that are utilized to illustrate the patterns and anatomy of crisis decision making; while the second, concerns simulations that consist as a powerful tool for generating awareness among participants [30]. In this paper crisis simulations are examined in the light of the first type.

2.1 Crisis Management Scenario Design

The crisis scenario forms the basis for the crisis simulation, and its design, requires respecting specific criteria. A crisis management scenario is a description of the conditions under which the crisis management system is to be designed, tested or evaluated in regards to performance assumptions [31]. The aforementioned classification of a crisis scenario is based on a more general definition recommended and discussed in [32].

The scenario design is the process in which the objects and events are defined, comprising the basis of the simulation content [33, 34]. The most critical consideration in the design of a scenario for a crisis simulation is the purpose that the simulation serves. There are four general aspects that must be addressed in detail, when designing a crisis scenario: (1) the time setting, (2) the environmental setting, (3) the level of detail and (4) the level of expertise. The first aspect is an important parameter due to that the aim of a crisis scenario is to determine how the current crisis response system works and what improvements can be done to it. Therefore, it is advised that for crisis simulations the time corresponds to the present. The second aspect incorporates features like geographic descriptions, demographics, as well as any information that it is believed the participants should know in order to arrive to an informed decision-making. Here, it is worth mentioning that the environmental setting of a scenario should incorporate as little change from the current world as possible. The third consideration carefully defines, clarifies and describes aspects covered over the second consideration - environmental setting, and advances the level of detail provided to the participants. Finally, the fourth takes into consideration the knowledge, experience and sophistication of the simulation participants. In regards to the level of expertise, attention should be given to the fact that the fewer skills, knowledge and background the participants bring into the simulation, the more detailed the scenario must be [34, 35].

2.2 Methodologies for Simulation Conceptualization

Scenario leads to Context - In practice, in order to meet specific requirements, a simulation is being conceptualized and formulated starting with the crisis. Then, working backwards, the aim is to create the most credible possible explanation for the crisis. After the formulation of a scenario that meets the research needs, world, regional and local contextual material consistent with the crisis, are being added. Finally, details in the various sections of the scenario are checked for internal and external consistency [34, 35].

Context leads to Scenario - Reversing the abovementioned methodology, another method for conceptualizing simulations, devotes more attention and study to the preparation of the context. This method supports the view that instead of starting with the crisis, detailed contexts should be prepared covering possible world, regional and local developments. After this has been done, one can more credibly deduce the kinds of crises which might occur under the defined context. Using this methodology, the scenario would emerge from the context, not the context from the scenario; and its credibility would rest on the consistency of the context [34, 35].

2.3 Scenario Modeling and Construction Practices

While each crisis has unique variables and outcomes, all exhibit similar characteristics: they generate urgency and time constraint [36]; they exhibit ambiguous and uncertain elements [37]; and they are unexpected [38]. In addition, they present a dilemma that needs immediate decision or judgment that is based on two distinct features: First, an emergency decision is usually made in a short period of time using partial or incomplete and inaccurate information, especially in the early stages of the disaster; Second, these decisions may have potentially serious outcomes [39–41]. Modeling scenarios for crisis simulations would therefore involve the reproduction of these characteristics and effects, for the crisis simulation. This can be achieved by implementing several types of injections with different consequences to the crisis scenario. The choice and arrangement of these injections allow obtaining the expected crisis characteristics.

An extensive literature review on the several stages needed to be defined when modeling a scenario, is provided among others in [42–44]. In a nutshell, these include: (1) writing a summary, (2) developing a list of possible events and setting the environment of the scenario, (3) gathering data in relation to the participants and the scenario environment, for instance the interconnections between multiple stakeholders involved in or affected by a crisis, based on the scenario, (4) recording a sequence of events, (5) connecting events by a script, adding if needed relevant details, (6) separating the scenario into phase (includes also regrouping in the same space of time various events), (7) developing sets of incidents and recording expected reactions, (8) examining consistency/reactions with the research objectives and the time sequence of events, (9) implementing post-crisis elements throughout the scenario covering a variety of societal spectrums.

3 Developing a Methodology for Simulated-Based Experiments

Based on the above literature an experiment methodology has been designed, to enable the exploration of indicators related to LoPHIEs management in the presence of CI decision making (RO1); and to examine the relation of CI-based Information Systems for the support of CI-enabled management of LoPHIEs (RO2). Using the *Context leads to Scenario* methodology, this section identifies and defines a credible context for a crisis scenario to fulfill the purposes of the experiment methodology. In addition, based on a real LoPHIE, Evangelos Florakis Naval Base explosion, it aims to form the foundations of a crisis scenario for simulated-based experiments.

The Context

The report on Global Risks for 2017 of the World Economic Forum, identifies 5 ruling trends most likely to determine global developments over the following 10 years. In addition, the report stresses on the risks that these global trends are connected to and provides a classification (economic, societal, geopolitical, environmental and technological). This paper is concerned with the study of events that have ***very low probabilities of occurring*** but may carry ***a huge impact upon their occurrence***. Therefore, through the analysis provided in the aforementioned report on global risks, the geopolitical sphere [45] is where most likely a LoPHIE might occur.

Weapons of mass destruction, identified as the risk with the highest impact, results mainly from trends associated with the international governance changing landscape, the rise of cyber dependency, the increasing national sentiment and polarization of societies and the shift of power (ex. from developed to emerging markets and developing economies). Failure of national governance on the other hand, which is identified as the lowest probability risk to develop into a crisis, results mainly from trends associated with the changing landscape of the international governance, the increasing national sentiment and polarization of societies, the shift of power as well as the rising income and wealth disparity [45].

Hence, the geopolitical sphere is identified as a credible context into which many different risks can be embedded for study and from which several scenarios of potential LoPHIEs, can be drawn for the conduction of a crisis simulation for the purposes of the experiment strategy.

The Experiment Frame

In order to evaluate the research objectives through a crisis scenario, two groups (Control Group and Experimental Group) need to be established per experiment. Each group will include action participants as well as participants that will take the role of experts. The objective of both groups (Control and Experimental) will be to address issues related to the management of a crisis/disaster occurring within the geopolitical sphere or being the result of a geopolitical matter. The experimental group will attempt to do this with the use of a CI enabled platform while this will not be available for the control group. This specific experiment frame, will allow to test and evaluate the crisis management system in the presence of CI decision making and in regard to relevant performance indicators and assumptions (see Table 1).

The Scenario

Evangelos Florakis Naval Base explosion, occurred on July 2011 in Cyprus, is an example of a LoPHIE that could form the foundations of a crisis scenario for simulated-based experiments within the geopolitical sphere. The incident ignited about 100 containers holding seized Iranian explosives grabbed by the United States Navy in 2009 after the interception of a Russian-owned, Cypriot-flagged vessel, travelling from Iran to Syria. Political pressure was applied so that Cyprus confiscates the shipment. The explosives were moved to Evangelos Florakis Naval Base and left in the open for over two years. The Cypriot government, fearing an adverse reaction from Syria, declined offers from several countries to remove or dispose the material [46, 47].

From the explosion 12 people were killed on the spot and 62 were injured. Houses in the nearby villages were damaged by the concussion wave of the blast, while the island's main power station was knocked out. This led to power cuts in several areas,

Table 1. Metrics and indicators

Indicators related to management of LoPHIEs	Metrics and Indicators mapping CI
DURING – event phase -Average time to respond -Ratio, accuracy and duplication of data -% of new activity developments (preparation, response plans) launched on time -% of ideas/response plans from outside the organization -% of risk issues exceeding defined risk tolerance without action plans -% of risk mitigation plans executed on time -% of escalated risk vulnerabilities -% of risk incident response plans with one or more open issues -Number of collected ideas/response plans that were implemented -% of unaccepted risk issues with action plan developed -Number of risk incident response plans with unresolved quality issues -Number of risk events with business impact due to delayed response plan execution -% of unaccepted risk issues without mitigation plans developed -% of risk issues exceeding defined risk tolerance without action plans -Number of ideas/response plans developed -Number of collected ideas that were developed further -% of ideas that are funded for development	*Key Metrics* -Ability to discover or elicit true responses -Accuracy of forecasts -% of problems solved -% of early discovery of problems -Quality, accuracy and frequency of contribution, and use of output in real situations -Number, quality and scope of unexpected uncovered issues -Access to difficult to obtain information, and minimization of damage infected by crisis *Key Indicators* -Access to talent, diversity of participants and participant engagement over time -Sample size and whether it is representative of market and participant engagement -Ability to track real quantities and participant engagement over time -Responsiveness to unsolved problems -Progress of testing and participant engagement -Models of communication utilized and participant engagement

taking months to bring the station fully back online. Paphos and Larnaca airports are reported to have turned on their generators, in an attempt to reduce power consumption to the minimum. At the same time, the public was asked to reduce water consumption as much as possible due to the fact desalination plants were taken offline as a result of power problems. Harsh criticism prompted the resignations of the country's defense minister and top military chief [46, 47].

Geopolitical risks are concerned at a great extent with the faith in collective security mechanisms [48]. The impact of the geopolitical crisis scenario described above, on economy, environment, society and political climate (national and international), strongly illustrates the interdependence and interconnection of risks.

Scenario Modeling and Construction

The sequence of events identified in the disaster at Evangelos Florakis Naval Base is:

i. minor blast of several containers,
ii. fire started,
iii. main/vast explosion,
iv. extensive damage in a wide area surrounding the blast,
v. severe damage to Vasilikos Power Station,
vi. electricity supply interruption to approximately half of Cyprus.

Modeling the above disaster scenario for the purposes of the experiment strategy, would involve separating the sequence of events into different phases and identifying the interconnections between the multiple stakeholders involved in or affected by the disaster. Characteristics and effects of the real crisis would also need to be reproduced within the context of our specific scenario by implementing relevant types of injections to create conditions of time constraints, urgency and uncertainty.

In relation to the above, quality improvement and innovation indicators related to the management of LoPHIEs as well as metrics and indicators mapping CI, can be used to test and evaluate the crisis management system in the presence of CI decision making (see Table 1).

4 Conclusion and Future Work

This paper identified the need for the development and conduction of a series of experiments, aiming to address the factors that qualify CI-enabled Information Systems with respect to their applicability towards support for LoPHIEs. It examined aspects of crisis scenario design, modelling and construction; and presented two opposing methodologies for conceptualizing simulations. In addition, using the *Context leads to Scenario* methodology, the paper identified and defined a credible context for a crisis scenario to fulfil the purposes of an experiment strategy within the field of study. Based on Evangelos Florakis Naval Base explosion, it established the foundations of a crisis scenario for simulated-based experiments. The purpose of the paper, was to propose an experiment framework as a methodology for scenario design in LoPHIEs settings. The authors acknowledge that this is an initial attempt to define a methodology for simulated based experiments in crisis scenarios, and that the proposed approach needs to be further evaluated in several contexts for identifying strengths, weaknesses and areas for improvement. Immediate future work will focus on evaluating the experiment frame within the defined context, using real users in virtual settings.

References

1. Hergert, M.: The effect of terrorist attacks on shareholder value: a study of United States international firms. Int. J. Manag. **21**(1), 25–28 (2004)
2. World Economic Forum: Global Risks 2014, 9th edn. World Economic Forum, Cologny (2014)
3. Coombs, W.T., Holladay, J.S.: PR Strategy and Application: Managing Influence. Blackwell, Chichester (2010)

4. Halder, B.: Approaches of humanitarian crisis management-associated risks with the ICT-based crowdsourcing paradigm (2010). http://ssrn.com/abstract=2568233, http://dx.doi.org/10.2139/ssrn.2568233
5. Coombs, W.T.: Ongoing Crisis Communication: Planning, Managing, and Responding, 2nd edn. Sage, Los Angeles (2007)
6. Diakou, C.-M., Kokkinaki, A.I.: Enabling sustainable development through networks of collective intelligence. In: 4th Conference of the IDRiM Society. Northumbria University, Newcastle, UK, 4–6 September 2013
7. Armstrong, J.S.: Findings from evidence-based forecasting: methods for reducing forecast error. Int. J. Forecast. **22**(3), 583–598 (2006)
8. Berg, J.E., Neumann, G.R., Rietz, T.A.: Searching for Google's value: using prediction markets to forecast market capitalization prior to an initial public offering. Manag. Sci. **55**(3), 348–361 (2009)
9. Fildes, R., Goodwin, P., Lawrence, M., Nikolopoulos, K.: Effective forecasting and judgmental adjustments: an empirical evaluation and strategies for improvement in supply-chain planning. Int. J. Forecast. **25**, 3–23 (2009)
10. Goodwin, P., Wright, G.: The limits of forecasting methods in anticipating rare events. Technol. Forecast. Soc. Change **77**, 355–368 (2010)
11. Jakoubi, S., Tjoa, S., Quirchmayr, G.: Rope: a methodology for enabling the risk-aware modelling and simulation of business processes. In: ECIS 2007 Proceedings, Paper 47 (2007)
12. Onkal, D., Gonul, M.S.: Judgmental adjustment: a challenge to providers and users of forecasts. Foresight **2**, 13–17 (2007)
13. Funston, F., Wagner, S.: Surviving and Thriving in Uncertainty: Creating the Risk Intelligent Enterprise. John Wiley & Sons, New Jersey (2012)
14. Hollis, N.: Crisis management: is a new prescription needed? Millward Brown Points of View, pp. 10–14. IT Governance Institute (2007). CombiT 4.1, The IT Governance Institute
15. Klein, M.: Achieving collective intelligence via large-scale on-line argumentation. Center for Collective Intelligence (2007)
16. Brabham, D.C.: Crowdsourcing. The MIT Press Essential Knowledge Series. Massachusetts Institute of Technology, Cambridge (2013)
17. Brabham, D.C.: Crowdsourcing in the Public Sector. Georgetown University Press, Washington DC (2015)
18. Howe, J.: Crowdsourcing: Why the POWER of the Crowd is Driving the Future of Business. Crown Business, New York (2008)
19. Liu, S.B.: Crisis crowdsourcing framework: designing strategic configurations of crowdsourcing for the emergency management domain. Comput. Support. Coop. Work (CSCW) **23**, 389–443 (2014). doi:10.1007/s10606-014-9204-3
20. Bonabeau, E.: Decisions 2.0: the power of collective intelligence. MITSloan Manag. Rev. **50**(2), 45–52 (2009)
21. Lyons, S.: The changing face of corporate defence in the 21st century. StrategicRISK (2008). http://ssrn.com/abstract=1288732
22. Malone, T.W., Laubacher, R., Dellarocas, C.: The collective intelligence genome. MITSloan Manag. Rev. **51**(3), 21–31 (2010)
23. Nickerson, J.V., Sakamoto, Y.: Crowdsourcing Creativity: Combining Ideas in Networks. In: Workshop on Information in Networks. Stevens Institute of Technology (2010)
24. Prpić, J., Jackson, P., Nguyen, T.: A computational model of crowds for collective intelligence. In: Collective Intelligence 2014. MIT Center for Collective Intelligence (2014). http://ssrn.com/abstract=2398206

25. Prpić, J.: Health care crowds: collective intelligence in public health. In: Collective Intelligence 2015. Center for the Study of Complex Systems, University of Michigan (2015). http://ssrn.com/abstract=2570593
26. Prpić, J., Taeihagh, A., Melton, J.: The fundamentals of policy crowdsourcing. Policy Internet **7**, 340–361 (2015). doi:10.1002/poi3.102
27. Dierks, E.: Intelligence-based business continuity part 1, 2, 3. Article of Continuity Insights (2012). http://www.continuityinsights.com
28. Kokkinaki, A.I.: The role of Information Systems for resilience under economic seize: the Cyprus template. In: Conference on Community Resilience (focus on Economic Resilience), University of Minho, Portugal, 12–13 July 2013
29. Zobel, C.W.: Decision support systems for more effective crisis management. In: Conference on Community Resilience (focus on Economic Resilience), University of Minho, Portugal, 12–13 July 2013
30. Limousin, P., Tixier, J., Bony-Dandrieux, A., Chapurlat, V., Sauvagnargues, S.: A new method and tools to scenarios design for crisis management exercises. Chem. Eng. Trans. **53**, 319–324 (2016). doi:10.3303/CET1653054
31. Walker, W.E., Giddings, J., Armstrong, S.: Training and learning for crisis management using a virtual simulation/gaming environment. Cogn. Tech. Work **13**, 163 (2011). doi:10.1007/s10111-011-0176-5
32. Quade, E.S.: Predicting the consequences: models and modeling, Chap. 7. In: Miser, H.J., Quade, E.S. (eds.) Handbook of Systems Analysis: Overview of Uses, Procedures, Applications, and Practice. Elsevier Science Publishing Co., Inc, New York (1985)
33. Kleiboer, M.: Simulation methodology for crisis management support. J. Conting. Crisis Manag. **5**(4), 198–206 (1997)
34. DeWeerd, H.A.: Political-military scenarios. The RAND Corporation, Santa Monica, p. 3535 (1967)
35. deLeon, P.: Scenario designs: an overview. Simul. Games **6**(1), 39–60 (1975)
36. Quarantelli, E.I.: Disaster crisis management: a summary of research findings. J. Manag. Stud. **25**, 373–385 (1988)
37. Lagadec, P.: Crisis management in the 21st century "unthinkable" events in "inconceivable" contexts, CECO-219 (2005)
38. Robert, B.: Nouvelles pratiques pour le pilotage des situations de crise: dix ruptures pour passer d'une logique de procédures à l'apprentissage de la surprise. Environnement, Risques and Santé **1**(1), 22–30 (2002)
39. Yu, L.A., Lai, K.K.: A distance-based group decision-making methodology for multi-person multi-criteria emergency decision support. Decis. Support Syst. **51**, 307–315 (2011)
40. Sun, B.Z., Ma, W.M., Zhao, H.Y.: A fuzzy rough set approach to emergency material demand prediction over two universes. Appl. Math. Model. **37**, 7062–7070 (2013)
41. Sayegh, L., Anthony, W.P., Perrewé, P.L.: Managerial decision-making under crisis: the role of emotion in an intuitive decision process. Human Resour. Manag. Rev. **14**, 179–199 (2004)
42. Bouget, G., Chapuis, J., Vincent, J.: Conception de scénarios d'attaque de systèmes complexes (2009). http://ws3sgs09.loria.fr/articles/vincent.pdf
43. Cannon-Bowers, J.A., Bowers, C.A.: Synthetic Learning Environments. University of Central Florida, Orlando, Florida, 12 (2007)
44. Walker, W. E.: The use of scenarios and gaming in crisis management planning and training, pp. 1–18. RAND Ed., Santa Monica (1995)
45. World Economic Forum: The Global Risks Report 2017, 12th edn. World Economic Forum, Cologny (2017)

46. Cyprus: Zygi naval base munitions blast kills 12 - BBC News. http://www.bbc.com/news/world-europe-14102253
47. Spencer, R.: Cyprus explosion knocks out island's electricity plant. http://www.telegraph.co.uk/news/worldnews/europe/cyprus/8629594/Cyprus-explosion-knocks-out-islands-electricity-plant.html
48. https://www.columbiathreadneedleus.com/content/columbia/pdf/GEOPOLITICAL.PDF

Multicriteria Transportation Problems with Fuzzy Parameters

Barbara Gładysz(✉)

Faculty of Computer Science and Management,
Wroclaw University of Science and Technology,
Wybrzeze Wyspianskiego 27, 50-370 Wroclaw, Poland
barbara.gladysz@pwr.edu.pl

Abstract. In the classical transportation problem, it is assumed that the transportation costs are known constants. In practice, however, transport costs depend on weather, road and technical conditions. The concept of fuzzy numbers is one approach to modeling the uncertainty associated with such factors. There have been a large number of papers in which models of transportation problems with fuzzy parameters have been presented. Just as in classical models, these models are constructed under the assumption that the total transportation costs are minimized. This article proposes two models of a transportation problem where decisions are based on two criteria. According to the first model, the unit transportation costs are fuzzy numbers. Decisions are based on minimizing both the possibilistic expected value and the possibilistic variance of the transportation costs. According to the second model, all of the parameters of the transportation problem are assumed to be fuzzy. The optimization criteria are the minimization of the possibilistic expected values of the total transportation costs and minimization of the total costs related to shortages (in supply or demand). In addition, the article defines the concept of a truncated fuzzy number, together with its possibilistic expected value. Such truncated numbers are used to define how large shortages are. Some illustrative examples are given.

1 Introduction

We present some elements of the theory of fuzzy sets. The concept of a fuzzy set was proposed by Zadeh (1965).

An interval fuzzy number $\tilde{X}$ is a family of intervals of real numbers $[\tilde{X}]_\lambda$, where $\lambda \in [0,1]$ such that: $\lambda_1 < \lambda_2 \Rightarrow [\tilde{X}]_{\lambda_1} \subset [\tilde{X}]_{\lambda_2}$ and $I \subseteq [0,1] \Rightarrow [\tilde{X}]_{\sup I} = \cap_{\lambda \in I}[\tilde{X}]_\lambda$. For a given $\lambda \in [0,1]$, the interval $[\tilde{X}]_\lambda$ is called the λ-level of the fuzzy number $\tilde{X}$. This interval will be denoted by $[\tilde{X}]_\lambda = [\underline{x}(\lambda), \overline{x}(\lambda)]$.

Dubois and Prade (1978) introduced the following useful definition of the class of L-R fuzzy numbers. A fuzzy number $\tilde{X}$ is called an L-R fuzzy number, if its membership function is given by

$$\mu_X(x) = \begin{cases} L\left(\frac{\underline{m}-x}{\alpha}\right) & \text{for} \quad x < \underline{m} \\ 1 & \text{for}\ \underline{m} \le x \le \overline{m} \\ R\left(\frac{x-\overline{m}}{\beta}\right) & \text{for} \quad x > \overline{m} \end{cases}, \tag{1}$$

N.T. Nguyen et al. (Eds.): ICCCI 2017, Part I, LNAI 10448, pp. 579–588, 2017.
DOI: 10.1007/978-3-319-67074-4_56

where $L(x)$ and $R(x)$ are continuous non-increasing functions and $x, \alpha, \beta > 0$.

The functions $L(x)$ and $R(x)$ are called the shape functions of the fuzzy numbers. The most commonly applied shape functions are $\max\{0, 1 - x^p\}$ and $\exp(-x^p), x \in [0, \infty), p \geq 1$. An interval fuzzy number for which $L(x) = R(x) = \max\{0, 1 - x^p\}$ and $\underline{m} = \overline{m} = m$ is called a triangular fuzzy number and will be denoted by (m, α, β).

Let $\tilde{X}$ and $\tilde{Y}$ be two fuzzy numbers with membership functions given by $\mu_X(x)$ and $\mu_Y(y)$, respectively. Based on Zadeh's extension principle (Zadeh 1965), the membership functions of the sum $\tilde{Z} = \tilde{X} + \tilde{Y}$ and the product $\tilde{V} = \tilde{X}\tilde{Y}$ are given by $\mu_Z(z) = \sup_{z=x+y}(\min(\mu_X(x), \mu_Y(y)))$; $\mu_V(v) = \sup_{v=xy}(\min(\mu_X(x), \mu_Y(y)))$.

Carlsson and Fullér (2001) defined the possibilistic expected value $E(\tilde{X})$ and variance $\mathrm{Var}(\tilde{X})$ of the fuzzy number $\tilde{X}$ as follows:

$$E(\tilde{X}) = \int_0^1 \frac{1}{2}\left(\underline{x}(\lambda) + \overline{x}(\lambda)\right) d\lambda \tag{2}$$

$$\mathrm{Var}(\tilde{X}) = \int_0^1 \frac{1}{4}\left(\overline{x}(\lambda) - \underline{x}(\lambda)\right)^2 d\lambda. \tag{3}$$

If $\tilde{X}$ is a triangular fuzzy number $\tilde{X} = (m, \alpha, \beta)$, the possibilistic expected value and the possibilistic variance are given by

$$E(\tilde{X}) = m + \frac{\beta - \alpha}{4} \tag{4}$$

$$\mathrm{Var}(\tilde{X}) = \frac{(\alpha + \beta)^2}{12}. \tag{5}$$

The possibilistic expected value has the following properties, see Carlsson and Fullér (2001):

$$E(a\tilde{X}) = aE(\tilde{X}), \tag{6}$$

$$E(\tilde{X} + \tilde{Y}) = E(\tilde{X}) + E(\tilde{Y}), \tag{7}$$

where a is a real number.

2 Truncated Interval Fuzzy Number

We now introduce the concept of a truncated interval fuzzy number and the possibilistic expected value of such a number.

Definition 1. *The interval fuzzy number $\tilde{X}_S$ is called the truncation of the fuzzy number $\tilde{X}$ ($[\tilde{X}]_\lambda$, where $\lambda \in [0, 1]$) on the crisp, closed (or semi-infinite) set $S \in \mathbb{R}$, if the corresponding λ-levels are given by $[\tilde{X}_S]_\lambda = [\underline{x}(\lambda), \overline{x}(\lambda)] \cap S = [\underline{x_S}(\lambda), \overline{x_S}(\lambda)]$.*

Definition 2. *The possibilistic expected value of a truncated fuzzy number $\tilde{X}_S$ is given by*

$$E(\tilde{X}_S) = E(\tilde{X}/\tilde{X} \in S) = \frac{1}{\lambda_{max}} \int_0^{\lambda_{max}} \frac{\underline{x_S}(\lambda) + \overline{x_S}(\lambda)}{2} d\lambda \tag{8}$$

where $\lambda_{max} = \max_\lambda\{\lambda : [\tilde{X}_S]_\lambda \neq \phi\}$.

For the triangular fuzzy number (m, α, β) truncated on the set $S = (-\infty, x_0]$, where $x_0 \leq m$ and $\mu(x_0) = \frac{x_0-(m-\alpha)}{\alpha}$:

$$E(\tilde{X}/\tilde{X} \in (-\infty, x_0]) = \frac{1}{2}(m + x_0)\mu(x_0) + \frac{\alpha}{4}\left(\mu^2(x_0) - 2\mu(x_0)\right).$$

For the triangular fuzzy number (m, α, β) truncated on the set $S = [x_0, \infty)$, where $x_0 \geq m$ and $\mu(x_0) = \frac{(m+\beta)-x_0}{\beta}$:

$$E(\tilde{X}/\tilde{X} \in [x_0, \infty)) = \frac{1}{2}(m + x_0)\mu(x_0) - \frac{\beta}{4}\left(\mu^2(x_0) - 2\mu(x_0)\right).$$

For the triangular fuzzy number (m, α, β) truncated on the set $S = [x_0, \infty)$, where $x_0 \leq m$ and $\mu(x_0) = \frac{x_0-(m-\alpha)}{\alpha}$:

$$E(\tilde{X}/\tilde{X} \in [x_0, \infty)) = \frac{1}{2}[m(3-2\mu(x_0))+x_0] + \frac{\beta-\alpha}{4}(2\mu^2(x_0) - 4\mu(x_0)+1) - \frac{\beta}{4}\left(\mu^2(x_0) - 2\mu(x_0)\right).$$

For the triangular fuzzy number (m, α, β) truncated on the set $S = (-\infty, x_0]$, where $x_0 \geq m$ and $\mu(x_0) = \frac{(m+\beta)-x_0}{\beta}$:

$$E(\tilde{X}/\tilde{X} \in (-\infty, x_0]) = \frac{1}{2}[m(3-2\mu(x_0))+x_0] + \frac{\beta-\alpha}{4}(2\mu^2(x_0) - 4\mu(x_0)+1) + \frac{\alpha}{4}(\mu^2(x_0) - 2\mu(x_0)).$$

3 Transportation Problem

The classical transportation problem involves transporting a uniform good from m suppliers to n customers. In a unit of time, supplier i can produce a_i units of the good and customer j demands b_j units. The unit cost of transporting a unit from supplier i to customer j is c_{ij}. The parameters of this problem (the capacities a_i, demands b_j and transportation costs c_{ij}) are all known constants. The objective is to select the transportation plan which minimizes the transportation costs while satisfying the demand of the customers, i.e.

$$\min \sum_{i=1}^{m} \sum_{j=1}^{n} c_{ij}x_{ij}, \tag{9}$$

subject to the constraints

$$\sum_{j=1}^{n} x_{ij} \leq a_i, \text{ for } i = 1, 2, \ldots m$$
$$\sum_{i=i}^{m} x_{ij} \geq b_j, \text{ for } j = 1, 2, \ldots n \tag{10}$$
$$x_{ij} \geq 0, \text{ for } i = 1, 2, \ldots m, j = 1, 2, \ldots n$$

The transportation problem given by (9)–(10) is a linear programming problem. When the capacities and demands are all integers, the algorithm proposed by Dantzig (1951) can be applied to find a solution where all the decision variables, the x_{ij}, are integers.

In reality, it is often the case that the parameters of the transportation problem (capacities, demands and transportation costs) are not known precisely. Applying the concept of probability theory or fuzzy logic is an approach to modeling uncertainty. Such an approach was used in many works. Probabilistic transportation problems are NP-hard problems, see Chaudhuri et al. (2013). The fuzzy models proposed in this article are linear and quadratic models.

Chanas and Kuchta (1996) assumed that the unit costs of transportation are fuzzy numbers, while supply and demand are given by crisp numbers. The object is to minimize the total transportation cost. In the optimal solution, the amounts to be transported along each route are crisp numbers, while the total transportation cost is a fuzzy number. This article proposes a two criterion approach based on minimizing both the possibilistic expected total transport cost and the variance of this cost. Consequently the total cost of optimal solution has a small diversity, see Model I.

Chanas and Kuchta (1988), Gupta and Kumar (2012) assumed that supply and demand are given by fuzzy numbers. The solution is given by the set of real numbers which determine how many units of the good are transported from each supplier to each customer. However, assuming that the capacities or the demands are not precisely known leads to the possibility that the realized values of the capacities and demands are such that the decision maker cannot find an appropriate solution, since e.g. there is not enough supply to satisfy the actual demand.

Pandian and Natarajan (2010), Narayanamoorthy et al. (2013), Salajapan and Jayaraman (2014), Hussain and Jayaraman (2014) assumed that the unit costs of transportation as well as supply and demand are given by fuzzy numbers. In the optimal solution the amounts of transportation, together with the total transportation cost, are also fuzzy numbers. Rita and Vimatka (2009) assumed that transport costs are crisp, while supply and demand are given by fuzzy numbers. The loads to be transported along each route are also fuzzy numbers. In this case, it is not clear to the decision maker what the specific loads should be, thus this approach is impractical. In fact, these loads can even takes negative values. This article proposes a model which considers the costs incurred due to shortages when the supply and demand are given by fuzzy numbers, see Model II.

In this article, we propose two models of the fuzzy transportation problem:

- Model I. Uncertainty regarding the transportation costs is modeled using fuzzy numbers and the decision is based on two criteria. The capacities and demands are known constants. The decision is based on two criteria (i) minimizing the possibilistic expected value of the total transportation costs and (ii) minimizing the possibilistic variance of these costs, since variance is a measure of risk (Sect. 4).
- Model II. Uncertainty regarding all the parameters (transportation costs, capacities and demands) is modeled using fuzzy numbers. The decision is based on two criteria: (i) minimizing the possibilistic expected value of the total transportation costs and (ii) minimizing the possibilistic expected costs of the shortages. We interpret both excess production (i.e. insufficient demand) and unsatisfied demand as shortages. These shortages are determined by the concrete realizations of the supply and the demand (Sect. 5).

In both models transportation costs are not known precisely. The first model can be applied when the supply and demand are deterministic. The second model we could be applied when supply and demand not known precisely. The solutions of both models are given by the set of real numbers which determine how many units of the good are transported from each supplier to each customer. Such structure of solutions is useful for decision maker.

4 Transportation Problem with Fuzzy Costs

Define the unit transportation cost on the route from the i-th supplier to the j-th customer in the transportation problem given by (9)–(10) to be the triangular fuzzy number $\tilde{C}_{ij} = (c_{ij}, \alpha_{ij}, \beta_{ij})$. According to Zadeh's extension principle, it follows that the total transportation cost is given by the following fuzzy number:

$$\tilde{C} = \sum_{i=1}^{m}\sum_{j=1}^{n} \tilde{C}_{ij}x_{ij} = (\sum_{i=1}^{m}\sum_{j=1}^{n} c_{ij}x_{ij}, \sum_{i=1}^{m}\sum_{j=1}^{n} \alpha_{ij}x_{ij}, \sum_{i=1}^{m}\sum_{j=1}^{n} \beta_{ij}x_{ij}). \tag{11}$$

Consider a transportation problem with the following objective functions:

- Minimization of the possibilistic expected value of the total costs of transport $F_1 : \min E(\tilde{C})$.
- Minimization of the possibilistic variance of the total costs of transport $F_2 :$ $\min \operatorname{Var}(\tilde{C})$.

Using Eqs. (4) and (11), we can write criterion F_1 in the following form:

$$F_1 : \min E(\tilde{C}) = \min \sum_{i=1}^{m}\sum_{j=1}^{n} E(\tilde{C}_{ij})x_{ij} = \min \sum_{i=1}^{m}\sum_{j=1}^{n} \left(c_{ij} + \frac{\beta_{ij} - \alpha_{ij}}{4} \right) x_{ij}. \tag{12}$$

It can be seen that the transportation problem based on the single optimality criterion F_1 with the constraints given by (10) is a linear programming problem. Hence, in order to solve it, we can use the simplex algorithm or Dantzig's algorithm (1951).

Now consider the problem in which the optimality criterion is the minimization of the variance of the total transportation costs. From Eqs. (5) and (11), we can write criterion F_2 in the following form:

$$F_2 : \min \operatorname{Var}(\tilde{C}) = \min \frac{1}{12}(\sum_{i=1}^{m}\sum_{j=1}^{n}(\alpha_{ij} + \beta_{ij})x_{ij})^2 \tag{13}$$

In this case, the objective function is quadratic. Hence, the transportation problem with objective function (13) and set of constraints given by (10) can be solved by quadratic programming.

Now consider the transportation problem with constraints given by (10) in which both F_1 and F_2 are used as optimality criteria. In order to solve such a problem, we can use e.g. the trade-off method. Assume that the decision maker wishes to find a transportation plan which minimizes the possibilistic variance of the total transportation costs while ensuring that the possibilistic expected costs of transportation are not greater than C. The measure of risk is defined to be the possibilistic variance of the total transportation costs. It follows that the appropriate transportation plan is the solution of the following optimization problem:

$$\min \frac{1}{12}(\sum_{i=1}^{m}\sum_{j=1}^{n}(\alpha_{ij} + \beta_{ij})x_{ij})^2 \tag{14}$$

subject to the constraints

$$\begin{aligned} \textstyle\sum_{i=1}^{m}\sum_{j=1}^{n}\left(c_{ij} + \frac{\beta_{ij}-\alpha_{ij}}{4}\right)x_{ij} \leq C \\ \textstyle\sum_{j=1}^{n} x_{ij} \leq a_i, \text{ for } i = 1, 2, \ldots m \\ \textstyle\sum_{i=i}^{m} x_{ij} \geq b_j, \text{ for } j = 1, 2, \ldots n \\ x_{ij} \geq 0, \text{ for } i = 1, 2, \ldots m, j = 1, 2, \ldots n \end{aligned} \tag{15}$$

A transportation problem formulated in this way has a quadratic objective function and a set of linear constraints.

Example 1. Consider the transportation problem with three suppliers and four customers defined by Liang et al. (2005). This problem was also analyzed in Kaur and Kumar (2011). In both of the articles, the authors assumed that the goal was to minimize the total transportation costs. The firm Dali, based in Taiwan, produces soft drinks and frozen foods. The firm wishes to extend its activities into the Chinese market. It plans to distribute its range of teas to four destinations: Taichung, Chiayi, Kaohsiung and Taipei. The production units are located in Changhua, Touliu and Hsinchu. The firm estimates the capacities of these units, together with the level of demand from the four destinations and

Table 1. Unit transportation costs, supply and demand for Example 1

Supplier	Customer				Supply (thou. of 12-packs)
	Taichung	Chiayi	Kaohsiung	Taipei	
Changhua	(10, 2, 0.8)	(20, 1.6, 2)	(c)	(20, 1.2, 2)	8
Touliu	(15, 1, 1)	(20, 1.8, 2)	(12, 2, 1)	(8, 2, 0.6)	14
Hsinchu	(20, 1.6, 1)	(12, 2.4, 1)	(10, 2.2, 0.8)	(15, 1, 1)	12
Demand (thousands of 12-packs)	7	10	8	9	

the transportation costs (see Table 1). The transportation costs are given in the form of triangular fuzzy numbers, since they are not known precisely. There are a number of factors which determine these transportation costs, such as weather, road and technical conditions. In the original paper (Liang et al. 2005), supply and demand were given in the form of fuzzy numbers. Here, they are given as constants, which are assumed to be the most likely values.

The possibilistic expected value and variance of the unit transportation costs can be derived from Eqs. (4) and (5).

The optimal transportation plans based on each objective individually: minimize the possibilistic expected value of the total transportation costs (F_1) and minimize the possibilistic variance of the total transportation costs (F_2) are given in Table 2. The only parts of these solutions which coincide are the use of two transportation routes, between Changhua and Kaoshiung and between Hsinchu and Chiayi.

Table 2. Optimal solutions to the transportation problems based on the criteria F_1 and F_2 and the associated costs

Criterion	F_1: min expected value				F_2: min variance				Multicriteria problem min F_2 with $F_1 \leq 360$			
Supplier	Customer				Customer				Customer			
	Taichung	Chiayi	Kaohsiung	Taipei	Taichung	Chiayi	Kaohsiung	Taipei	Taichung	Chiayi	Kaohsiung	Taipei
Changhua	7	0	1	0	0	0	8	0	3	0	5	0
Touliu	0	0	5	9	7	7	0	0	4	0	1	9
Hsinchu	0	10	2	0	0	3	0	9	0	10	2	0
Cost $\tilde{C}$ [thou. $]	(352, 72.4, 30)				(496, 51.8, 37.8)				(364, 68.4, 29.2)			

In the case of minimizing the expected value of the transportation costs, the total transportation costs are given by the triangular fuzzy number (352, 72.4, 30) [in thousands of $] with expected value $ 341.4 thou. and dispersion $ 29.6 thou. This is the same solution as the one obtained using the algorithms proposed by Kaur and Kumar (2011) and Liang et al. (2005). In the case of minimizing the variance of the transportation costs, the total transportation costs are given by the triangular fuzzy number (496, 51.8, 37.8) [in thousands of $] with expected value $ 492.5 thou. and dispersion $ 25.9 thou. The variance of the costs are somewhat smaller. However, the expected costs are $ 148.73 thou. greater. One advantage of this transportation plan lies in the fact that only five routes are used.

Now we consider the model of the transportation problem with objective functions (14) and set of constraints given by (15). Assume that the decision maker is interested in a transportation plan which minimizes the variance of the transportation costs given that the expected value of the total transportation costs is less than $ 360 thou. The optimal transportation plan for this problem is described in Table 2. The total transportation costs are given by the triangular fuzzy number (364, 68.4, 29.2) [in thousands of $], which has expected value $ 354.2 thou. and dispersion $ 28.67 thou.

5 Transportation Problem with Fuzzy Costs, Demand and Supply

Consider a fuzzy transportation problem in which all the parameters (the unit transportation costs, supply and demand) are all given by fuzzy numbers. Such problems are common in practice, especially in the case of goods which have seasonal demand (which commonly depends on the weather). Similarly, supply can also be uncertain. Given such a model, we consider the following objective functions: minimization of the possibilistic expected value of the total transportation costs, minimization of the sum of the possibilistic expected value of the total costs associated with shortages. The unit transportation costs are given by triangular fuzzy numbers of the form $\tilde{C}_{ij} = (c_{ij}, \alpha_{ij}, \beta_{ij})$. In addition, let the capacities and demands be given by the set of triangular fuzzy numbers $\tilde{A}_i = (a_i, \gamma_i, \delta_i)$ for $i = 1, 2, \ldots, m$ and $\tilde{B}_j = (b_j, \epsilon_j, \theta_j)$ for $j = 1, 2, \ldots, n$. Let the unit cost of purchasing (producing) the good at the last moment at the i-th point of supply be PA_i, $i = 1, 2, \ldots, m$ and the unit penalty for not delivering a product to the j-th customer be PB_j, $j = 1, 2, \ldots, n$. In the case when $PA_i = PB_j$ for all i and j, the objective is to minimize the possibilistic expected value of the sum of the shortage in supply and the shortage in demand. The appropriate model of a multicriteria fuzzy transportation problem is given by

$$F_1 \min E(\tilde{C}) = \min \sum_{i=1}^{m} \sum_{j=1}^{n} E(\tilde{C}_{ij}) x_{ij} \tag{16}$$

$$F_2 \min \left[\sum_{i=1}^{m} PA_i \cdot \left[\sum_{j=1}^{n} x_{ij} - E\left(\tilde{A}_i / \tilde{A}_i \in (-\infty, \sum_{j=1}^{n} x_{ij}] \right) \right] + \right. \tag{17}$$

$$\left. + \sum_{j=1}^{n} PB_j \cdot \left[E\left(\tilde{B}_j / \tilde{B}_j \in [\sum_{i=1}^{m} x_{ij}, \infty) \right) - \sum_{i=1}^{m} x_{ij} \right] \right]$$

subject to the conditions

$$a_i - \gamma_i \le \sum_{j=1}^{n} x_{ij} \le a_i + \delta, \text{ for } i = 1, 2, \ldots m$$

$$b_j - \epsilon_j \le \sum_{i=i}^{m} x_{ij} \le b_j + \theta_j, \text{ for } j = 1, 2, \ldots n \tag{18}$$

$$x_{ij} \ge 0, \text{ for } i = 1, 2, \ldots m, j = 1, 2, \ldots n.$$

Example 2. We return to the transportation problem considered in Example 1. However, supply and demand are assumed to be given by the following triangular fuzzy numbers: $\tilde{A}_1 = (8, 0.8, 0.8), \tilde{A}_2 = (14, 2, 2), \tilde{A}_3 = (12, 1.8, 1.8), \tilde{B}_1 = (7, 0.8, 0.8)$, $\tilde{B}_2 = (10, 1.4, 1.4), \tilde{B}_3 = (8, 1.5, 1.5), \tilde{B}_4 = (9, 1.2, 1.2)$, see Liang et al. (2005). The optimality function is taken to be combination (sum) of the two functions defined above, $F = F_1 + F_2$, where $PA_i = PB_j = 10000$. This means that the F_2 criterion is to minimize the possibilistic expected shortage (in supply or demand). Optimal solution is:

$$x_{11} = 4, x_{13} = 3, x_{21} = 2, x_{23} = 2, x_{24} = 8, x_{31} = 1, x_{32} = 9, x_{33} = 2$$

The expected shortage is 3.73 thousand 12-packs, the transportation cost is the triangular fuzzy number (336, 63.6, 27) [in thou. $] and the possibilistic expected value of the transport costs are $ 326.85 thou. Lets now compare our solution with the solutions of this problem obtained by other methods: the generalized fuzzy methods (GFNWCM, GFLCM, GFVAM) proposed by Kaur and Kumar (2011) and the method proposed by Liang et al. (2005). The optimal solution obtained by each of this method is the same as our solution in the example 1 when the criterion function is minimization of the possibilistic expected value of transportation costs. The optimal transportation plan and fuzzy transportation cost are given in Table 2 in column 2. The possibilistic expected value of the transportation costs are $ 341.4 thou. the expected shortage is 2.375 thousand 12-packs. So the expected transport cost in the solution proposed by our method is smaller and the expected shortage is greater.

6 Conclusion

This article has presented two models of a multicriteria transportation problem. Under the first model, it is assumed that the unit transportation costs are fuzzy numbers. The two optimality criteria considered were: (a) minimization of the possibilistic expected value of the total transportation costs and (b) minimization of the possibilistic variance of the total transportation costs. Under the second model, all of the parameters of the problem are fuzzy numbers. The two optimality criteria considered here were: (a) minimization of the possibilistic expected value of the total transportation costs and (b) minimization of the expected costs resulting from shortages in supply and demand. The first model enables the derivation of transportation plans which achieve "stable" (minimum variance) costs. Applying the second model, we take into account both transportation costs as well as losses resulting from unsatisfied demand or excessive production. This approach uses the concept of truncated fuzzy numbers, as proposed in this article, as well as the possibilistic expected value. Two illustrative examples were given.

References

Carlsson, C., Fullér, R.: On possibilistic mean value and variance of fuzzy numbers. Fuzzy Sets Syst. **122**, 315–326 (2001)

Chanas, S., Kuchta, D.: A concept of the optimal solution of the transportation problem with fuzzy cost coefficients. Fuzzy Sets Syst. **82**, 299–305 (1996)

Chanas, S., Kuchta, D.: Fuzzy integer transportation problem. Fuzzy Sets Syst. **98**, 291–298 (1988)

Chaudhuri, A., De, K., Subhas, N.: A comparative study of transportation problems under probabilistic and fuzzy uncertainties. GANIT: J. Bangladesh Math. Soc. (in press). arxiv:1307.1891v1

Dantzig, G.B.: Application of the simplex method to a transportation problem. In: Koopmans, T.C. (ed.) Activity Analysis of Production and Allocation. Wiley, New York (1951)

Dubois, D., Prade, H.: Algorithmes de plus courts Chemins pour traiter des données floues. RAIRO Rech. Opérationnelle/Oper. Res. **12**, 213–227 (1978)

Gupta, A., Kumar, A.: A new method for solving linear multi-objective transportation problems with fuzzy parameters. Appl. Math. Modell. **36**, 1421–1430 (2012)

Hussain, R.J., Jayaraman, P.: Fuzzy transportation problem using improved fuzzy Russell's method. Int. J. Math. Trends Technol. **5**, 50–59 (2014)

Kaur, A., Kumar, A.: A new method for solving fuzzy transportation problem using ranking function. Appl. Math. Modell. **35**, 5652–5661 (2011)

Liang, T.F., Chiu, C.S., Heng, H.W.: Using possibilistic linear programming for fuzzy transportation planning decision. Hsiuping J. **11**, 93–112 (2005)

Narayanamoorthy, S., Saranya, S., Maheswari, S.: A method for solving fuzzy transportation problem (FTP) using fuzzy Russell's method. Int. J. Intell. Syst. Appl. **2**, 71–75 (2013)

Pandian, P., Natarajan, G.: A new algorithm for finding a fuzzy optimal solution for fuzzy transportation problems. Appl. Math. Sci. **4**, 79–90 (2010)

Ritha, W., Vinotha, J.M.: Multi-objective two-stage transportation problem. J. Phys. Sci. **13**, 107–120 (2009)

Solaiappan, S., Jeyaraman, D.K.: A new optimal solution method for trapezoidal fuzzy transportation problem. Int. J. Adv. Res. **2**(1), 933–942 (2014)

Zadeh, L.A.: Fuzzy sets. Inf. Control **8**, 338–353 (1965)

Author Index

Abduali, Balzhan II-491
Abedin, Mahmudul I-298
Achilleos, Achilleas P. I-484
Adnan, Foysal Amin II-479
Ahmed, Muyeed II-479
Albishry, Nabeel II-469
Alimzhanov, Yermek II-509
Aloui, Nadia I-233
Amirova, Dina II-491
Aubakirov, Sanzhar II-509
Augustynek, Martin II-541

Bac, Maciej I-113
Bădică, Costin I-331, I-381
Bąk, Jarosław I-93
Balabanov, Kristiyan I-223, I-361
Barbar, Kablan II-590
Bartuskova, Aneta I-63, I-452, I-462
Basher, Sheikh Faisal I-298
Bernas, Marcin II-119
Blecha, Petr I-548
Blinkiewicz, Michał I-93
Bobrowski, Leon I-73
Boryczka, Urszula II-76, II-107
Bosse, Tibor I-125
Bourgne, Gauvain I-202
Bregulla, Markus II-195, II-205, II-249
Bruha, Radek II-305
Bryjova, Iveta II-182, II-541
Bytniewski, Andrzej I-34

Cerri, Stefano A. I-212
Chabchoub, Yousra II-590
Charfi, Nesrine I-538
Chau, Nguyen Hai I-266
Chebil, Raoudha I-212
Chiky, Raja I-137, II-590
Chłopaś, Łukasz II-292
Chohra, Amine II-32
Chojnacka-Komorowska, Anna I-342
Choroś, Kazimierz II-569
Choudhury, Deboshree I-307
Cimler, Richard I-528, II-315, II-335, II-345
Cimr, Dalibor I-528
Ciorbaru, Vicentiu-Marian I-192
Crick, Tom II-469
Cupek, Rafał II-238, II-272, II-282, II-292

Danicek, Matej II-345
Deepthi, P.S. II-129
Demerjian, Jacques II-590
Diakou, Chrysostomi Maria I-569
Djaghloul, Younes I-172
Dolezal, Rafael II-171
Doroz, Rafal II-161
Draszawka, Karol II-438
Drewniak, Marek II-195, II-227, II-272, II-282, II-292
Du Nguyen, Van I-83
Du, Phuong-Hanh I-148
Duda, Jakub II-292
Dworak, Kamil II-107
Dzięjdziel, Grzegorz II-292

El Sibai, Rayane II-590
Ellouze, Mehdi I-172
Esmaili, Parvaneh I-497

Faltynova, Kamila II-541
Fathalla, Said I-14
Fernandes de Mello Araújo, Eric II-386
Fietz, Robinson Guerra I-223
Filonenko, Alexander II-549, II-558
Fojcik, Marcin II-249, II-272, II-282
Formolo, Daniel I-160
Fougères, Alain-Jérôme I-389
Foulonneau, Muriel I-172
Franke, Annelore II-386

Gargouri, Faiez I-233
Gataullin, Ramil II-519, II-529
George, K.M. II-417
Georgiou, Kyriaki I-484
Gilmullin, Rinat II-519, II-529
Gładysz, Barbara I-579
Gogoglou, Antonia I-244
Gouider, Mohamed Salah II-448

Grzechca, Damian II-215, II-260
Guinand, Frédéric I-422, I-442
Gumede, Andile M. I-257
Gwetu, Mandlenkosi V. I-257
Gwizdałła, Tomasz M. II-66

Hajdú-Szücs, Katalin I-401
Hamem, Sihem II-448
Haron, Habibollah I-497
Hasan, Syeda Shabnam I-288
Hernes, Marcin I-34, I-113, I-342
Himel, Ahsan Habib I-298
Hlioui, Fedia I-233
Hoang, Dinh Tuyen I-182
Hosain, Rukshar Wagid II-386
Hwang, Dosam I-182

Imtiaz, Mir Tahsin II-479
Ivanović, Mirjana I-381
Ivković, Jovana I-381

Jankowski, Jarosław II-579
Jędrzejowicz, Joanna II-3, II-357
Jędrzejowicz, Piotr II-3
Jo, Kang-Hyun II-549, II-558
Jodłowiec, Marcin I-24
Juanals, Brigitte II-376

Kadery, Ivan II-479
Kannot, Yaman I-14
Karaskova, Natalie II-171
Karibayeva, Aidana II-491
Karwowski, Jan I-518
Kempa, Olgierd I-317
Khakimov, Bulat II-529
Khan, Haymontee I-288
Khan, Raiyan II-479
Khisha, Joytu I-307
Khusainov, Aidar II-407
Kiss, Attila I-401
Kłak, Sebastian II-249
Kleanthous, Styliani I-569
Klein, Michel C.A. I-473
Kokkinaki, Angelika I. I-569
Kolar, Karel II-171
Komarek, Ales II-325
Konstantinidis, Andreas I-484
Korczak, Jerzy I-113
Kotelnikova, Anastasiia I-433
Kovarnik, Jaroslav I-548
Kowalczyk, Mateusz I-411
Kozierkiewicz-Hetmańska, Adrianna I-44, I-103
Krejcar, Ondrej II-335, II-345
Krenek, Jiri II-171
Kriz, Pavel II-305
Krótkiewicz, Marek I-24
Kubicek, Jan II-182, II-541
Kuca, Kamil II-171, II-182
Kuhnova, Jitka I-528
Kurnianggoro, Laksono II-549, II-558
Kutrzyński, Marcin I-317

Labidi, Mohamed II-459
Laki, Sándor I-401
Lasota, Tadeusz I-317
Latif, Asiful Haque II-479
Le, Hong-Quang II-22
Lejouad Chaari, Wided I-212
Leon, Florin I-331
Lis, Robert I-433
Logofătu, Doina I-223, I-361
Lu, Dang-Nhac II-22
Luckner, Marcin I-518
Lupu, Andrei-Ştefan I-331

Madani, Kurosh II-32
Madiyeva, Aigerim II-501
Madiyeva, Gulmira II-509
Maia, Nuno I-558
Małecki, Krzysztof II-56
Maleszka, Bernadetta II-428
Maleszka, Marcin I-54
Maltsevskaya, Nadezhda V. II-171
Mannan, Noel I-288
Manolopoulos, Yannis I-244
Mansurova, Madina II-509
Manzoor, Adnan I-473
Maraoui, Mohsen II-459
Maresova, Petra II-541
Mariano, Manuel I-558
Markides, Christos I-484
Marreiros, Goreti I-558
Mars, Ammar II-448
Mashbu, Ruhul I-298
Matouk, Kamal I-342
Matyska, Jan II-335, II-345
Medeiros, Lenin I-125

Melikova, Michaela II-171
Meltzer, Flavian II-205
Merayo, Mercedes G. I-83
Mercik, Jacek II-13
Mercl, Lubos II-325
Métais, Elisabeth I-137
Meziane, Farid I-137
Mikitiuk, Artur I-411, I-422
Minel, Jean-Luc II-376
Mollee, Julia S. I-473
Molnarova, Kristyna II-182
Moni, Jebun Nahar I-288
Morozkin, Pavel II-600
Mukhamedshin, Damir II-407

Nachazel, Tomas I-371
Nalepa, Marek II-238
Nemcova, Zuzana I-371
Neves, José I-558
Nevzorova, Olga II-407
Ngo, Thi-Thu-Trang II-22
Nguyen, Manh-Hai II-22
Nguyen, Ngoc Thanh I-83
Nguyen, Ngoc-Hoa I-148
Nowak-Brzezińska, Agnieszka II-139, II-150
Nowakowski, Arkadiusz II-45

Ogorodnikov, Nikita II-398

Pałka, Dariusz II-97
Papadopoulos, George A. I-484
Paryani, Jyotsna II-417
Paszek, Krzysztof II-260
Paul, Anirudha II-479
Pavlik, Jakub II-325
Penhaker, Marek II-182, II-541
Peter, Lukas II-182
Pham, Hai-Dang I-148
Piekarz, Jakub II-195
Pietranik, Marcin I-44
Płaczek, Bartłomiej II-119
Poloczek, Dawid II-260
Porwik, Piotr II-161
Pozo, Manuel I-137
Procházka, Jan I-351
Przybyła-Kasperek, Małgorzata II-139, II-150
Pscheidl, Pavel II-315

Racakova, Veronika II-171
Rahman, Rashedur M. I-288, I-298, I-307, II-479
Rahman, Rashida I-288
Rakhimova, Diana II-501
Rebedea, Traian I-192
Rędziński, Michał II-227
Rybka, Paweł II-215
Rybotycki, Tomasz II-150

Safaverdi, Hossein II-161
Sapek, Alicja II-119
Schmitt, Ulrich I-3
Schrittenloher, Sebastian II-195
Sec, David II-335
Seredynski, Franciszek I-442
Siemiński, Andrzej I-277
Sikder, Tonmoy I-298
Simiński, Roman II-139, II-150
Sitarczyk, Mateusz I-103
Skinderowicz, Rafał II-87
Sobeslav, Vladimir II-325
Sobol, Gil I-361
Solaiman, Basel I-538
Soldano, Henry I-202
Soukal, Ivan I-63, I-452, I-462
Stamate, Daniel I-361
Štekerová, Kamila I-351
Stepan, Jan II-335, II-345
Strąk, Łukasz II-45
Suder, Tomasz II-569
Suleymanov, Dzhavdet II-529
Sundetova, Aida II-491
Swynghedauw, Marc II-600
Szuba, Tadeusz I-507
Szwarc, Krzysztof II-76
Szymański, Julian II-438

T.K., Ashwin Kumar II-417
Tamanna, Nusrat Jahan I-298
Telec, Zbigniew I-317
Thampi, Sabu M. II-129
Tokarz, Krzysztof II-227, II-260
Tomášková, Hana I-528, II-315
Toshimasa, Yamanaka II-367
Tran, Thi-Thu-Hien II-22
Tran, Van Cuong I-182
Trawiński, Bogdan I-317
Trejbal, Jan II-171
Tretyakova, Antonina I-442

Trichili, Hanene I-538
Trocan, Maria II-600
Trojanowski, Krzysztof I-411, I-422
Tryfonas, Theo II-469
Tucnik, Petr I-371, I-548
Tukeyev, Ualsher II-491
Turki, Slim I-172

van der Wal, C. Natalie I-160
van Halteren, Aart T. I-473
Vanin, Artem I-433
Veillon, Lise-Marie I-202
Vicente, Henrique I-558
Viriri, Serestina I-257

Wahyono, II-549
Wąs, Jarosław II-97
Wątróbski, Jarosław II-56, II-579
Wesolowski, Tomasz Emanuel II-161
Wieczorek, Wojciech II-45
Wojtkiewicz, Krystian I-24
Wolski, Waldemar II-56
Wrobel, Krzysztof II-161
Wypych, Michał I-422

Xanat, Vargas Meza II-367

Yermekov, Zhantemir II-509

Zakrzewska, Magdalena II-357
Zerin, Naushaba I-307
Zhumanov, Zhandos II-501
Zięiński, Adam II-215, II-238, II-249, II-272, II-282
Ziemba, Paweł II-579
Zonenberg, Dariusz II-292
Zrigui, Mounir II-459

Printed in the United States
By Bookmasters